ROBOTICS

Fundamental Concepts and Analysis

ROBOTICS

Fundamental Concepts and Analysis

ASHITAVA GHOSAL

Professor
Indian Institute of Science
Bangalore

OXFORD
UNIVERSITY PRESS

OXFORD
UNIVERSITY PRESS

Oxford University Press is a department of the University of Oxford. It furthers the University's objective of excellence in research, scholarship, and education by publishing worldwide. Oxford is a registered trademark of Oxford University Press in the UK and in certain other countries

Published in India by
Oxford University Press
22 Workspace, 2nd Floor, 1/22 Asaf Ali Road, New Delhi 110002, India

First published 2006
13th impression 2023
Digitally Printed in 2024

ISBN-13: 978-0-19-567391-3
ISBN-10: 0-19-567391-3

Typeset in Times Roman
by Archetype, New Delhi 110063

Printed at Manipal Technologies Limited, Manipal

To Aruna Ghosal
and
In memory of Satya Brata Ghosal

Preface

The subject of robotics has fascinated many engineers, scientists, and the general public. This is perhaps due to the desire of humans to create 'human-like' machines that can perform tasks that humans do not like to perform or cannot do them efficiently at a stretch. Although major efforts are underway across the world towards creating intelligent robots capable of performing at the behest of humans, the goal is far from being achieved. The field of robotics itself is fairly new—the word 'robot' was coined only in the 1920s and the first patent obtained on a robot was only about 50 years back. In the past two and half decades, however, the field of robotics has witnessed an explosive activity in terms of research and application in areas ranging from factory environments to outer space and, recently, to new areas such as homes, robotic surgery, and micro-electromechanical systems.

Robotics is a vast and interdisciplinary field that incorporates concepts and topics from mathematics, physics, and various branches of engineering and computer science. Study courses at undergraduate and postgraduate levels in mechanical engineering, electrical engineering, and computer science entail topics such as kinematics, dynamics, control, planning, sensing, intelligence, locomotion, design, and applications of robots. This book covers the topics of kinematics, dynamics, and control of robot manipulators in detail. It is an attempt to provide a more updated view of the available tools and techniques for kinematic, dynamic, and control system modelling and analyses of various kinds of robot manipulators. It provides a unified treatment for the modelling and analysis of serial, parallel, and hybrid manipulators. This textbook introduces the relatively advanced topic of modelling and analysis of flexible manipulators, wherein the rigidity assumption in links and joints of typical robot manipulators is relaxed. The textbook also introduces modelling and analysis of wheeled mobile robots capable of traversing flat or uneven terrains where the notion of non-holonomic constraints at the wheel–ground contact points is used.

Chapters 1 through 8 cover the topics of kinematics, dynamics, trajectory planning, and control of mechanical manipulators. The contents of these chapters can be used for an undergraduate and/or an introductory postgraduate course in robotics. The sections marked with '*' (e.g., Section 2.8) present a more advanced or abstract treatment of the topic in question and can be skipped in the undergraduate course. Chapters 9 and 10 deal with two advanced topics, namely, modelling and analysis of flexible manipulators and wheeled mobile robots. These topics are more suitable for postgraduate students and starting researchers in the field of robotics.

In the area of robotics, a more thorough understanding on a topic is obtained by solving problems and obtaining results in analytical and closed-form expressions (as opposed to purely numerical simulations). To obtain closed-form and analytical expressions, the symbolic manipulation software MAPLE has been used and several examples on serial and parallel manipulators have been presented throughout the textbook. At the end of each chapter, exercise problems which require use of MAPLE for obtaining symbolic expressions (MATHEMATICA can also be used) and/or MATLAB for numerical simulations have been marked with '†'. Students are encouraged to try out these and other exercise problems at the end of each chapter.

The contents of this textbook have evolved over several years of teaching a postgraduate level course in robotics at the Indian Institute of Science (IISc), Bangalore, to students from mechanical, electrical, aerospace, and computer science background. A large portion of the material has also been used for continuing education courses for industry and other local undergraduate college students in Bangalore. These courses are typically for 45 hours spread over approximately four months. In addition to the contents in this textbook, the students in these courses were encouraged to read current research papers and other material available in journals, conference proceedings, other textbooks, and the Internet. Towards this end, at the end of each chapter, an attempt has been made to list significant and relevant reference material on the topics covered in the chapter. I hope that interested students at undergraduate and postgraduate levels will benefit from them.

I would like to thank all my research students, past and present, at the Robotics and CAD Lab at the Mechanical Engineering Department of IISc, Bangalore, for their help during the preparation and writing of this textbook. In particular, I would like to thank Sangamesh Deepak R. and Sandipan Bandyopadhyay for help in obtaining several of the numerical results, plots, figures, and MAPLE results presented in this textbook. I would like to thank Rex J. Theodore, whose PhD work forms a major portion of the contents of Chapter 9. I would like to thank Nilanjan Chakraborty, whose research resulted in the bulk of the contents of Chapter 10. I would like to thank T.A. Dwarakanath, Balakrishna R., Dheeman Basu, L. Shrinivas, A. Ravishankar, M. Chandra Shaker, B.P. Nagaraj, B.S. Reddy, and R. Ranganath—I had fruitful collaboration with all of them over the past several years. I would like to thank the students of the robotics course ME246 at IISc Bangalore, on whom the contents of this textbook have been tested. I would like to thank Prof. A. Krothapalli of Florida State University for allowing and encouraging me, during my visit in the summer/fall of 2004, to collect material and make preparations for this textbook. I acknowledge the effort and commitment of the editorial team at Oxford University Press—without their prodding and help this book may not have seen the light of day. Finally, I would like to thank Oishee and Mithu for their patience and understanding during the period of writing this textbook.

Ashitava Ghosal

Contents

1

Introduction

1.1 Introduction

There has been a rapidly growing and widespread interest in robots, mechanical manipulators and hands, mobile platforms, walking machines, and many other so-called robotic devices and 'intelligent' systems. These robotic technologies combined with rapid advances in electronics, controls, vision and other forms of sensing, and computing have been widely recognized for their potential applications in almost all areas where machines enter our society. In fact, it has been claimed that we are in the midst of a robotics revolution, and this revolution will have more influence on human society than had the Industrial Revolution of the 17th and 18th centuries. Although the number of robots and their applications are proliferating, the underlying science and, to some extent, the technology are moving at a much slower pace. In this chapter, we present a brief history of robots, the various types and classifications of robots, and the underlying science and technology in the field of robotics.

1.2 Brief History

The word 'robot' came into English language in 1923 from the translation of a 1921 Czech play *R.U.R (Rossum's Universal Robot)* by Karel Capek (Capek 1975). It is derived from the Czech word 'robota' meaning slave labour. The 'robots' in the play are designed to replace human workers and are depicted as very efficient and indistinguishable from humans except for their lack of emotions. In the play the robots rebel against their human masters and destroy the entire human race except one man so that he can continue making robots! Unfortunately, the formula gets lost in the destruction.

In contrast to the horrors of mechanization in Capek's play, in 1942, the science fiction writer Issac Asimov (Asimov 1970) in a story titled '*Runaround*' coined the word '*robotics*' to describe the study of robots and gave the three laws of robotics. In many of his stories dealing with robots,

Fig. 1.1 A PUMA 560 robot

always designed on the basis of these three laws, Asimov portrays robots as harmless and totally under the control of human beings.

The modern industrial robot, as it first appeared, bore little resemblance to the science-fiction-inspired vision of a robot. Figures 1.1 and 1.2 show a PUMA (Programmable Universal Machine for Assembly) and a $T3$ (The Tomorrow Tool) robot manufactured by Unimation Inc. and Cincinnati Milacron Inc., respectively, and seen extensively in textbooks, industry, and academic and research institutions. Clearly, these are not even remotely similar to the human-like robots depicted in science fiction. The modern industrial robot was patented by George C. Devol in 1954 and he called it a 'programmable articulated transfer device'. J. Engelberger and George C. Devol founded the world's first robot company, Unimation Inc., in 1956, and the first industrial robot, called Unimate, was purchased by General Motors and installed at an automobile plant in New Jersey, USA, in 1961.

What is a robot in the modern sense of the word? There are several definitions. The Webster dictionary defines a robot as '.... an automatic device that performs functions normally ascribed to humans or a machine in the form of a human'. A more formal and restrictive definition from the

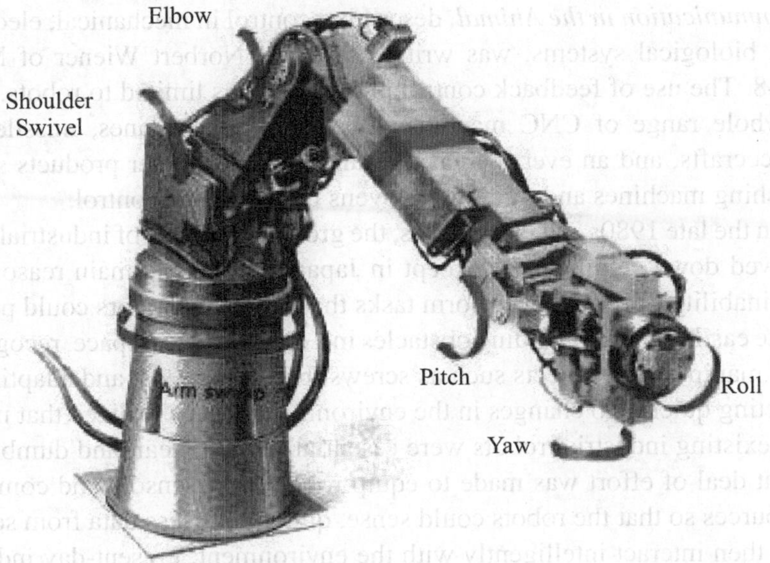

Fig. 1.2 A Cincinatti Milacron $T3$ robot

Robot Institute of America (1969) defines robot as '.... a re-programmable, multi-functional manipulator designed to move materials, parts, tools or specialized devices through various programmed motions for the performance of a variety of tasks'. The key word in the definition is 're-programmable' and this word has closely linked the development of robots to the rapid development of the digital computer and developments in the art and science of computing. A computer or a microprocessor that allows running of different robot programs for the various applications is an essential component of a robotic system. One of the first computers, ENIAC, was developed in the University of Pennsylvania in 1946 and in 1959 a programmable lathe was first demonstrated at MIT. It may be mentioned that the word 're-programmable' also distinguishes a robot from computer numerically controlled (CNC) machines since the level and sophistication of re-programmability is significantly higher in an industrial robot.

In addition to the digital computer, the other key ingredient in the development of the modern industrial robot is the concept and implementation of feedback control. Feedback control allows the execution of the programmed or desired motion (chosen by a robot operator or another program called a *task planner*) with the required accuracy in spite of 'small' changes in the robot or the environment, and thus improve the performance of robots. The first textbook on feedback control, *Cybernetics or Control and*

Communication in the Animal, describing control in mechanical, electronic, and biological systems, was written by Prof. Norbert Wiener of MIT in 1948. The use of feedback control is by no means limited to robots. Today a whole range of CNC machines, automobiles, airplanes, missiles, and spacecrafts, and an ever increasing number of consumer products such as washing machines and microwave ovens use feedback control.

In the late 1980s and early 1990s, the growth in the use of industrial robots slowed down significantly, except in Japan. One of the main reasons was the inability of robots to perform tasks that human operators could perform quite easily, such as avoiding obstacles in a cluttered workspace, recognizing and manipulating objects such as screws, bolts, and nuts, and adapting and reacting quickly to changes in the environment. It was realized that most of the existing industrial robots were essentially blind, deaf, and dumb, and a great deal of effort was made to equip robots with sensors and computing resources so that the robots could sense, quickly process data from sensors, and then interact intelligently with the environment. Present-day industrial robots are often equipped with sensors to detect the presence or absence of the object to be manipulated, measure applied forces and moments, and obtain the position and orientation of objects in its environment. Present-day industrial robots also come with a wide variety of end-effectors, hands, and grippers (which are often equipped with sensing elements) to grasp and manipulate a wide variety of tools and objects. With the advancement in sensing and computing, the modern industrial robot is easier to program and use, more flexible, and more intelligent. The late 1990s have seen a renewed interest in the use of robots.

Modern industrial robots are used in a variety of places and situations. These can be broadly classified into three categories. The first typical area is in an environment that is hazardous for humans to operate in, or an environment where the cost of protecting humans is very high. Examples are in handling of fuel and radioactive material in nuclear power plants, and in space and underwater operations. For example, the satellites and experimental payloads in the space shuttle are often removed from the cargo bay by a mechanical manipulator operated by an astronaut from the safety of the cabin. The pictures of the ocean liner *Titanic* and retrieval of portions of the Air India's Kanishka airplane, from great ocean depths, were taken by unmanned submersible robots. The exploration of Mars was done by mobile robots Sojourner in 1997 and Spirit and Opportunity in 2004, which not only beamed back spectacular pictures of the Martian landscape but also the data and the images sent by them allowed researchers to infer the presence of large amounts of water on the Martian surface in the past. Robots are also being used in environments not hazardous to human beings but where human beings are hazardous to the product—robots are being increasingly used in ultraclean rooms in the electronic industry as it is expensive to keep off dust and other foreign material carried and generated by human beings.

The second area where modern industrial robots have been employed in large numbers is in tasks that are repetitive, back-breaking, and also boring for human beings. In these tasks, human beings cannot maintain the required accuracy or quality because of the monotonous and tedious nature of the task. Typical applications can be seen in automotive industries where robots are often used for spray painting and welding of car bodies, in general manufacturing where robots are used for loading and unloading of material, parts, and tools from other machines, and more recently for assembly of components such as electric motors and computer peripherals.

The third area where robots are used is in manufacturing of consumer products where the number of items is not very large and the product or the model is frequently changing. Typical examples are television sets, cameras, and other audio/video consumer products. Robots are ideally suited for these industries because of their ease of re-programmability. Reprogramming the robots to handle different parts of newer models can be done more easily than expensive re-tooling and changes in the assembly line and this allows a manufacturer to keep pace with the changing consumer tastes and stay competitive.

With the maturing of the technology, robots are finding their way into many other areas of human activity. To name a few, robots are and have been used extensively for entertainment as evidenced by the *Star Wars* and *Terminator* series movies. The robot dog Aibo from Sony and the LEGO Mindstorms robot construction kit is very popular among children and adults. The humanoid robot P3 and Asimo from Honda have appeared in advertising for a variety of products. A robotic system called Da Vinci has been used for heart surgery, where it is programmed to follow the physician's hand movements very accurately and with no tremors (the tremors and unwanted movement can be removed with the help of a computer and the robot controller), and thus can be used to perform very delicate bypass and heart valve surgeries. Finally, robots are finding their way into human homes as robotic vacuum cleaners and lawn mowers which can clean the house or mow the lawns on their own when the occupant is away!

Japan, accounting for more than half the number of robots installed worldwide, is the largest user of robots, followed by the European Union and USA. To give an idea of the explosive growth in the use of robots in Japan, the number of robots in use in Japan went up from approximately 5,500 in 1980 to over 65,000 in 1985 and about 400,000 in 1995. Although the number of robots in use in Japan has not increased much during the last decade due to the retiring of older robots and the depressed economy, the number of robots in use have gone up in Europe and USA. It is forecast that there will be a dramatic increase, in tens of thousands, of robots for domestic and medical uses. In India (and other developing countries) too robots have found their way into a few industries such as for spray painting and spot welding of automobile bodies, handling of molten metals and other hazardous substances, and in nuclear waste handling. The use, however, is insignificant compared to that in Japan, USA, and Europe.

Some of the important dates in the history of robotics are given below.

- 1770—Mechanism-driven life-like machines that can draw, play instruments, and clocks made in Germany and Switzerland.
- 1830—Cam programmable lathe invented.
- 1923—Karel Capek's play *R.U.R*.
- 1942—Asimov coins the word 'robotics' and gives his three laws of robotics.
- 1946—ENIAC, the first electronic computer, developed at the University of Pennsylvania.
- 1947—The first servo electric-powered tele-operated robot at MIT.
- 1948—Book on feedback control, *Cybernetics*, written by Prof. Norbert Weiner of MIT.
- 1948—Transistor invented at Bell Laboratories.
- 1952—IBM's first commercial computer, IBM 701, marketed.
- 1954—First programmable robot patented and designed by Devol.
- 1955—Paper by J. Denavit and R. S. Hartenberg (Denavit and Hartenberg, 1955) provides a notation to describe links and joints in a manipulator.
- 1959—Unimation Inc. founded by Engelberger; CNC lathe demonstrated at MIT.
- 1961—General Motors buys and installs the first Unimate at a plant in New Jersey to tend a die casting machine.
- 1968—Shakey, the first mobile robot with vision capability, made at SRI.
- 1970—The Stanford Arm designed with electrical actuators and controlled by a computer.
- 1973—Cincinnati Milacron's (*T*3) electrically actuated, mini-computer controlled industrial robot.

- 1976—Viking II lands on Mars and an arm scoops Martian soil for analysis.
- 1978—Unimation develops PUMA, which can still be seen in many research labs.
- 1981—*Robot Manipulators* by R. Paul, one of the first textbooks on robotics.
- 1982—First educational robots introduced by Microbot and Rhino.
- 1983—Adept Technology, maker of SCARA robot, started.
- 1995—Intuitive Surgical formed to design and market surgical robots.
- 1997—Sojourner robot sends back pictures of Mars; the Honda P3 humanoid robot, started in 1986, unveiled.
- 2000—Honda demonstrates Asimo humanoid robot.
- 2001—Sony releases second generation Aibo robot dog.
- 2004—Spirit and Opportunity explore Mars surface and detect evidence of past water.

1.3 Types of Robots

Robots are generally classified according to their number of degrees of freedom or axes. The degrees of freedom (DOF) of a robot roughly indicate the capability of a robot. A general task consisting of arbitrarily positioning and orienting an object or a tool can be achieved only by a six-DOF or a six-axes robot. Painting and simple welding can be done by a five-axes robot, and often assembly robots have only four degrees of freedom. A five- or a six-DOF welding robot is often mounted on a three-axes gantry, giving rise to an eight- or nine-axes robotic system for larger operating volume and flexibility. We will discuss the concept of *degree of freedom* in detail in Chapter 3.

Based on its configuration, a robot may be a Cartesian, spherical, or cylindrical robot, since the motion of a point after the first three joints in the robot is best described by the use of Cartesian, spherical, or cylindrical coordinates. The term 'anthropomorphic' or 'articulate' is used for a robot because of its 'similarity' to a human arm, and a SCARA (selective compliance adaptive robot arm) design is based on a folding door. In most manipulators, there are two or three additional joints after the first three joints, which form a wrist. The first three joints are typically used to position an object or a tool in the workspace of the manipulator, whereas the wrist joints are used to orient the tool or the object being manipulated.

All the above configurations are called *serial* manipulators since they have one fixed end, a free end which carries the end-effector or tool, and no closed loops. Many present-day robots have one or more joints fixed to the ground and one or more closed loops. These are called *parallel* manipulators or robots.[1] The Stewart-Gough platform, shown in Fig. 1.3, is one of the most famous examples and has found use in many applications such as in a flight simulator, a six-DOF manipulator, and a six-component force–torque sensor. In Chapters 3 and 4, we present the kinematic analysis of serial and parallel manipulators.

Fig. 1.3 The original Stewart-Gough platform (Stewart 1965)

According to the mode of operation in a playback robot, a robot is physically taken through each step of the desired motion by an operator and these recorded positions are simply played back by the robot on being signalled to do so. In a computer-controlled robot, the desired motion is obtained from a computer after computations according to specified algorithms. An intelligent robot is equipped with sensors and processors and is capable of performing tasks such as avoiding obstacles, taking simple decisions based on sensor inputs and even 'learn' about the environment in which it operates. The topics of feedback control of robots and sensors used in robots form the content of Chapter 8.

[1] Parallel manipulators are often further classified as fully parallel and hybrid. In a fully parallel manipulator, as shown in Fig. 2.22, all the connections between the end-effector and the ground are by means of two links and a single actuated joint. In a hybrid manipulator, as shown in Fig. 2.23, the end-effector can be connected to the ground by several links and actuated joints in a series. In this text, we will use the term 'parallel' for all configurations which are not serial, i.e., those having one or more closed loops.

In addition to the above classification, we also have other types of robots and robotic devices. To name a few, there exist multi-DOF walking machines, robots mounted on two- or three-DOF mobile platforms or automated guided vehicles (AGVs), and multi-DOF mechanical hands with fingers attached to a robot. In Chapter 10, we will discuss mobile robots in more details. For other kinds of robots, the reader is referred to the literature mentioned at the end of the chapter.

1.4 Technology of Robots

The technology and hardware of robots are changing continuously, and we can, at best, describe briefly and qualitatively some of the main components of a robot. A typical robot consists of mechanical components, actuators, power transmission devices, sensors, an electronic controller, and computers. The main mechanical components of a robot are links connected by joints. As mentioned earlier, in a serial manipulator, the links are arranged sequentially, starting from the base and ending in the end-effector with no loops. In a parallel manipulator, on the other hand, there can be one or more loops. Links are assumed to be rigid in most of this text (except in Chapter 9, where we discuss modelling and analysis of flexible link manipulators) and are generally made of metal such as steel or aluminium (cast or machined). It is desirable that links be as lightweight as possible so that torque (force) requirements from an actuator are low, and at the same time the links must have rigidity to achieve positioning accuracy. The joints allow relative rotation or translation between the connecting links, and various types of bearings are used to ensure relatively free and smooth motion. The end-effector carries the tool and is application specific. In painting or welding, the paint gun or the welding tool is fixed on the end-effector and the arrangement is made to continuously supply the paint or the welding wire. In material handling, often a two-fingered gripper is used to grasp objects, as shown in Fig. 1.4.

The links are moved by actuators, which are electric motors or pneumatic and hydraulic cylinders. Electric motors can be DC or AC servo-motors, or sometimes stepper motors. The motors required for robots should have ideally a low rpm (of less than 100), be lightweight and have high torque. Most lightweight DC servo-motors, however, run at a high speed of 3000 rpm or more[2] and a suitable transmission has to be used to bring down the speed. Usual speed reduction approaches using standard spur gears, chains, and

[2] Lightweight, high-torque, and low-rpm motors or direct drive motors are difficult to design and are expensive.

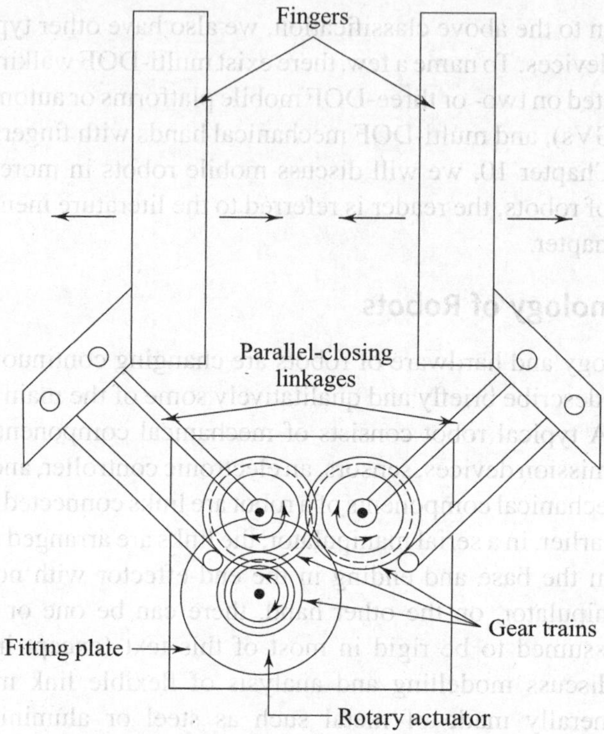

Fig. 1.4 A two-fingered gripper

sprockets or belts cannot be used since the accuracy is lost due to backlash or slippage. Special low-backlash gear sets, harmonic drives, and ball screws are used for transmission of power and to reduce speed. Often brakes are present to stop the motion or to hold the robot when it is not in motion.

The speed and repeatability requirement vary with the nature of the task and the load-carrying capacity. The tip of a large welding robot can move as fast as 2 m/sec (although the welding process speed is much smaller at about 1–5 m/min) carrying a payload of 5–10 kg, and can have a repeatability of less than 1 mm and a reach of 2–3 m. Typical speeds of material-handling robots are 1 m/sec carrying a payload of about 10 kg, with a repeatability of about 0.1 mm (thickness of a human hair) and a reach of 1–2 m. Finally, there are small electronic assembly robots which can have maximum speeds of more than 5 m/sec, a payload of less than 1 kg, a repeatability of 0.01 mm or less, and a reach less than 1 m.

Most robots have sensors at joints, which measure rotation or translation at joints for feedback control. The angular rotation at a joint is measured by optical encoders, and the angular velocity can be measured by tachometers. Translation and linear velocities can be measured by linear variable differential transformers (LVDTs) and video cameras. Force at the end-effector or at the links can be accurately measured by force–torque sensors, which use strain gauges.

Most controllers are implemented digitally using microprocessors and contain circuitry for analog to digital (A/D) and digital to analog (D/A) conversions, memory, and other electronics. Typically, the measured signals and the processing are digital (0/5 V and currents in milliamperes) whereas the input to the actuators are higher voltages (often 12 or 24 V) and currents (in amperes) in analog form, and hence carefully designed servo-amplifiers are used to drive actuators. The technology of controllers for robots is quite complex, difficult, and expensive to develop, and together with computers also makes up the bulk of the cost of a robotic system—the controller may cost as much as 60% of the entire robotic system.

In addition to the sensors required for control, most industrial robots have sensors required for the task and application for which the robot is being used. For example, an arc-welding robot will have sensors to maintain the arc length while the robot is moving along the required weld path. A few robots are equipped with simple vision systems. They are used to inspect or pick components not oriented in the desired way. Several research robots, at universities and research laboratories, have full-fledged vision systems or sonars which allow them to avoid obstacles and navigate in cluttered environments.

Two of the most important components in a robotic system are the computers and the software or the programs residing in them. Often there are two kinds of computers in a robotic system: one set performs the task of controlling the actuators in a robot and the other is a supervisory or a master computer where application programs can be developed and stored, fault detection, diagnosis, and corrective actions can be taken, or where a high-level task planner or an expert system can reside. At the actuator level, the programs are simple and the processing has to be very fast (typical sampling rates are about 50–100 Hz), and an assembly language programming is often used for achieving the fast processing requirements. At the supervisory level, the processing speed is much slower, but the programs are more complex. At this level, a user-friendly environment and a standard high-level language, such as C, or a robot programming language is available for developing application programs. There are two-way communication channels between the actuator control computers and the supervisory computer.

Finally, an industrial robot is a complex and expensive machine. It is important to provide a user-friendly operator interface so that the robot operator can learn to use it easily and quickly. The commands to the robot to accomplish a given task must be simple and straight forward to the operator. The operator interface is normally through a teach pendant or through a computer. A large amount of careful programming, often including graphics, is required to build this user interface.

1.5 Basic Principles in Robotics

To understand how and why robots work, we have to understand the scientific principles that form the basis of robotics. Robotics is an interdisciplinary subject drawing ideas and tools from mathematics, physics, engineering, and computing. We take a brief look at some of the ingredients of robotics, namely, kinematics, dynamics, controls, sensing, and intelligence.

Kinematics deals with the motion of rigid bodies (motion of the rigid links of a robot) in a three-dimensional (3D) space. A rigid body moving in a three-dimensional space has six degrees of freedom, or, in other words, a rigid body requires six independent parameters to be fully specified. Three of the six parameters specify the position of a point of interest, which could be a point on the end-effector, or the tip of the paint or welding tool, or the centre of gravity of the part being moved. Three other parameters are required to specify the orientation of the same object. In order to achieve these six degrees of freedom, a robot must have at least six independently actuated joints. In kinematics of robots we study the functional relationships between the motion at the joints and the motion of the end-effector or tool without reference to the cause (external forces and moments) of the motion. One can find and study, quantitatively, the motion at the end-effector for a given motion at the joints (the direct kinematics problem), the motion at the joints for a required motion of the end-effector (the inverse kinematics problem), the workspace or the volume in 3D space which a point of interest on the robot can reach, and other issues related to the time derivatives of the position and orientation of the links of a robot, i.e., the velocity and acceleration. In Chapter 2, we discuss in detail the representation of the rigid body position and orientation and the representation of links and joints of a serial or a parallel robot. In Chapters 3 and 4, we discuss the kinematics of serial and parallel robots, respectively, and in Chapter 5, we discuss velocity and acceleration.

In dynamics, we study the motion of the links and the end-effector under the action of external forces and torques from the actuators. The methodology is to first obtain the mass and inertia of the moving links and the end-effector, then obtain the dynamic equations of motion by the application of the well-known Newton's laws of motion, or by the Lagrangian formulation, or by the use of Kane's equations. There are two problems of interest. In the so-called direct problem, the differential equations of motion are solved for given initial conditions, and the time evolution of the variables which describe the motion of the complete robot system is obtained. The differential equations of motion are non-linear and coupled, and hence they can be solved only numerically on a computer. In the so-called inverse problem, we compute the actuator forces and torques required to achieve a desired motion of the robot. The direct problem is useful for simulation, whereas the inverse problem is useful for design (or choice) of actuators and links and for model-based control. Robot dynamics is discussed in detail in Chapter 6.

To ensure that a robot follows a desired motion, the paradigm of feedback control is used. In feedback control, as applied to a robot, the actual motion either at the joints or of the end-effector (or tool) is measured by means of sensors, and these measurements together with the known desired motions are used as inputs to a controller. The outputs of a controller are the torques and forces acting at the actuators, and they act in a way so as to reduce the errors between the desired and the actual motion. In a very common controller, the so-called proportional integral plus derivative (PID) controller, the output is proportional to the error (defined as the difference between the desired motion and the measured or estimated motion), the rate of change of the error and the integral of the error. One can show, for a linear system, that by a proper choice of the proportionality constants, also known as the controller gains, the errors can be driven to zero in an asymptotic manner. A PID controller also works reasonably well for common industrial robots in spite of them being non-linear systems. There are other advanced controllers which use a model (essentially the dynamic equations of motion) of the robot system to achieve better performance in terms of the accuracy and speed of response.

In many applications, in particular robotic assembly, position control is not suitable. In situations where the robot end-effector is in contact with the environment it is useful to control the force which the robot applies on the environment. Although force and position cannot be controlled in the same direction, there are advanced hybrid position/force control schemes which allow the user to switch between position and force control depending on the application. In Chapters 7 and 8, we deal with the generation of desired trajectories, position and force control in a robot, respectively.

A robot without sensors is like a human being without eyes, ears, and sense of touch or smell. It is also believed that to make robots more intelligent in interacting with the external environment and taking simple decisions, they need to be equipped with sensors. The single most important sensor in humans is the eye. It is also the most complex sensor and requires the maximum processing by the brain—to process the vast amount of information coming from the eyes quickly, unlike the other sense organs, the eye is directly connected to the brain by means of the optic nerve. Efforts have been made and are still continuing to endow robots with vision—a video camera acts as an eye and the processing is done by specialized electronic and high-speed computers.

Sensors which can measure force and simulate the sense of touch in humans have also been used in robots. These sensors use the phenomenon of change of electrical resistance of some material under pressure or contact forces. These sensors enable a robot to apply the correct amount of force to grip delicate objects, such as an egg or a paper cup, and detect objects and obstacles that come in its path. The reader interested in sensors used in robots is referred to Fu, Gonzalez, and Lee (1987) and other related refernces listed at the end of the chapter.

A robot equipped with vision and touch sensors that can adapt to various changes in its environment can be said to be intelligent. The foundation of this ability lies in the field of artificial intelligence or AI in short. The goal of AI is to produce systems that can imitate human performance in a large variety of tasks considered to be intelligent. A class of AI systems known as expert systems have a knowledge base consisting of 'if ..., then ...,' rules, rules of logic, heuristics and data about the current domain of interest, and algorithms to manipulate the data and the rules (it can learn and add new rules). These expert systems can obtain solutions to the problem at hand or extract useful information about the external world from sensory data, and serve as an interface with a human user or a robot. Other AI systems[3] for extracting useful data about a scene viewed by a TV camera, to avoid obstacles while moving in cluttered spaces, and to assemble blocks and other objects have been devised and used in robots. The present level of intelligence in a robot is minuscule—even an insect can move around with more agility and adapt to the changing environment with more ease than an advanced, expensive, state-of-the-art research robot. The topic of intelligence is outside the scope of this text, but interested readers can refer to some of the references listed at the end.

[3] AI systems are not restricted to robotics—they are used in widely varying areas such as in medicine to diagnose diseases, in organic chemistry to determine composition from spectroscopic data, prospecting of oils and minerals, and even in the design of computers.

1.6 Notation

The following notations will be used throughout the text.

Symbols such as a, x, P will be used to denote scalars. Boldfaced symbols such as **p**, **q** will denote vectors. Often the components of a vector are required, and we will denote the components of a 3D vector **p** by $(p_x, p_y, p_z)^T$ or $(p_1, p_2, p_3)^T$. Vectors are described with respect to a coordinate system, and the leading superscript A as in ${}^A\mathbf{p}$ denotes the coordinate system in which **p** is described. The subscript with a vector is used for distinguishing one object out of many, for example, the vector ${}^0\mathbf{O}_1$ denotes a point $\mathbf{O}_1$ with respect to the coordinate system $\{0\}$. The first and second derivatives of vector **q** with respect to time will be denoted by $\dot{\mathbf{q}}$ and $\ddot{\mathbf{q}}$, respectively.

We will use several coordinate systems. The symbols such as $\{A\}$ or $\{Tool\}$ will be used to denote the coordinate system named A or $Tool$. The origin of the coordinate system $\{A\}$ will be denoted by O_A and it is located by a vector $\mathbf{O}_A$. The unit vectors along the three coordinate axes of $\{A\}$ will be denoted by $\widehat{\mathbf{X}}_A$, $\widehat{\mathbf{Y}}_A$, $\widehat{\mathbf{Z}}_A$. A rigid body i, in the context of kinematics, is equivalent to a coordinate system $\{i\}$, and will be used interchangeably.

Symbols enclosed in square brackets such as $[\mathbf{J}]$ or $[T]$ will denote matrices. Matrices are often associated with coordinate systems, for example, the orientation of a rigid body or a coordinate system with respect to another rigid body or a coordinate system can be defined with the help of a rotation matrix ${}^A_B[R]$. The matrix ${}^A_B[R]$ will be used to denote the rotation matrix describing $\{B\}$ relative to $\{A\}$. The inverse and transpose of a matrix $[R]$ will be denoted by the usual $[R]^{-1}$ and $[R]^T$, respectively. The identity matrix will be denoted by $[U]$ (for unitary) to differentiate from the inertia matrix $[I]$. Other matrices, as and when they appear, will be explained in the text.

We will use trigonometric functions $\sin(\cdot)$ and $\cos(\cdot)$ very frequently. For convenience, we will use the symbols c_1, s_{12}, etc. to denote $\cos\theta_1$ and $\sin(\theta_1 + \theta_2)$, etc., respectively, throughout the text.

Other symbols will be explained as and when they are introduced in the text.

1.7 Symbolic Computation and Numerical Analysis

Before we end this chapter, a word about computations is in order. In this text, we will be making significant use of computers in two main ways. Firstly, in this text, several problems and exercises are provided at the end of each chapter which enhance the understanding of the topics covered in that chapter. Some of the exercise problems require a significant amount of numerical computation, and a software such as MATLAB (Mathwork

1992) or MATHEMATICA (Wolfram 1999), on a PC or a Unix machine, is adequate. MATLAB or MATHEMATICA provide in-built and very easy to use functions and routines for almost all mathematical operations, plotting of numerical data, and other scientific computations. The use of MATLAB or MATHEMATICA is preferred over writing from scratch, in a language such as C or C++, error free codes for numerical computations.

Secondly, we have used extensive symbolic computations to derive and simplify many of the analytical results presented in this text. We have used the software MAPLE[4] (Heck 2003) which has an added advantage of being compatible with MATLAB, and symbolic expressions obtained in MAPLE can be easily ported to MATLAB for numerical computations and other post-processing. Several exercise problems at the end of the chapters require symbolic computations for derivations of analytical expressions. Readers are urged to obtain familiarity with MAPLE and MATLAB. Readers interested in knowing more about symbolic computations are referred to the book by Cohen (2002).

Exercises

1.1 Visit some of the websites mentioned in the text, for example *http://www.jpl.nasa.gov* and links to Mars Robots, and *http://world.honda.com/ASIMO/*, and learn about the robots.

1.2 From the Internet find out the most recent numbers of robots in use worldwide. What is the cost of a typical welding robot? What is the cost of a surgical robot? What is the largest robot in use? How much payload can it carry and what is its weight?

1.3 In later chapters, there will be many exercises where a computer will be necessary. We will use MAPLE or MATHEMATICA. Familiarize yourself with one of these software packages by
- Assignment of symbolic or numeric expressions to identifiers.
- Dot and cross-product of two vectors.
- Manipulating trigonometric expressions and simplification of expressions using known trigonometric and other identities.
- Symbolic multiplication of two 4×4 matrices.
- Symbolic inverse of a 3×3 matrix.

1.4 Test expressions obtained from symbolic computations numerically by choosing random numeric values for the variables.

1.5 Familiarize yourself with MATLAB by
- trying out mathematical operations on scalars, vectors, and matrices.
- plotting graphs for some common functions such as $x = \sin\theta, \theta \in [-\pi, \pi]$.
- writing and using a simple 'm'-functions.

[4] MATHEMATICA can also be used.

References and Suggested Additional Reading

Asimov, I. 1970, *I, Robot*, Fawcett Greenwich.

Capek, K. 1975, 'R.U.R', in the Brothers Capel, *R. U. R and The Insect Play*, Oxford University Press.

Cohen, J.S. 2002, *Computer Algebra and Symbolic Computation: Elementary Algorithms*, A K Peters.

Denavit, J. and R.S. Hartenberg 1955, 'A kinematic notation for lower-pair mechanisms based on matrices', *J. Appl. Mech.*, pp. 215–21.

Heck, A. 2003, *Introduction to MAPLE*, 3rd edn, Springer-Verlag.

MATLAB Users Manual, Mathwork Inc, USA 1992.

Stewart, D. 1965, 'A platform with six degrees of freedom', *Proc. Inst. Mech. Eng.*, vol. 180, no. 1, pp. 371–86.

Wolfram, S. 1999, *The Mathematica Book*, 4th edn, Cambridge University Press.

Other References and Textbooks on Robotics

Asada, H. and J.J. Slotine 1986, *Robot Analysis and Control*, Wiley.

Brady, M. et al. 1983, *Robot Motion*, MIT Press.

Craig, J.J. 1989, *Introduction to Robotics: Mechanics and Control*, 2nd edn, Addison-Wesley.

Dorf, R.C. (ed.) 1988, *International Encyclopedia of Robotics*, John Wiley, vols 1–3.

Fu, K., R. Gonzalez, and C.S.G. Lee 1987, *Robotics: Control, Sensing, Vision and Intelligence*, McGraw-Hill.

Hunt, V. 1983, *Industrial Robotics Handbook*, Industrial Press.

Koren, Y. 1985, *Robotics for Engineers*, McGraw-Hill.

Murray, R.M., Z. Li, and S.S. Sastry 1994, *A Mathematical Introduction to Robotic Manipulation*, CRC Press.

Paul, R. 1981, *Robot Manipulators*, MIT Press.

General Reference Journals and Magazines on Robotics

IEEE Transactions on Robotics (originally *Robotics and Automation*)

IEEE Transactions on System, Man and Cybernetics

International Journal of Robotics & Automation

Journal of Robotic Systems

Mechanism and Machine Theory

Robotics Today

The Industrial Robot

The International Journal of Robotics Research

Transactions of ASME, Journal of Dynamical Systems, Measurement and Control

Transactions of ASME, Journal of Mechanical Design

Some Useful and Interesting Robot Related Websites

http://www.jpl.nasa.gov and links to Mars Robots.

http://world.honda.com/ASIMO/

http://www.sony.net/Products/aibo/index.html

http://www.legomindstorms.com

http://www.howstuffworks.com

http://www.ai.mit.edu

http://robotics.stanford.edu

http://www.frc.ri.cmu.edu/robotics-faq/

2

Mathematical Representation of Robots

2.1 Introduction

Before we can analyse serial or parallel manipulators or other types of robots, we need to develop a description for them which is amenable to mathematical analysis. As mentioned earlier, for most of this textbook, we will assume that a manipulator is made up of rigid links connected by joints. In this chapter, we will first address the representation of a rigid body in three-dimensional space in terms of a position vector and its orientation. Then we will discuss the representation of typical joints present in serial and parallel manipulators and present the constraints introduced by a joint connecting two rigid bodies. Next we discuss the representation of the links of a manipulator in terms of the well-known Denavit–Hartenberg parameters and the concept of link transformation matrices, which form the basis of all kinematic analyses of serial or parallel manipulators. The chapter will end with examples of several serial and parallel manipulators.

2.2 Position and Orientation of a Rigid Body

A rigid body in three-dimensional space, $\Re^3$, has six degrees of freedom (DOF) and is described by its position and orientation. By 'position' of a rigid body, we mean the position of any point on the rigid body. In most situations, we will choose a special point such as the centre of mass of a link, the mid-point of a two fingered hand, the point where the last three joint axes intersect, or some other point of interest on the rigid body. It must be noted that often the point of interest may not be actually on any physical object.

The position of a point on the rigid body is best described by a vector drawn from a chosen reference point, located in another rigid body, to the point under consideration, as shown in Fig. 2.1. The position vector has three scalar components since we are considering points in $\Re^3$. The components of the vector have different meanings depending on the type of the coordinate system in use. There are many types of coordinate systems, such as Cartesian, cylindrical, and spherical; however, in this text, we will be using mainly the Cartesian coordinate system. The Cartesian coordinate system consists of an origin O and three coordinate axes, $\widehat{\mathbf{X}}$, $\widehat{\mathbf{Y}}$, and $\widehat{\mathbf{Z}}$ which form an orthonormal[1] right-handed basis. Since we will be introducing many Cartesian coordinate systems, we will identify them as $\{A\}$, $\{B\}$, $\{Tool\}$, etc. to keep track of them. We will also attach leading superscript and subscript, A, B, etc., to the vectors and a subscript, A, B, etc., to the origins to signify the coordinate system they are related to. Hence, in Fig. 2.1, the origin of $\{A\}$ is denoted by O_A, and the coordinate axes are denoted by $\widehat{\mathbf{X}}_A$, $\widehat{\mathbf{Y}}_A$, and $\widehat{\mathbf{Z}}_A$. The position vector of the point P, in the rigid body B, with reference to the coordinate system $\{A\}$ or the rigid body A, is denoted by ${}^A\mathbf{p}$. It may be noted that a coordinate system and a rigid body are one and the same and will be used interchangeably. In situations where the coordinate system is clearly understood, we will drop the superscript and subscript for convenience. We assume that there exists a coordinate system relative to which everything can be described.

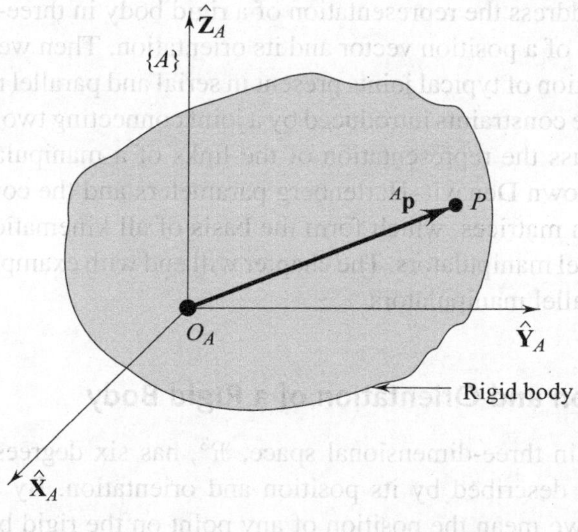

Fig. 2.1 Position of point P denoted by ${}^A\mathbf{p}$

In a chosen Cartesian coordinate system, denoted by $\{A\}$, the components of the position vector of the point P, denoted by $(p_x, p_y, p_z)^T$, are the distances along the three axes measured from the origin, O_A. The three

[1] The vectors $\widehat{\mathbf{X}}$, $\widehat{\mathbf{Y}}$, and $\widehat{\mathbf{Z}}$ are unit vectors, and are orthogonal to one another.

components can also be thought of as projection of the vector ${}^{A}\mathbf{p}$ onto the three corresponding axes. Hence we can write

$$ {}^{A}\mathbf{p} = p_x\widehat{\mathbf{X}}_A + p_y\widehat{\mathbf{Y}}_A + p_z\widehat{\mathbf{Z}}_A \tag{2.1} $$

where $\widehat{\mathbf{X}}_A$, $\widehat{\mathbf{Y}}_A$, and $\widehat{\mathbf{Z}}_A$ are unit vectors associated with $\{A\}$.

Obtaining the position of any one point on the rigid body is clearly not enough to completely describe[2] the rigid body in $\Re^3$. We need to know the orientation of the rigid body. The orientation of a rigid body, B, is obtained by attaching a coordinate system, $\{B\}$, to the rigid body, and then obtaining a description of $\{B\}$ with respect to another rigid body or coordinate system $\{A\}$ (see Fig. 2.2). It may be noted that as far as orientation is concerned, the origin of $\{B\}$ can be chosen to coincide with the origin of $\{A\}$.

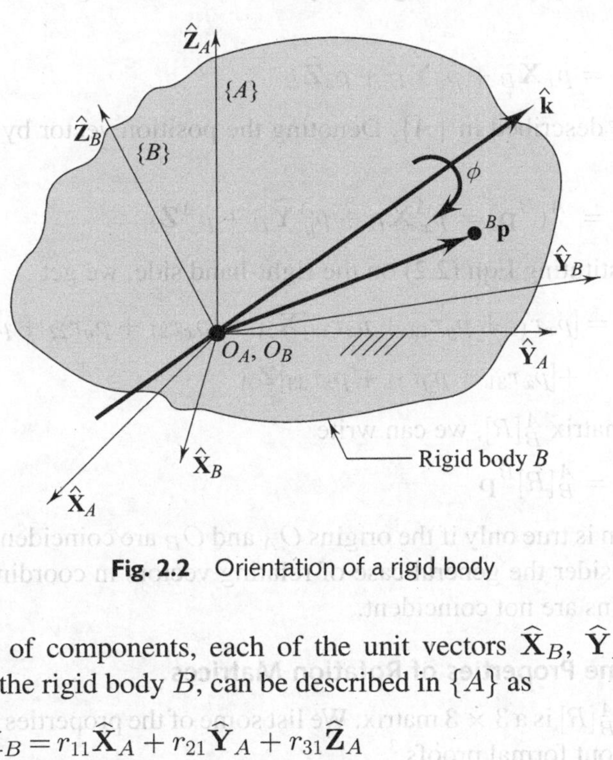

Fig. 2.2 Orientation of a rigid body

In terms of components, each of the unit vectors $\widehat{\mathbf{X}}_B$, $\widehat{\mathbf{Y}}_B$, and $\widehat{\mathbf{Z}}_B$, attached to the rigid body B, can be described in $\{A\}$ as

$$ \begin{aligned} {}^{A}\widehat{\mathbf{X}}_B &= r_{11}\widehat{\mathbf{X}}_A + r_{21}\widehat{\mathbf{Y}}_A + r_{31}\widehat{\mathbf{Z}}_A \\ {}^{A}\widehat{\mathbf{Y}}_B &= r_{12}\widehat{\mathbf{X}}_A + r_{22}\widehat{\mathbf{Y}}_A + r_{32}\widehat{\mathbf{Z}}_A \\ {}^{A}\widehat{\mathbf{Z}}_B &= r_{13}\widehat{\mathbf{X}}_A + r_{23}\widehat{\mathbf{Y}}_A + r_{33}\widehat{\mathbf{Z}}_A \end{aligned} \tag{2.2} $$

[2] It is possible to completely describe a rigid body in $\Re^3$ by obtaining the position vectors of three points not lying on a straight line.

where r_{ij} (i, j being 1, 2, 3) are the components. In Eqn (2.2), the leading superscript in $A(\widehat{\cdot})B$ signifies that we are describing the unit vectors $(\widehat{\cdot})_B$ in $\{A\}$.

We define a 3×3 matrix whose columns are ${}^A\widehat{\mathbf{X}}_B$, ${}^A\widehat{\mathbf{Y}}_B$, and ${}^A\widehat{\mathbf{Z}}_B$ and denote it by ${}^A_B[R]$. We call ${}^A_B[R]$ the rotation matrix—the leading superscript and subscript signify that the matrix relates $\{B\}$ to $\{A\}$. From Eqn (2.2), the elements of ${}^A_B[R]$ are r_{ij} (where $i, j = 1, 2, 3$), and they completely describe the three coordinate axes of $\{B\}$ relative to $\{A\}$. Recalling that the orientation of rigid body B with reference to $\{A\}$ is equivalent to a description of the coordinate axis of $\{B\}$ with reference to $\{A\}$, we can say that the rotation matrix ${}^A_B[R]$ gives the orientation of the rigid body B in $\{A\}$.

An arbitrary point in the rigid body B, as shown in Fig. 2.2 and described by

$$
{}^B\mathbf{p} = p_x\widehat{\mathbf{X}}_B + p_y\widehat{\mathbf{Y}}_B + p_z\widehat{\mathbf{Z}}_B \tag{2.3}
$$

can now be described in $\{A\}$. Denoting the position vector by ${}^A\mathbf{p}$, we can write

$$
{}^A\mathbf{p} = {}^A({}^B\mathbf{p}) = p_x{}^A\widehat{\mathbf{X}}_B + p_y{}^A\widehat{\mathbf{Y}}_B + p_z{}^A\widehat{\mathbf{Z}}_B
$$

and on substituting Eqn (2.2) on the right-hand side, we get

$$
\begin{aligned}
{}^A\mathbf{p} = & [p_x r_{11} + p_y r_{12} + p_z r_{13}]\widehat{\mathbf{X}}_A + [p_x r_{21} + p_y r_{22} + p_z r_{23}]\widehat{\mathbf{Y}}_A \\
& + [p_x r_{31} + p_y r_{32} + p_z r_{33}]\widehat{\mathbf{Z}}_A
\end{aligned}
$$

Using the matrix ${}^A_B[R]$, we can write

$$
{}^A\mathbf{p} = {}^A_B[R]{}^B\mathbf{p} \tag{2.4}
$$

This relation is true only if the origins O_A and O_B are coincident. In Section 2.3, we consider the general case of relating vectors in coordinate systems whose origins are not coincident.

2.2.1 Some Properties of Rotation Matrices

The matrix ${}^A_B[R]$ is a 3×3 matrix. We list some of the properties of a rotation matrix without formal proofs.[3]

1. The columns of the rotation matrix are unit vectors, orthogonal to one another. Hence,

$$
{}^A\widehat{\mathbf{X}}_B \cdot {}^A\widehat{\mathbf{X}}_B = {}^A\widehat{\mathbf{Y}}_B \cdot {}^A\widehat{\mathbf{Y}}_B = {}^A\widehat{\mathbf{Z}}_B \cdot {}^A\widehat{\mathbf{Z}}_B = 1
$$

[3] Rigorous treatment of properties of orthonormal matrices can be found in any textbook on linear algebra [see, for example, the textbook by Strang (1976)].

and

$${}^{A}\widehat{\mathbf{X}}_B \cdot {}^{A}\widehat{\mathbf{Y}}_B = {}^{A}\widehat{\mathbf{Y}}_B \cdot {}^{A}\widehat{\mathbf{Z}}_B = {}^{A}\widehat{\mathbf{Z}}_B \cdot {}^{A}\widehat{\mathbf{X}}_B = 0$$

The above relationships also imply ${}_{B}^{A}[R]^T\, {}_{B}^{A}[R] = [U]$, where $[U]$ is a 3×3 unit (or identity) matrix.

2. The determinant of the rotation matrix is $+1$.
   From properties (1) and (2), it follows that the inverse of the rotation matrix is equal to its transpose.

$${}_{A}^{B}[R] = {}_{B}^{A}[R]^{-1} = {}_{B}^{A}[R]^T \tag{2.5}$$

3. From Eqn (2.2), the element r_{23} can be obtained by taking the dot product of $\widehat{\mathbf{Y}}_A$ and $\widehat{\mathbf{Z}}_B$. Since $\widehat{\mathbf{Y}}_A$ and $\widehat{\mathbf{Z}}_B$ are unit vectors, the dot product is the cosine of the angle between $\widehat{\mathbf{Y}}_A$ and $\widehat{\mathbf{Z}}_B$. Likewise, all the elements r_{ij} (where i, j = 1, 2, 3) can be similarly obtained by taking appropriate dot products, and all the elements are cosines of angles. For this reason, the elements of the rotation matrix are also called the direction cosines.
4. One of the eigenvalues of ${}_{B}^{A}[R]$ is unity $(+1)$. The other two are complex conjugate pairs of the form $e^{\pm\iota\phi}$, where

$$\phi = \cos^{-1}\frac{r_{11} + r_{22} + r_{33} - 1}{2} \tag{2.6}$$

   and ι denotes $\sqrt{-1}$.
   The eigenvector corresponding to the unity eigenvalue is given by

$$\widehat{\mathbf{k}} = (1/2\sin\phi)[r_{32} - r_{23},\ r_{13} - r_{31},\ r_{21} - r_{12}]^T \tag{2.7}$$

We can make the following observations about the unit vector $\widehat{\mathbf{k}}$ and the angle ϕ, assuming $\phi \neq 0$ or $\pm\pi$:[4]

(a) For the rotation matrix ${}_{B}^{A}[R]$ and its eigenvector corresponding to the $+1$ eigenvalue, we have

$${}^{A}\widehat{\mathbf{k}} = {}_{B}^{A}[R]\,{}^{B}\widehat{\mathbf{k}} = 1\,{}^{B}\widehat{\mathbf{k}} \tag{2.8}$$

The first equality follows from Eqn (2.4), and the second from the definition of an eigenvector. This implies that the components of $\widehat{\mathbf{k}}$ are the same in $\{A\}$ and $\{B\}$, and hence we can drop the leading superscript. It also implies that $\widehat{\mathbf{k}}$ is fixed in both $\{A\}$ and $\{B\}$, and a line parallel to $\widehat{\mathbf{k}}$ stays at the same place during the rotation of the rigid body B from $\{A\}$ to $\{B\}$. A single

[4] If $\phi = 0$, then there is no change in orientation and ${}_{B}^{A}[R]$ is an identity matrix. The case of $\phi = \pm\pi$ is a special case and needs to be handled carefully—when $\phi = \pm\pi$, the matrix ${}_{B}^{A}[R]$ has all three real roots ($+1$, -1, and -1) and the eigenvector corresponding to $+1$ can still be obtained.

rotation about $\widehat{\mathbf{k}}$ by an angle of ϕ, shown schematically in Fig. 2.2, can take the rigid body coincident with $\{A\}$ to $\{B\}$.

(b) If the components of $\widehat{\mathbf{k}}$ are denoted by $(k_x, k_y, k_z)^T$ [given in Eqn (2.7)], then the elements r_{ij} (where $i, j = 1, 2, 3$) of ${}^A_B[R]$ can be written as

$$\begin{aligned}
r_{11} &= k_x^2(1-\cos\phi)+\cos\phi \\
r_{12} &= k_x k_y(1-\cos\phi)-k_z\sin\phi \\
r_{13} &= k_z k_x(1-\cos\phi)+k_y\sin\phi \\
r_{21} &= k_x k_y(1-\cos\phi)+k_z\sin\phi \\
r_{22} &= k_y^2(1-\cos\phi)+\cos\phi \\
r_{23} &= k_y k_z(1-\cos\phi)-k_x\sin\phi \\
r_{31} &= k_z k_x(1-\cos\phi)-k_y\sin\phi \\
r_{32} &= k_y k_z(1-\cos\phi)+k_x\sin\phi \\
r_{33} &= k_z^2(1-\cos\phi)+\cos\phi
\end{aligned} \tag{2.9}$$

(c) If $\widehat{\mathbf{k}}$ is parallel to $\widehat{\mathbf{X}}_A$ and hence to $\widehat{\mathbf{X}}_B$, the rotation matrix can be written as

$${}^A_B[R] = [R(\widehat{\mathbf{X}},\phi)] = \begin{pmatrix} 1 & 0 & 0 \\ 0 & \cos\phi & -\sin\phi \\ 0 & \sin\phi & \cos\phi \end{pmatrix} \tag{2.10}$$

The rotation matrix, given in Eqn (2.10), will be denoted by $[R(\widehat{\mathbf{X}},\phi)]$ to signify that the rotation axis[5] is $\widehat{\mathbf{X}}$ and the rotation is by angle ϕ. The rotation about $\widehat{\mathbf{X}}$ is shown in Fig. 2.3.

We can similarly obtain the rotation matrices for rotation about $\widehat{\mathbf{Y}}$ and $\widehat{\mathbf{Z}}$, denoted by $[R(\widehat{\mathbf{Y}},\phi)]$ and $[R(\widehat{\mathbf{Z}},\phi)]$, respectively. They are given in Eqns (2.11) and (2.12).

$$[R(\widehat{\mathbf{Y}},\phi)] = \begin{pmatrix} \cos\phi & 0 & \sin\phi \\ 0 & 1 & 0 \\ -\sin\phi & 0 & \cos\phi \end{pmatrix} \tag{2.11}$$

$$[R(\widehat{\mathbf{Z}},\phi)] = \begin{pmatrix} \cos\phi & -\sin\phi & 0 \\ \sin\phi & \cos\phi & 0 \\ 0 & 0 & 1 \end{pmatrix} \tag{2.12}$$

The rotation matrices given in Eqns (2.10) through (2.12), as we will see in Section 2.2.3, are used to obtain a representation of the orientation of rigid

[5] No subscript, A or B, is required since $\widehat{\mathbf{X}}_B$ is coincident with $\widehat{\mathbf{X}}_A$.

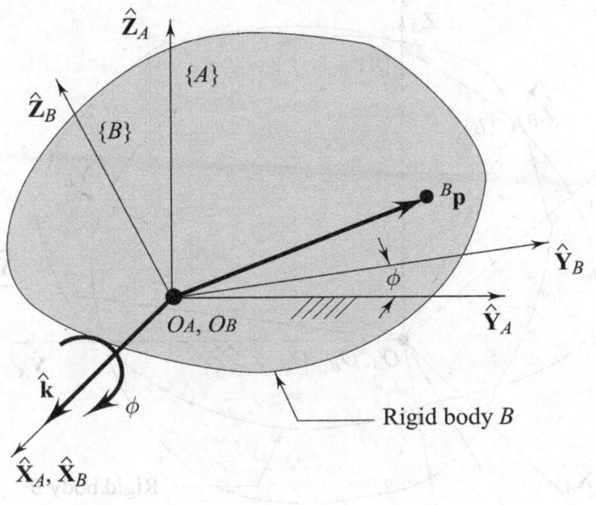

Fig. 2.3 Rotation about $\hat{\mathbf{X}}$ axes by angle ϕ

bodies in terms of Euler angles. But, first we look at a rigid body undergoing successive rotations.

2.2.2 Successive Rotations of a Rigid Body

Consider a rigid body B after it has undergone two successive rotations in the following manner.

1. Initially B is coincident with $\{A\}$.
2. The first rotation is relative to $\{A\}$, and, after the first rotation, the rigid body is described by $\{B_1\}$.
3. The second rotation is relative to $\{B_1\}$, which is now considered the fixed frame. After the second rotation, the rigid body is described by $\{B\}$.

The coordinate systems $\{A\}$, $\{B_1\}$, and $\{B\}$ with axes are shown in Fig. 2.4. We wish to obtain a single equivalent rotation matrix which describes the orientation of B in $\{A\}$ after the two successive rotations, i.e., we wish to obtain $^A_B[R]$. The rotation matrix $^A_B[R]$ can be obtained as follows.

The axes of $\{B_1\}$ can be written in terms of the axes of $\{A\}$ as

$$
\begin{aligned}
{}^A\hat{\mathbf{X}}_{B_1} &= p_{11}\hat{\mathbf{X}}_A + p_{21}\hat{\mathbf{Y}}_A + p_{31}\hat{\mathbf{Z}}_A \\
{}^A\hat{\mathbf{Y}}_{B_1} &= p_{12}\hat{\mathbf{X}}_A + p_{22}\hat{\mathbf{Y}}_A + p_{32}\hat{\mathbf{Z}}_A \\
{}^A\hat{\mathbf{Z}}_{B_1} &= p_{13}\hat{\mathbf{X}}_A + p_{23}\hat{\mathbf{Y}}_A + p_{33}\hat{\mathbf{Z}}_A
\end{aligned}
$$

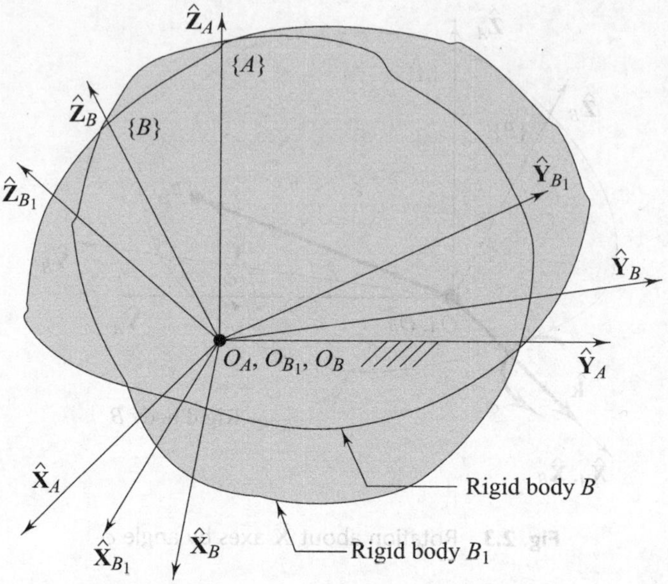

Fig. 2.4 Successive rotation of a rigid body

and the axes of $\{B\}$ can be written in terms of the axes of $\{B_1\}$ as

$$\begin{aligned}
{}^{B_1}\widehat{\mathbf{X}}_B &= q_{11}\widehat{\mathbf{X}}_{B_1} + q_{21}\widehat{\mathbf{Y}}_{B_1} + q_{31}\widehat{\mathbf{Z}}_{B_1} \\
{}^{B_1}\widehat{\mathbf{Y}}_B &= q_{12}\widehat{\mathbf{X}}_{B_1} + q_{22}\widehat{\mathbf{Y}}_{B_1} + q_{32}\widehat{\mathbf{Z}}_{B_1} \\
{}^{B_1}\widehat{\mathbf{Z}}_B &= q_{13}\widehat{\mathbf{X}}_{B_1} + q_{23}\widehat{\mathbf{Y}}_{B_1} + q_{33}\widehat{\mathbf{Z}}_{B_1}
\end{aligned}$$

where p_{ij} and q_{ij} $(i, j = 1, 2, 3)$ are direction cosines of the matrices ${}^{A}_{B_1}[R]$ and ${}^{B_1}_{B}[R]$, respectively. The columns of the matrix ${}^{A}_{B}[R]$, by definition, are the vectors ${}^A({}^{B_1}\widehat{\mathbf{X}}_B)$, ${}^A({}^{B_1}\widehat{\mathbf{Y}}_B)$, and ${}^A({}^{B_1}\widehat{\mathbf{Z}}_B)$, and are given as

$$\begin{aligned}
{}^A\widehat{\mathbf{X}}_B &= {}^A({}^{B_1}\widehat{\mathbf{X}}_B) = q_{11}\,{}^A\widehat{\mathbf{X}}_{B_1} + q_{21}\,{}^A\widehat{\mathbf{Y}}_{B_1} + q_{31}\,{}^A\widehat{\mathbf{Z}}_{B_1} \\
{}^A\widehat{\mathbf{Y}}_B &= {}^A({}^{B_1}\widehat{\mathbf{Y}}_B) = q_{12}\,{}^A\widehat{\mathbf{X}}_{B_1} + q_{22}\,{}^A\widehat{\mathbf{Y}}_{B_1} + q_{32}\,{}^A\widehat{\mathbf{Z}}_{B_1} \\
{}^A\widehat{\mathbf{Z}}_B &= {}^A({}^{B_1}\widehat{\mathbf{X}}_B) = q_{13}\,{}^A\widehat{\mathbf{X}}_{B_1} + q_{23}\,{}^A\widehat{\mathbf{Y}}_{B_1} + q_{33}\,{}^A\widehat{\mathbf{Z}}_{B_1}
\end{aligned}$$

Substituting expressions for ${}^A\widehat{\mathbf{X}}_{B_1}$, ${}^A\widehat{\mathbf{Y}}_{B_1}$, and ${}^A\widehat{\mathbf{Z}}_{B_1}$, and denoting the elements of the matrix ${}^{A}_{B}[R]$ by r_{ij} (where $i, j = 1, 2, 3$), we find that $r_{11} = p_{11}q_{11} + p_{12}q_{21} + p_{13}q_{31}$, etc. In matrix form, we can write

$$ {}^{A}_{B}[R] = {}^{A}_{B_1}[R]\;{}^{B_1}_{B}[R] \tag{2.13}$$

One can also show by a similar procedure that the resultant rotation matrix after n successive rotations is given by

$$ {}^{A}_{B}[R] = {}^{A}_{B_1}[R]\;{}^{B_1}_{B_2}[R]\quad \ldots \quad {}^{B_{n-1}}_{B}[R] \tag{2.14}$$

It may be mentioned that matrix multiplication is generally not commutative, and, except for special situations, ${}^{A}_{B_1}[R]\ {}^{B_1}_{B}[R] \neq {}^{B_1}_{B}[R]\ {}^{A}_{B_1}[R]$. This implies that finite rotations, in general, do not commute and the order of rotations is important.

2.2.3 Representation of Orientation by Three Angles

In the last section we have seen two representations of the orientation of a rigid body B in a reference frame $\{A\}$—in terms of direction cosines and in terms of the unit vector $\hat{\mathbf{k}}$ and angle ϕ. In the first case, we have nine quantities and six constraints (see property 1 of ${}^{A}_{B}[R]$), and in terms of the unit vector $\hat{\mathbf{k}}$ and angle ϕ, we have four quantities and one unit vector constraint. It is also possible to describe or specify the orientation of a rigid body by successive rotations of three angles about two or three axes. These are called Euler angles and these can be of two kinds—the three rotations can be about fixed axes of $\{A\}$, or they can be about the axes attached to the moving rigid body, i.e., about the axes of $\{B\}$. In either case, one can also have rotations about two or three distinct axes. In robotics, rotations about axes fixed in $\{B\}$ are of more interest and this is described, very briefly, next. An excellent treatment of Euler angles is given in the textbook by Kane, Likins, and Levinson (1983)

Rotations about Three Distinct Moving Axes

To bring the rigid body B or the coordinate system $\{B\}$ to a desired orientation, we subject B to successive rotations about the axes $\hat{\mathbf{X}}_B$, $\hat{\mathbf{Y}}_B$, and $\hat{\mathbf{Z}}_B$ fixed to the moving rigid body. In particular, we can subject B to a rotation θ_1 about $\hat{\mathbf{X}}_B$, then a rotation θ_2 about $\hat{\mathbf{Y}}_B$, and finally a rotation θ_3 about $\hat{\mathbf{Z}}_B$. The angles θ_1, θ_2, and θ_3 are called 1-2-3 or the X-Y-Z Euler angles. The three rotations are schematically shown in Fig. 2.5.

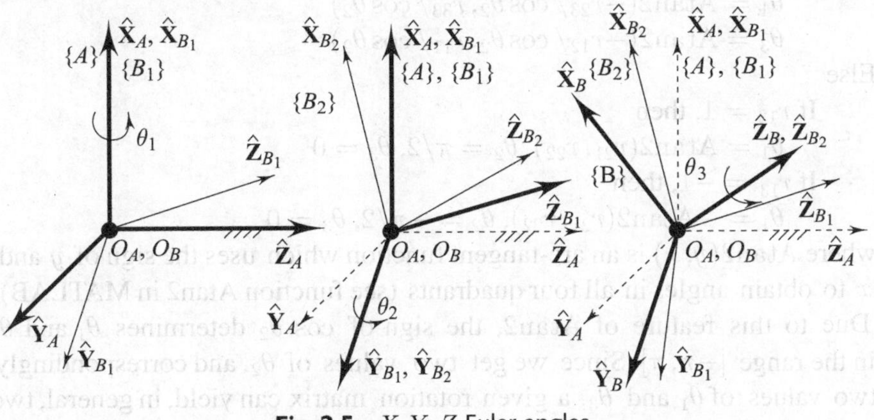

Fig. 2.5 X-Y-Z Euler angles

To obtain an equivalent single resultant matrix, we use Eqns (2.10) through (2.12) and (2.14). For the case described above, we have

$$ {}^{A}_{B_1}[R] = [R(\widehat{\mathbf{X}}, \theta_1)] = \begin{pmatrix} 1 & 0 & 0 \\ 0 & \cos\theta_1 & -\sin\theta_1 \\ 0 & \sin\theta_1 & \cos\theta_1 \end{pmatrix} $$

$$ {}^{B_1}_{B_2}[R] = [R(\widehat{\mathbf{Y}}, \theta_2)] = \begin{pmatrix} \cos\theta_2 & 0 & \sin\theta_2 \\ 0 & 1 & 0 \\ -\sin\theta_2 & 0 & \cos\theta_2 \end{pmatrix} $$

$$ {}^{B_2}_{B}[R] = [R(\widehat{\mathbf{Z}}, \theta_3)] = \begin{pmatrix} \cos\theta_3 & -\sin\theta_3 & 0 \\ \sin\theta_3 & \cos\theta_3 & 0 \\ 0 & 0 & 1 \end{pmatrix} $$

and from Eqn (2.14), we get

$$ {}^{A}_{B}[R] = {}^{A}_{B_1}[R]\, {}^{B_1}_{B_2}[R]\, {}^{B_2}_{B}[R] $$

$$ = \begin{pmatrix} c_2c_3 & -c_2s_3 & s_2 \\ s_1s_2c_3 + s_3c_1 & -s_1s_2s_3 + c_3c_1 & -s_1c_2 \\ -c_1s_2c_3 + s_3s_1 & c_1s_2s_3 + c_3s_1 & c_1c_2 \end{pmatrix} \quad (2.15) $$

It must be stressed that the rotation matrix in Eqn (2.15) will be different if the orders of successive rotations are different from θ_1 about $\widehat{\mathbf{X}}_B$, θ_2 about $\widehat{\mathbf{Y}}_B$, and θ_3 about $\widehat{\mathbf{Z}}_B$.

To determine θ_1, θ_2, and θ_3, given the rotation matrix, we can use the following algorithm [obtained from inspection of Eqn (2.15)].

Algorithm $r_{ij} \Rightarrow \theta_i$ for body-fixed X-Y-Z rotations

If $r_{13} \neq \pm 1$, then

$$ \theta_2 = \text{Atan2}[r_{13}, \pm\sqrt{(r_{11}^2 + r_{12}^2)}] $$
$$ \theta_1 = \text{Atan2}(-r_{23}/\cos\theta_2, r_{33}/\cos\theta_2) $$
$$ \theta_3 = \text{Atan2}(-r_{12}/\cos\theta_2, r_{11}/\cos\theta_2) $$

Else

If $r_{13} = 1$, then

$$ \theta_1 = \text{Atan2}(r_{21}, r_{22}),\ \theta_2 = \pi/2,\ \theta_3 = 0 $$

If $r_{13} = -1$, then

$$ \theta_1 = -\text{Atan2}(r_{21}, r_{22}),\ \theta_2 = -\pi/2,\ \theta_3 = 0 $$

where Atan2(y, x) is an arc-tangent function which uses the sign of y and x to obtain angles in all four quadrants (see function Atan2 in MATLAB). Due to this feature of Atan2, the sign of $\cos\theta_2$ determines θ_1 and θ_3 in the range $[-\pi, \pi]$. Since we get two values of θ_2, and correspondingly two values of θ_1 and θ_3, a given rotation matrix can yield, in general, two sets of X-Y-Z Euler angles $(\theta_1, \theta_2, \theta_3)$. If θ_2 is $\pm\pi/2$, then we cannot find θ_1

can only find the sum or the difference of θ_1 and θ_3, and we choose θ_3 as zero. These are called singularities, and all Euler angle representations are known to have singularities.

Rotations about Two Distinct Moving Axes

In robotic applications, such as in the commonly used three intersecting axes wrists (see Fig. 2.19), rotations are best modelled by the use of three rotations about the two distinct (in the case of Fig. 2.19 $\hat{\mathbf{Z}}$ and $\hat{\mathbf{Y}}$) axes. These are called the 3-2-3 or Z-Y-Z Euler angles.

***Z*-*Y*-*Z* Euler angles**

In this scheme, as shown in Fig. 2.6, the first rotation is by an amount θ_1 about $\hat{\mathbf{Z}}_B$, the second rotation is by an amount θ_2 about $\hat{\mathbf{Y}}_B$, and finally the third rotation is by an amount θ_3 about $\hat{\mathbf{Z}}_B$. The resultant rotation matrix is given by

$$
{}^A_B[R] = \begin{pmatrix} \cos\theta_1 & -\sin\theta_1 & 0 \\ \sin\theta_1 & \cos\theta_1 & 0 \\ 0 & 0 & 1 \end{pmatrix} \begin{pmatrix} \cos\theta_2 & 0 & \sin\theta_2 \\ 0 & 1 & 0 \\ -\sin\theta_2 & 0 & \cos\theta_2 \end{pmatrix} \begin{pmatrix} \cos\theta_3 & -\sin\theta_3 & 0 \\ \sin\theta_3 & \cos\theta_3 & 0 \\ 0 & 0 & 1 \end{pmatrix}
$$

$$
= \begin{pmatrix} c_1c_2c_3 - s_1s_3 & -c_1c_2s_3 - s_1c_3 & c_1s_2 \\ s_1c_2c_3 + c_1s_3 & -s_1c_2s_3 + c_1c_3 & s_1s_2 \\ -s_2c_3 & s_2s_3 & c_2 \end{pmatrix} \qquad (2.16)
$$

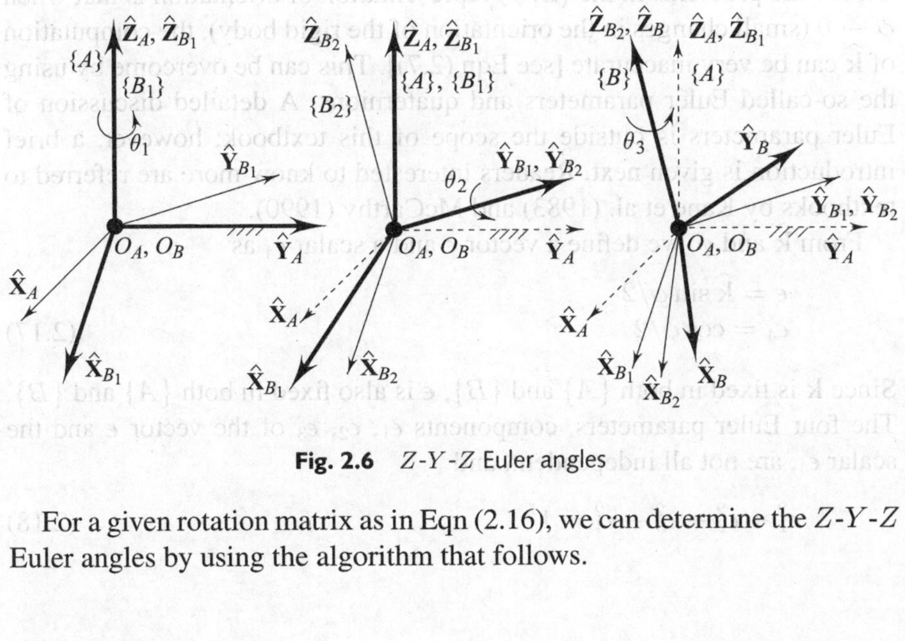

Fig. 2.6 Z-Y-Z Euler angles

For a given rotation matrix as in Eqn (2.16), we can determine the Z-Y-Z Euler angles by using the algorithm that follows.

Algorithm $r_{ij} \Rightarrow Z$-Y-Z Euler angles

If $r_{33} \neq \pm 1$, then

$$\theta_2 = \text{Atan2}[\pm\sqrt{(r_{31}^2 + r_{32}^2)}, r_{33}]$$
$$\theta_1 = \text{Atan2}(r_{23}/\sin\theta_2, r_{13}/\sin\theta_2)$$
$$\theta_3 = \text{Atan2}(r_{32}/\sin\theta_2, -r_{31}/\sin\theta_2)$$

Else

If $r_{33} = 1$, then

$$\theta_1 = \theta_2 = 0, \theta_3 = \text{Atan2}(-r_{12}, r_{11})$$

If $r_{33} = -1$, then

$$\theta_1 = 0, \theta_2 = \pi, \theta_3 = \text{Atan2}(r_{12}, -r_{11})$$

As in the case of X-Y-Z Euler angles, we have two possible sets of Z-Y-Z Euler angles $(\theta_1, \theta_2, \theta_3)$ for a given rotation matrix. It may be noted that $r_{33} = \pm 1$ corresponds to a singularity and again only the sum or the difference of θ_1 and θ_3 can be found. To obtain unique θ_1 and θ_3 when $r_{33} = \pm 1$, we choose θ_1 as zero.

It must be stressed that, as in the case of the Euler angles about three distinct axes, the rotation matrix will be different if instead of a 3-2-3 sequence some other sequence is followed. It may also be noted that, in rotations about two distinct axes, the consecutive axes cannot be the same. Considering all possible combinations of rotations about two or three distinct axes attached to $\{B\}$ or attached to $\{A\}$, there are 24 possible Euler angle sets.

2.2.4 Other Representations of Orientation

One of the problems in the $(\widehat{\mathbf{k}}, \phi)$ representation of orientation is that when $\phi \to 0$ (small changes in the orientation of the rigid body), the computation of $\widehat{\mathbf{k}}$ can be very inaccurate [see Eqn (2.7)]. This can be overcome by using the so-called Euler parameters and quaternions. A detailed discussion of Euler parameters is outside the scope of this textbook; however, a brief introduction is given next. Readers interested to know more are referred to textbooks by Kane et al. (1983) and McCarthy (1990).

From $\widehat{\mathbf{k}}$ and ϕ, we define a vector ϵ and a scalar ϵ_4 as

$$\begin{aligned} \epsilon &= \widehat{\mathbf{k}} \sin\phi/2 \\ \epsilon_4 &= \cos\phi/2 \end{aligned} \tag{2.17}$$

Since $\widehat{\mathbf{k}}$ is fixed in both $\{A\}$ and $\{B\}$, ϵ is also fixed in both $\{A\}$ and $\{B\}$. The four Euler parameters, components ϵ_1, ϵ_2, ϵ_3 of the vector ϵ and the scalar ϵ_4, are not all independent, and

$$\epsilon_1^2 + \epsilon_2^2 + \epsilon_3^2 + \epsilon_4^2 = 1 \tag{2.18}$$

One can also find ϵ and ϵ_4 directly from the elements r_{ij} (where $i, j = 1, 2, 3$) of ${}^A_B[R]$ as

$$\epsilon_4 = (1/2)(1 + r_{11} + r_{22} + r_{33})^{1/2}$$
$$\epsilon = \frac{1}{4\epsilon_4}[r_{32} - r_{23}, r_{13} - r_{31}, r_{21} - r_{12}]^T \tag{2.19}$$

It may be noted that for small ϕ (small rotations), ϵ_4 is well defined. It may appear from Eqn (2.19) that when $\phi \longrightarrow \pm\pi$, the computation of ϵ can be inaccurate. However, in the limit $\phi \longrightarrow \pm\pi$ [see Eqn (2.17)], we get

$$\epsilon_4 = \pm\delta\phi/2$$
$$\epsilon = \pm[k_x, k_y, k_z]^T \tag{2.20}$$

where $\delta\phi$ is the small deviation of ϕ from $\pm\pi$, and hence the Euler parameters are always well defined.

Another way to represent orientation is by means of quaternions. Quaternions were invented by Hamilton (1866) and they are also called hyper-complex numbers since they can be viewed as extensions of ordinary complex numbers. A general quaternion, H, can be described in terms of four parameters, c_0, c_1, c_2, c_3, and four 2×2 matrices, closely related to the so-called Pauli spin matrices combined with an identity matrix [see p. 185 in Arfken (1985)], as

$$H = c_0\begin{pmatrix} 1 & 0 \\ 0 & 1 \end{pmatrix} + c_1\begin{pmatrix} \imath & 0 \\ 0 & -\imath \end{pmatrix} + c_2\begin{pmatrix} 0 & 1 \\ -1 & 0 \end{pmatrix} + c_3\begin{pmatrix} 0 & \imath \\ \imath & 0 \end{pmatrix} \tag{2.21}$$

where $\imath = \sqrt{-1}$. Denoting the four 2×2 matrices associated with the four parameters c_0, c_1, c_2, c_3 by $\mathcal{U}, \mathcal{I}, \mathcal{J}$, and $\mathcal{K}$, respectively, we can verify that

$$\mathcal{I}^2 = \mathcal{J}^2 = \mathcal{K}^2 = -1 \tag{2.22}$$

indicating that the matrices are 'similar' to complex numbers. It can be also verified that

$$\mathcal{I}\mathcal{J} = -\mathcal{J}\mathcal{I} = \mathcal{K}$$
$$\mathcal{J}\mathcal{K} = -\mathcal{K}\mathcal{J} = \mathcal{I} \tag{2.23}$$
$$\mathcal{K}\mathcal{I} = -\mathcal{I}\mathcal{K} = \mathcal{J}$$

indicating that the product of two matrices, from $\mathcal{I}, \mathcal{J}, \mathcal{K}$, are 'similar' to the rules of the cross product of two vectors in $\Re^3$. A general quaternion can also be interpreted as a 'sum' of a scalar c_0 and a vector with components $(c_1, c_2, c_3)^T$ as

$$C = c_0 + c_1\widehat{\mathbf{i}} + c_2\widehat{\mathbf{j}} + c_3\widehat{\mathbf{k}} \tag{2.24}$$

where $\widehat{\mathbf{i}}$, $\widehat{\mathbf{j}}$, and $\widehat{\mathbf{k}}$ are the unit vectors in $\Re^3$. In this form the product of two quaternions, which is also a quaternion, can be easily computed as

$$C \cdot D = (c_0, \mathbf{c}) \cdot (d_0, \mathbf{d}) = (c_0 d_0 - \mathbf{c} \cdot \mathbf{d}, c_0 \mathbf{d} + d_0 \mathbf{c} + \mathbf{c} \times \mathbf{d}) \quad (2.25)$$

In addition, a quaternion, unlike a vector in $\Re^3$, has an inverse, C^*, similar to a conjugate in complex numbers. The conjugate C^* is given by

$$C = c_0 - c_1 \widehat{\mathbf{i}} - c_2 \widehat{\mathbf{j}} - c_3 \widehat{\mathbf{k}} \quad (2.26)$$

One can also show that if $c_0^2 + c_1^2 + c_2^2 + c_3^2 = 1$, then the quaternion, called a unit quaternion, represents the orientation of a rigid body in $\Re^3$. In particular, the quaternion

$$Z = \cos\phi/2 + k_x \sin\phi/2\, \widehat{\mathbf{i}} + k_y \sin\phi/2\, \widehat{\mathbf{j}} + k_z \sin\phi/2\, \widehat{\mathbf{k}} \quad (2.27)$$

clearly represents an orientation of a rigid body in $\Re^3$ since c_0, c_1, c_2, and c_3 are the four Euler parameters.

The algebra of quaternions is well known; however, quaternions and their advantages in representing the rigid body orientation, over other representations, is outside the scope of this text.

Let us now sum up the discussion so far. A rigid body in three-dimensional space, $\Re^3$, has six degrees of freedom and can be described by a position and an orientation. The position is described by a vector which has three components in $\Re^3$. The orientation of the rigid body can be described in several ways. The orthogonal rotation matrix ${}^A_B[R]$ is most suitable for analysis (it will be used to obtain angular velocity in Chapter 4) but not very useful for motion planning since one has to carry around nine variables and six constraints. With the axis-angle, $(\widehat{\mathbf{k}}, \phi)$, form one has to deal with a lesser number of parameters and only one constraint. The axis-angle form (and the Euler parameters or quaternions) is more suitable for motion planning. The three Euler angle representations are minimal; however, they suffer from the problem of singularities and one needs to be careful in doing computations with them. It may be noted that we can easily go from one representation to another as shown by the closed-form expressions and algorithms developed in this section.

2.3 Transformation Between Coordinate Systems

In Section 2.2, we introduced the concept of describing a vector $\mathbf{p}$ in $\{A\}$ and $\{B\}$ when the origins O_A and O_B are coincident. To recapitulate, a vector ${}^B\mathbf{p}$ can be described in $\{A\}$ as

$${}^A\mathbf{p} = {}^A_B[R]\,{}^B\mathbf{p}$$

where the rotation matrix ${}^A_B[R]$ gives the orientation of B or the coordinate system $\{B\}$ in $\{A\}$. Figure 2.7 shows two coordinate systems $\{A\}$ and $\{B\}$ as in Fig. 2.2, but now the origins O_A and O_B are not coincident— the origin of $\{B\}$ is at the point described by the vector ${}^A\mathbf{O}_B$. The leading superscript denotes the fact that the vector is described in $\{A\}$. The vector ${}^B\mathbf{p}$ can be described in $\{A\}$ in terms of ${}^B\mathbf{p}$, the rotation matrix ${}^A_B[R]$, and ${}^A\mathbf{O}_B$ as

$$
{}^A\mathbf{p} = {}^A\mathbf{O}_B + {}^A_B[R]\,{}^B\mathbf{p} \tag{2.28}
$$

Equation (2.28) is simply a result of the addition of two vectors described in the same coordinate system. It represents a general displacement of a rigid body consisting of a rotation and a translation. A rotation is a displacement in which one point on the rigid body is fixed and the rigid body changes its orientation (in our discussion on orientation, we had kept the origins, O_A of $\{A\}$ and O_B of $\{B\}$, coincident). A translation is a displacement in which all points in the rigid body move equal distances along a parallel line and the rotation matrix ${}^A_B[R]$ is identity. Rotations in $\Re^3$ form a group called the special orthogonal group denoted by $\mathbf{SO}(3)$. General rigid body displacements also form a group, called the special Euclidean group in $\Re^3$, and is denoted by $\mathbf{SE}(3)$. There exists a large amount of literature in the kinematics of rigid body displacements and their relationships to the properties of groups $\mathbf{SO}(3)$ and $\mathbf{SE}(3)$. These are beyond the scope of this text and readers interested to know more are referred to the textbook by Murray et al. (1994).

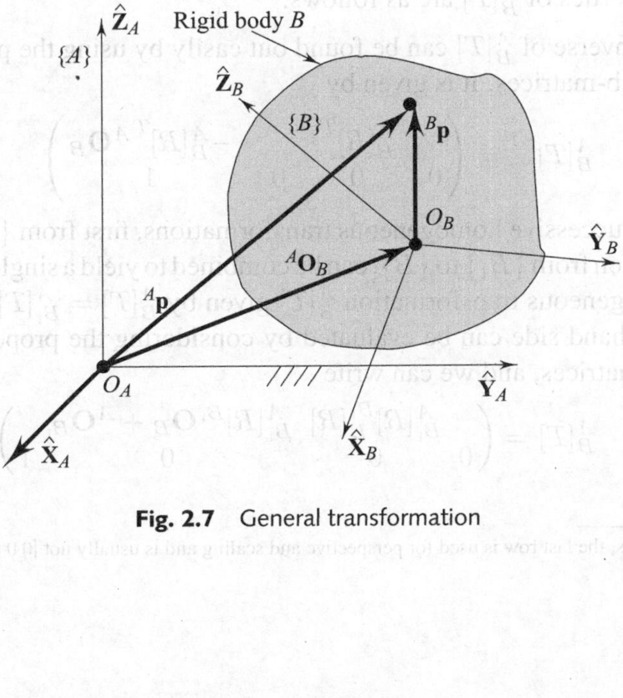

Fig. 2.7 General transformation

2.3.1 Homogeneous Transformation

Equation (2.28) can be written in a compact form as

$$^{A}\mathbf{P} = {}^{A}_{B}[T]\,^{B}\mathbf{P} \tag{2.29}$$

where $^{A}\mathbf{P}$ and $^{B}\mathbf{P}$ are 4×1 vectors constructed by concatenating a '1' to $^{A}\mathbf{p}$ and $^{B}\mathbf{p}$ as

$$^{A}\mathbf{P} = [^{A}\mathbf{p} \mid 1]^{T}$$
$$^{B}\mathbf{P} = [^{B}\mathbf{p} \mid 1]^{T}$$

The 4×1 vectors are called homogeneous coordinates (see Section 2.8 for a brief discussion on homogeneous coordinates).

The matrix ${}^{A}_{B}[T]$ is formed as

$${}^{A}_{B}[T] = \begin{pmatrix} & {}^{A}_{B}[R] & & {}^{A}\mathbf{O}_{B} \\ 0 & 0 & 0 & 1 \end{pmatrix} \tag{2.30}$$

The 4×4 matrix in Eqn. (2.30) is called the homogeneous transformation matrix and, in addition to robotics, is used in fields such as computer graphics and computer vision.[6] As shown in Eqn (2.30), the top left 3×3 matrix is the rotation matrix ${}^{A}_{B}[R]$. This will be an identity matrix for a pure translation of the rigid body. If the top right 3×1 vector is zero, then ${}^{A}_{B}[T]$ describes a pure rotation.

2.4 Properties of ${}^{A}_{B}[T]$

Some properties of ${}^{A}_{B}[T]$ are as follows.

- The inverse of ${}^{A}_{B}[T]$ can be found out easily by using the properties of the sub-matrices. It is given by

$${}^{A}_{B}[T]^{-1} = \begin{pmatrix} & {}^{A}_{B}[R]^{T} & & -{}^{A}_{B}[R]^{T}\,{}^{A}\mathbf{O}_{B} \\ 0 & 0 & 0 & 1 \end{pmatrix} \tag{2.31}$$

- Two successive homogeneous transformations, first from $\{A\}$ to $\{B_1\}$ and then from $\{B_1\}$ to $\{B\}$, can be combined to yield a single equivalent homogeneous transformation ${}^{A}_{B}[T]$ given by ${}^{A}_{B}[T] = {}^{A}_{B_1}[T]\;{}^{B_1}_{B}[T]$. The right-hand side can be evaluated by considering the properties of the sub-matrices, and we can write

$${}^{A}_{B}[T] = \begin{pmatrix} & {}^{A}_{B_1}[R]\,{}^{B_1}_{B}[R] & & {}^{A}_{B_1}[R]\,{}^{B_1}\mathbf{O}_{B} + {}^{A}\mathbf{O}_{B_1} \\ 0 & 0 & 0 & 1 \end{pmatrix} \tag{2.32}$$

[6] In these fields, the last row is used for perspective and scaling and is usually not [0 0 0 1].

Successive transformations need not be limited to two and, for n successive transformations, we can write

$$ {}^{A}_{B}[T] = {}^{A}_{B_1}[T]\; {}^{B_1}_{B_2}[T] \quad \ldots \quad {}^{B_{n-1}}_{B}[T] \tag{2.33}$$

It may be mentioned again that the matrix multiplication is generally not commutative, and, except for special situations, ${}^{A}_{B_1}[T]\;{}^{B_1}_{B}[T] \neq {}^{B_1}_{B}[T]\;{}^{A}_{B_1}[T]$.

- As in the case of ${}^{A}_{B}[R]$, we can try to obtain the eigenvalues and eigenvectors of ${}^{A}_{B}[T]$. From the definition of ${}^{A}_{B}[T]$ in Eqn (2.30), one of the eigenvalues is clearly $+1$. In fact, one can show that there are two repeated real eigenvalues of $+1$ and two complex conjugate pairs of the form $e^{\pm\imath\phi}$, where ϕ can be obtained from the trace of ${}^{A}_{B}[R]$ as before. To understand the meaning of the eigenvectors corresponding to $+1$ eigenvalue, we first recognize that the eigenvectors must be four dimensional since the matrix ${}^{A}_{B}[T]$ is a 4×4 matrix, and the eigenvalue $+1$ is repeated or has *algebraic* multiplicity 2.

The 4×1 eigenvectors, $\mathbf{X}$, associated with the repeated eigenvalue $+1$ are obtained from

$$\begin{pmatrix} {}^{A}_{B}[R] - [U] & {}^{A}\mathbf{O}_B \\ 0\,0\,0 & 0 \end{pmatrix} \mathbf{X} = \mathbf{0} \tag{2.34}$$

where $[U]$ is the 3×3 identity matrix.

Clearly $\mathbf{X} = (\widehat{\mathbf{k}}, 0)^T$, where $\widehat{\mathbf{k}}$ is the eigenvector of ${}^{A}_{B}[R]$ corresponding to eigenvalue $+1$, is one of the solutions to Eqn (2.34). To show that there cannot be any other solutions to Eqn (2.34) for arbitrary ${}^{A}\mathbf{O}_B$, we proceed as follows: Assume that $(\mathbf{Y}, 1)^T$ is a solution to Eqn (2.34). Then we get

$$([U] - {}^{A}_{B}[R])\mathbf{Y} = {}^{A}\mathbf{O}_B \tag{2.35}$$

However, the matrix $([U] - {}^{A}_{B}[R])$ is singular and, hence, $\mathbf{Y}$ cannot exist for arbitrary ${}^{A}\mathbf{O}_B$.

The above reasoning implies that the geometric multiplicity cannot be 2 and the matrix ${}^{A}_{B}[T]$ cannot be diagonalized and at best we can reduce it to a Jordan form [see Strang (1976)].

- The matrix ${}^{A}_{B}[T]$ represents the general motion of a rigid body in $\Re^3$, and it is known that the general motion of a rigid body can be described in terms of six parameters. One can also describe the general motion of a rigid body as a twist defined as a rotation and a translation about and along a line in $\Re^3$ (see Section 2.8 for a brief discussion on lines, screws, and twists). Rotation and translation are described by two scalars and a line is located in $\Re^3$ by means of four parameters, and hence the six parameters associated with a twist contain the same information as ${}^A_B[T]$. The expressions for the rotation axis, $\widehat{\mathbf{k}}$ and the amount of rotation, ϕ, were obtained from the eigen analysis of ${}^A_B[R]$ [see Eqns (2.6) and (2.7)]. To locate the line along which the rigid body rotates and translates and the amount of translation for a given ${}^A_B[T]$, we proceed as follows.

We know from Eqn (2.35) that $\mathbf{Y}$ cannot exist for arbitrary ${}^A\mathbf{O}_B$. We can, however, consider the existence of $\mathbf{Y}$ for the vector ${}^A\mathbf{O}_B^*$ given by

$$ {}^A\mathbf{O}_B^* = {}^A\mathbf{O}_B - ({}^A\mathbf{O}_B \cdot \widehat{\mathbf{k}})\widehat{\mathbf{k}} \tag{2.36} $$

It may be noted that the vector ${}^A\mathbf{O}_B^*$ is the projection of the translation vector ${}^A\mathbf{O}_B$ on to a plane whose normal is the rotation axis $\widehat{\mathbf{k}}$ and it can be written as $\widehat{\mathbf{k}} \times ({}^A\mathbf{O}_B \times \widehat{\mathbf{k}})$.

It can be verified [see Exercise 2.12 and Sangamesh & Ghosal (2006)] that $\mathbf{Y}$ given by

$$ \mathbf{Y} = \frac{([U] - {}^A_B[R]^T)\,{}^A\mathbf{O}_B}{2(1-\cos\phi)} \tag{2.37} $$

satisfies

$$ ([U] - {}^A_B[R])\mathbf{Y} = {}^A\mathbf{O}_B^* \tag{2.38} $$

It can be also verified that $\mathbf{Y}$ is normal to $\widehat{\mathbf{k}}$ and the four independent parameters in the vector pair $(\widehat{\mathbf{k}}, \mathbf{Y} \times \widehat{\mathbf{k}})$ determine the line along the twist axis (see Section 2.8). Finally, the translation along the line is given by $({}^A\mathbf{O}_B \cdot \widehat{\mathbf{k}})$.

It may be noted that the vector $\mathbf{Y}$ is not defined when $\phi = 0$. This is to be expected since $\phi = 0$ implies pure translation. In this case, any unit vector parallel to ${}^A\mathbf{O}_B$ can be chosen as $\widehat{\mathbf{k}}$; the amount of translation is simply the magnitude of ${}^A\mathbf{O}_B$; and $\mathbf{Y}$ could be chosen as $(0,0,0)^T$.

2.5 Representation of Joints

A joint connects two or more[7] rigid bodies or links in a way as to permit relative motion between them. When two rigid bodies are connected by a joint, the joint imposes constraints on this system of two rigid bodies. For example, if two rigid bodies are connected by a hinged joint, the two rigid bodies no longer have 12 (6 + 6) degrees of freedom. The first rigid body can be described by six parameters as before, but the second connected rigid

[7] In serial manipulators, a joint connects two links; whereas in parallel and hybrid manipulators, a joint may connect more than two links.

body can now be described by an additional single variable. This is because a hinge only allows relative rotation between the two bodies, and a hinge is said to introduce five constraints between two connected rigid bodies or, alternatively, a hinge is said to have one degree of freedom. Formally, the degree of freedom of a joint is the number of independent variables, also called joint variables, required to describe the relative motion, between the two connected rigid bodies, allowed by the joint. The degree of freedom of a joint is also the same as $6 - m$, where m is number of constraints imposed by a joint.

Pair	Symbol	DOF	Representation
Screw	H	1	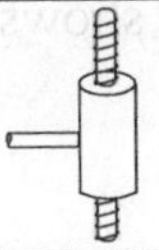
Revolute	R	1	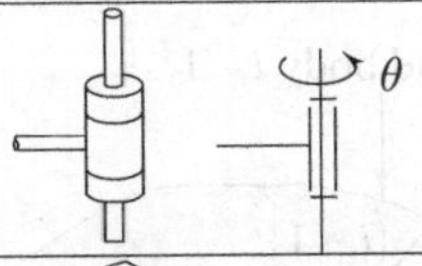
Prismatic	P	1	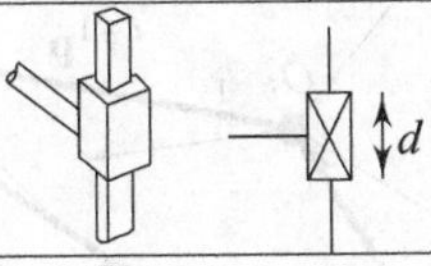
Cylindrical	C	2	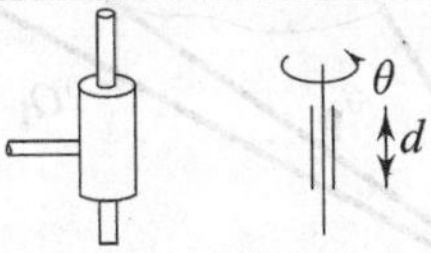
Spherical	S	3	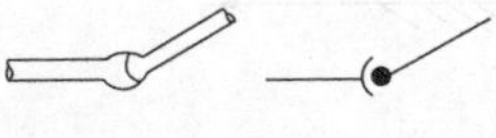
Planar pair	E	3	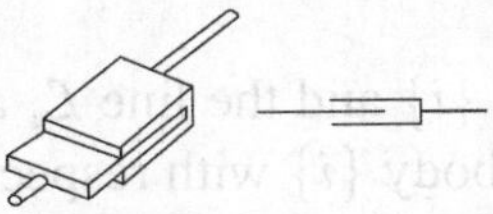
Hooke joint	T	2	

Fig. 2.8 Types of joints

Figure 2.8 shows some of the common joints—the hinge or revolute (R), sliding or prismatic (P), and screw (H) joints have one degree of freedom; the cylindrical (C) joint has two degrees of freedom; and the spherical (S) joint has three degrees of freedom. The last column in Fig. 2.8 shows the symbolic representation of the joints and the joint variable symbol(s) typically used for that joint. To mathematically represent joints, we present the constraints associated with a joint starting with the rotary (R) joint.

2.5.1 Rotary Joint

Associated with a rotary (R) joint is a line[8] about which the relative rotational motion of the two connected rigid bodies is allowed. The line is also called the joint axis. Figure 2.9 shows, schematically, two rigid bodies denoted by

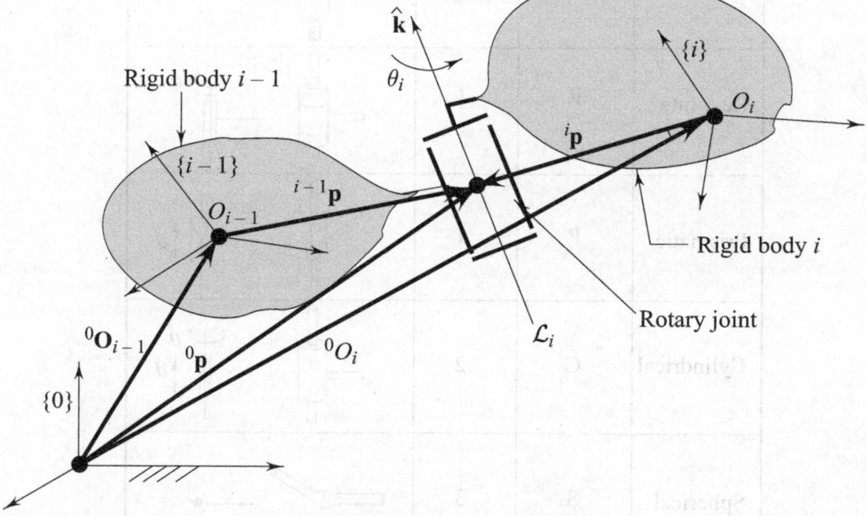

Fig. 2.9 A rotary joint

$\{i-1\}$ and $\{i\}$ and the line $\mathcal{L}_i$ along the rotary joint axis. The orientation of the rigid body $\{i\}$ with respect to $\{i-1\}$ is given by

$$ {}^{0}_{i}[R] = {}^{0}_{i-1}[R]\; {}^{i-1}_{i}[R(\widehat{\mathbf{k}}, \theta_i)] \tag{2.39}$$

where $\widehat{\mathbf{k}}$ is a unit vector along the line $\mathcal{L}_i$ and θ_i is the rotation about the joint axis. The rotation axis $\widehat{\mathbf{k}}$ (along $\mathcal{L}_i$) is fixed in $\{i-1\}$ and $\{i\}$. In addition, for the point P on the rotary joint axis, we can write

$$ {}^{0}\mathbf{p} = {}^{0}\mathbf{O}_{i-1} + {}^{0}_{i-1}[R]^{i-1}\mathbf{p} = {}^{0}\mathbf{O}_i + {}^{0}_{i}[R]^{i}\mathbf{p} \tag{2.40}$$

[8] See Section 2.8 for mathematical representation of lines in $\Re^3$.

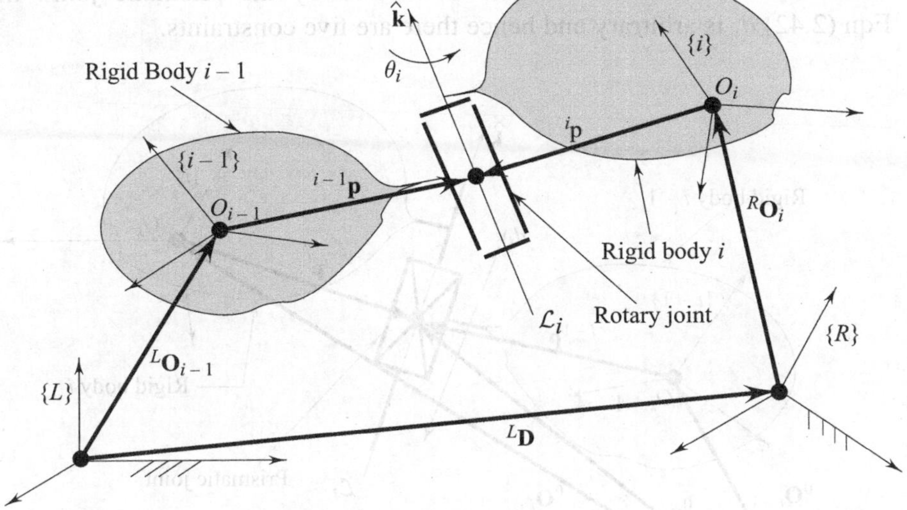

Fig. 2.10 A rotary joint in a loop

where ${}^{i-1}\mathbf{p}$ and ${}^{i}\mathbf{p}$ are constant vectors locating the point P on the rotary joint axis in $\{i-1\}$ and $\{i\}$, respectively. The six equations in Eqns (2.39) and (2.40) yield five constraints since θ_i is arbitrary.

Now consider the rotary joint as a part of a loop in a parallel manipulator, as shown schematically in Fig. 2.10. The two fixed ends of the loop are denoted $\{L\}$ and $\{R\}$; the transformation matrix relating $\{R\}$ to $\{L\}$ is described by ${}^{L}_{R}[R]$; and the vector locating $\{R\}$ with respect to $\{L\}$ is given by ${}^{L}\mathbf{D}$. Similar to Eqns (2.39) and (2.40), we can write

$$
\begin{aligned}
{}^{L}_{i}[R] &= {}^{L}_{i-1}[R]\; {}^{i-1}_{i}[R(\widehat{\mathbf{k}}, \theta_i)] = {}^{L}_{R}[R]\,{}^{R}_{i}[R] \\
{}^{L}\mathbf{p} &= {}^{L}\mathbf{O}_{i-1} + {}^{L}_{i-1}[R]\,{}^{i-1}\mathbf{p} = {}^{L}\mathbf{D} + {}^{R}\mathbf{O}_i + {}^{R}_{i}[R]\,{}^{i}\mathbf{p}
\end{aligned}
\tag{2.41}
$$

Again, the rotary joint introduces five constraints since θ_i can be arbitrary. As we will see in Chapter 4, the equations derived above can be used for the kinematic analysis of parallel manipulators.

2.5.2 Prismatic Joint

For a prismatic (P) joint we can also associate a line about which there is relative translation between the two connected rigid bodies. Figure 2.11 shows two rigid bodies $\{i-1\}$ and $\{i\}$ and the line $\mathcal{L}_i$ along the prismatic joint axis. Since the prismatic joint only allows relative translation, we have

$$
\begin{aligned}
{}^{0}_{i}[R] &= {}^{0}_{i-1}[R] \\
{}^{0}\mathbf{O}_{i-1} + {}^{0}_{i-1}[R]\,{}^{i-1}\mathbf{p} + d_i\widehat{\mathbf{k}} &= {}^{0}\mathbf{O}_i + {}^{0}_{i}[R]\,{}^{i}\mathbf{p}
\end{aligned}
\tag{2.42}
$$

where d_i is the relative translation allowed by the prismatic joint. In Eqn (2.42) d_i is arbitrary and hence there are five constraints.

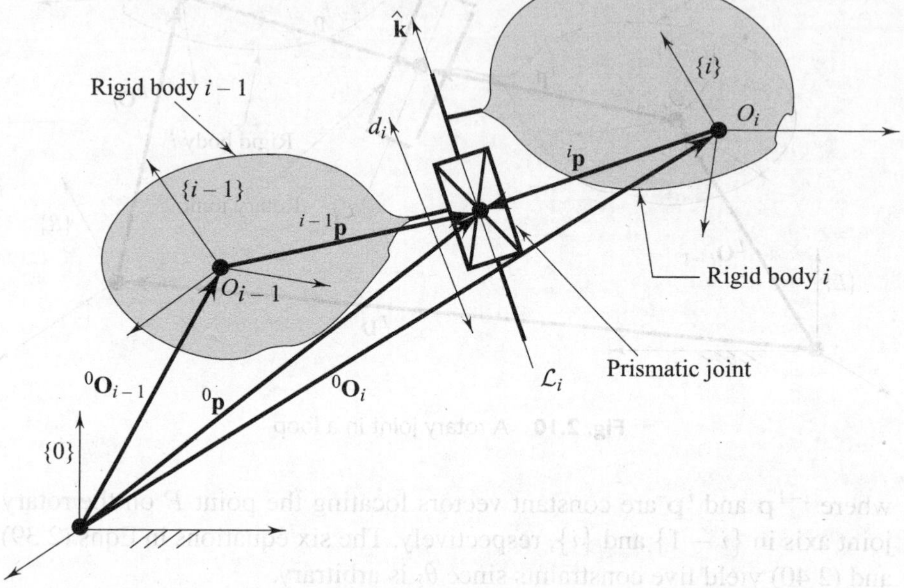

Fig. 2.11 A prismatic joint

If the prismatic joint is a part of a loop in a parallel manipulator, similar to the case of a rotary joint, we can write

$$
\begin{aligned}
&{}^L_i[R] = {}^L_{i-1}[R] = {}^L_R[R]{}^R_i[R] \\
&{}^L\mathbf{O}_{i-1} + {}^L_{i-1}[R]{}^{i-1}\mathbf{p} + d_i\widehat{\mathbf{k}} = {}^L\mathbf{D} + {}^R\mathbf{O}_i + {}^R_i[R]{}^i\mathbf{p}
\end{aligned} \tag{2.43}
$$

where the two fixed ends of the loop are denoted $\{L\}$ and $\{R\}$ and the transformation matrix relating $\{R\}$ to $\{L\}$ is described by ${}^L_R[R]$ and ${}^L\mathbf{D}$.

2.5.3 Screw Joint

A screw joint allows relative rotation and translation between $\{i-1\}$ and $\{i\}$. However, the translation d_i is related to the rotation θ_i by a constant called the pitch p. The constraint equations for a screw joint are given by

$$
\begin{aligned}
&{}^0_i[R] = {}^0_{i-1}[R]\,{}^{i-1}_i[R(\widehat{\mathbf{k}}, \theta_i)] \\
&{}^0\mathbf{O}_{i-1} + {}^0_{i-1}[R]{}^{i-1}\mathbf{p} + p\theta_i\widehat{\mathbf{k}} = {}^0\mathbf{O}_i + {}^0_i[R]{}^i\mathbf{p}
\end{aligned} \tag{2.44}
$$

In the case of a screw joint, there are five constraints as θ_i can be arbitrary.

2.5.4 Cylindrical Joint

Figure 2.12 shows a cylindrical joint between two rigid bodies. In this case the rigid body $\{i\}$ can rotate and translate along $\mathbf{k}$ with respect to $\{i-1\}$. The constraint equations for a cylindrical joint are

$$
\begin{aligned}
{}^0_i[R] &= {}^{\ 0}_{i-1}[R]\ {}^{i-1}_{\ \ i}[R(\widehat{\mathbf{k}}, \theta_i)] \\
{}^0\mathbf{O}_{i-1} + {}^{\ 0}_{i-1}[R]^{i-1}\mathbf{p} + d_i\widehat{\mathbf{k}} &= {}^0\mathbf{O}_i + {}^0_i[R]^i\mathbf{p}
\end{aligned} \tag{2.45}
$$

where d_i and θ_i are independent and arbitrary. Equation (2.45) yields four constraint equations, and the cylindrical joint has two degrees of freedom.

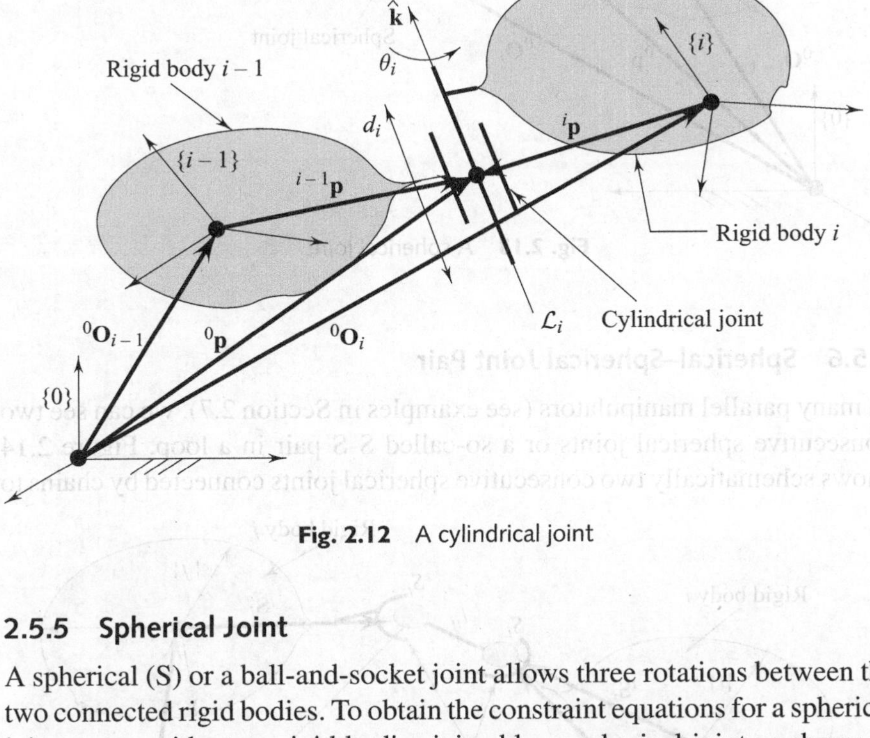

Fig. 2.12 A cylindrical joint

2.5.5 Spherical Joint

A spherical (S) or a ball-and-socket joint allows three rotations between the two connected rigid bodies. To obtain the constraint equations for a spherical joint, we consider two rigid bodies joined by a spherical joint as shown in Figure 2.13. In this case the constraint equation is given by

$$
{}^0\mathbf{p} = {}^0\mathbf{O}_{i-1} + {}^{\ 0}_{i-1}[R]^{i-1}\mathbf{p} = {}^0\mathbf{O}_i + {}^0_i[R]^i\mathbf{p} \tag{2.46}
$$

where ${}^0\mathbf{p}$ is the position vector of the centre of the spherical joint. The vector equation above gives rise to three constraints and the spherical joint has three degrees of freedom. It may be noted that we can write

$$
{}^0_i[R] = {}^{\ 0}_{i-1}[R]^{i-1}_{\ \ i}[R] \tag{2.47}
$$

with the arbitrary ${}^{i-1}_{i}[R]$ defined in terms of three angles (see *X-Y-Z* Euler angles in Section 2.2.3). In this sense, it is possible to represent a spherical joint by means of three intersecting rotary joints.

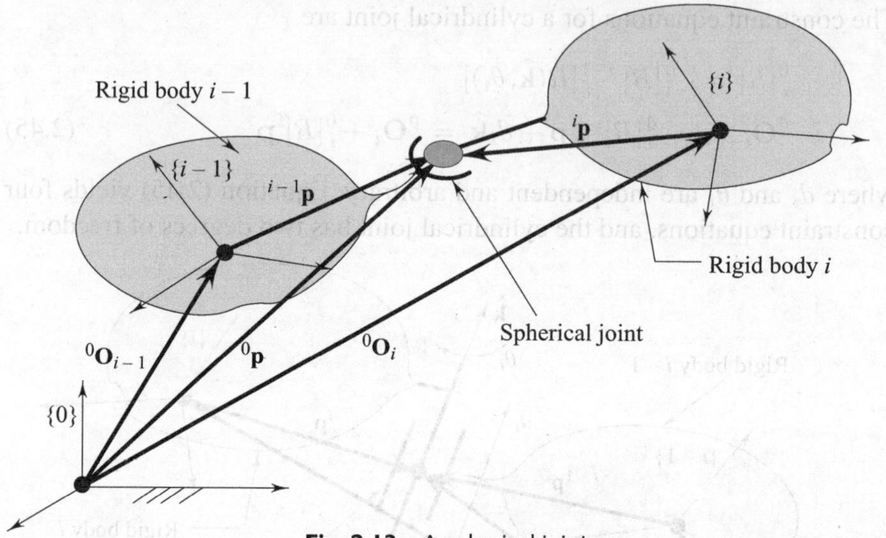

Fig. 2.13 A spherical joint

2.5.6 Spherical–Spherical Joint Pair

In many parallel manipulators (see examples in Section 2.7), we can see two consecutive spherical joints or a so-called S-S pair in a loop. Figure 2.14 shows schematically two consecutive spherical joints connected by chains to

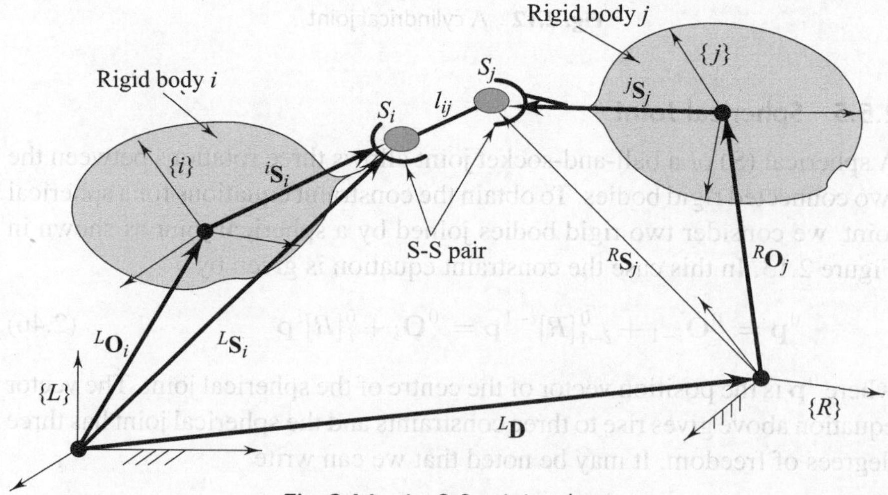

Fig. 2.14 An S-S pair in a loop

two fixed ends. The constraint equation for such an S-S pair can be written as

$$[{}^{L}\mathbf{S}_i - ({}^{L}\mathbf{D} + {}^{R}\mathbf{S}_j)] \cdot [{}^{L}\mathbf{S}_i - ({}^{L}\mathbf{D} + {}^{R}\mathbf{S}_j)] = l_{ij}^2 \tag{2.48}$$

where l_{ij} is a constant, the two fixed ends are denoted $\{L\}$ and $\{R\}$, ${}^{L}\mathbf{D}$ locates the origin of $\{R\}$ with respect to $\{L\}$, and ${}^{(\cdot)}\mathbf{S}_{(\cdot)}$ denotes the position vector of the centre of the spherical joints.

2.5.7 Other Joints

Additionally, there are the two-DOF Hooke (T) joint [also called the universal (U) joint] and the three-DOF planar (E) joint (Fig. 2.8). By following the above development, one can obtain the constraint equations for these joints.

The key feature of all the constraints developed in this subsection is that they are holonomic, i.e., they involve position or rotational variables and not their derivatives. In Chapter 10, we analyse wheeled mobile robots and the rolling of a wheel on uneven terrain. We model the contact between the wheel and the ground as a non-holonomic joint wherein the constraint equations contain position and rotational variables and their derivatives.

2.6 Representation of Links Using Denavit–Hartenberg Parameters

We recall that the links of a manipulator are rigid bodies which are connected by joints. Also, in general, a rigid body or a link can be described with respect to another coordinate system with six independent parameters—three for the position vector of the origin and three angles for the orientation. As first mentioned by Denavit and Hartenberg in 1955 [later in a textbook (Denavit & Hartenberg 1964)], when links are connected by rotary (R) or prismatic (P) joints, it is possible to use only four parameters to perform a kinematic analysis of closed-loop mechanisms, serial and parallel manipulators by carefully choosing the origins and the coordinate systems. These are the so-called Denavit–Hartenberg (DH) parameters.[9] Unfortunately, there are several ways to make these choices and to number the links, and these lead to several conventions. In this textbook we follow one of the commonly used conventions [see Craig (1989)], however, it must be kept in mind that there are other conventions (see Paul 1981 and Fu et al. 1987 listed in the references of Chapter 1).

In the convention used in this textbook, the coordinate system $\{i\}$ is attached to the link i, and the origin of $\{i\}$ lies on the joint axis i—the link

[9] When two- or three-DOF joints are present, they can be modelled as two or three one-DOF joints, respectively, and the Denavit–Hartenberg parameters can be obtained.

i is after the joint i. In serial manipulators, the notion of 'after' is clear—we start from the fixed link denoted $\{0\}$, and link 1, denoted $\{1\}$, is after the joint 1 connecting $\{0\}$ and $\{1\}$. In parallel and hybrid manipulators, with one or more loops, one needs to be careful since the loop can be traversed in more than one way.

Figure 2.15 shows three intermediate rotary joints in a manipulator. The axes of the joints $i-1$, i, and $i+1$ are labelled $\widehat{\mathbf{Z}}_{i-1}$, $\widehat{\mathbf{Z}}_i$ and $\widehat{\mathbf{Z}}_{i+1}$, respectively, and the links $i-2$, $i-1$, i, and $i+1$ are connected to the joints as shown in Fig. 2.15. The coordinate systems $\{i-1\}$ and $\{i\}$ are attached to links $i-1$ and i, respectively, and the origins O_{i-1} and O_i are located on the joint axes $\widehat{\mathbf{Z}}_{i-1}$ and $\widehat{\mathbf{Z}}_i$, respectively, as per our convention. For the coordinate system $\{i-1\}$, the axis $\widehat{\mathbf{X}}_{i-1}$ is chosen along the common perpendicular between the lines along the joint axes $\widehat{\mathbf{Z}}_{i-1}$ and $\widehat{\mathbf{Z}}_i$. The origin of $\{i-1\}$, O_{i-1}, is the point of intersection of the mutual perpendicular line and the line along the joint axis $i-1$. Likewise, the coordinate axis $\widehat{\mathbf{Z}}_i$ is along the joint axis i, $\widehat{\mathbf{X}}_i$ is along the common perpendicular between $\widehat{\mathbf{Z}}_i$ and $\widehat{\mathbf{Z}}_{i+1}$, and the origin of $\{i\}$, O_i, is the point of intersection of the line along $\widehat{\mathbf{X}}_i$ and the line along $\widehat{\mathbf{Z}}_i$ (see Fig. 2.15). For clarity, $\widehat{\mathbf{Y}}_{i-1}$ and $\widehat{\mathbf{Y}}_i$, formed by the vector cross product of $\widehat{\mathbf{Z}}$ and $\widehat{\mathbf{X}}$, are not shown in Fig. 2.15.

2.6.1 Link Parameters for Intermediate Links

The first parameter for link i is the twist angle α_{i-1}. The twist angle is defined as the angle between the lines along joints $i-1$ and i measured about the common perpendicular $\widehat{\mathbf{X}}_{i-1}$ according to the right-hand rule (see Fig. 2.15). The twist angle is a signed quantity between 0 and $\pm\pi$ rad.

The second parameter for link i is the link length a_{i-1}. a_{i-1} is the distance between the lines along joints $i-1$ and i along the common perpendicular (see Fig. 2.15) and it is always a positive quantity.

The third parameter is the link offset d_i. It is the distance along $\widehat{\mathbf{Z}}_i$ from the line parallel to $\widehat{\mathbf{X}}_{i-1}$ to the line parallel to $\widehat{\mathbf{X}}_i$. If the joint i is a rotary joint, then d_i is constant; whereas if joint i is prismatic, then d_i is the joint variable (see Fig. 2.15). d_i can be positive or negative.

The fourth parameter is the link rotation angle θ_i. It is the angle between $\widehat{\mathbf{X}}_{i-1}$ and $\widehat{\mathbf{X}}_i$ measured about $\widehat{\mathbf{Z}}_i$ according to the right-hand rule. If the joint i is prismatic, then θ_i is constant (assumed zero for simplicity); whereas if joint i is rotary, then θ_i is the joint variable (see Fig. 2.15). θ_i is a signed quantity between 0 and $\pm\pi$ rad.

If the axes of two consecutive joints $i-1$ and i are parallel, then we have infinitely many common perpendiculars. In this case the twist angle α_{i-1} is 0 or π, the link length a_{i-1} is the distance along any of the common perpendiculars (all of them are equal). If joints $i-1$ and i are parallel and

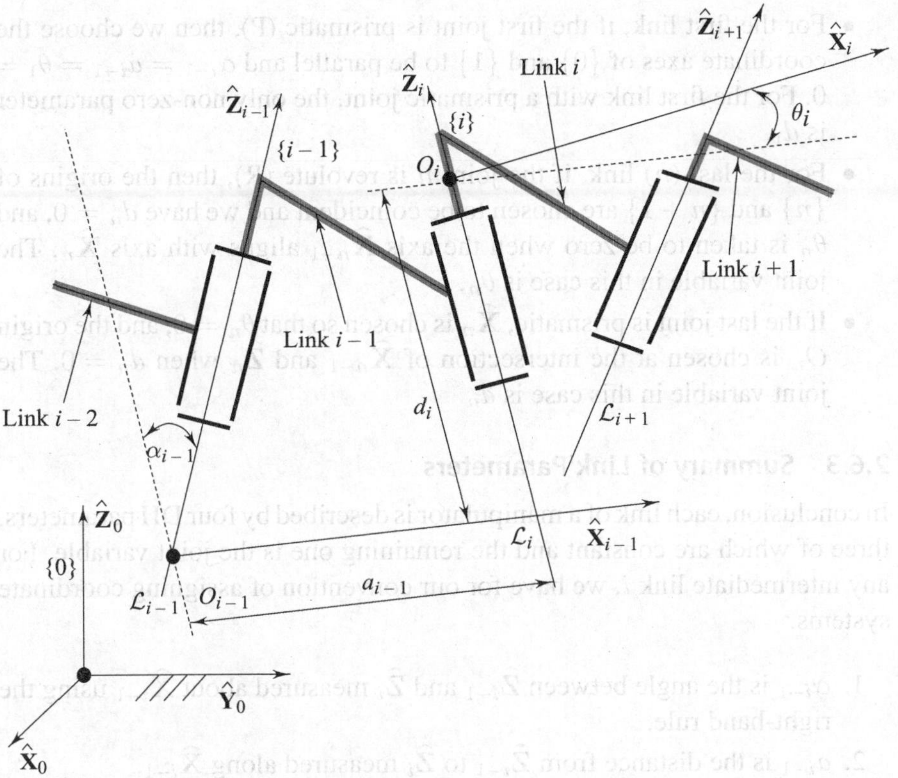

Fig. 2.15 Intermediate links and DH parameters

joint i is rotary, then d_i is taken as zero. If joint i is prismatic, then θ_i is taken as zero. It may be noted that if the two consecutive joints are prismatic and parallel, then the two joint variables are not independent.

If two consecutive joints are intersecting, then there are two choices for the direction of the common perpendicular (or the normal to the plane formed by the lines along the joints), and a_{i-1} is clearly zero.

2.6.2 First and Last Links

The above assignment of origins and coordinate axes fails for the first and last link. For the first link, the choice of $\widehat{\mathbf{Z}}_0$ and thereby $\widehat{\mathbf{X}}_1$ is arbitrary; and for the last link, $\widehat{\mathbf{Z}}_{n+1}$ does not exist. As a consequence, for the first link, we cannot obtain the twist angle α_{i-1} and the link length a_{i-1}; and for the last link we cannot obtain the link offset d_n and the rotation angle θ_n. In order to overcome this problem, we follow the following convention.

- For the first link, if the first joint is rotary (R), then we choose $\{0\}$ and $\{1\}$ to be coincident with $\alpha_{i-1} = a_{i-1} = 0$. This also implies $d_1 = 0$ if first joint is rotary and the only non-zero variable is θ_1.

- For the first link, if the first joint is prismatic (P), then we choose the coordinate axes of $\{0\}$ and $\{1\}$ to be parallel and $\alpha_{i-1} = a_{i-1} = \theta_1 = 0$. For the first link with a prismatic joint, the only non-zero parameter is d_1.
- For the last (n) link, if the joint n is revolute (R), then the origins of $\{n\}$ and $\{n+1\}$ are chosen to be coincident and we have $d_n = 0$, and θ_n is taken to be zero when the axis $\widehat{\mathbf{X}}_{n-1}$ aligns with axis $\widehat{\mathbf{X}}_n$. The joint variable in this case is θ_n.
- If the last joint is prismatic, $\widehat{\mathbf{X}}_n$ is chosen so that $\theta_n = 0$, and the origin O_n is chosen at the intersection of $\widehat{\mathbf{X}}_{n-1}$ and $\widehat{\mathbf{Z}}_n$ when $d_n = 0$. The joint variable in this case is d_n.

2.6.3 Summary of Link Parameters

In conclusion, each link of a manipulator is described by four DH parameters, three of which are constant and the remaining one is the joint variable. For any intermediate link i, we have for our convention of assigning coordinate systems.

1. α_{i-1} is the angle between $\widehat{\mathbf{Z}}_{i-1}$ and $\widehat{\mathbf{Z}}_i$ measured about $\widehat{\mathbf{X}}_{i-1}$ using the right-hand rule.
2. a_{i-1} is the distance from $\widehat{\mathbf{Z}}_{i-1}$ to $\widehat{\mathbf{Z}}_i$ measured along $\widehat{\mathbf{X}}_{i-1}$.
3. d_i is the distance from $\widehat{\mathbf{X}}_{i-1}$ to $\widehat{\mathbf{X}}_i$ measured along $\widehat{\mathbf{Z}}_i$. d_i is constant for a rotary joint and is the joint variable for a prismatic joint.
4. θ_i is the angle between $\widehat{\mathbf{X}}_{i-1}$ and $\widehat{\mathbf{X}}_i$ measured about $\widehat{\mathbf{Z}}_i$ using the right-hand rule. θ_i is constant for a prismatic joint and is the joint variable for the rotary joint.

a_{i-1} is always positive, but the other three could be positive or negative.

For the first link $\alpha_0 = 0$ and $a_0 = 0$, $d_1 = 0$ if joint 1 is rotary, and $\theta_1 = 0$ if joint 1 is prismatic. For the last link, if the joint is revolute, $d_n = 0$ and $\theta_n = 0$ if the joint is prismatic.

As a consequence of the convention, two of the four parameters of the link i, α_{i-1} and a_{i-1}, have subscripts $i-1$ and two of them, d_i and θ_i, have subscript i. Another consequence of the convention is that the link length a_n and the twist angle α_n need not be defined. The link n is the end-effector or the tool of the manipulator and, to represent the tool or the end-effector, we assign a separate coordinate system $\{Tool\}$ on the tool. Usually, this end-effector or tool coordinate system has the same orientation as $\{n\}$ and its origin is at some point of interest along $\{Tool\}$. For example, in the case of a parallel jaw gripper shown in Fig. 2.16, the origin of $\{Tool\}$ is at the mid-point of the jaws, and it is at a distance l from the origin of the last frame along $\widehat{\mathbf{Z}}_n$. This is explained in detail in the examples in Section 2.7.

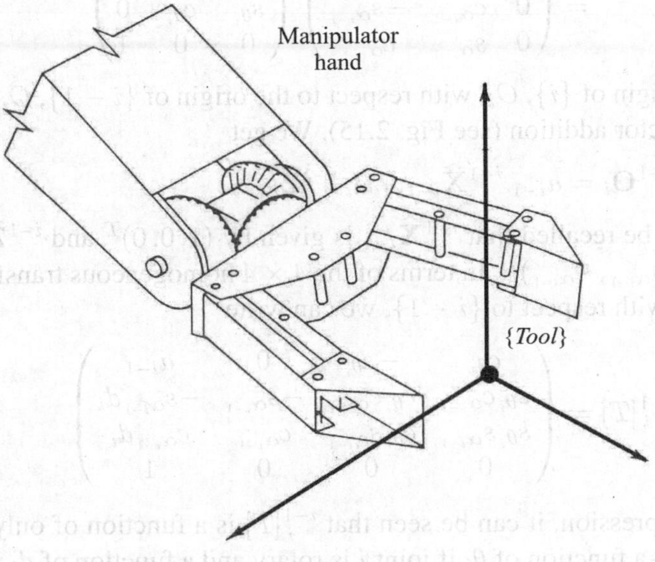

Fig. 2.16 Parallel jaw gripper with $\{Tool\}$ frame

The above convention and the Denavit–Hartenberg parameters are ideally suited for manipulators with one-DOF joints. In serial robots, the commonly used joints are one-DOF rotary and prismatic joints since these can be easily actuated by electric motors or other actuators. Two- and three-DOF joints are much harder to actuate, and they are commonly used in parallel and hybrid manipulators but as a passive joint. For the analysis of parallel and hybrid manipulators, as we show in Chapter 4, it is easier to use the joint constraints developed in Section 2.5.

Finally, in the case of intersecting or parallel joints (which maybe 'special' cases for an analyst but most commonly used by robot designers), there are choices in assigning frames, and thereby one can get different link parameters depending on the choice.

2.7 Link Transformation Matrices

The link parameters allow us to describe link i with respect to link $i-1$. The rotation matrix ${}^{i-1}_{i}[R]$ can be obtained by two successive rotations about

$\widehat{\mathbf{X}}_{i-1}$ by angle α_{i-1} and by angle θ_i about $\widehat{\mathbf{Z}}_i$ (see Fig. 2.15). Hence we get

$$
\begin{aligned}
{}^{i-1}_{i}[R] &= [R(\widehat{\mathbf{X}}_{i-1},\alpha_{i-1})][R(\widehat{\mathbf{Z}}_{i},\theta_i)] \\
&= \begin{pmatrix} 1 & 0 & 0 \\ 0 & c_{\alpha_{i-1}} & -s_{\alpha_{i-1}} \\ 0 & s_{\alpha_{i-1}} & c_{\alpha_{i-1}} \end{pmatrix} \begin{pmatrix} c_{\theta_i} & -s_{\theta_i} & 0 \\ s_{\theta_i} & c_{\theta_i} & 0 \\ 0 & 0 & 1 \end{pmatrix}
\end{aligned}
\tag{2.49}
$$

The origin of $\{i\}$, O_i, with respect to the origin of $\{i-1\}$, O_{i-1} is given by the vector addition (see Fig. 2.15). We get

$$
{}^{i-1}\mathbf{O}_i = a_{i-1}\,{}^{i-1}\widehat{\mathbf{X}}_{i-1} + d_i\,{}^{i-1}\widehat{\mathbf{Z}}_i \tag{2.50}
$$

It may be recalled that ${}^{i-1}\widehat{\mathbf{X}}_{i-1}$ is given by $(1,0,0)^T$ and ${}^{i-1}\widehat{\mathbf{Z}}_i$ is given by $(0,\ -s_{\alpha_{i-1}},\ c_{\alpha_{i-1}})^T$. In terms of the 4×4 homogeneous transformations of link i with respect to $\{i-1\}$, we can write

$$
{}^{i-1}_{i}[T] = \begin{pmatrix} c_{\theta_i} & -s_{\theta_i} & 0 & a_{i-1} \\ s_{\theta_i}c_{\alpha_{i-1}} & c_{\theta_i}c_{\alpha_{i-1}} & -s_{\alpha_{i-1}} & -s_{\alpha_{i-1}}d_i \\ s_{\theta_i}s_{\alpha_{i-1}} & c_{\theta_i}s_{\alpha_{i-1}} & c_{\alpha_{i-1}} & c_{\alpha_{i-1}}d_i \\ 0 & 0 & 0 & 1 \end{pmatrix} \tag{2.51}
$$

In this expression, it can be seen that ${}^{i-1}_{i}[T]$ is a function of only one joint variable—a function of θ_i if joint i is rotary and a function of d_i if joint i is prismatic.

To obtain the transformation matrix of a link i in any other frame, we can use the properties described in Section 2.3. For example, the link i can be described in $\{0\}$ as

$$
{}^{0}_{i}[T] = {}^{0}_{1}[T]\,{}^{1}_{2}[T] \cdots {}^{i-1}_{i}[T] \tag{2.52}
$$

Since each of ${}^{i-1}_{i}[T]$ is a function of only one joint variable, the right-hand side of Eqn (2.52) will be a function of i joint variables with the other $3i$ Denavit–Hartenberg parameters fixed.

Next, we discuss, with the help of examples, assigning coordinate systems, determining the Denavit–Hartenberg parameters for several serial, parallel, and hybrid manipulators, and finally obtaining link transforms. We start with a simple planar 3R manipulator.

Example 2.1: The planar 3R manipulator

Figure 2.17 shows a planar three-link manipulator with three rotary joints and a parallel jaw gripper as the end-effector. All the joint axes are parallel and are pointing out of the paper. To assign the coordinate systems and determine the Denavit–Hartenberg parameters, we proceed as follows.

We start by choosing the fixed or reference coordinate system. $\{0\}$ is chosen with its $\widehat{\mathbf{Z}}_0$ coming out of the paper, and $\widehat{\mathbf{X}}_0$ and $\widehat{\mathbf{Y}}_0$ pointing to the

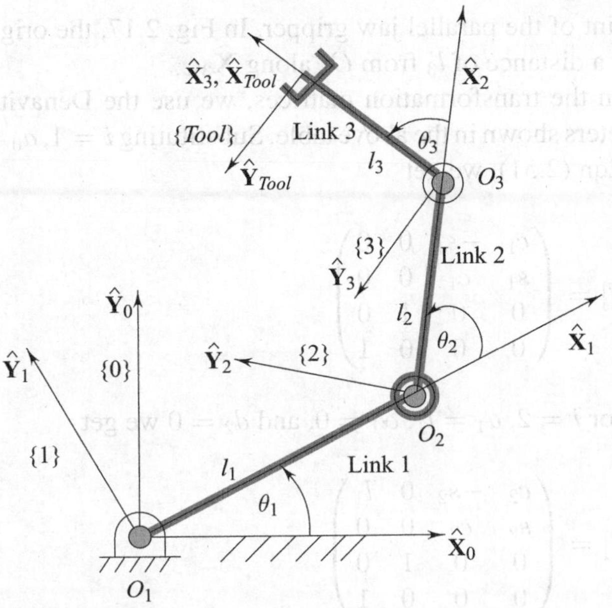

Fig. 2.17 The planar 3R manipulator

right and top, respectively. For the first coordinate system, the origin O_1 and $\hat{\mathbf{Z}}_1$ are coincident with O_0 and $\hat{\mathbf{Z}}_0$, and $\hat{\mathbf{X}}_1$ and $\hat{\mathbf{Y}}_1$ are coincident with $\hat{\mathbf{X}}_0$ and $\hat{\mathbf{Y}}_0$ when θ_1 is zero. $\hat{\mathbf{X}}_1$ is also along the mutual perpendicular between $\hat{\mathbf{Z}}_1$ and $\hat{\mathbf{Z}}_2$. Similarly, $\hat{\mathbf{X}}_2$ is along the mutual perpendicular between $\hat{\mathbf{Z}}_2$ and $\hat{\mathbf{Z}}_3$. For the last frame, $\hat{\mathbf{X}}_3$ is aligned to $\hat{\mathbf{X}}_2$ when $\theta_3 = 0$. The origin O_2 is located at the intersection of the mutual perpendicular along $\hat{\mathbf{X}}_2$ and $\hat{\mathbf{Z}}_2$. The origin O_3 is chosen such that d_3 is zero. The origins and the axes of $\{1\}$, $\{2\}$, and $\{3\}$ are shown in Fig. 2.17.

Once the axes and origins have been assigned, we can write the Denavit–Hartenberg parameters in a tabular form as follows.

i	α_{i-1}	a_{i-1}	d_i	θ_i
1	0	0	0	θ_1
2	0	l_1	0	θ_2
3	0	l_2	0	θ_3

In this table, l_1 and l_2 are the link lengths as shown in Fig. 2.17. It may be noted that the length of the end-effector does not appear in the table. To describe the end-effector, we attach a tool frame, $\{Tool\}$, aligned to $\{3\}$ at the mid-point of the parallel jaw gripper. In Fig. 2.17, the origin of $\{Tool\}$ is shown at a distance of l_3 from O_3 along $\widehat{\mathbf{X}}_3$.

To obtain the transformation matrices, we use the Denavit–Hartenberg link parameters shown in the above table. Substituting $i = 1, a_0 = 0, \alpha_0 = 0, d_1 = 0$ in Eqn (2.51), we get

$$ {}^{0}_{1}[T] = \begin{pmatrix} c_1 & -s_1 & 0 & 0 \\ s_1 & c_1 & 0 & 0 \\ 0 & 0 & 1 & 0 \\ 0 & 0 & 0 & 1 \end{pmatrix} \tag{2.53} $$

Similarly, for $i = 2$, $a_1 = l_1$, $\alpha_1 = 0$, and $d_2 = 0$ we get

$$ {}^{1}_{2}[T] = \begin{pmatrix} c_2 & -s_2 & 0 & l_1 \\ s_2 & c_2 & 0 & 0 \\ 0 & 0 & 1 & 0 \\ 0 & 0 & 0 & 1 \end{pmatrix} \tag{2.54} $$

and finally, for $i = 3$, $a_2 = l_2$, $\alpha_2 = 0$, and $d_3 = 0$ we get

$$ {}^{2}_{3}[T] = \begin{pmatrix} c_3 & -s_3 & 0 & l_2 \\ s_3 & c_3 & 0 & 0 \\ 0 & 0 & 1 & 0 \\ 0 & 0 & 0 & 1 \end{pmatrix} \tag{2.55} $$

To find the transformation matrix ${}^{3}_{Tool}[T]$, we recall that the orientation of $\{Tool\}$ is the same as the orientation of $\{3\}$ and the origin is at a distance l_3 along $\widehat{\mathbf{X}}_3$. Hence we have

$$ {}^{3}_{Tool}[T] = \begin{pmatrix} 1 & 0 & 0 & l_3 \\ 0 & 1 & 0 & 0 \\ 0 & 0 & 1 & 0 \\ 0 & 0 & 0 & 1 \end{pmatrix} \tag{2.56} $$

To find the transformation matrix ${}^{0}_{3}[T]$, we simply multiply ${}^{0}_{1}[T]\ {}^{1}_{2}[T]\ {}^{2}_{3}[T]$ and get

$$ {}^{0}_{3}[T] = \begin{pmatrix} c_{123} & -s_{123} & 0 & l_1c_1 + l_2c_{12} \\ s_{123} & c_{123} & 0 & l_1s_1 + l_2s_{12} \\ 0 & 0 & 1 & 0 \\ 0 & 0 & 0 & 1 \end{pmatrix} \tag{2.57} $$

Finally, to obtain ${}^{0}_{Tool}[T]$, we multiply ${}^{0}_{3}[T]\ {}^{3}_{Tool}[T]$ and get

$$
{}^{0}_{Tool}[T] = \begin{pmatrix} c_{123} & -s_{123} & 0 & l_1c_1 + l_2c_{12} + l_3c_{123} \\ s_{123} & c_{123} & 0 & l_1s_1 + l_2s_{12} + l_3s_{123} \\ 0 & 0 & 1 & 0 \\ 0 & 0 & 0 & 1 \end{pmatrix} \tag{2.58}
$$

Example 2.2: The PUMA 560 manipulator

The PUMA 560 is a six-DOF manipulator with all rotary joints. A schematic drawing of the manipulator is shown in Figs 2.18 and 2.19 with the assigned coordinate systems.

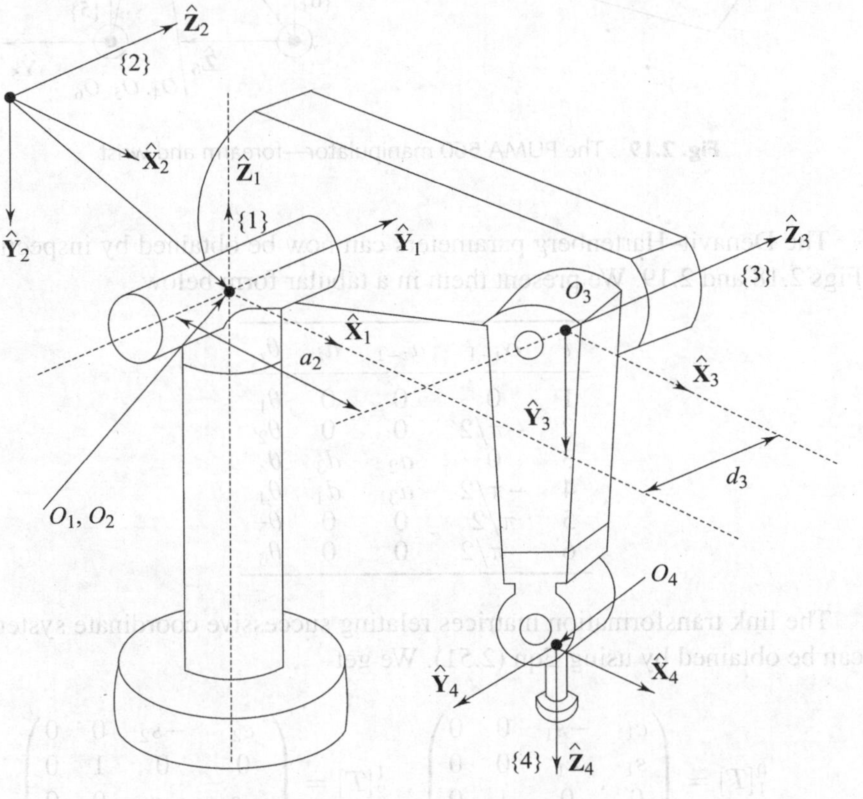

Fig. 2.18 The PUMA 560 manipulator

The coordinate systems {0}, {1}, and {2} have the same origin. In many industrial manipulators, the last three joint axes intersect at a point called the *wrist*. For the PUMA, the last three joint axes intersect, and the origins of the coordinate systems {4}, {5}, and {6} are located at this wrist point.

The coordinate systems have been assigned according to our convention, as discussed previously.

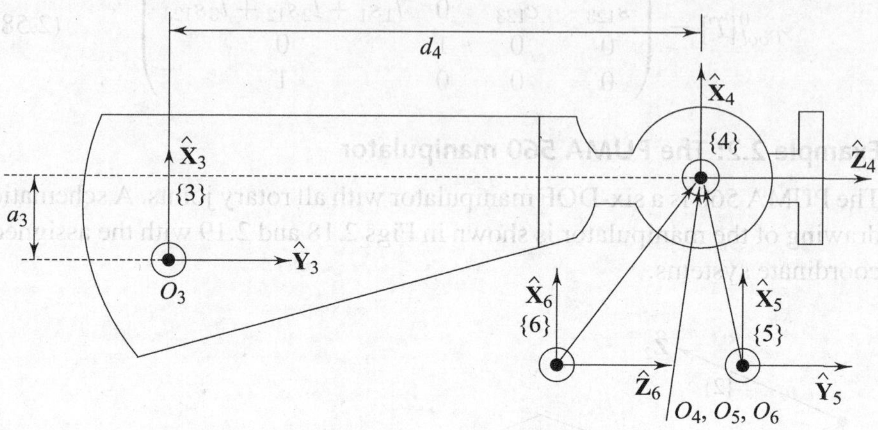

Fig. 2.19 The PUMA 560 manipulator—forearm and wrist

The Denavit–Hartenberg parameters can now be obtained by inspecting Figs 2.18 and 2.19. We present them in a tabular form below.

i	α_{i-1}	a_{i-1}	d_i	θ_i
1	0	0	0	θ_1
2	$-\pi/2$	0	0	θ_2
3	0	a_2	d_3	θ_3
4	$-\pi/2$	a_3	d_4	θ_4
5	$\pi/2$	0	0	θ_5
6	$-\pi/2$	0	0	θ_6

The link transformation matrices relating successive coordinate systems can be obtained by using Eqn (2.51). We get

$$ {}^0_1[T] = \begin{pmatrix} c_1 & -s_1 & 0 & 0 \\ s_1 & c_1 & 0 & 0 \\ 0 & 0 & 1 & 0 \\ 0 & 0 & 0 & 1 \end{pmatrix}, \quad {}^1_2[T] = \begin{pmatrix} c_2 & -s_2 & 0 & 0 \\ 0 & 0 & 1 & 0 \\ -s_2 & -c_2 & 0 & 0 \\ 0 & 0 & 0 & 1 \end{pmatrix} $$

$$ {}^2_3[T] = \begin{pmatrix} c_3 & -s_3 & 0 & a_2 \\ s_3 & c_3 & 0 & 0 \\ 0 & 0 & 1 & d_3 \\ 0 & 0 & 0 & 1 \end{pmatrix}, \quad {}^3_4[T] = \begin{pmatrix} c_4 & -s_4 & 0 & a_3 \\ 0 & 0 & 1 & d_4 \\ -s_4 & -c_4 & 0 & 0 \\ 0 & 0 & 0 & 1 \end{pmatrix} $$

$$
{}^4_5[T] = \begin{pmatrix} c_5 & -s_5 & 0 & 0 \\ 0 & 0 & -1 & 0 \\ s_5 & c_5 & 0 & 0 \\ 0 & 0 & 0 & 1 \end{pmatrix}, \quad {}^5_6[T] = \begin{pmatrix} c_6 & -s_6 & 0 & 0 \\ 0 & 0 & 1 & 0 \\ -s_6 & -c_6 & 0 & 0 \\ 0 & 0 & 0 & 1 \end{pmatrix}
$$

We can obtain the resultant transformation matrix ${}^0_6[T]$ by multiplying the transformation matrices. We will obtain ${}^0_6[T]$ in two steps as the intermediate results will be useful later during inverse kinematics. We first obtain ${}^0_3[T]$ by multiplying ${}^0_1[T]\,{}^1_2[T]\,{}^2_3[T]$.

$$
{}^0_3[T] = \begin{pmatrix} c_1c_{23} & -c_1s_{23} & -s_1 & a_2c_1c_2 - d_3s_1 \\ s_1c_{23} & -s_1s_{23} & c_1 & a_2s_1c_2 + d_3c_1 \\ -s_{23} & -c_{23} & 0 & -a_2s_2 \\ 0 & 0 & 0 & 1 \end{pmatrix} \tag{2.59}
$$

The matrix ${}^3_6[T]$ is obtained by multiplying ${}^3_4[T]\,{}^4_5[T]\,{}^5_6[T]$.

$$
{}^3_6[T] = \begin{pmatrix} c_4c_5c_6 - s_4s_6 & -c_4c_5s_6 - s_4c_6 & -c_4s_5 & a_3 \\ s_5c_6 & -s_5s_6 & c_5 & d_4 \\ -s_4c_5c_6 - c_4s_6 & s_4c_5s_6 - c_4c_6 & s_4s_5 & 0 \\ 0 & 0 & 0 & 1 \end{pmatrix} \tag{2.60}
$$

The transformation matrix ${}^0_6[T]$ is obtained by multiplying ${}^0_3[T]$ and ${}^3_6[T]$ given in Eqns (2.59) and (2.61).

Example 2.3: A SCARA manipulator

A manipulator with a SCARA configuration is very popular for robotic assembly due to its compliance and rigidity in desired directions. A SCARA manipulator has four degrees of freedom with three rotary joints and the third joint is prismatic. Figure 2.20 shows a schematic drawing of a SCARA manipulator and the assigned coordinate systems.

The coordinate systems $\{0\}$ and $\{1\}$ have the same origin. The origins of $\{3\}$ and $\{4\}$ are chosen at the base of the parallel jaw gripper. The directions of $\widehat{\mathbf{Z}}_3$ have been chosen pointing upwards, as shown in Fig. 2.20. It may be noted that we can choose $\widehat{\mathbf{Z}}_3$ in the opposite direction and O_3 at some other point. In an actual SCARA manipulator, the translation at the third joint may be realized by means of a (rotary) motor and a ball–screw. We will, however, assume that there is a prismatic joint.

The Denavit–Hartenberg parameters can now be obtained by inspecting Fig. 2.20. We present them in a tabular form as follows.

i	α_{i-1}	a_{i-1}	d_i	θ_i
1	0	0	0	θ_1
2	0	a_1	0	θ_2
3	0	a_2	$-d_3$	0
4	0	0	0	θ_4

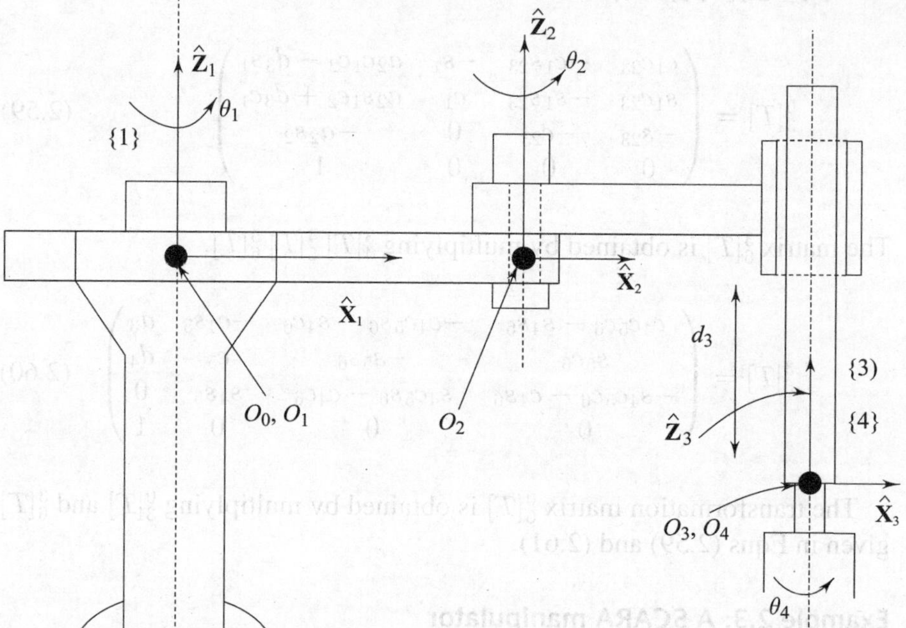

Fig. 2.20 A SCARA manipulator

The link transformation matrices relating successive coordinate systems can be obtained by using Eqn (2.51). Hence we get

$$
{}^0_1[T] = \begin{pmatrix} c_1 & -s_1 & 0 & 0 \\ s_1 & c_1 & 0 & 0 \\ 0 & 0 & 1 & 0 \\ 0 & 0 & 0 & 1 \end{pmatrix}, \quad {}^1_2[T] = \begin{pmatrix} c_2 & -s_2 & 0 & a_1 \\ s_2 & c_2 & 0 & 0 \\ 0 & 0 & 1 & 0 \\ 0 & 0 & 0 & 1 \end{pmatrix}
$$

$$
{}^2_3[T] = \begin{pmatrix} 1 & 0 & 0 & a_2 \\ 0 & 1 & 0 & 0 \\ 0 & 0 & 1 & -d_3 \\ 0 & 0 & 0 & 1 \end{pmatrix}, \quad {}^3_4[T] = \begin{pmatrix} c_4 & -s_4 & 0 & 0 \\ s_4 & c_4 & 0 & 0 \\ 0 & 0 & 1 & 0 \\ 0 & 0 & 0 & 1 \end{pmatrix}
$$

The transformation matrix ${}^0_4[T]$ is obtained as

$$
{}^0_4[T] = {}^0_1[T]\,{}^1_2[T]\,{}^2_3[T]\,{}^3_4[T]
$$

$$
= \begin{pmatrix} c_{124} & -s_{124} & 0 & a_1c_1 + a_2c_{12} \\ s_{124} & c_{124} & 0 & a_1s_1 + a_2s_{12} \\ 0 & 0 & 1 & -d_3 \\ 0 & 0 & 0 & 1 \end{pmatrix} \tag{2.61}
$$

Example 2.4: The planar four-bar mechanism

Consider the closed-loop four-bar mechanism shown in Fig. 2.21. The rotary joints 1 and 4 are fixed to the ground as shown. One way to represent the three links is to consider the four-bar mechanism as two serial manipulators connected by a rotary joint at joint 3. Similar to the example of a planar 3R manipulator, the DH parameters of the 2R manipulator are given by

i	α_{i-1}	a_{i-1}	d_i	θ_i
1	0	0	0	θ_1
2	0	l_1	0	ϕ_2

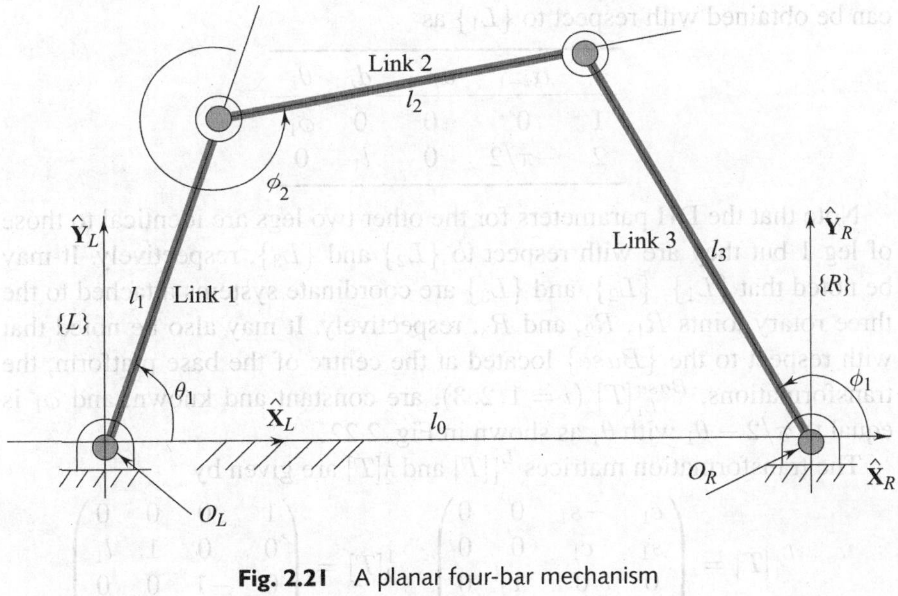

Fig. 2.21 A planar four-bar mechanism

and the DH parameters of the 1R manipulator are given by

i	α_{i-1}	a_{i-1}	d_i	θ_i
1	0	0	0	ϕ_1

It may be noted that the first set of DH parameters is with respect to the coordinate system $\{L\}$ and the second set is with respect to $\{R\}$ as shown in Fig. 2.21. The transformation matrix ${}^L_R[T]$ is a known constant matrix.

From the above DH parameters, we can obtain the individual transformation matrices. As we will see in Chapter 4, from the individual transformation matrices, ${}^L_R[T]$, and the constraints imposed by the rotary joint 3 (see Section 2.5), we can obtain two relationships between the three joint variables θ_1, ϕ_2, and ϕ_1. We can eliminate ϕ_2 from the two relationships and get the so-called input–output equation for a four-bar mechanism.

Example 2.5: A three-DOF parallel manipulator

Figure 2.22 shows a spatial three-DOF parallel manipulator. It consists of a top, moving platform connected to a bottom base platform with three 'legs'. Each of the legs consists of a rotary, a prismatic, and a spherical joint (R-P-S). Since, it is a three-DOF manipulator, only three joints are actuated. We assume that the three prismatic joints are actuated and all other joints are passive. It may be noted that this is a fully parallel manipulator, since in each leg only one joint (prismatic) is actuated. It was first proposed as a parallel wrist by Lee and Shah (1988). The DH parameters of the first leg can be obtained with respect to $\{L_1\}$ as

i	α_{i-1}	a_{i-1}	d_i	θ_i
1	0	0	0	ϕ_1
2	$-\pi/2$	0	l_1	0

Note that the DH parameters for the other two legs are identical to those of leg 1 but they are with respect to $\{L_2\}$ and $\{L_3\}$, respectively. It may be noted that $\{L_1\}$, $\{L_2\}$, and $\{L_3\}$ are coordinate systems attached to the three rotary joints R_1, R_2, and R_3, respectively. It may also be noted that with respect to the $\{Base\}$ located at the centre of the base platform, the transformations, ${}^{Base}_{L_1}[T]$ $(i = 1, 2, 3)$, are constant and known, and ϕ_1 is equal to $\pi/2 - \theta_1$ with θ_1 as shown in Fig. 2.22.

The transformation matrices ${}^{L_1}_{1}[T]$ and ${}^1_2[T]$ are given by

$$
{}^{L_1}_{1}[T] = \begin{pmatrix} c_1 & -s_1 & 0 & 0 \\ s_1 & c_1 & 0 & 0 \\ 0 & 0 & 1 & 0 \\ 0 & 0 & 0 & 1 \end{pmatrix}, \quad {}^1_2[T] = \begin{pmatrix} 1 & 0 & 0 & 0 \\ 0 & 0 & 1 & l_1 \\ 0 & -1 & 0 & 0 \\ 0 & 0 & 0 & 1 \end{pmatrix}
$$

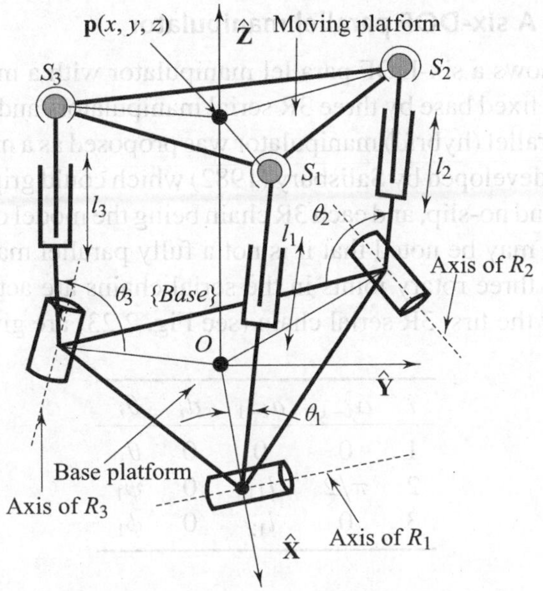

Fig. 2.22 A three-DOF parallel manipulator

The transformation matrix ${}^{2}_{S_1}[T]$ is an identity matrix since $\{S_1\}$ is located at the centre of the spherical joint and parallel to $\{2\}$.

The transformation matrix ${}^{L_1}_{S_1}[T]$ is obtained by multiplying the above three transformation matrices. The transformation matrix ${}^{Base}_{S_1}[T]$ is given by ${}^{Base}_{L_1}[T]{}^{L_1}_{S_1}[T]$ and we can obtain the position vector of the spherical joint by extracting the rightmost 3×1 column vector of ${}^{Base}_{S_1}[T]$. This is given by

$$^{Base}\mathbf{S}_1 = (b - l_1 \cos\theta_1, 0, l_1 \sin\theta_1)^T \tag{2.62}$$

where b is the distance of the rotary joint R_1 from the origin O of $\{Base\}$ (see Fig. 2.22). Assuming that all the rotary joints are at the vertices of an equilateral triangle, we can obtain the position vectors of other two spherical joints with respect to $\{Base\}$ as

$$^{Base}\mathbf{S}_2 = \left(-\frac{b}{2} + \frac{1}{2} l_2 \cos\theta_2, \frac{\sqrt{3}b}{2} - \frac{\sqrt{3}l_2}{2}\cos\theta_2, l_2 \sin\theta_2\right)^T$$

$$^{Base}\mathbf{S}_3 = \left(-\frac{b}{2} + \frac{1}{2} l_3 \cos\theta_3, -\frac{\sqrt{3}b}{2} + \frac{\sqrt{3}l_3}{2}\cos\theta_3, l_3 \sin\theta_3\right)^T \tag{2.63}$$

As we will see in Chapter 4, by using the S-S pair joint constraint (see Section 2.5.6), we can obtain three expressions relating all the joint variables θ_i, l_i $(i = 1, 2, 3)$.

Example 2.6: A six-DOF parallel manipulator

Figure 2.23 shows a six-DOF parallel manipulator with a moving platform connected to a fixed base by three 3R serial manipulators and three spherical joints. This parallel (hybrid) manipulator was proposed as a model of a three-fingered hand developed by Salisbury (1982) which could grip an object with point contact and no-slip, and each 3R chain being the model of a finger (Hunt et al. 1991). It may be noted that it is not a fully parallel manipulator since two out of the three rotary joints in the serial chains are actuated. The DH parameters for the first 3R serial chain (see Fig. 2.23) are given as

i	α_{i-1}	a_{i-1}	d_i	θ_i
1	0	0	0	θ_1
2	$\pi/2$	l_{11}	0	ψ_1
3	0	l_{12}	0	ϕ_1

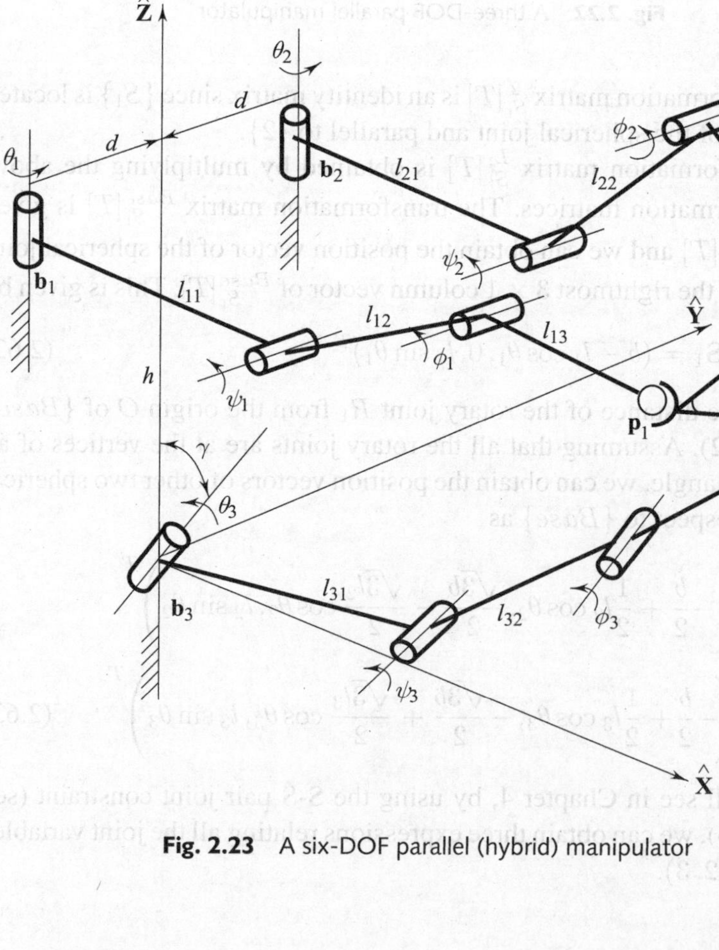

Fig. 2.23 A six-DOF parallel (hybrid) manipulator

As in the case of the planar 3R manipulator, the length of the last link l_{13} does not appear in the DH table. The transformation matrices ${}^0_1[T]$, ${}^1_2[T]$, and ${}^2_3[T]$ can be obtained by using Eqn (2.51) and the DH table given above. The DH parameters of the three fingers with respect to a coordinate system chosen at the first rotary joint, $\{F_i\}$ (where $i = 1, 2, 3$), are identical. The position vectors of the spherical joint $\mathbf{p}_i$ with respect to $\{F_i\}$ all have the same form:

$$^{F_i}\mathbf{p}_i = \begin{pmatrix} \cos\theta_i[l_{i1} + l_{i2}\cos\psi_i + l_{i3}\cos(\psi_i + \phi_i)] \\ \sin\theta_i[l_{i1} + l_{i2}\cos\psi_i + l_{i3}\cos(\psi_i + \phi_i)] \\ l_{i2}\sin\psi_i + l_{i3}\sin(\psi_i + \phi_i) \end{pmatrix} \tag{2.64}$$

where the index $i = 1, 2, 3$ denotes the three fingers.

With respect to $\{Base\}$, the locations of $\{F_i\}$ (where $i = 1, 2, 3$), are known and constant [see Fig. (2.23)], and are given as

$$\begin{aligned} {}^{Base}\mathbf{b}_1 &= (0, -d, h)^T \\ {}^{Base}\mathbf{b}_2 &= (0, d, h)^T \\ {}^{Base}\mathbf{b}_3 &= (0, 0, 0)^T \end{aligned} \tag{2.65}$$

and the orientation of $\{F_i\}$ $(i = 1, 2, 3)$, with respect to $\{Base\}$ are also known—$\{F_1\}$ and $\{F_2\}$ are parallel to $\{Base\}$ and $\{F_3\}$ is rotated by γ about the $\hat{\mathbf{Y}}$. The transformation matrices from $\{F_1\}$ to $\{p_1\}$ can be obtained by multiplying ${}^{Base}_{F_1}[T]{}^0_1[T]{}^1_2[T]{}^2_3[T]{}^3_{p_1}[T]$ where the last transformation takes into account the length l_{13}. We can extract the position vector ${}^{Base}\mathbf{p}_1$ from the last column of the matrix ${}^{Base}_{F_1}[T]$, and this is given as

$$\begin{aligned} {}^{Base}\mathbf{p}_1 &= {}^{Base}\mathbf{b}_1 + {}^{F_1}\mathbf{p}_1 \\ &= \begin{pmatrix} \cos\theta_1[l_{11} + l_{12}\cos\psi_1 + l_{13}\cos(\psi_1 + \phi_1)] \\ -d + \sin\theta_1[l_{11} + l_{12}\cos\psi_1 + l_{13}\cos(\psi_1 + \phi_1)] \\ h + l_{12}\sin\psi_1 + l_{13}\sin(\psi_1 + \phi_1) \end{pmatrix} \end{aligned} \tag{2.66}$$

where h and d are as shown in Fig. 2.23.

Similar to first finger, we can also find the position vector of the other two spherical joints and these are given as

$$^{Base}\mathbf{p}_2 = \begin{pmatrix} \cos\theta_2[l_{21} + l_{22}\cos\psi_2 + l_{23}\cos(\psi_2 + \phi_2)] \\ d + \sin\theta_2[l_{21} + l_{22}\cos\psi_2 + l_{23}\cos(\psi_2 + \phi_2)] \\ h + l_{22}\sin\psi_2 + l_{23}\sin(\psi_2 + \phi_2) \end{pmatrix} \tag{2.67}$$

and

$$^{Base}\mathbf{p}_3 = [R(\hat{\mathbf{Y}}, \gamma)] \begin{pmatrix} \cos\theta_3[l_{31} + l_{32}\cos\psi_3 + l_{33}\cos(\psi_3 + \phi_3)] \\ \sin\theta_3[l_{31} + l_{32}\cos\psi_3 + l_{33}\cos(\psi_3 + \phi_3)] \\ l_{32}\sin\psi_3 + l_{33}\sin(\psi_3 + \phi_3) \end{pmatrix} \tag{2.68}$$

As will be shown in detail in Chapter 4, the above equations together with the S-S pair joint constraint (see Section 2.5.6) can be used for the kinematic analysis of this six-DOF manipulator.

2.8* Homogeneous Coordinates, Lines, Screws, and Twists

In this chapter, we have used homogeneous coordinates, lines in $\Re^3$, and twists to describe the general motion of a rigid body. We have also used lines to represent joint axes. In this section, we present a brief description of homogeneous coordinates, mathematical representation of lines, screws, and twists using Plücker coordinates and also present expressions for the angle and distance between two lines [for details, the reader is referred to Brand (1947)]. We start with the concept of homogeneous coordinates.

Let (x, y) denote the Cartesian coordinates of a point in the Euclidean plane $\mathbf{E}^2$; then the homogeneous coordinates of the point are given by $(x, y, w) \in \mathbf{E}^3$ with $w \neq 0$. One of the key properties of homogeneous coordinates is that scaling does not matter e.g., the coordinates (x, y, w) and $(\lambda x, \lambda y, \lambda w)$, where λ is a non-zero constant, represent the same point.

From elementary mathematics, we know that any point (x, y, z) on a line passing through two points, say ${}^A(x_0, y_0, z_0)^T$ and ${}^A(x_1, y_1, z_1)^T$, in $\{A\}$ satisfies

$$\frac{x - x_0}{x_0 - x_1} = \frac{y - y_0}{y_0 - y_1} = \frac{z - z_0}{z_0 - z_1} = c \tag{2.69}$$

where c is an arbitrary non-zero constant. In Eqn (2.69), if we consider a line through the origin, i.e., $(x_0, y_0, z_0) = \mathbf{0}$, we get $(x, y, z) = -c(x_1, y_1, z_1)$. If z is considered the same as w, then the equation of a line through the origin is equivalent to scaling. Hence, homogenous coordinates represent a point in $\mathbf{E}^2$ by a line through the origin in $\mathbf{E}^3$. Likewise, a line in $\mathbf{E}^2$ is a plane through the origin of $\mathbf{E}^3$.

To go from homogeneous coordinates to Cartesian coordinates, we simply extract from (x, y, w) the quantities $(x/w, y/w)$ and set w to 1. This implies that the Euclidean plane $\mathbf{E}^2$ with points (x, y) can be embedded as a $w = 1$ plane and the ordinary Euclidean point [with Cartesian coordinates (x, y)] can be thought of as a line through the origin intersecting the $w = 1$ plane.

In addition to the ordinary Euclidean points, it is possible to have homogeneous coordinates of the form $(x, y, 0)$. These are lines through the origin of $\mathbf{E}^3$ parallel to $w = 1$ plane. These are called ideal points or points at infinity which can be shown to form a line called the *line at infinity*. The set of lines through the origin of $\mathbf{E}^3$ defines the projective plane $\mathbf{P}^2$. This plane can be thought of as the Euclidean plane $\mathbf{E}^2$ to which we have added points at infinity. In this form, the projective plane has the interesting property of duality, which states that in every axiom we can replace 'point' by 'line' and still make perfect sense without any exceptions. For example, we can say 'two points determine a line' or 'two lines determine a point'. We can also state two parallel lines meet at infinity without any mathematical problem. The projective space $\mathbf{P}^2$ is one of the fundamental concepts in geometry and, as we will see in Chapter 3, allows us to 'correctly' count the number of solutions of non-linear equations.

The concept of a projective space is also useful in theoretical kinematics since the 4×1 homogeneous coordinates and the 4×4 transformation matrices of Section 2.3 can be put on a more formal footing. The 4×1 vector obtained by appending a '1' to $(x, y, z)^T$ is obtained from the homogeneous coordinates $(x, y, z, w)^T$ by setting $w = 1$. Similar to the discussion above, we can also have points at infinity for $w = 0$. One difference between $\mathbf{P}^2$ earlier and $\mathbf{P}^3$ is that now the axioms and the notion of duality involve points, lines, and planes.

In Eqn (2.69), instead of three equations, we can also represent the line as

$$\mathcal{L} = {}^A(x_0, y_0, z_0)^T + t\widehat{\mathbf{Q}}_A \tag{2.70}$$

where t is an arbitrary constant and $\widehat{\mathbf{Q}}_A$ is a unit vector from ${}^A(x_0, y_0, z_0)^T$ to ${}^A(x_1, y_1, z_1)^T$ in the coordinate system $\{A\}$. In addition, the point ${}^A(x_0, y_0, z_0)^T$ need not have three independent parameters. Since the line extends to infinity in both directions, along the line, we can choose the point ${}^A(x_0, y_0, z_0)^T$ as the point where the line intersects any of the three coordinate planes ($x = 0$ or $y = 0$ or $z = 0$). Hence, a line in $\Re^3$ can be described by four independent parameters.

We will represent a line by a pair of vectors of the form $({}^A\mathbf{Q};\ {}^A\mathbf{Q}_0)$ where ${}^A\mathbf{Q}$ is the direction vector and ${}^A\mathbf{Q}_0$ is the moment vector given by

$${}^A\mathbf{Q}_0 = {}^A\mathbf{r} \times {}^A\mathbf{Q} \tag{2.71}$$

where ${}^A\mathbf{r}$ locates a point on the line. It can be shown that ${}^A\mathbf{Q}_0$ is independent of the chosen point on the line.

The vector pair $({}^A\mathbf{Q};\ {}^A\mathbf{Q}_0)$ are the six Plücker coordinates of a line in $\Re^3$. It may be noted that there are only four independent parameters since ${}^A\mathbf{Q} \cdot {}^A\mathbf{Q}_0 = 0$ and $c({}^A\mathbf{Q};\ {}^A\mathbf{Q}_0)$ $(c \in \Re^1 \neq 0)$ is the same line as $({}^A\mathbf{Q};\ {}^A\mathbf{Q}_0)$. Since, the Plücker coordinates of a line are unchanged by scaling, they are homogeneous coordinates. Hence, similarly to choosing $w = 1$ in the case of points, as long as $|{}^A\mathbf{Q}| \neq 0$, we can represent lines in

$\Re^3$ by a unit vector and its moment as

$$\widehat{\mathbf{Q}}_A = \frac{{}^A\mathbf{Q}}{|{}^A\mathbf{Q}|}$$

$${}^A\widehat{\mathbf{Q}}_0 = {}^A\mathbf{r} \times \widehat{\mathbf{Q}}_A \tag{2.72}$$

Note that the vector pair $(\widehat{\mathbf{Q}}_A;\ {}^A\widehat{\mathbf{Q}}_0)$ has four independent parameters.

The Denavit–Hartenberg parameters, discussed in Section 2.6, are based on the notion of the distance and the angle between lines in three-dimensional space. In the following, we present analytical expressions for the common perpendicular, angle, and distance between two lines in $\Re^3$. These expressions can be used to compute numerical values of DH parameters from a CAD model of a robot.

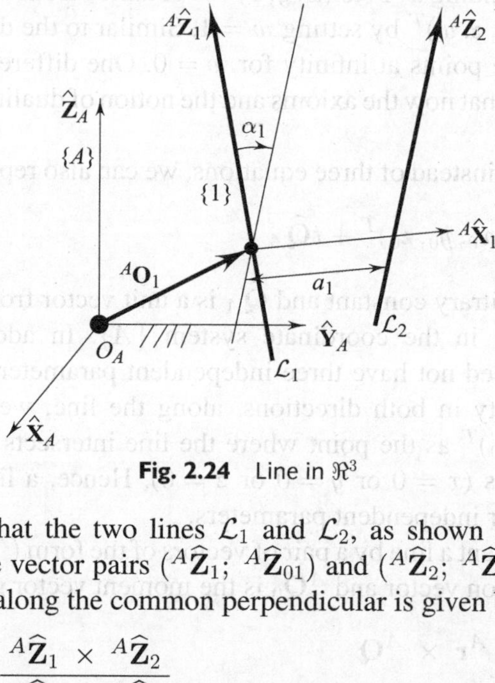

Fig. 2.24 Line in $\Re^3$

We assume that the two lines $\mathcal{L}_1$ and $\mathcal{L}_2$, as shown in Fig. 2.24, are described by the vector pairs $({}^A\widehat{\mathbf{Z}}_1;\ {}^A\widehat{\mathbf{Z}}_{01})$ and $({}^A\widehat{\mathbf{Z}}_2;\ {}^A\widehat{\mathbf{Z}}_{02})$, respectively. The unit vector along the common perpendicular is given by

$${}^A\widehat{\mathbf{X}}_1 = \frac{{}^A\widehat{\mathbf{Z}}_1 \times {}^A\widehat{\mathbf{Z}}_2}{|{}^A\widehat{\mathbf{Z}}_1 \times {}^A\widehat{\mathbf{Z}}_2|} \tag{2.73}$$

It may be noted that the unit vector ${}^A\widehat{\mathbf{X}}_1$ is from $\mathcal{L}_1$ to $\mathcal{L}_2$. If $\mathcal{L}_1$ and $\mathcal{L}_2$ intersect, then the unit vector ${}^A\widehat{\mathbf{X}}_1$ is normal to the plane formed by the two intersecting lines, and there are two choices for the direction of ${}^A\widehat{\mathbf{X}}_1$. If the lines are parallel, the common perpendicular is not unique and the vector cross product in Eqn (2.73) is zero. For this case, any line perpendicular to $\mathcal{L}_1$ and $\mathcal{L}_2$ is a common perpendicular.

The angle between the two lines is

$$\alpha_1 = \cos^{-1}({}^A\widehat{\mathbf{Z}}_1 \cdot {}^A\widehat{\mathbf{Z}}_2), \quad 0 \le \alpha_1 \le \pi \tag{2.74}$$

The angle α_1 can also be negative, $(-\pi \leq \alpha_1 \leq 0)$, and we can choose the correct sign by ensuring that the angle is measured from ${}^A\widehat{\mathbf{Z}}_1$ to ${}^A\widehat{\mathbf{Z}}_2$ about ${}^A\widehat{\mathbf{X}}_1$ using the right-hand rule. If the two lines are parallel, the angle is 0 or π.

The shortest distance is along the common perpendicular and is given by

$$a_1 = \frac{{}^A\widehat{\mathbf{Z}}_1 \cdot {}^A\widehat{\mathbf{Z}}_{02} + {}^A\widehat{\mathbf{Z}}_2 \cdot {}^A\widehat{\mathbf{Z}}_{01}}{|{}^A\widehat{\mathbf{Z}}_1 \times {}^A\widehat{\mathbf{Z}}_2|} \tag{2.75}$$

If the lines $\mathcal{L}_1$ and $\mathcal{L}_2$ intersect, then a_1 is zero. If the lines are parallel, the length of any of the common perpendiculars is a_1. The point of intersection of the common perpendicular line with $\mathcal{L}_1$ can be obtained by solving simultaneously the equations of line $\mathcal{L}_1$ and the plane formed by line $\mathcal{L}_2$ and the common perpendicular line. Denoting the point by the vector ${}^A\mathbf{O}_1$, we have

$${}^A\mathbf{O}_1 = \frac{({}^A\widehat{\mathbf{X}}_1 \cdot {}^A\widehat{\mathbf{Z}}_{02}){}^A\widehat{\mathbf{Z}}_1 - {}^A\widehat{\mathbf{Z}}_{01} \times ({}^A\widehat{\mathbf{Z}}_2 \times {}^A\widehat{\mathbf{X}}_1)}{|{}^A\widehat{\mathbf{Z}}_1 \times {}^A\widehat{\mathbf{Z}}_2|} \tag{2.76}$$

The moment vector ${}^A\widehat{\mathbf{X}}_{01}$ can be obtained by noting that the distances between $\mathcal{L}_1, \mathcal{L}_2$, and the common perpendicular are zero, and ${}^A\widehat{\mathbf{X}}_{01} \cdot {}^A\widehat{\mathbf{X}}_1 = 0$. The vector ${}^A\widehat{\mathbf{X}}_{01}$ is given by

$$\begin{aligned} {}^A\widehat{\mathbf{X}}_{01} = {} & [\{\cos\alpha_1 {}^A\widehat{\mathbf{Z}}_{02} - {}^A\widehat{\mathbf{Z}}_{01}\} \cdot {}^A\widehat{\mathbf{X}}_1]{}^A\widehat{\mathbf{Z}}_1 \\ & +[\{\cos\alpha_1 {}^A\widehat{\mathbf{Z}}_{01} - {}^A\widehat{\mathbf{Z}}_{02}\} \cdot {}^A\widehat{\mathbf{X}}_1]{}^A\widehat{\mathbf{Z}}_2 \end{aligned} \tag{2.77}$$

The lines $({}^A\widehat{\mathbf{X}}_1; {}^A\widehat{\mathbf{X}}_{01}), ({}^A\widehat{\mathbf{Z}}_1; {}^A\widehat{\mathbf{Z}}_{01})$ and their point of intersection, ${}^A\mathbf{O}_1$, completely determine the coordinate system $\{1\}$ with respect to the fixed coordinate system $\{A\}$.

A screw $\mathcal{S}$ with respect to $\{A\}$ can be specified by a line and a pitch denoted by h. The screw coordinates denoted by $({}^A\mathbf{S}; {}^A\mathbf{S}_0)$ are defined as

$$\begin{aligned} {}^A\mathbf{S} &= {}^A\mathbf{Q} \\ {}^A\mathbf{S}_0 &= {}^A\mathbf{Q}_0 + h{}^A\mathbf{Q} \end{aligned} \tag{2.78}$$

The pitch can be obtained from a given $({}^A\mathbf{S}; {}^A\mathbf{S}_0)$ by

$$h = \frac{{}^A\mathbf{S} \cdot {}^A\mathbf{S}_0}{{}^A\mathbf{S} \cdot {}^A\mathbf{S}} \tag{2.79}$$

Since a line has four independent parameters, a screw has five independent parameters. If the pitch h is zero, the screw coordinates are the same as the line coordinates. A screw has infinite pitch if ${}^A\mathbf{S} = \mathbf{0}$. A screw is an element of $\mathbf{P}^5$.

A twist is a six-dimensional entity which completely describes the motion of a rigid body in $\Re^3$. It can be thought of as a screw with a magnitude. Denoting a twist by $\mathcal{V}$, we can write

$$\mathcal{V} = c\left(\frac{{}^A\mathbf{S}}{|{}^A\mathbf{S}|};\ \frac{{}^A\mathbf{S}_0}{|{}^A\mathbf{S}|}\right), \quad c \in \Re^1 \tag{2.80}$$

In terms of (normalized) line coordinates, we can write

$$\mathcal{V} = c(\widehat{\mathbf{Q}}_A;\ {}^A\widehat{\mathbf{Q}}_0 + h\widehat{\mathbf{Q}}_A), \quad c \in \Re^1 \tag{2.81}$$

The six independent parameters are the four in normalized line coordinates $(\widehat{\mathbf{Q}}_A;\ {}^A\widehat{\mathbf{Q}}_0)$, the pitch h, and the magnitude c. A twist of zero pitch is pure rotation and is of the form $\theta(\widehat{\mathbf{Q}}_A;\ {}^A\widehat{\mathbf{Q}}_0)$, where θ is the amount of rotation. A twist of infinite pitch is a pure translation and is of the form $(\mathbf{0};\ d\widehat{\mathbf{Q}}_A)$, where d is the amount of translation.

Exercises

2.1 The position vectors of three points, not lying on a line, on a rigid body $\{B\}$ are given by ${}^A\mathbf{p_1}$, ${}^A\mathbf{p_2}$, and ${}^A\mathbf{p_3}$ with respect to a fixed coordinate system $\{A\}$. Obtain the rotation matrix ${}^A_B[R]$ in terms of the three position vectors.

2.2 What are the eigenvectors corresponding to complex eigenvalues $e^{\pm\imath\phi}$? What is the geometrical significance of these eigenvectors?

†2.3 Figure 2.25 shows two orientations of a commonly used dice. The frames $\{A\}$ and $\{B\}$ are as shown and, for clarity, the origins are not shown coincident. Estimate (a) the rotation matrix ${}^A_B[R]$, (b) the equivalent $(\widehat{\mathbf{k}}, \phi)$, and (c) the X-Y-Z Euler angles which can take $\{A\}$ to $\{B\}$. For part (a), use of a 3D CAD software to draw the dice in the two orientation may be helpful.

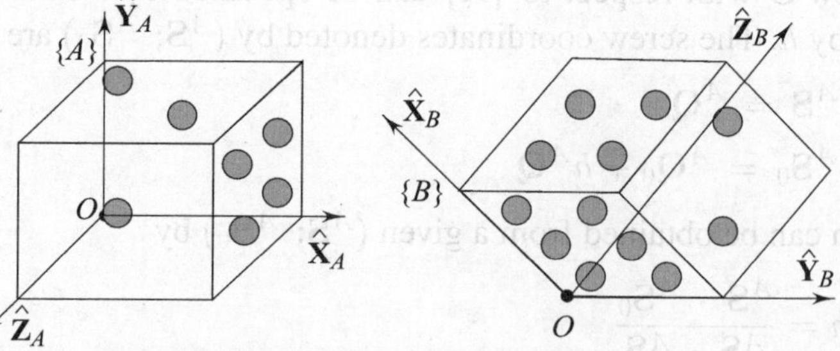

Fig. 2.25 Two orientations of a dice

2.4 Consider a vector ${}^A\mathbf{Q}$ being rotated about the vector ${}^A\widehat{\mathbf{k}}$ by an angle θ to form the vector ${}^A\mathbf{Q}'$. Show

$${}^A\mathbf{Q}' = {}^A\mathbf{Q}\cos\theta + \sin\theta({}^A\widehat{\mathbf{k}} \times {}^A\mathbf{Q}) + (1 - \cos\theta)({}^A\widehat{\mathbf{k}} \cdot {}^A\mathbf{Q})\,{}^A\widehat{\mathbf{k}}$$

This is known as Rodrigues' formula.

2.5 Use Rodrigues' formula to derive the expressions given in Eqn (2.9).

2.6 Consider the skew symmetric matrix $[\mathcal{K}]$ given by

$$[\mathcal{K}] = \begin{pmatrix} 0 & -k_z & k_y \\ k_z & 0 & -k_x \\ -k_y & k_x & 0 \end{pmatrix}$$

where $(k_x, k_y, k_z)^T$ is the eigenvector corresponding to $+1$ eigenvalue of ${}^A_B[R]$. Show that $e^{[\mathcal{K}]\phi}$ is the same as ${}^A_B[R]$ where the exponential of a matrix $[A]$ is defined as

$$e^{[A]} = [U] + [A] + (1/2)[A][A] + (1/6)[A][A][A] + \cdots$$

where $[U]$ is an identity matrix.

2.7 Consider three successive rotations of $[{}^A\widehat{\mathbf{X}}, \theta_1]$, $[{}^A\widehat{\mathbf{Y}}, \theta_2]$, and $[{}^A\widehat{\mathbf{Z}}, \theta_3]$—note that the rotations are about the axis of $\{A\}$. Derive the rotation matrix ${}^A_B[R]$. [***Hint***: Transform the second and third rotation axes, $\widehat{\mathbf{Y}}_A$ and $\widehat{\mathbf{Z}}_A$, to the moving $\{B\}$ and use Eqn (2.9)].

2.8 Show that the rotation matrix obtained in Exercise 2.7 is the same as 3-2-1 (or Z-Y-X) Euler angles about the moving axis. Explain why this is so.

2.9 If the rotation angle is small, i.e., $\sin\phi \cong \phi$ and $\cos\phi \cong 1$ holds, then obtain ${}^A_B[R]$ from Eqn (2.9). Using this result, show two infinitesimally small rotations commute.

2.10 Obtain expressions for r_{ij} (where $i = 1, 2, 3$) in terms of the four Euler parameters, ϵ_1, ϵ_2, ϵ_3, and ϵ_4.

†2.11 Figure 2.26 shows a commonly used dice in two orientations. The origin of $\{B\}$ with respect to O_A is as shown in Fig. 2.26. The frames $\{A\}$ and $\{B\}$ are as shown. Estimate ${}^A_B[T]$. Use of a 3D CAD software may be helpful in visualization.

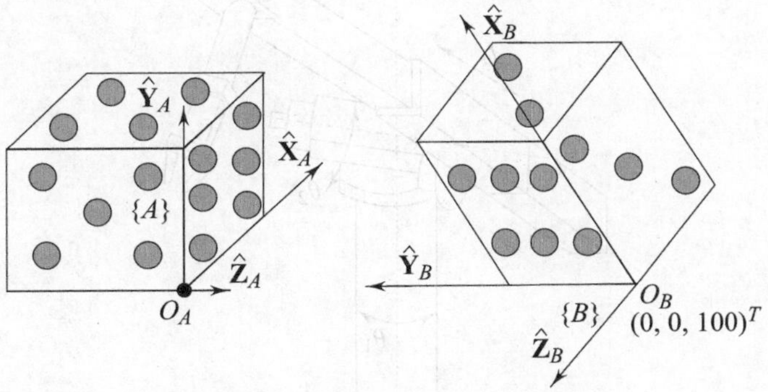

Fig. 2.26 Two orientations of a dice

2.12 Derive the expression for $\mathbf{Y}$ given in Eqn (2.37). Show that

$$\mathbf{Y} = (1/2)\left(\frac{\mathbf{b} \times {}^A\mathbf{O}_B^*}{\mathbf{b} \cdot \mathbf{b}} + {}^A\mathbf{O}_B^*\right)$$

where $\mathbf{b} = \widehat{\mathbf{k}} \tan(\phi/2)$, and $\widehat{\mathbf{k}}$ and ϕ denote the axis and amount of rotation, respectively.

2.13 For the problem given in 2.11 compute the screw parameters.

2.14 Derive the constraint equations for a Hooke (T) [also called universal (U)] joint.

2.15 Assign coordinate systems and obtain the DH parameters for the robot shown in Fig. 2.27. The arrangement of the non-intersecting joints at the wrist is shown on the right-hand side of the figure. The arrangement of the first three joints is similar to the PUMA 560 manipulator shown in Figs 2.18 and 2.19.

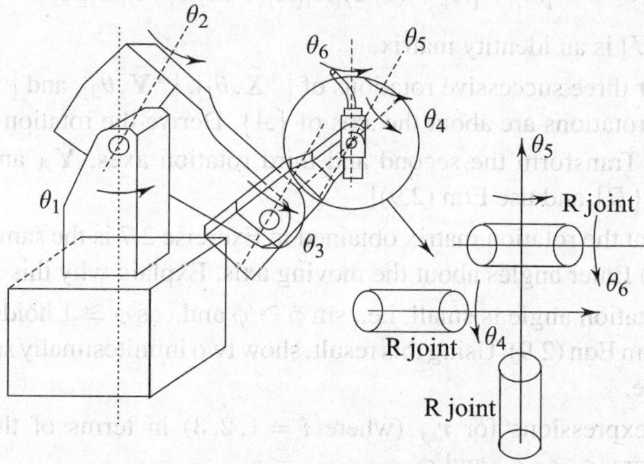

Fig. 2.27 IGM robot with non-intersecting wrist

2.16 Obtain the DH parameters for the Stanford arm shown in Fig. 2.28.

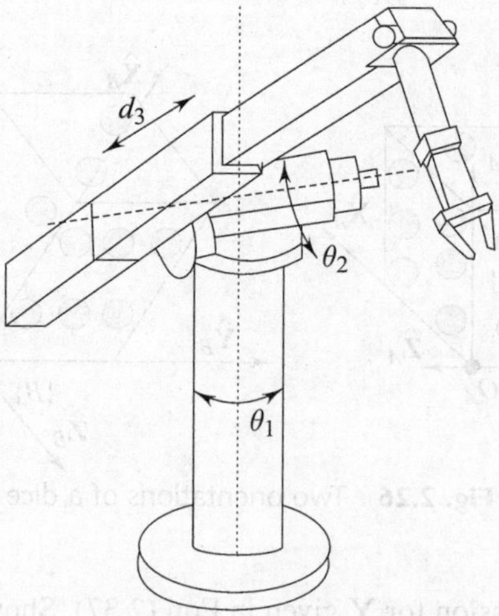

Fig. 2.28 Stanford arm

2.17 Obtain the transformation matrices for the robots given in Exercises 2.15 and 2.16.

2.18 In Example 2.5 assume that the top and bottom platforms are both equilateral triangles of sides a and b, respectively. Consider $\{Top\}$ fixed at the centroid of the top (moving) platform and the $\{Base\}$ fixed at the centroid of the bottom (fixed) platform as discussed in the example. Obtain ${}^{Base}_{Top}[T]$ in terms of a, b and the DH parameters of each leg as given in Example 2.5. [***Hint***: The centroid is given by $(1/3)({}^{Base}\mathbf{S}_1 + {}^{Base}\mathbf{S}_2 + {}^{Base}\mathbf{S}_3)$].

†2.19 In Problem 2.18, there is possibility of making mistakes if it is done manually. Write a MAPLE program to obtain symbolically ${}^{i-1}_{\ \ i}[T]$ given a DH table. Test the programs with the known example of the PUMA robot (Example 2.2).

†2.20 For the three-DOF parallel manipulator Example 2.5, derive and check the symbolic expressions for the position vectors given in Eqns (2.62) and (2.63) using MAPLE. Check numerically with numerical values to see whether the expressions make sense.

†2.21 For the six-DOF manipulator Example 2.6, derive and check the expressions given in Eqns (2.66)–(2.68) with MAPLE.

†2.22 For Example 2.6, obtain ${}^{Base}_{Top}[T]$ using MAPLE. Again assume the geometry of the gripped object as an equilateral triangle of side s as shown in Fig. 2.23.

References and Suggested Additional Reading

Arfken, G. 1985, *Mathematical Methods for Physicists*, 3rd edn, Academic Press.

Brand, L. 1947, *Vector and Tensor Analysis*, John Wiley.

Craig, J.J. 1989, *Introduction to Robotics: Mechanics and Control*, 2nd edn. Addison-Wesley.

Hamilton, W.R. 1866, *Elements of Quaternions*, Longmans Green.

Hartenberg, R.S. and J. Denavit 1964, *Kinematic Synthesis of Linkages*, McGraw-Hill.

Hunt, K.H., A. Samuel, and P. McAree 1991, 'Special configurations of multi-finger multi-freedom grippers—A kinematic study, *Int. J. Robot. Res.*, vol. 10, pp. 123–34.

Kane, T.R., P.W. Likins, and D.A. Levinson 1983, *Spacecraft Dynamics*, McGraw-Hill.

Lee, K.M. and D. Shah 1988, 'Kinematic analysis of a three-degrees-of-freedom in-parallel actuated manipulator', *IEEE Trans. Robotic Autom.*, vol. 14, no. 4, pp. 354–60.

McCarthy, J.M. 1990, *An Introduction to Theoretical Kinematics*, MIT Press.

Murray, R.M., Z. Li, and S.S. Sastry 1994, *A Mathematical Introduction to Robotic Manipulation*, CRC Press.

Salisbury, J.K. 1982, *Kinematics and force analysis of articulated hands*, PhD Thesis, Stanford University.

Sangamesh, Deepak R. and A. Ghosal 2006, 'A note on the diagonalizablity and the Jordon form of the 4×4 homogeneous transformation matrix', *Trans. ASME, J. Mech. Design.* vol. 128, no. 6, pp. 1343–48.

Strang, G. 1976, *Linear Algebra and its Application*, Academic Press.

3

Kinematics of Serial Manipulators

3.1 Introduction

In kinematics of manipulators, we study the motion of the links without considering the forces and torques which cause the motion of the links. A manipulator, it may be recalled, is broadly classified as serial or parallel and consists of rigid links and joints. In a serial manipulator, one end is fixed and, after a finite number of links connected one after another by joints, we arrive at a link, also called an end-effector, which is free. In a serial manipulator, as opposed to a parallel manipulator, there are no closed loops. In this chapter, we study the kinematics of serial manipulators. The manipulator geometries are described by the use of DH parameters discussed in Chapter 2, and we present and discuss two well-known problems—the direct and inverse kinematics problems. We look at the associated concept of workspace of a serial manipulator. In Chapter 4, we will study the kinematics of parallel manipulators. We start this chapter with a discussion on the key concept of degrees of freedom of a manipulator/mechanism.

3.2 Degrees of Freedom of a Manipulator

The degrees of freedom (DOF) of a manipulator or a mechanism can be obtained from the Grübler criterion

$$\text{dof} = \lambda(N - J - 1) + \sum_{i=1}^{J} F_i \tag{3.1}$$

where dof is the degrees of freedom of the mechanism/manipulator chain, N is the total number of links including the fixed link (or base), J is the total number of joints connecting only two links (if a joint connects three links, then it must be counted as two joints), F_i is the degrees of freedom at the ith joint, and

$$\lambda = \begin{cases} 6 & \text{for spatial manipulators and mechanisms} \\ 3 & \text{for planar manipulators and mechanisms} \end{cases}$$

For a serial manipulator such as the PUMA, $N = 7$ (6 moving links and 1 fixed base), $J = 6$ with each rotary joint having one degree of freedom, $\lambda = 6$ since motion is in $\Re^3$, and therefore dof $= \sum_{i=1}^{J} F_i = 6$. For a four-bar mechanism (see Fig. 2.21 in Chapter 2), $N = 4$ (3 moving links and 1 fixed base), $J = 4$ with each rotary joint having one degree of freedom, $\lambda = 3$ since motion is in a plane, and, hence, dof $= 1$. It may be noted that although for a vast majority of mechanisms and manipulators, the above formula correctly gives the degrees of freedom, there exist several mechanisms that can move even though the dof according to Eqn (3.1) is less than 1. These are called over-constrained mechanisms and they are still a subject of kinematics research [see Mavroidis and Roth (1995) and Gan and Pellegrino (2003)].

The quantity dof obtained from Eqn (3.1) is the number of independent actuators that can be put in a mechanism or a manipulator. It also describes, in the broadest sense, the capability of a mechanism or a manipulator with respect to λ which denote the dimension of the ambient space in which the motion is taking place. We have the following possibilities.

1. dof $= \lambda$—In this case, an end-effector of a manipulator can be positioned and oriented arbitrarily. As mentioned earlier, λ is 3 for planar motion and 6 for motion in $\Re^3$.
2. dof $< \lambda$—In this case, the arbitrary position and orientation of the end-effector is not achievable. There exist ($\lambda-$ dof) functional relationships containing the position and orientation variables.
3. dof $> \lambda$—There exist extra degrees of freedom in the manipulator and any position and orientation of the end-effector can be obtained in ∞ ways. Such manipulators are called *redundant manipulators*.

In serial manipulators with a fixed base, a free end-effector and two links connected by a joint, as in the example of a PUMA manipulator, $N = J + 1$ and dof $= \sum_{i=1}^{J} F_i$. If all the actuated joints are one-DOF joints, then $J =$ dof.

If $J >$ dof, as in parallel manipulators, then ($J-$ dof) of the joints are not actuated and are called passive joints. Passive joints need not be one-DOF joints. Spherical (S) and cylindrical (C) joints, with $F_i = 3$ and 2, respectively, are often used in spatial mechanisms and parallel manipulators. For parallel manipulators/mechanisms, $N \leq J$. We will look at the kinematic analysis of parallel manipulators in Chapter 4.

If $J <$ dof, then one or more of the actuated joints are multi-DOF joints, and this is not used in mechanical manipulators.[1]

[1] Due to space restrictions and the available type of actuators, it is difficult to actuate more than one-DOF joints. In biological systems, such as in a human shoulder joint, muscles can easily actuate multi-DOF joints.

In manipulators, the J joint variables (θ or d for a one-DOF R and P joint) form the joint space. The variables describing the position and orientation of a link or the end-effector are called the task space variables and these are less than or equal to 6 for motions in 3D space and less than or equal to 3 for motion in a plane. Finally, there are often mechanical linkages, gears, etc. between actuators and joints. The space of all actuator variables is called the actuator space. If the dimension of the actuator space is more than 3 for planar motion and more than 6 for 3D motion, the manipulator is called *redundant*. If the dimension of the actuator space is less than the degree of freedom, then the manipulator is called *under-actuated*.

For serial manipulators with N links, there are $3 \times (N-1)$ constant Denavit–Hartenberg parameters and J joint variables.

With the above definitions, we can now consider the two main problems in serial manipulator kinematics, namely, the direct and inverse kinematics problems.

3.3 Direct Kinematics of Serial Manipulators

The direct kinematics problem of a serial manipulator can be stated as follows: given the link parameters and the joint variable, a_{i-1}, α_{i-1}, d_i, and θ_i, find the position and orientation of the last link in the fixed or reference coordinate system.

The above is the simplest possible problem in manipulator kinematics. It follows directly from the notion of the link transformation matrix of Section 2.7. If the fixed coordinate system is $\{0\}$ and the coordinate system of the end-effector is $\{n\}$, we can write

$$ {}^{0}_{n}[T] = {}^{0}_{1}[T]{}^{1}_{2}[T] \cdots {}^{n-1}_{n}[T] \tag{3.2} $$

It may be recalled that the link transformation matrices on the right-hand side are functions of a_{i-1}, α_{i-1}, d_i, and θ_i. In the direct kinematics problem, all these are known and hence all the 4×4 matrices on the right-hand side are known. The direct kinematics problem for serial manipulators can be solved by simple matrix multiplication and extraction of the rotation matrix, ${}^{0}_{n}[R]$, and the origin, ${}^{0}\mathbf{O}_n$, of $\{n\}$.

In many situations, the orientation of the tool or end-effector and the position of a point on the tool or end-effector is of more practical importance. In addition, instead of $\{0\}$, the chosen reference frame may be $\{Base\}$ (see Fig. 3.1). Denoting the transformation between $\{0\}$ and $\{Base\}$ by ${}^{Base}_{0}[T]$ and the transformation between $\{n\}$ and $\{Tool\}$ by ${}^{n}_{Tool}[T]$, we can get

$$ {}^{Base}_{Tool}[T] = {}^{Base}_{0}[T]\,{}^{0}_{n}[T]\,{}^{n}_{Tool}[T] \tag{3.3} $$

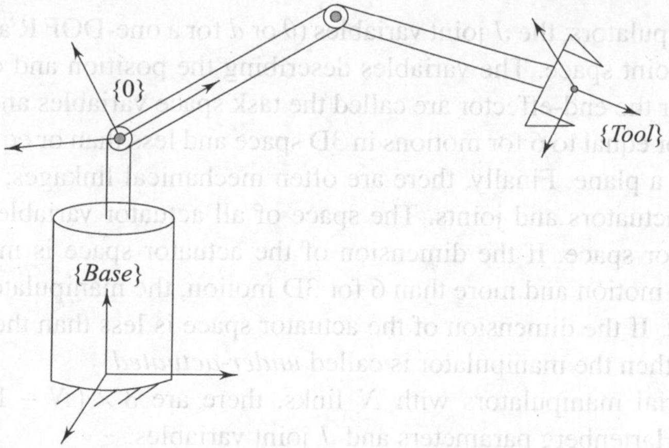

Fig. 3.1 The {*Base*} and {*Tool*} frames

The direct kinematics problems for three serial manipulators are solved in the following examples.

Example 3.1 The planar 3R manipulator

In the case of the planar 3R manipulator (see Fig. 2.17), the orientation of the tool or the gripper can be described by an angle ϕ as shown in Fig. 2.17. From Eqn (2.58), one can directly write the position, x, y, and orientation of the tool as

$$\begin{aligned} x &= l_1 c_1 + l_2 c_{12} + l_3 c_{123} \\ y &= l_1 s_1 + l_2 s_{12} + l_3 s_{123} \\ \phi &= \theta_1 + \theta_2 + \theta_3 \end{aligned} \tag{3.4}$$

where, as before, $c_{(\cdot)}$ and $s_{(\cdot)}$ represent $\cos(\cdot)$ and $\sin(\cdot)$, respectively.

Example 3.2 The PUMA 560 manipulator

The transformation matrix ${}^0_6[T]$ for the PUMA manipulator is obtained by multiplying ${}^0_3[T]$ and ${}^3_6[T]$ given in Eqns (2.59) and (2.60). Denoting the elements of the rotation matrix ${}^0_6[R]$ by r_{ij} (where $i, j = 1, 2, 3$), and the components of the origin of the coordinate system {6}, ${}^0\mathbf{O}_6$, by $(O_{6x}, O_{6y}, O_{6z})^T$, we can write

$$\begin{aligned} r_{11} &= c_1\{c_{23}(c_4c_5c_6 - s_4s_6) - s_{23}s_5c_6\} + s_1(s_4c_5c_6 + c_4s_6) \\ r_{21} &= s_1\{c_{23}(c_4c_5c_6 - s_4s_6) - s_{23}s_5c_6\} - c_1(s_4c_5c_6 + c_4s_6) \\ r_{31} &= -s_{23}(c_4c_5c_6 - s_4s_6) - c_{23}s_5c_6 \end{aligned}$$

$$\begin{aligned} r_{12} &= c_1\{c_{23}(-c_4c_5s_6 - s_4c_6) + s_{23}s_5s_6\} + s_1(-s_4c_5s_6 + c_4c_6) \\ r_{22} &= s_1\{c_{23}(-c_4c_5s_6 - s_4c_6) + s_{23}s_5s_6\} - c_1(-s_4c_5s_6 + c_4c_6) \\ r_{32} &= -s_{23}(c_4c_5s_6 - s_4c_6) + c_{23}s_5s_6 \end{aligned}$$

$$
\begin{aligned}
r_{13} &= -c_1(c_{23}c_4s_5 + s_{23}c_5) - s_1s_4s_5 \\
r_{23} &= -s_1(c_{23}c_4s_5 + s_{23}c_5) + c_1s_4s_5 \\
r_{33} &= s_{23}c_4s_5 - c_{23}c_5 \\
O_{6x} &= c_1(a_2c_2 + a_3c_{23} - d_4s_{23}) - d_3s_1 \\
O_{6y} &= s_1(a_2c_2 + a_3c_{23} - d_4s_{23}) + d_3c_1 \\
O_{6z} &= -a_2s_2 - a_3s_{23} - d_4c_{23}
\end{aligned} \tag{3.5}
$$

To obtain the position and orientation of the tool, we have to multiply ${}^0_6[T]$ with a known transform ${}^{\ 6}_{Tool}[T]$.

Example 3.3 A SCARA manipulator

For the SCARA manipulator, the matrix ${}^0_4[T]$ is given in Eqn (2.61). The orientation of the $\{4\}$ can be described by the angle ϕ (see Fig. 2.20), and the position (x, y, z) and orientation of $\{4\}$ can be given as

$$
\begin{aligned}
x &= a_1c_1 + a_2c_{12} \\
y &= a_1s_1 + a_2s_{12} \\
z &= -d_3 \\
\phi &= \theta_1 + \theta_2 + \theta_4
\end{aligned} \tag{3.6}
$$

It may be noted that the direct kinematics problem for a serial manipulator is well defined, easily solvable, and unique for any number of links. The last statement means that as long as all the Denavit–Hartenberg parameters are given, we can obtain ${}^0_n[T]$ for any n.

3.4 Inverse Kinematics of Serial Manipulators

The inverse kinematics problem for serial manipulators can be stated as follows: given the link parameters and the position and orientation of $\{n\}$ with respect to the fixed frame $\{0\}$, find the joint variables. For the planar 3R manipulator example, the inverse kinematics problem would be to obtain θ_i (where $i = 1, 2, 3$), given the position of the end-effector x, y and its orientation ϕ. In the case of the spatial PUMA manipulator, the transformation matrix ${}^0_6[T]$ is given and the aim is to find the joint angles θ_i (i being $1, \ldots, 6$). In general, we are given six task space variables (three for position and three for orientation) for 3D motion (three task space variables for planar motion) and the goal is to find the n joint variables which make up ${}^0_n[T]$. We can have the following cases.

1. $n = 6$ for motion in 3D or $n = 3$ for a planar motion. In this case, we have the required number of equations for the unknowns.

2. $n < 6$ for motion in 3D or $n < 3$ for a planar motion. In this case, the number of task space variables is more than the number of equations and hence, for solutions to exists, there must be $6 - n$ ($3 - n$ for planar case) relationships involving the task space variables.
3. $n > 6$ for motion in 3D or $n > 3$ for a planar motion. This is the case of more unknowns than equations and hence we will have infinite number of solutions. These are called redundant manipulators.

We start by considering the inverse kinematics of serial manipulators with $n = 6$ ($n = 3$ for planar). It can be seen from the direct kinematics equations of the planar 3R or the spatial PUMA manipulator [see Eqns (3.4) and (3.5)], the position and orientation of the end-effector are related to the joint variables by means of non-linear transcendental equations. The solution to the inverse kinematics problem involves 'inverting' the direct kinematics equations and solving the non-linear equations. Our aim is to obtain closed-form expressions[2] for the joint variables in terms of the given position and orientation of the last link. Like most sets of non-linear equations, there were no known general methods for solving the inverse kinematics problem for an arbitrary serial manipulator till the recent work of Raghavan and Roth (1993). Before we look at this general method, it is instructive to obtain the inverse kinematics equations of a few simple serial manipulators.

Example 3.4 The planar 3R manipulator

For the planar 3R manipulator shown in Fig. 2.17, the direct kinematics equations are given in Eqn (3.4) and are given below for convenience. They are

$$\begin{aligned} x &= l_1 c_1 + l_2 c_{12} + l_3 c_{123} \\ y &= l_1 s_1 + l_2 s_{12} + l_3 s_{123} \\ \phi &= \theta_1 + \theta_2 + \theta_3 \end{aligned}$$

where l_i (i being $1, 2$), are the link lengths for the 3R manipulator, and l_3 is the length of the end-effector. To solve for θ_i (where $i = 1, 2, 3$), given x, y, ϕ, we proceed as follows.

Define $X = x - l_3 c_\phi$ and $Y = y - l_3 s_\phi$. X and Y are known since x, y, ϕ, and l_3 are known. Squaring and adding, we get

$$X^2 + Y^2 = l_1^2 + l_2^2 + 2 l_1 l_2 c_2 \tag{3.7}$$

[2] One can also solve the inverse kinematics problem numerically. However, numerical solution procedures may not give all the solutions, depend critically on initial guess, and are not very suitable to obtain the workspace or the singularities of a manipulator.

and we can find

$$\theta_2 = \pm \cos^{-1}\left(\frac{X^2 + Y^2 - l_1^2 - l_2^2}{2l_1 l_2}\right) \tag{3.8}$$

Once θ_2 is known, we can find θ_1 using the four quadrant arc tangent formula as

$$\theta_1 = \text{Atan2}(Y, X) - \text{Atan2}(k_2, k_1) \tag{3.9}$$

where $k_2 = l_2 s_2$ and $k_1 = l_1 + l_2 c_2$. Finally, θ_3 is obtained from

$$\theta_3 = \phi - \theta_1 - \theta_2 \tag{3.10}$$

We define the workspace of the planar 3R manipulator as the set of values of $\{x, y, \phi\}$ for which the inverse kinematics solution exists. From Eqn (3.8), we can obtain the workspace of the planar 3R manipulator. We know that

$$-1 \leq \left(\frac{X^2 + Y^2 - l_1^2 - l_2^2}{2l_1 l_2}\right) \leq +1$$

This implies that

$$(l_1 - l_2)^2 \leq (X^2 + Y^2) \leq (l_1 + l_2)^2 \tag{3.11}$$

where $X = x - l_3 c_\phi$ and $Y = y - l_3 s_\phi$. Figure 3.2 shows a 3D plot of the region in $\{x, y, \phi\}$ space where the inequalities in Eqn (3.11) are satisfied and the inverse kinematics solution exists. A projection of the workspace on the $\hat{X}_0$-$\hat{Y}_0$ plane is shown in Fig. 3.3 for $l_1 > l_2 > l_3$. We have four circles of radii $l_1 + l_2 + l_3$, $l_1 + l_2 - l_3$, $l_1 - l_2 + l_3$, and $l_1 - l_2 - l_3$.

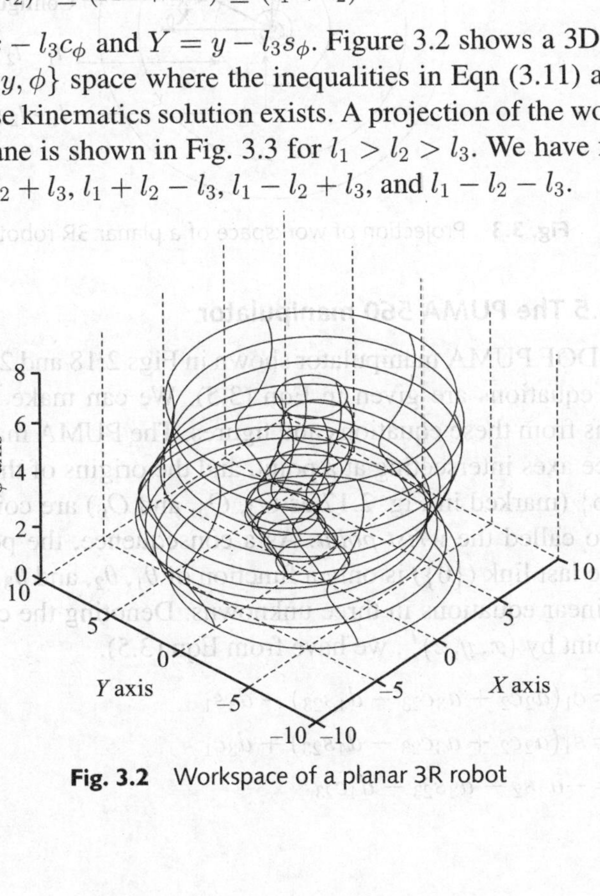

Fig. 3.2 Workspace of a planar 3R robot

It can be clearly seen that the maximum reach of the planar 3R manipulator are points on the circle of radius $l_1 + l_2 + l_3$ and the closest it can reach from the origin are points on the circle of radius $l_1 - l_2 - l_3$. This annular region is called the reachable workspace. Points in-between the other two circles of radii $l_1 + l_2 - l_3$ and $l_1 - l_2 + l_3$ can be reached with any ϕ (see Exercise 3.4). This region was termed as dexterous workspace by Kumar and Waldron (1980; 1981). It can be seen that as l_3 increases, the reachable workspace increases whereas the dexterous workspace decreases.

Given any point in the workspace, we can see from the inverse kinematics procedure that we can get two sets of values of θ_1, θ_2, and θ_3. This is shown schematically in Fig. 3.3—the given X and Y can be reached by the two configurations. As with any non-linear equation, the solutions are non-unique and we could have several solutions. In the planar 3R manipulator, the above analysis shows that we can have two sets of solutions.

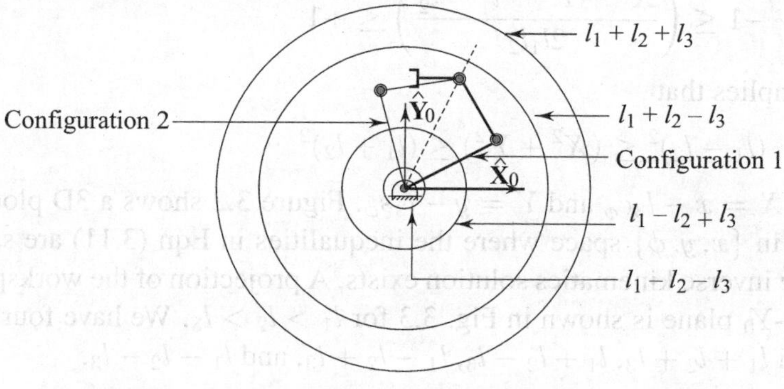

Fig. 3.3 Projection of workspace of a planar 3R robot

Example 3.5 The PUMA 560 manipulator

For the six-DOF PUMA manipulator shown in Figs 2.18 and 2.19, the direct kinematics equations are given in Eqn (3.5). We can make the following observations from these equations and figures. The PUMA manipulator has the last three axes intersecting at a point, and the origins of the frames $\{4\}$, $\{5\}$, and $\{6\}$ (marked in Fig. 2.17 as O_4, O_5, and O_6) are coincident. This point is also called the *wrist point*. As a consequence, the position of the origin of the last link ($\{6\}$) is only a function of θ_1, θ_2, and θ_3, and we have three non-linear equations in three unknowns. Denoting the coordinates of the wrist point by $(x, y, z)^T$, we have from Eqn (3.5),

$$
\begin{aligned}
x &= c_1(a_2c_2 + a_3c_{23} - d_4s_{23}) - d_3s_1 \\
y &= s_1(a_2c_2 + a_3c_{23} - d_4s_{23}) + d_3c_1 \qquad (3.12) \\
z &= -a_2s_2 - a_3s_{23} - d_4c_{23}
\end{aligned}
$$

We can observe from the first two equations that

$$-s_1x + c_1y = d_3 \tag{3.13}$$

The above can be solved for θ_1 by making the substitution[3]

$$x_1 = \tan\frac{\theta_1}{2}$$
$$c_1 = \frac{1-x_1^2}{1+x_1^2},\quad s_1 = \frac{2x_1}{1+x_1^2} \tag{3.14}$$

which converts the transcendental equation (3.13) to the quadratic

$$x_1^2(d_3+y) + 2xx_1 + (d_3 - y) = 0 \tag{3.15}$$

This quadratic can be solved in closed form for x_1, and we can get

$$\theta_1 = 2\tan^{-1}\left(\frac{-x \pm \sqrt{x^2+y^2-d_3^2}}{y+d_3}\right) \tag{3.16}$$

It may be noted that $\tan^{-1}$ gives an angle between 0 and π and hence we get $0 \le \theta_1 \le 2\pi$. However, due to the $\pm$ sign before the square root, we can get two possible values of θ_1.

To obtain θ_3 we can observe that

$$x^2+y^2+z^2 = d_3^2 + a_2^2 + a_3^2 + d_4^3 + 2a_2a_3c_3 - 2a_2d_4s_3 \tag{3.17}$$

leading to a single equation in sine and cosine of θ_3 as in the case of θ_1. We can solve for θ_3 by again making the tangent half-angle substitutions and obtain

$$\theta_3 = 2\tan^{-1}\left(\frac{-d_4 \pm \sqrt{d_4^2+a_3^2-K^2}}{K+a_3}\right) \tag{3.18}$$

where the constant K is given as $(1/2a_2)(x^2+y^2+z^2-d_3^2-a_2^2-a_3^2-d_4^2)$. Again, $\tan^{-1}$ gives an angle between 0 and π and hence we get $0 \le \theta_3 \le 2\pi$. Due to the $\pm$ sign, we can get two possible values of θ_3.

Finally, to obtain an expression for θ_2, we observe that the Z component of the wrist point is only a function of θ_2 and θ_3. We can rewrite the third equation in Eqn (3.12) as

$$-s_2(a_2 + a_3c_3 - d_4s_3) + c_2(-a_3s_3 - d_4c_3) = z \tag{3.19}$$

[3] These are known as the tangent half-angle rules in trigonometry.

We can solve the above for θ_2 by making the tangent half-angle substitutions, solving the resulting quadratic, and get

$$\theta_2 = 2\tan^{-1}\left(\frac{-a_2 - a_3c_3 + d_4s_3 \pm \sqrt{a_2^2 + a_3^2 + d_4^2 + 2a_2(a_3c_3 - d_4s_3) - z^2}}{z - (a_3s_3 + d_4c_3)}\right) \tag{3.20}$$

It may again be noted that we get $0 \leq \theta_2 \leq 2\pi$, and we can get two possible values of θ_2 due to the $\pm$ sign. Since θ_3 appears on the right-hand side of Eqn (3.20) in c_3 and s_3, and we can have two possible values of θ_3 from Eqn (3.18), we can get four possible values of θ_2.

To obtain θ_4, θ_5, and θ_6, we make use of Eqn (2.60) which gives ${}^3_6[T]$. The top 3×3 sub-matrix gives the rotation matrix ${}^3_6[R]$, which is given as

$${}^3_6[R] = \begin{pmatrix} c_4c_5c_6 - s_4s_6 & -c_4c_5s_6 - s_4c_6 & -c_4s_5 \\ s_5c_6 & -s_5s_6 & c_5 \\ -s_4c_5c_6 - c_4s_6 & s_4c_5s_6 - c_4c_6 & s_4s_5 \end{pmatrix} \tag{3.21}$$

We can also write

$${}^3_6[R] = {}^0_3[R]^T \, {}^0_6[R] \tag{3.22}$$

and since θ_1, θ_2, and θ_3 are now known, we can obtain the right-hand side. Let the right-hand side matrix have elements r_{ij}, where $i, j = 1, 2, 3$. By comparing, we can write the following algorithm to obtain θ_4, θ_5, and θ_6.

Algorithm $r_{ij} \Rightarrow \theta_4, \theta_5,$ and θ_6

If $r_{23} \neq \pm 1$, then
$\theta_5 = \text{Atan2}[\pm\sqrt{(r_{21}^2 + r_{22}^2)}, r_{23}]$
$\theta_4 = \text{Atan2}(r_{33}/s_5, -r_{13}/s_5),$
$\theta_6 = \text{Atan2}(-r_{22}/s_5, r_{21}/s_5)$

Else
If $r_{23} = 1$, then
$\theta_4 = 0$
$\theta_5 = 0$
$\theta_6 = \text{Atan2}(-r_{12}, r_{11}),$
If $r_{23} = -1$, then
$\theta_4 = 0$
$\theta_5 = \pi$
$\theta_6 = -\text{Atan2}(r_{12}, -r_{11}),$

The above inverse kinematics procedure gives two values of θ_1, two values of θ_3, and four values of θ_2. We also get two sets of values for θ_4, θ_5, and θ_6. Hence a PUMA 560 manipulator can have eight possible configurations.

The workspace of a PUMA manipulator is the set of values of the position and orientation of the last link or tool for which the inverse kinematics solution exists. This is clearly much harder to imagine or describe since it is six dimensional. It is, however, possible to describe the possible positions of the wrist point. The position vector of the wrist point is given by

$$\begin{aligned} x &= c_1(a_2c_2 + a_3c_{23} - d_4s_{23}) - d_3s_1 \\ y &= s_1(a_2c_2 + a_3c_{23} - d_4s_{23}) + d_3c_1 \\ z &= -a_2s_2 - a_3s_{23} - d_4c_{23} \end{aligned} \tag{3.23}$$

Equations (3.23) describe a 3D solid. The equation of the bounding surface(s) of the solid can be obtained as follows.

Squaring and adding the three equations in Eqn (3.23), we get Eqn (3.17), which can be re-arranged and written as

$$R^2 = x^2 + y^2 + z^2 = K_1 + K_2c_3 - K_3s_3$$

where K_1, K_2, and K_3 are constants. To obtain the envelope of this family of surfaces, we must have

$$\frac{\partial R^2}{\partial \theta_3} = 0$$

which gives

$$K_2s_3 + K_3c_3 = 0$$

Eliminating θ_3 from these two equations, and denoting $a_3^2 + d_4^2$ by l^2, we can get, after simplification,

$$\{x^2 + y^2 + z^2 - [(a_2 + l)^2 + d_3^2]\}\{x^2 + y^2 + z^2 - [(a_2 - l)^2 + d_3^2]\} = 0 \tag{3.24}$$

which implies that the bounding surfaces are spheres.

At every point in the solid, all possible orientations make up the orientation workspace of the PUMA. To describe the orientation workspace, we recognize that the last three rotations in the PUMA, namely, θ_4, θ_5, and θ_6, are like a set of three Euler angles about two distinct axes (see Chapter 2, Section 2.2.3). In particular, they are Z-Y-Z rotations, except that the second Y rotation is $-\theta_5$. From Chapter 2, we know that, except for two special 'singular' configurations, we can determine the three Euler angles given an arbitrary rotation matrix. Hence, at every point in the solid, the last link can be oriented arbitrarily, except at two singular configurations which are also called the wrist singularities. This intuitive reasoning can be verified from the algorithm given above to determine θ_4, θ_5, and θ_6 given above and we can see that the only problem cases are when $r_{23} = \pm 1$.

Numerical example of a PUMA

For the PUMA 560, the Denavit–Hartenberg parameters are given by

i	α_{i-1} degrees	a_{i-1} m	d_i m	θ_i degrees
1	0	0	0	45
2	−90	0	0	60
3	0	0.4318	0.125	135
4	−90	0.019	0.432	30
5	90	0	0	−45
6	−90	0	0	120

In the above, we have chosen arbitrarily the values of θ_i, where $i = 1, ..., 6$. Using the direct kinematics Eqn (3.5), we obtain the transformation matrix ${}^0_6[T]$ as

$$ {}^0_6[T] = \begin{bmatrix} 0.9749 & -0.2192 & -0.0388 & 0.1304 \\ 0.1643 & 0.8262 & -0.5388 & 0.3071 \\ 0.1502 & 0.5190 & 0.8415 & 0.0482 \\ 0 & 0 & 0 & 1 \end{bmatrix} $$

The above ${}^0_6[T]$ can be used as an input to the inverse kinematics algorithm. Using the inverse kinematics equations, we get eight sets of solutions. They are given in a tabular form below with the angles given in degrees.

i	θ_1	θ_2	θ_3	θ_4	θ_5	θ_6
1	−91	120	50.04	177.51	−42.65	105.34
2	−91	120	50.04	−2.49	42.65	−74.66
3	45	−77.73	50.04	85.25	−159.22	−132.87
4	45	−77.73	50.04	−94.75	159.22	47.13
5	−91	−102.27	135	92.28	−178.31	15.79
6	−91	−102.27	135	−87.72	178.31	−164.21
7	45	60	135	30	−45	120
8	45	60	135	210	45	300

As expected, one of the solutions (set 7) matches the chosen values of θ_i, where $i = 1, ..., 6$, in the direct kinematics.

The workspace of the wrist point of the PUMA for the numerical values assumed in this numerical example is shown in Fig. 3.4. It may be noted that the workspace of an actual PUMA robot is only a subset due to joint limits.

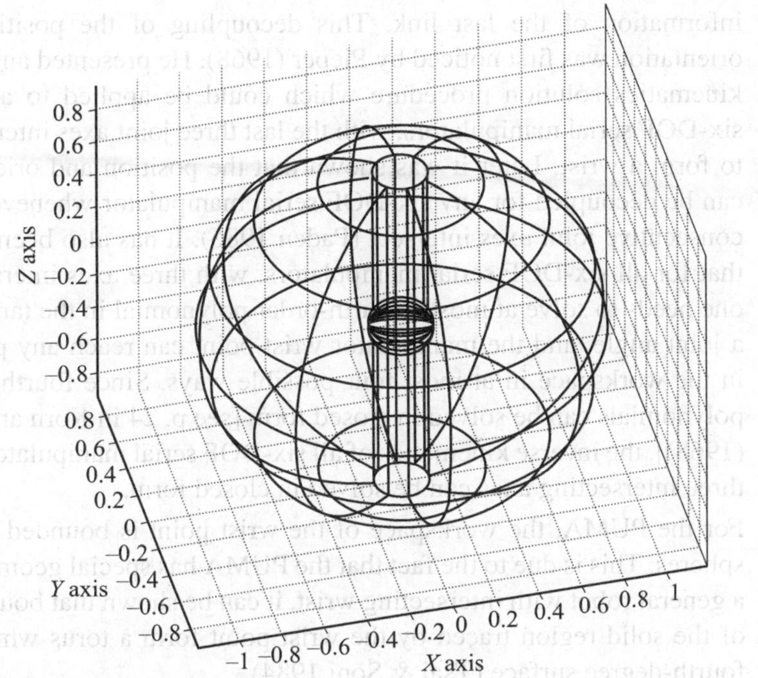

Fig. 3.4 Workspace of the wrist point of the PUMA

We can conclude the following from these two examples.

- Transcendental equations can be converted to polynomial equations by the tangent half-angle substitution. However, the polynomial equation is of a higher degree—an equation linear in $\sin(\theta)$ or $\cos(\theta)$ gives rise to a quadratic containing x^2 and lower powers of x with $x = \tan(\theta/2)$ [see Eqn (3.14)].
- To solve the inverse kinematics problem analytically, we need to eliminate joint variables from sets of non-linear equations in several joint variables to finally arrive at a single equation in one joint variable, which can then be solved. In the planar 3R example, we started with three equations in three joint variables and first reduced to two equations in θ_1 and θ_2 and then obtained one equation, Eqn (3.7), in θ_2 alone. This single equation was solved for θ_2 and then we went backward to solve for θ_1 and θ_3.
- In the case of the spatial PUMA, we were able to obtain three equations in three joint variables from the position information of the last link. This allowed us to solve for the first three joint variables, and then the last three joint variables were solved by considering the orientation

information of the last link. This decoupling of the position and orientation was first noticed by Pieper (1968). He presented an inverse kinematics solution procedure which could be applied to arbitrary six-DOF serial manipulators, with the last three joint axes intersecting to form a wrist. Later it was shown that the position and orientation can be decoupled for any six-DOF serial manipulator whenever three consecutive joint axes intersect (Paden 1986). It has also been shown that for all six-DOF serial manipulators, with three axes intersecting, one needs to solve at most a fourth-order polynomial in the tangent of a joint angle, and the manipulator wrist point can reach any position in the workspace in at most four possible ways. Since fourth-degree polynomials can be solved in closed form [see p. 24 in Korn and Korn (1968)], the inverse kinematics of all six-DOF serial manipulators with three intersecting axes can be solved in closed form.

- For the PUMA, the workspace of the wrist point is bounded by two spheres. This is due to the fact that the PUMA has special geometry. In a general robot with intersecting wrist, it can be shown that boundaries of the solid region traced by the wrist point form a torus which is a fourth-degree surface (Tsai & Soni 1984).

For manipulators where three consecutive axes do not intersect, the elimination procedure to solve the inverse kinematics problem is much more complex. Several researchers worked on this problem. Duffy and Crane (1980) first derived a 32nd-order polynomial, and it was finally demonstrated by Raghavan and Roth (1993) that the inverse kinematics of an arbitrary 6R serial manipulator required the solution of at most a sixteenth-degree polynomial. Before we discuss the general algorithm by Raghavan and Roth for a 6R manipulator, we obtain the inverse kinematics equations of a robot with only the last two joints intersecting or with a non-intersecting wrist.

3.5 Manipulator With Non-intersecting Wrist

The three-axes-intersecting wrist, as in the PUMA manipulator, is a fairly complicated design and also difficult to manufacture. In addition, there is always some manufacturing tolerance that makes it impossible to have three axes intersecting at a point. Figure 2.27 shows a six-DOF robot manufactured by IGM and often used for robotic welding. In this robot, as shown in the figure, the last three axes do not intersect and there is an offset d_5. The rest of the Denavit–Hartenberg parameters are the same as in a PUMA and are shown below. We use this example to discuss the inverse kinematics of a manipulator with a non-intersecting wrist.

i	α_{i-1}	a_{i-1}	d_i	θ_i
1	0	0	0	θ_1
2	$-\pi/2$	0	0	θ_2
3	0	a_2	d_3	θ_3
4	$-\pi/2$	a_3	d_4	θ_4
5	$\pi/2$	0	d_5	θ_5
6	$-\pi/2$	0	0	θ_6

From the above DH table, we can compute ${}^0_1[T], \ldots, {}^5_6[T]$ and then multiplying the transformation matrices we can compute ${}^0_6[T]$. From the last column of ${}^0_6[T]$, we get

$$\begin{aligned}
x &= c_1(a_2c_2 + a_3c_{23} - d_4s_{23}) - d_3s_1 + d_5(s_1c_4 - c_1s_4c_{23}) \\
y &= s_1(a_2c_2 + a_3c_{23} - d_4s_{23}) + d_3c_1 - d_5(c_1c_4 + s_1s_4c_{23}) \\
z &= -a_2s_2 - a_3s_{23} - d_4c_{23} - d_5s_4s_{23}
\end{aligned} \tag{3.25}$$

Equations (3.25) are in four joint variables $(\theta_1, \theta_2, \theta_3, \theta_4)$ in terms of known x, y, and z. We need one more equation without introducing any more unknown joint variable. This can be obtained by considering the rotation matrices. Noting that the inverse of a rotation matrix is the same as its transpose, we can write

$${}^3_6[R] = {}^0_3[R]^T\,{}^0_6[R] \tag{3.26}$$

where ${}^0_6[R]$ have elements r_{ij} (where $i, j = 1, 2, 3$), and are given. Equation (3.26) can be expanded to obtain

$$\begin{pmatrix} c_4c_5c_6 - s_4s_6 & -c_4c_5s_6 - s_4c_6 & -c_4s_5 \\ s_5c_6 & -s_5s_6 & c_5 \\ -s_4c_5c_6 - c_4s_6 & s_4c_5s_6 - c_4c_6 & s_4s_5 \end{pmatrix} = \begin{pmatrix} c_1c_{23} & s_1c_{23} & s_{23} \\ -c_1s_{23} & -s_1s_{23} & -c_{23} \\ -s_1 & c_1 & 0 \end{pmatrix} \times \begin{pmatrix} r_{11} & r_{12} & r_{13} \\ r_{21} & r_{22} & r_{23} \\ r_{31} & r_{32} & r_{33} \end{pmatrix} \tag{3.27}$$

Dividing the $(1, 3)$ and the $(3, 3)$ terms of the above matrix equation, we get for $\theta_5 \neq 0, \pi$,

$$s_4(r_{13}c_1c_{23} + r_{23}s_1c_{23} + r_{33}s_{23}) = c_4(r_{13}s_1 - r_{23}c_1) \tag{3.28}$$

We can solve for $(\theta_1, \theta_2, \theta_3, \theta_4)$ from Eqns (3.25) and (3.28) by using elimination theory as described in Section 3.9 (see Exercise 3.9), and we can also obtain the joint variables numerically. We discuss, briefly, the issues involved in the numerical solution of the inverse kinematics of a non-intersecting wrist robot.

We assume that the numerical values of the DH parameters are the same as that of a PUMA and d_5 is chosen to be 20 mm. For the ${}^0_6[T]$ used for the PUMA example, namely,

$$
{}^0_6[T] = \begin{bmatrix} 0.9749 & -0.2192 & -0.0388 & 0.1304 \\ 0.1643 & 0.8262 & -0.5388 & 0.3071 \\ 0.1502 & 0.5190 & 0.8415 & 0.0482 \\ 0 & 0 & 0 & 1 \end{bmatrix}
$$

we get, by using the non-linear equation solving routine, `fsolve` in MATLAB,

$$
\theta_1 = 41.82,\ \theta_2 = 60.43, \theta_3 = 135.33, \theta_4 = 31.96
$$

Once $(\theta_1, \theta_2, \theta_3)$ are solved, we can solve for θ_4, θ_5, and θ_6 by considering left- and right-hand sides of matrix (3.27). For $\theta_5 \neq 0, \pi$, following a Z–$(-Y)$–Z inverse Euler angle algorithm, we get

$$
\begin{aligned}
\theta_5 &= \text{Atan2}[\pm\sqrt{(1,3)^2 + (3,3)^2}, (2,3)] \\
\theta_4 &= \text{Atan2}[(3,3)/s_5, -(1,3)/s_5] \\
\theta_6 &= \text{Atan2}[-(2,2)/s_5, (2,1)/s_5]
\end{aligned}
$$

where the terms $(3,3)$, $(2,2)$, etc. are from the right-hand side of Eqn (3.27) and are known once θ_1, θ_2, and θ_3 are known. For the numerical values chosen above, we get two sets of values, namely, $\theta_4 = 31.96, -148.04$, $\theta_5 = -45.22, +45.22$, and $\theta_6 = 121.57, -58.43$. It can be verified that one value of θ_4 is the same as the value obtained from the numerical iterative procedure. It may be noted that $\theta_5 = 0$ or π is a singular configuration for the non-intersecting wrist and, at a singular configuration, the inverse Euler angles algorithms (see Chapter 2) can at most give the sum or difference of θ_4 and θ_6.

In the above example, by using the numerical iterative method we could easily solve the inverse kinematics for a robot with non-intersecting wrists. Iterative techniques can be used for any serial manipulator [see Goldenberg, et al (1985)] and efficient numerical techniques, using a Newton–Raphson technique and the easily available Jacobian matrices (see Chapter 5), can solve the inverse kinematics of a serial manipulator extremely quickly in very few iterations (typically 2 or 3). The two main problems associated with any numerical technique are the choice of the initial guess and effort required to find all solutions. In practice, the choice of the initial guess is not a problem since the initial guess can be the previous available or stored joint values.[4] The second difficulty is not so easy to overcome even though continuation methods (Tsai & Morgan 1985) can obtain all the solutions. Analytical expressions have the advantage of finding all possible inverse kinematics solutions and also give insight into the nature of the workspace. This has been a motivating reason for finding general algorithms for the inverse kinematics of arbitrary manipulators. This topic is discussed in the following section.

3.6* Inverse Kinematics of a General 6R Robot

In this section, we discuss the salient features of the algorithm of Raghavan and Roth (1993) which solves the inverse kinematics of an arbitrary 6R serial manipulator. The manipulator is arbitrary in the sense that none of the fixed Denavit–Hartenberg parameters (link lengths, twist angles, or link offsets) have special values, such as 0, $\pi/2$, or π, which results in simpler equations and easier elimination of one or more joint variables. We will use the symbols and the Denavit–Hartenberg convention used in this text, but the approach follows the paper by Raghavan and Roth (1993).

We start by considering the direct kinematics equations for a general 6R manipulator given symbolically as

$$ {}^0_6[T] = {}^0_1[T]{}^1_2[T]{}^2_3[T]{}^3_4[T]{}^4_5[T]{}^5_6[T] \tag{3.29} $$

where ${}^{i-1}_{i}[T]$ is given in terms of Denavit–Hartenberg parameters in Section 2.7 [see Eqn (2.51)]. It may be recalled that a 4×4 transformation matrix ${}^{i-1}_{i}[T]$ is a function of only one joint variable θ_i and the other three Denavit–Hartenberg parameters are constants. For the inverse kinematics problem, the left-hand side ${}^0_6[T]$ is given and we have to find the six joint variables in each of ${}^{i-1}_{i}[T]$, where $i = 1, 2, ..., 6$.

The first step is to recognize that ${}^{i-1}_{i}[T]$ can be written as a product of two matrices, $({}^{i-1}_{i}[T])_{st}({}^{i-1}_{i}[T])_{jt}$. The first matrix $({}^{i-1}_{i}[T])_{st}$ is a function of a_{i-1} and α_{i-1} and is constant. The second matrix $({}^{i-1}_{i}[T])_{jt}$ is a function of the joint variables θ_i (for a rotary joint) or d_i (for a prismatic joint). The rotation part follows directly from Eqn (2.49) and the position follows from Eqns (2.50) and (2.51), and we can obtain

$$ {}^{i-1}_{i}[T] = ({}^{i-1}_{i}[T])_{st}({}^{i-1}_{i}[T])_{jt} $$

[4] Typically, for good control, a manipulator end-effector is commanded to move by a small amount from its present position and orientation. The joint values at the future commanded position and orientation are expected to be close to the known current values and these current values of the joint variables serve as a very good initial guess for the iterative numerical procedure.

$$= \begin{pmatrix} 1 & 0 & 0 & a_{i-1} \\ 0 & c_{\alpha_{i-1}} & -s_{\alpha_{i-1}} & 0 \\ 0 & s_{\alpha_{i-1}} & c_{\alpha_{i-1}} & 0 \\ 0 & 0 & 0 & 1 \end{pmatrix} \begin{pmatrix} c\theta_i & -s\theta_i & 0 & 0 \\ s\theta_i & c\theta_i & 0 & 0 \\ 0 & 0 & 1 & d_i \\ 0 & 0 & 0 & 1 \end{pmatrix} \qquad (3.30)$$

The next step is to rewrite Eqn (3.29) as

$$({}^2_3[T])_{jt}\ {}^3_4[T]{}^4_5[T]({}^5_6[T])_{st} = ({}^2_3[T])^{-1}_{st}({}^1_2[T])^{-1}({}^0_1[T])^{-1}\ {}^0_6[T]({}^5_6[T])^{-1}_{jt} \qquad (3.31)$$

It is clear from this matrix equation that the left-hand side is only a function of $(\theta_3, \theta_4, \theta_5)$ and the right-hand side is only a function of $(\theta_1, \theta_2, \theta_6)$. On expanding the left- and right-hand sides, we can observe, in addition, that the six scalar equations obtained by equating the top three elements of columns 3 and 4 on both sides of Eqn (3.31) do not contain θ_6. The third column, however, has the constraint that its magnitude is unity. The third and fourth columns are called **p** and **l** for convenience, and we can rewrite the six equations as

$$[A](s_4s_5\ s_4c_5\ c_4s_5\ c_4c_5\ s_4\ c_4\ s_5\ c_5\ 1)^T = [B](s_1s_2\ s_1c_2\ c_1s_2\ c_1c_2\ s_1\ c_1\ s_2\ c_2)^T \qquad (3.32)$$

where $[A]$ is 6×9 matrix whose elements are linear in s_3, c_3, 1, and $[B]$ is 6×8 matrix of constants.

The next step is to eliminate four of the remaining five variables, θ_1, θ_2, θ_3, θ_4, and θ_5 in Eqn (3.32), by obtaining the minimal set of equations. As pointed out by Raghavan and Roth (1993), the minimal set of equations are 14 in number. These are the three equations from **p**, three equations from **l**, one scalar equation from the scalar dot product $\mathbf{p} \cdot \mathbf{p}$, one scalar equation from the scalar dot product $\mathbf{p} \cdot \mathbf{l}$, three equations from the vector cross product $\mathbf{p} \times \mathbf{l}$, and three scalar equations from $(\mathbf{p} \cdot \mathbf{p})\mathbf{l} - (2\mathbf{p} \cdot \mathbf{l})\mathbf{p}$. It is shown that all these 14 equations contain the same variables as in Eqn (3.32) with no new variables. They can be written as

$$[P](s_4s_5\ s_4c_5\ c_4s_5\ c_4c_5\ s_4\ c_4\ s_5\ c_5\ 1)^T = [Q](s_1s_2\ s_1c_2\ c_1s_2\ c_1c_2\ s_1\ c_1\ s_2\ c_2)^T \qquad (3.33)$$

where $[P]$ is a 14×9 matrix whose elements are linear in c_3, s_3, 1, and $[Q]$ is a 14×8 matrix of constants. To eliminate four out of the five joint rotations, we first use any eight of the 14 equations in Eqn (3.33) and solve for the eight variables $s_1s_2, s_1c_2, c_1s_2, c_1c_2, s_1, c_1, s_2, c_2$. Note this is always possible since we are only solving eight linear equations in eight unknowns. Once this is done we can substitute the eight variables in the rest of the six equations and get an equation of the form

$$[R]\,(s_4s_5\ s_4c_5\ c_4s_5\ c_4c_5\ s_4\ c_4\ s_5\ c_5\ 1)^T = \mathbf{0} \qquad (3.34)$$

where $[R]$ is a 6×9 matrix whose elements are linear in s_3 and c_3.

In the next step, the tangent half-angle formulas [see Eqn (3.14)] are introduced for s_3, c_3, s_4, c_4, s_5, and c_5. After simplifying, we can write

$$[S]\left(x_4^2x_5^2\ x_4^2x_5\ x_4^2\ x_4x_5^2\ x_4x_5\ x_4\ x_5^2\ x_5\ 1\right)^T = \mathbf{0} \tag{3.35}$$

where $[S]$ is a 6×9 matrix and $x_{(.)} = \tan(\theta./2)$.

Next x_4 and x_5 are eliminated using Sylvester's dialytic method (see Section 3.9). Six additional equations are generated by multiplying equations in Eqn (3.35) by x_4. In the process three additional 'linearly' independent variables, namely, $x_4^3x_5^2$, $x_4^3x_5$, and x_4^3, are generated, and we have a system of 12 equations in 12 unknowns. The equations can be written as

$$\begin{pmatrix} S & 0 \\ 0 & S \end{pmatrix} \begin{pmatrix} x_4^3x_5^2 \\ x_4^3x_5 \\ x_4^3 \\ x_4^2x_5^2 \\ x_4^2x_5 \\ x_4^2 \\ x_4x_5^2 \\ x_4x_5 \\ x_4 \\ x_5^2 \\ x_5 \\ 1 \end{pmatrix} = \mathbf{0} \tag{3.36}$$

Following Sylvester's dialytic method, we set the determinant of the coefficient matrix to zero, which gives a 16th-degree polynomial in x_3. The roots of this polynomial can be solved numerically and then we can find $\theta_3 = 2\tan^{-1}(x_3)$. Since there can be 16 solutions to the polynomial, the inverse kinematics problem for the general 6R serial manipulator has 16 possible solutions.

Once θ_3 is known, by using standard tools of linear algebra we can find x_4 and x_5 from Eqn (3.36), which will give θ_4 and θ_5. Once θ_3, θ_4, and θ_5 are known, we can go back to Eqn (3.33) and solve for the right-hand side variables s_1s_2, s_1c_2, ..., s_2, c_2 from eight linearly independent equations and then obtain unique θ_1 and θ_2. Finally, to obtain θ_6, we go back to Eqn (3.29) and rewrite it as

$$^5_6[T] = {}^4_5[T]^{-1}\,{}^3_4[T]^{-1}\,{}^2_3[T]^{-1}\,{}^1_2[T]^{-1}\,{}^0_1[T]^{-1}\,{}^0_6[T] \tag{3.37}$$

Since θ_i (i being $1, 2, ..., 5$) are known, the $(1, 1)$ and $(2, 1)$ elements give two equations in s_6 and c_6 which give unique values of θ_6.

It may be noted that if the 6R manipulator has special geometry, i.e., some Denavit–Hartenberg parameters are 0, $\pi/2$, or π, the 16th-degree polynomial in x_3 can be of lower order. In addition, if one or more joints are prismatic, then the inverse kinematics becomes simpler since the prismatic joint variable is not in terms of sines or cosines.

Finally, a word about the workspace of a general 6R manipulator. Obtaining the workspace of a general 6R serial manipulator has been studied by several researchers [see, for example, papers by Sugimoto and Duffy (1981a, b)] and conditions for a manipulator to be at its extreme reach were known before the 16th-degree polynomial was obtained. As in the case of the planar 3R manipulator, the workspace of a general 6R serial manipulator can be obtained from the solution of the 16th-degree polynomial. If all the roots of the 16th-degree polynomial are complex, then ${}^0_6[T]$ is not in the workspace of the manipulator. However, since there are no closed-form solutions for a 16th-degree polynomial, it is not possible to obtain analytical conditions to check whether a given ${}^0_6[T]$ lies in the workspace or not.

All the inverse kinematics solutions may not be achievable due to the presence of joint limits associated with any manipulator hardware, and, in any robot software, the choice of the inverse kinematics solution out of all the possible solutions is dictated by the presence of joint limits in that particular robot. In addition, due to the joint limits the entire workspace is not available and this must be taken into account for any application of a robot [see also Rastegar and Deravi (1987) and Dwarakanath et al. (1992)].

3.7 Inverse Kinematics for Manipulators with $n < 6$

In robotic gas or arc welding or painting, there is no need to rotate the welding torch or the paint gun about its own axis. To decrease cost and to simplify the design and manufacture, many welding or painting robots have five degrees of freedom or five actuators. Likewise, in the assembly of electronic components on a printed circuit board, it is simpler for the robot to assemble from one direction. Hence an assembly robot such as the ADEPT SCARA robot has only four degrees of freedom. In these industrial robots $n < 6$, and the tool or the end-effector cannot be positioned and oriented arbitrarily in 3D space. To solve for the inverse kinematics of such robots, we must have $6 - n$ ($3 - n$ for planar manipulators) relationships or constraint equations involving the position and orientation variables, and the given ${}^0_n[T]$ must satisfy these relationships or constraints. Most of the time, the constraints or the relationships are obvious since the manipulator was designed with the constraint in mind. For example, in the case of the SCARA manipulator (Example 3.3), the tool can be positioned arbitrarily inside the workspace but its orientation capabilities are restricted to rotations about $\widehat{\mathbf{Z}}_4$ (see Fig. 2.20). Hence, the constraints would be $\eta = \psi = 0$, where η and ψ

are rotations about $\widehat{\mathbf{X}}_4$ and $\widehat{\mathbf{Y}}_4$, respectively.[5] For the SCARA manipulator, given x, y, z, and ϕ, the inverse kinematics is described by

$$\begin{aligned}
\theta_2 &= \pm\cos^{-1}\left(\frac{x^2+y^2-l_1^2-l_2^2}{2l_1l_2}\right)\\
\theta_1 &= \mathrm{Atan2}(y,x) - \mathrm{Atan2}(l_2s_2, l_1+l_2c_2)\\
d_3 &= -z\\
\theta_4 &= \phi - \theta_1 - \theta_2
\end{aligned} \tag{3.38}$$

The constraints or the relationships between the position and orientation variables in any serial manipulator can be, in principle, determined from the direct kinematics equations. For example, in a five-DOF robot, we can get six independent equations, representing the position and orientation of the end-effector, from ${}^0_5[T]$ in terms of the five joint variables. Following the theory of elimination, described in Section 3.9, it is possible to eliminate the five joint variables from the six equations and obtain a single equation in terms of the position and orientation variables of the tool. This single equation would be the constraint and would represent a five-dimensional subspace of the reachable workspace. A manipulator with four joints would lead to two expressions involving the position and orientation variables and thus we would have four-dimensional subspace. Obtaining explicit expressions for the constraints is, however, difficult in practice (see the example of a four-bar mechanism in Chapter 4) due to the difficulties in elimination of variables from non-linear equations.

3.8 Inverse Kinematics of Redundant Manipulators

In the case of a redundant manipulator, the number of joint variables is more than the number of equations. One of the simplest examples is a planar 3R robot shown in Fig. 2.17. Its inverse kinematics was considered in Section 3.4. Let us assume we are not interested in the orientation of the last link. In such a situation, the (x, y) coordinates of the end-effector can be written as

$$\begin{aligned}
x &= l_1c_1 + l_2c_{12} + l_3c_{123}\\
y &= l_1s_1 + l_2s_{12} + l_3s_{123}
\end{aligned} \tag{3.39}$$

For the inverse kinematics, (x, y) are given and our task is to find θ_1, θ_2, and θ_3. Since there are only two equations, we can have an ∞ of θ_1, θ_2, and θ_3 for a given (x, y). In order to obtain unique θ_1, θ_2, and θ_3, we need one more equation or a constraint involving θ_1, θ_2, and θ_3. One can impose a simple constraint, such as θ_3 equals constant (the third joint is locked), but that

[5] The X-Y-Z Euler angles, η, ψ, and ϕ, could be chosen to represent the orientation of $\{4\}$.

defeats the purpose of designing and building manipulators with more than required joints and actuators. The availability of an extra joint can be used for optimization. Several researchers have suggested the use of redundancy to minimize joint rotations, joint velocities, and accelerations. Several others have also suggested the use of redundancy for avoiding singularities and avoiding obstacles (Nakamura 1991). Obtaining meaningful and useful equations is known as the resolution of redundancy, and this is the key issue in inverse kinematics of redundant robots. The resolution of redundancy can be achieved at various levels, such as position, velocity, accelerations, and torques. In Chapter 5, we will discuss resolution at the level of joint velocities. In this chapter, we use the example of the redundant planar 3R manipulator to illustrate the minimization of joint rotations.

A candidate function for optimization is $\theta_1^2 + \theta_2^2 + \theta_3^2$. Minimization of $\theta_1^2 + \theta_2^2 + \theta_3^2$ subject to constraints given in Eqn (3.39) results in the planar 3R manipulator following a given trajectory with least rotation of the joints. We solve this problem using the method of Lagrange multipliers. We have

$$\text{Minimize} f(\boldsymbol{\theta}) = \theta_1^2 + \theta_2^2 + \theta_3^2$$

subject to

$$\begin{aligned} g_1(\boldsymbol{\theta}) &= -x + l_1 c_1 + l_2 c_{12} + l_3 c_{123} = 0 \\ g_2(\boldsymbol{\theta}) &= -y + l_1 s_1 + l_2 s_{12} + l_3 s_{123} = 0 \end{aligned}$$

where $\boldsymbol{\theta}$ denotes the three joint variables $(\theta_1, \theta_2, \theta_3)$, and (x, y) are points on the trajectory of the end-effector. Following classical optimization theory, using Lagrange multipliers we form the function

$$F(\boldsymbol{\theta}) = f(\boldsymbol{\theta}) - \lambda_1 g_1(\boldsymbol{\theta}) - \lambda_2 g_2(\boldsymbol{\theta}) \tag{3.40}$$

and equating the derivatives of $F(\boldsymbol{\theta})$ to zero, we get three equations

$$\begin{aligned} \frac{\partial f}{\partial \boldsymbol{\theta}} &= \lambda_1 \frac{\partial g_1}{\partial \boldsymbol{\theta}} + \lambda_2 \frac{\partial g_2}{\partial \boldsymbol{\theta}} \\ g_1(\boldsymbol{\theta}) &= 0 \\ g_2(\boldsymbol{\theta}) &= 0 \end{aligned} \tag{3.41}$$

To solve the above three equations, we first eliminate λ_1 and λ_2 by rewriting the first equation as

$$\begin{pmatrix} \frac{\partial f}{\partial \theta_1} & \frac{\partial g_1}{\partial \theta_1} & \frac{\partial g_2}{\partial \theta_1} \\ \frac{\partial f}{\partial \theta_2} & \frac{\partial g_1}{\partial \theta_2} & \frac{\partial g_2}{\partial \theta_2} \\ \frac{\partial f}{\partial \theta_3} & \frac{\partial g_1}{\partial \theta_3} & \frac{\partial g_2}{\partial \theta_3} \end{pmatrix} \begin{pmatrix} 1 \\ -\lambda_1 \\ -\lambda_2 \end{pmatrix} = 0 \tag{3.42}$$

For non-trivial λ_1 and λ_2 we must have the determinant of the 3×3 matrix as zero. For the planar 3R manipulator, we get

$$l_1 l_2 \theta_3 s_2 + l_2 l_3 (\theta_1 - \theta_2) s_3 + l_3 l_1 (\theta_3 - \theta_2) s_{23} = 0 \tag{3.43}$$

We can solve Eqn (3.43) together with $g_1(\boldsymbol{\theta}) = 0$ and $g_2(\boldsymbol{\theta}) = 0$ numerically. Figure 3.5 shows the plot of θ_1, θ_2, θ_3, and $f(\boldsymbol{\theta})$ (l_1, l_2, and l_3 are chosen to be 5, 3, and 1, respectively) when the end-effector of the planar 3R manipulator traces a straight line parallel to the Y axis as shown in the bottom of Fig. 3.5.

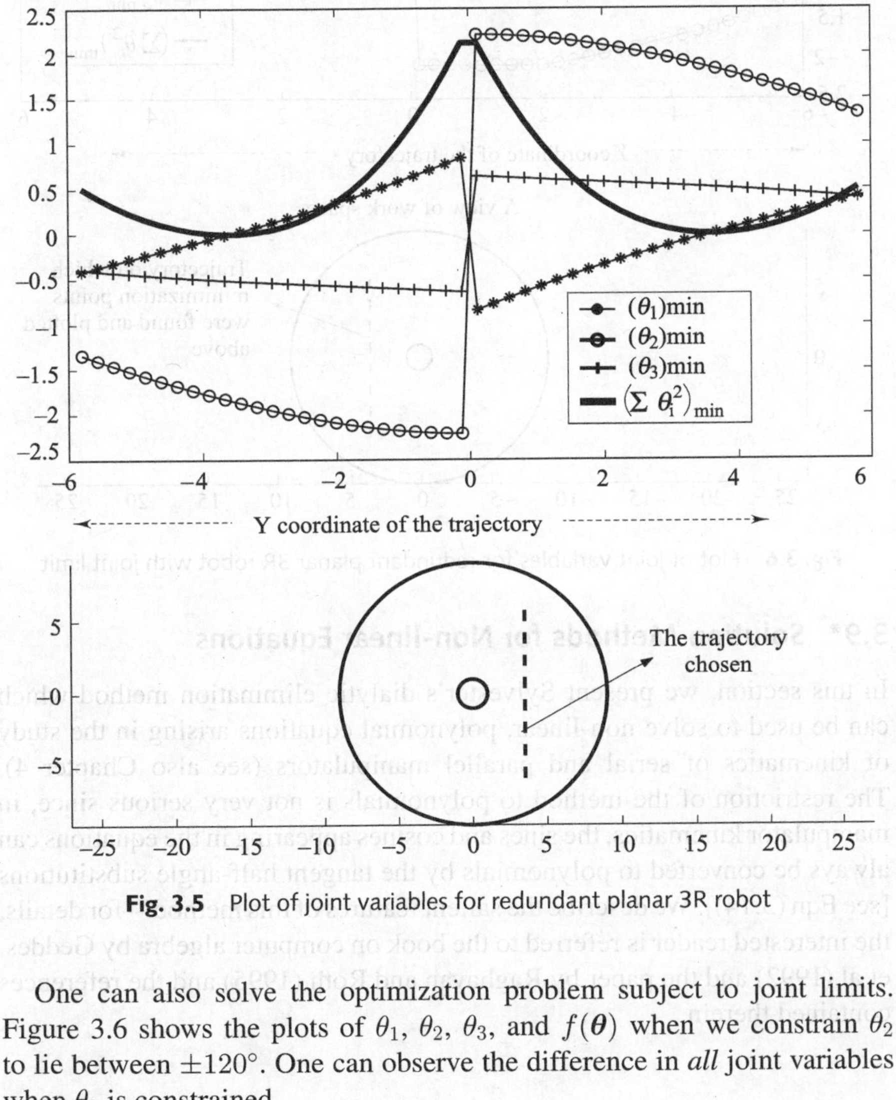

Fig. 3.5 Plot of joint variables for redundant planar 3R robot

One can also solve the optimization problem subject to joint limits. Figure 3.6 shows the plots of θ_1, θ_2, θ_3, and $f(\boldsymbol{\theta})$ when we constrain θ_2 to lie between $\pm 120°$. One can observe the difference in *all* joint variables when θ_2 is constrained.

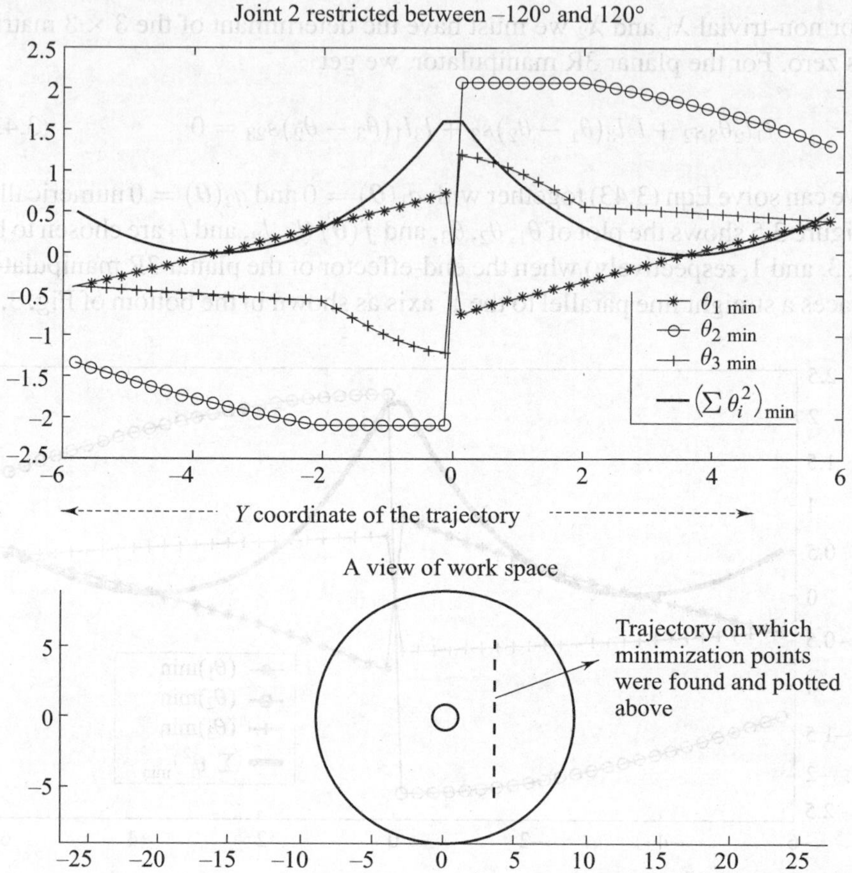

Fig. 3.6 Plot of joint variables for redundant planar 3R robot with joint limit

3.9* Solution Methods for Non-linear Equations

In this section, we present Sylvester's dialytic elimination method which can be used to solve non-linear, polynomial equations arising in the study of kinematics of serial and parallel manipulators (see also Chapter 4). The restriction of the method to polynomials is not very serious since, in manipulator kinematics, the sines and cosines appearing in the equations can always be converted to polynomials by the tangent half-angle substitutions [see Eqn (3.14)]. We describe the salient features of this method—for details, the interested reader is referred to the book on computer algebra by Geddes, et al (1992) and the paper by Raghavan and Roth (1995) and the references contained therein.

Consider two non-linear polynomial equations $f(x, y) = 0$ and $g(x, y) = 0$, of degree m and n,[6] respectively. It is known from the Bézout theorem (Semple & Roth 1949) that there are a maximum of $m \times n$ (x, y) values that satisfy both the equations. This is an upper bound and includes real, complex conjugate and solutions at infinity. For example, a circle given by $x^2 + y^2 = 1$ and a line $y - x = 0$ are satisfied by two sets of values, namely, $\pm(\frac{1}{\sqrt{2}}, \frac{1}{\sqrt{2}})$. The fact that there may not be any real solutions can be seen if the line $y = x$ is replaced by the line $y = x + 2$. A sketch shows that in this case the line $y = x + 2$ does not intersect the circle. In this case, however, we will get two complex conjugate values of (x, y) which satisfy the two equations. It is also possible to get one real solution if we consider the line $y = x + \sqrt{2}$. In this case, the line is a tangent to the circle and we count the point of tangency as two coincident points. If we consider two circles, $(x - a_1)^2 + (y - b_1)^2 = r_1^2$ and $(x - a_2)^2 + (y - b_2)^2 = r_2^2$, then we never get four sets of values of (x, y), although it is possible to get four solution sets for two ellipses or a circle and an ellipse or any other quadratic curve. It can be shown, by using homogenous coordinates (see Section 2.8), that for two circles there are two additional points at infinity (see Exercise 3.12).

The Bézout theorem is useful in giving an upper bound on the number of solutions for two polynomials, and a generalization states that two m- and n-order manifolds intersect in a $m \times n$ order sub-manifold. For example, a sphere $x^2 + y^2 + z^2 = 1$ intersects a plane $x = 0$ in a second-order curve, namely, a circle $y^2 + z^2 = 1$. The Bezout theorem is, however, of not much use in obtaining the solutions of the two polynomials since it is not a constructive theorem. One such constructive method is Sylvester's dialytic elimination method (Salmon 1964).

Consider two polynomials of degrees m and n with $m \geq n > 0$ denoted by $P(x) = \sum_{i=0}^{m} a_i x^i$ and $Q(x) = \sum_{i=0}^{n} b_i x^i$. The Sylvester matrix of $P(x)$ and $Q(x)$ is the $(m + n) \times (m + n)$ matrix given by

$$[SM] = \begin{bmatrix} a_m & a_{m-1} & \dots & a_1 & a_0 & & & \\ & a_m & a_{m-1} & \dots & a_1 & a_0 & & \\ & & & \dots & \dots & \dots & & \\ & & & a_m & \dots & \dots & a_0 & \\ b_n & b_{n-1} & \dots & b_1 & b_0 & & & \\ & b_n & b_{n-1} & \dots & b_1 & b_0 & & \\ & & & \dots & \dots & \dots & & \\ & & & b_n & \dots & \dots & b_0 & \end{bmatrix} \tag{3.44}$$

[6] The degree of a polynomial is defined as the degree of its highest degree term and the degree of a term is the sum of the exponents in that term.

where the unfilled entries are 0. It may be noted that the ith row of the top half of the matrix are the coefficients of $P(x) \times x^i$ for $i = n-1, n-2, ..., 1$ and the ith row in the bottom half of the matrix are the coefficients of $Q(x) \times x^i$ for $i = m-1, m-2, ..., 1$. According to the Sylvester criterion,[7] $P(x)$ and $Q(x)$ have a non-trivial common factor if and only if $\det[SM] = 0$.[8] The Sylvester criterion follows from the analogy with linear homogeneous equations. A system of $m+n$ linear equations in $m+n$ unknowns, $x^{m+n-1}, x^{m+n-2}, ..., x^1, x^0$ (note that all powers of x including the constant term x^0 are treated as linearly independent variables), given by

$$[SM](x^{m+n-1}, x^{m+n-2}, ..., x^1, x^0)^T = \mathbf{0} \tag{3.45}$$

can have a non-trivial solution if and only if $\det[SM] = 0$. The determinant of the Sylvesters matrix $[SM]$ is also called the resultant of $P(x)$ and $Q(x)$ and is denoted by $\text{res}(P, Q)$.

The above results can be used to obtain the solution to two polynomials, $f(x,y) = 0$ and $g(x,y) = 0$ of degrees m and n, respectively, as follows.

1. Rewrite given $f(x,y) = 0$ and $g(x,y) = 0$ as $P(x) = \sum_{i=0}^{m} a_i x^i$ and $Q(x) = \sum_{i=0}^{n} b_i x^i$, i.e., only one variable is explicitly shown and the coefficients a_i and b_i are functions of the other suppressed variable y or constants.
2. The second step is to obtain the set of linear equations, Eqn (3.45) and Sylvester's matrix as described in Eqn (3.44) and obtain the equation $\det[SM] = 0$ which is a function of all the a_i's and b_i's. Since a_i's and b_i's are only functions of y, we get a polynomial in y alone.
3. Solve $\det[SM](y) = 0$ for all roots analytically (if possible) or numerically.
4. The system of linear equations, obtained in step 2, can be solved by standard linear algebra techniques for the 'linearly' independent unknowns $x^{m+n-1}, x^{m+n-2}, ..., x^1, x^0$. The integrity of the numerical procedure can be verified by checking that the values of variables x^1 and say x^2 are related by $(x^1)^2 = x^2$.

To illustrate this procedure, consider two polynomial equations in the required form:

$$f_1(x,y) = a_2(y)x^2 + a_1(y)x + a_0(y) = 0$$
$$f_2(x,y) = b_2(y)x^2 + b_1(y)x + b_0(y) = 0 \tag{3.46}$$

[7] Sylvester and Trudi worked in the late 19th century on the theory of equations, later called the theory of algebraic curves, which forms the foundation of algebraic geometry.
[8] We will use the notation $\det[\cdot]$ to denote 'determinant of a square matrix $[\cdot]$' throughout this textbook.

where a_i and b_i ($i = 0, 1, 2$) are arbitrary polynomials in y or constants.

Sylvester's matrix is given by

$$[SM] = \begin{bmatrix} a_2 & a_1 & a_0 & 0 \\ 0 & a_2 & a_1 & a_0 \\ b_2 & b_1 & b_0 & 0 \\ 0 & b_2 & b_1 & b_0 \end{bmatrix} \tag{3.47}$$

and $\det[SM](y) = 0$ reduces to

$$(a_2b_1 - b_2a_1)(a_1b_0 - b_1a_0) - (a_2b_0 - b_2a_0)^2 = 0 \tag{3.48}$$

where the dependence of the coefficients on y has been dropped for convenience.

The expression for x can be obtained from the set of 'linear equations'

$$\begin{bmatrix} a_2 & a_1 & a_0 & 0 \\ 0 & a_2 & a_1 & a_0 \\ b_2 & b_1 & b_0 & 0 \\ 0 & b_2 & b_1 & b_0 \end{bmatrix} \begin{pmatrix} x^3 \\ x^2 \\ x^1 \\ x^0 \end{pmatrix} = \mathbf{0} \tag{3.49}$$

as

$$x^1 = -\frac{a_1b_0 - b_1a_0}{a_2b_0 - b_2a_0} = \frac{a_2b_0 - b_2a_0}{a_1b_2 - a_2b_1} \tag{3.50}$$

Sylvester's matrix is $(m+n) \times (m+n)$ and the computation of $\det[SM]$ [or the resultant $\text{res}(P, Q)$] can be computationally quite expensive. Bézout in the 18th century proposed a method where the $\text{res}(P, Q)$ can be computed as a determinant of order $\max(m, n)$. Consider two polynomial equations $P(x) = \sum_{i=0}^{m} a_i x^i = 0$ and $Q(x) = \sum_{i=0}^{n} b_i x^i = 0$ with $m > n$. We can eliminate x^m from $P(x) = 0$ and $x^{m-n}Q(x) = 0$ by writing

$$\frac{a_m}{b_n} = \frac{a_{m-1}x^{m-1} + \cdots + a_0}{b_{n-1}x^{m-1} + \cdots + b_0x^{m-n}}$$

On simplification, we get

$$\begin{aligned}(a_{m-1}b_n - a_mb_{n-1})x^{m-1} &+ (a_{m-2}b_n - a_mb_{n-2})x^{m-2} \\ &+ \cdots + a_0b_n = 0\end{aligned} \tag{3.51}$$

We could also eliminate x^m by writing

$$\frac{a_mx + a_{m-1}}{b_nx + b_{n-1}} = \frac{a_{m-2}x^{m-2} + \cdots + a_0}{b_{n-2}x^{m-2} + \cdots + b_0x^{m-n}}$$

and this on simplification will give

$$(a_{m-2}b_n - b_{n-2}a_m)x^{m-1} + [(a_{m-3}b_n - b_{n-3}a_m) + (a_{m-2}b_{n-1} - b_{n-2}a_{m-1})]x^{m-2} + \cdots + a_0 b_{n-1} = 0 \tag{3.52}$$

The above steps could be repeated to obtain n equations, with the nth equation given by

$$(a_{m-n}b_n - a_m b_0)x^{m-1} + (a_{m-n-1}b_n + a_{m-n}b_{n-1} - a_{m-1}b_0)x^{m-2} + \cdots + a_0 b_1 = 0 \tag{3.53}$$

In addition, we construct $m - n$ equations

$$\begin{aligned} x^{m-n-1}Q(x) &= b_n x^{m-1} + b_{n-1}x^{m-2} + \cdots + b_0 x^{m-n-1} = 0 \\ x^{m-n-2}Q(x) &= b_n x^{m-2} + \cdots + b_0 x^{m-n-2} = 0 \\ \cdots &= 0 \\ Q(x) &= b_n x^n + \cdots + b_0 = 0 \end{aligned} \tag{3.54}$$

The Bézout matrix is given as

$$[BM] = \begin{bmatrix} a_{m-1}b_n - a_m b_{n-1} & a_{m-2}b_n - a_m b_{n-2} & \cdots\cdots\cdots & a_0 b_n \\ \cdots & \cdots & \cdots\cdots\cdots & \cdots \\ a_{m-n}b_n - a_m b_0 & a_{m-n-1}b_n + a_{m-n}b_{n-1} - a_{m-1}b_0 & \cdots & a_0 b_1 \\ b_n & b_{n-1} & \cdots & b_0 \\ & b_n & \cdots\cdots & b_0 \\ & \cdots & \cdots\cdots\cdots & \cdots \\ & & b_n \quad \cdots \quad \cdots & b_0 \end{bmatrix} \tag{3.55}$$

where the unfilled entries are 0's.

The criterion for $P(x)$ and $Q(x)$ to have a non-trivial common factor is given by $\det[BM] = 0$. It may be noted that if $m = n$, then the $m - n$ equations, Eqn (3.54), are not needed. In Eqns (3.51) through (3.53), we already have the required set of n 'linearly independent equations' in n unknowns $x^{n-1}, \cdots, x^0$. As in Sylvester's method, we can solve for the unknowns by standard linear algebra techniques.

As an illustration of the Bézout method, consider two cubics of the form

$$\begin{aligned} a_3x^3 + a_2x^2 + a_1x + a_0 &= 0 \\ b_3x^3 + b_2x^2 + b_1x + b_0 &= 0 \end{aligned} \tag{3.56}$$

where a_i, b_i $(i = 0, ..., 4)$ are arbitrary polynomials in y or constants. The Bézout matrix is given by

$$[BM] = \begin{bmatrix} b_3a_2 - a_3b_2 & b_3a_1 - a_3b_1 & b_3a_0 - a_3b_0 \\ b_3a_1 - a_3b_1 & (b_3a_0 - a_3b_0) + (b_2a_1 - a_2b_1) & b_2a_0 - a_2b_0 \\ b_3a_0 - a_0b_3 & b_2a_0 - a_2b_0 & b_1a_0 - a_1b_0 \end{bmatrix} \tag{3.57}$$

It can be seen that $[BM]$ is 3×3 as opposed to Sylvester's matrix which would be 6×6 for this case.

The matrices $[SM]$ and $[BM]$ are related as can be seen from the following example. Consider $P(x) = a_3x^3 + a_2x^2 + a_1x + a_0 = 0$ and $Q(x) = b_2x^2 + b_1x + b_0 = 0$. Sylvester's matrix is given by

$$[SM] = \begin{bmatrix} a_3 & a_2 & a_1 & a_0 & 0 \\ 0 & a_3 & a_2 & a_1 & a_0 \\ b_2 & b_1 & b_0 & 0 & 0 \\ 0 & b_2 & b_1 & b_0 & 0 \\ 0 & 0 & b_2 & b_1 & b_0 \end{bmatrix} \tag{3.58}$$

The Bézout matrix is given as

$$[BM] = \begin{bmatrix} a_1b_2 - a_3b_0 & a_0b_2 + a_1b_1 - a_2b_0 & a_0b_1 \\ a_2b_2 - a_3b_1 & a_1b_2 - a_3b_0 & a_0b_2 \\ b_2 & b_1 & b_0 \end{bmatrix} \tag{3.59}$$

If we pre-multiply $[SM]$ by

$$[A] = \begin{bmatrix} 1 & 0 & 0 & 0 & 0 \\ 0 & 1 & 0 & 0 & 0 \\ b_2 & b_1 & -a_3 & -a_2 & 0 \\ 0 & b_2 & 0 & -a_3 & 0 \\ 0 & 0 & 0 & 0 & 1 \end{bmatrix} \tag{3.60}$$

we get

$$[B] = \begin{bmatrix} a_3 & a_2 & a_1 & a_0 & 0 \\ 0 & a_3 & a_2 & a_1 & a_0 \\ 0 & 0 & a_1b_2 - a_3b_0 & a_0b_2 + a_1b_1 - a_2b_0 & a_0b_1 \\ 0 & 0 & a_2b_2 - a_3b_1 & a_1b_2 - a_3b_0 & a_0b_2 \\ 0 & 0 & b_2 & b_1 & b_0 \end{bmatrix} \tag{3.61}$$

and we can observe that

$$\begin{aligned} \det[A]\det[SM] &= (a_3)^2 \det[SM] = \det[B] \\ &= (a_3)^2 \det[BM] \end{aligned} \tag{3.62}$$

which shows that $\det[SM] = \det[BM]$.

Exercises

3.1 Determine the expression for the area of the workspace of 2R manipulator with l_1 and l_2. Assume that l_1 and l_2 can be changed with $l_1 + l_2 =$ constant. Show that $l_1 = l_2$ for maximum workspace.

3.2 Sketch the workspace of planar 2R manipulator with $-\pi/2 \leq \theta_i \leq \pi/2$, where $i = 1, 2$.

3.3 If the joint motions are limited as in $\theta_{i\min} \leq \theta_i \leq \theta_{i\max}$, what is the expression for the area of the workspace of a planar 2R manipulator. Show that maximum workspace may not occur for $l_1 = l_2$ when joint rotation is limited.

3.4 For the planar 3R manipulator, discussed in Example 3.4, verify numerically that for a point (x, y) chosen in the dexterous region, i.e., between circles of radii $l_1 + l_2 - l_3$, and $l_1 - l_2 + l_3$, the inverse kinematics can be solved with arbitrary ϕ. Use $l_1 = 5$, $l_2 = 3$, and $l_3 = 1$.

3.5 Obtain the DH parameters for the RRR manipulator shown in Fig. 3.7. Derive expressions for the (x, y, z) coordinates of the point P on the manipulator, with respect to $\{Base\}$, as a function of $(\theta_1, \theta_2, \theta_3)$. Derive the expressions for θ_1, θ_2, and θ_3 for a given $(x, y, z)^T$.

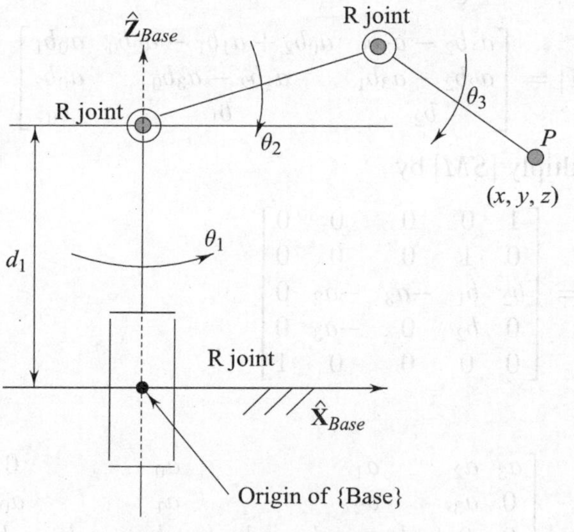

Fig. 3.7 A RRR manipulator

3.6 Assign coordinate systems, obtain DH parameters and derive expressions for the (x, y, z) coordinates of the point P, on the manipulator, shown in Fig. 3.8, as a function of $(\theta_1, \theta_2, \theta_3)$. Derive the expressions for θ_1, θ_2, and θ_3 for a given $(x, y, z)^T$.

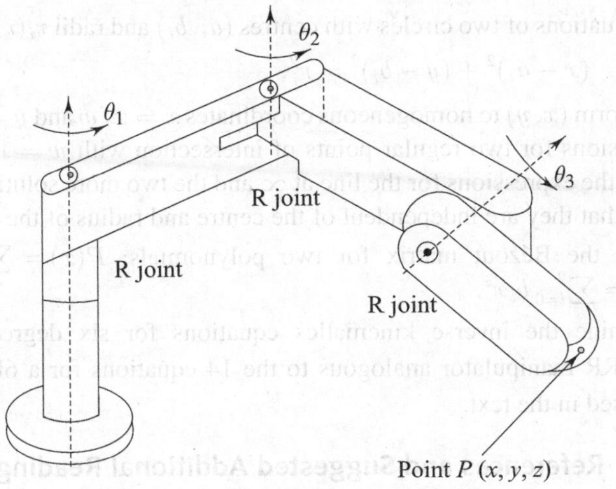

Fig. 3.8 A different RRR manipulator

3.7 Derive the direct and inverse kinematics for the Stanford arm shown in Fig. 2.28.

3.8 Show that the expressions for θ_1, θ_2 and θ_3 for a PUMA 560 robot can also be obtained in the form

$$\theta_1 = \text{Atan2}(O_{6y}, O_{6x}) - \text{Atan2}(d_3, \pm\sqrt{O_{6x}^2 + O_{6y}^2 - d_3^2})$$
$$\theta_3 = \text{Atan2}(a_3, d_4) - \text{Atan2}(K, \pm\sqrt{a_3^2 + d_4^2 - K^2})$$
$$\theta_2 = \text{Atan2}[(-a_3 - a_2c_3)O_{6z} - (c_1O_{6x} + s_1O_{6y})(d_4 - a_2s_3), (a_2s_3 - d_4)O_{6z} + (c_1O_{6x} + s_1O_{6y})(a_3 + a_2c_3)] - \theta_3$$

where

$$K = (1/2a_2)(O_{6x}^2 + O_{6y}^2 + O_{6z}^2 - a_2^2 - a_3^2 - d_3^2 - d_4^2)$$

and $\text{Atan2}(y, x)$ is the four-quadrant $\tan^{-1}$ function.

†3.9 Obtain other inverse kinematics solutions by using different initial guesses for the ${}^0_6[T]$ given for the non-intersecting wrist IGM robot.

†3.10 Using the theory of elimination, derive expressions for the inverse kinematics equations for IGM and check with numerical values.

†3.11 Derive the inverse kinematics equations of the SCARA manipulator shown in Eqn (3.38) and sketch the workspace of the SCARA manipulator.

†3.12 Take a straight line path in the workspace and plot θ_i when joint rotations are limited for the redundant planar 3R manipulator. Assume $-120° \leq \theta_i \leq +120°$, where $i = 1, 2, 3$, and the link lengths are 5, 3, and 1 units, respectively.

†3.13 Take a straight line path in the workspace and evaluate θ_i for a redundant planar 4R manipulator. Assume $-120° \leq \theta_i \leq +120°$, where $i = 1, 2, 3, 4$, and the link lengths are 5, 2.5, 1, and 0.5 units, respectively. Plot the θ_i's and compare with Exercise 3.12.

3.14 The equations of two circles with centres (a_i, b_i) and radii $r_i (i$ being $1, 2)$ are

$$(x - a_i)^2 + (y - b_i)^2 = r_i^2,$$

Transform (x, y) to homogeneous coordinates $x = x/w$ and $y = y/w$. Obtain expressions for two regular points of intersection with $w = 1$. With $w = 0$, obtain the expressions for the line at ∞ and the two more solutions at infinity. Show that they are independent of the centre and radius of the circles.

†3.15 Obtain the Bézout matrix for two polynomials, $P(x) = \sum_{i=0}^{4} a_i x^i$ and $Q(x) = \sum_{i=0}^{2} b_i x^i$.

†3.16 Determine the inverse kinematics equations for six degree of freedom RRPRRR manipulator analogous to the 14 equations for a 6R manipulator discussed in the text.

References and Suggested Additional Reading

Dwarakanath, T.A., A. Ghosal, and U. Shrinivasa 1992, 'Kinematic analysis and design of articulated manipulators with joint motion constraints,' *Trans. ASME J. Mech. Des.*, vol. 116, pp. 969–72.

Duffy, J. and C. Crane 1980, 'A displacement analysis of the general spatial 7R mechanism,' *Mech. Mach. Theory*, vol. 15, pp. 153–69.

Gan, W.W. and S. Pellegrino 2003, 'Closed-loop deployable structures', in *Proceedings of the 44th AIAA Structures, Structural Dynamics and Material Conference*, Paper 2003-1450, pp. 1–9.

Geddes, K.O., S.R. Czapor, and G. Labahn 1992, *Algorithms for Computer Algebra*, Kluwer Academic.

Goldenberg, A.A., B. Benhabib, and R.G. Fenton 1985, 'A complete generalised solution to the inverse kinematics of robots,' *IEEE Trans. J. Robotic. Autom.*, RA-1(1), pp. 14–20.

Korn, G.A. and T.M. Korn 1968, *Mathematical Handbook for Scientists and Engineers*, 2nd edn, McGraw-Hill.

Kumar, A. and K.J. Waldron 1980, 'The dexterous workspace', *ASME Paper no. 80-DET-108*.

Kumar, A. and K.J. Waldron 1981, 'The workspace of a mechanical manipulator,' *Trans. ASME J. Mech. Des.*, vol. 103, pp. 665–72.

Mavroidis, C. and B. Roth 1995, 'Analysis of overconstrained mechanisms,' *Trans. ASME J. Mech. Des.*, vol. 117, pp. 75–81.

Nakamura, Y. 1991, *Advanced Robotics: Redundancy and Optimization*, Addison-Wesley.

Paden, B. 1986, 'Kinematics and Control of Robot Manipulators,' PhD Thesis, Dept. of Mechanical Engineering, University of California, Berkeley.

Pieper, D.L. 1968, 'The kinematics of manipulators under computer control,' PhD Thesis, Dept. of Mechanical Engg., Stanford University.

Raghavan, M. and B. Roth 1993, 'Inverse kinematics of the general 6R manipulator and related linkages,' *Trans. ASME J. Mech. Des.*, vol. 115, pp. 502–08.

Raghavan, M. and B. Roth 1995, 'Solving polynomial systems for kinematic analysis and synthesis of mechanisms and robot manipulators,' *Trans. ASME J. Mech. Des.*, vol. 117, pp. 71–79.

Rastegar, J. and P. Deravi 1987, 'The effect of joint motion constraints on the workspace and number of configurations of manipulators,' *Mech. Mach. Theory*, vol. 22, pp. 401–09.

Salmon, G. 1964, *Lessons Introductory to Modern Higher Algebra*, Chelsea Publishing Co.

Semple, J.G. and L. Roth 1949, *Introduction to Algebraic Geometry*, Oxford University Press.

Sugimoto, K. and J. Duffy 1981a, 'Determination of extreme distances of a robot hand—Part 1 A general theory,' *Trans. ASME J. Mech. Des.*, vol. 103, pp. 631–36.

Sugimoto, K. and J. Duffy 1981b, 'Determination of extreme distances of a robot hand—Part 2 Robot arms with special geometry,' *Trans. ASME J. Mech. Des.*, vol. 103, pp. 776–83.

Tsai, L.-W. and A. Morgan 1985, 'Solving the kinematics of the most general six- and five-degree-of-freedom manipulators by continuation methods,' *Trans. ASME J. Mech., Transm. Autom. Des.*, vol. 107, pp. 189–200.

Tsai, Y.C. and A.H. Soni 1984, 'The effect of link parameters on the working space of the general 3R robot arms,' *Mech. Mach. Theory*, vol. 19, pp. 9–16.

4

Kinematics of Parallel Manipulators

4.1 Introduction

In Chapter 3 we studied the kinematics of serial manipulators consisting of a fixed end and a sequence of rigid links and joints ending in a free link also called an end-effector. In this chapter, we study the kinematics of parallel manipulators which are characterized by the existence of one or more loops wherein a moving rigid link is connected to the fixed ground with more than one arrangement of links and joints. In a parallel manipulator, as discussed briefly in Section 3.2, the number of joints is more than the degrees of freedom, and as a results not all joints are actuated. Since there can be un-actuated or passive joints, in contrast to serial manipulators, a parallel manipulator or a closed-loop mechanism often has multi-DOF spherical and universal (Hooke) joints. The presence of passive joints and multi-DOF joints makes the kinematic analysis very different from the kinematic analysis of serial manipulators. As we will see in this chapter, the direct kinematics problem, unlike in a serial manipulator, is much harder and involves elimination of passive joint variables. The inverse kinematics problem, on the other hand, is simpler for parallel manipulators and closed-loop mechanisms. In serial manipulators, the existence of inverse kinematic solutions led to the notion of a workspace. In contrast, in parallel manipulators and closed-loop mechanisms, as we will see in this chapter, a more important notion is that of mobility, which follows from the direct kinematics problem. Unlike the workspace in serial manipulators, mobility deals with assemblability of a parallel manipulator or a closed-loop mechanism. These concepts are illustrated, in this chapter, with the help of several planar and spatial multi-DOF parallel manipulators and closed-loop mechanisms. We start this chapter with a recap of the notion of degrees of freedom and a few definitions.

4.2 Degrees of Freedom

As mentioned in Section 3.2, the degrees of freedom of a parallel manipulator or a closed-loop mechanism can be obtained from the Grübler criterion:

$$\text{dof} = \lambda(N - J - 1) + \sum_{i=1}^{J} F_i$$

where the symbols are as explained after Eqn (3.1). The quantity dof obtained above is the number of independent actuators that can be put in a closed-loop mechanism or a manipulator. For example, in a four-bar mechanism (see Fig. 2.20), we have $\lambda = 3$, $N = 4$, $J = 4$ and all F_i are 1, thereby giving a dof of 1. Hence in a four-bar mechanism, only one of the four rotary (R) joints can be actuated.

In general, for parallel manipulators and closed-loop mechanisms, $J >$ dof and $J -$ dof of the joints are not actuated. We will denote the actuated joint variables by θ_i, where $i = 1, ..., n$ or the $n \times 1$ vector $\boldsymbol{\theta}$. It may be noted that n is the same as dof for a fully actuated parallel manipulator. If the parallel manipulator is under-actuated, then $n \leq$ dof, and, except for redundant parallel manipulators, $n \leq 6$. The actuated joints are usually one-DOF joints and could be rotary (R) or prismatic (P).

In closed-loop mechanisms and parallel manipulators, the un-actuated or passive joints need not be single-DOF joints and we can have spherical (S), cylindrical (C), Hooke (U), or other multi-DOF joints. The number of passive joint variables can at most be $J-$ dof (or $J - n$).[1] However, as we will see in detail in the next section, not all the passive joint variables may appear during the kinematic analysis of closed-loop mechanisms and parallel manipulators. The passive joint variables required for the purpose of kinematic analysis will be denoted by ϕ_i, where $i = 1, \ldots, m$ or by the $m \times 1$ vector $\boldsymbol{\phi}$.

Finally, the vector $\mathbf{q}$ will be used to denote the configuration space of a closed-loop mechanism or a parallel manipulator. The vector $\mathbf{q}$ is the set $(\boldsymbol{\theta}, \boldsymbol{\phi})$ and is of dimension $n + m$. Together with the loop closure or other constraint equations, $\mathbf{q}$ completely describes the configuration of the closed-loop mechanism or the parallel manipulator. It may be noted $n + m \leq J$ (see footnote for calculating J for multi-DOF joints).

4.3 Loop-closure Constraint Equations

Out of the $n + m$ configuration variables, n are actuated. To completely specify the configuration of a closed-loop mechanism or a parallel

[1] For the purpose of counting the number of joint variables J in a closed-loop mechanism or a parallel manipulator, multi-DOF joints must be considered as equivalent single-DOF joints.

manipulator, we need m additional constraint equations $\eta_i(q_1, \ldots, q_{n+m}) = 0$ (where $i = 1, \ldots, m$), which can be used to obtain the values of the passive variables ϕ_i, where $i = 1, \ldots, m$ for given values of actuated variables $\theta_i (i$ being $1, \ldots, n)$. These holonomic constraint equations are called loop-closure equations in this text.

As the name implies, loop closure implies going around a loop in the closed-loop mechanism or a parallel manipulator. Consider the four-bar mechanism shown in Fig. 4.1. There is only one loop, and as shown we have assigned frames $\{L\}$, $\{R\}$. For clarity, Fig. 4.1 shows only the origins and the $\hat{\mathbf{X}}$ of frames $\{1\}$, $\{2\}$, $\{3\}$, and $\{Tool\}$. Note $\{L\}$ and $\{R\}$ have the same orientation and $\{R\}$ is translated by l_0 along the X axis. The portion O_L-O_1-O_2-O_3-O_{Tool} of the loop can be thought of as a planar 3R manipulator whose DH parameters are given as

i	α_{i-1}	a_{i-1}	d_i	θ_i
1	0	0	0	θ_1
2	0	l_1	0	ϕ_2
3	0	l_2	0	ϕ_3

From the above DH table we can find the 4×4 transformations, ${}^L_1[T]$, ${}^1_2[T]$, ${}^2_3[T]$, and, similar to the planar 3R, for the tool of length l_3, we can find ${}^{3}_{Tool}[T]$. In addition, as shown in Fig. 4.1, the transformation matrix ${}^{Tool}_{R}[T]$ can be obtained in terms of ϕ_1 (note that the angle between $\{Tool\}$ and $\{R\}$ is $180 + \phi_1$) and is given by

$$
{}^{Tool}_{R}[T] = \begin{pmatrix} -\cos\phi_1 & -\sin\phi_1 & 0 & 0 \\ \sin\phi_1 & -\cos\phi_1 & 0 & 0 \\ 0 & 0 & 1 & 0 \\ 0 & 0 & 0 & 1 \end{pmatrix} \tag{4.1}
$$

The loop-closure equation for the four-bar mechanism is simply given as

$$
{}^L_1[T]{}^1_2[T]{}^2_3[T]{}^{3}_{Tool}[T]{}^{Tool}_{R}[T] = {}^L_R[T] \tag{4.2}
$$

Since the loop is planar, only the X and Y components of the position vector and the rotation about the Z axis are useful, and we can write

$$
\begin{aligned}
l_1 \cos\theta_1 + l_2 \cos(\theta_1 + \phi_2) + l_3 \cos(\theta_1 + \phi_2 + \phi_3) &= l_0 \\
l_1 \sin\theta_1 + l_2 \sin(\theta_1 + \phi_2) + l_3 \sin(\theta_1 + \phi_2 + \phi_3) &= 0 \\
\theta_1 + \phi_2 + \phi_3 + (\pi - \phi_1) &= 4\pi
\end{aligned} \tag{4.3}
$$

where the last equation is derived from the fact that $\{L\}$ is parallel to $\{R\}$ and the angles θ_1, ϕ_1, ϕ_2, and ϕ_3 are as shown in Fig. 4.1. Since the sum of the interior angles of a planar quadrilateral is 2π, we can see that the right-hand side must be 4π.

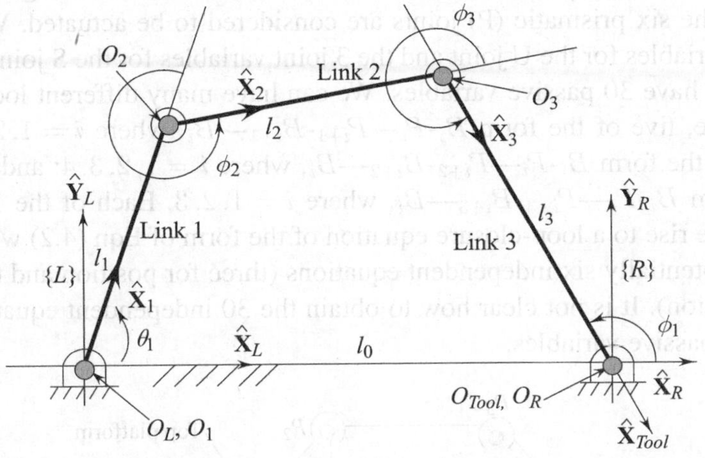

Fig. 4.1 The four-bar mechanism

In this approach of deriving loop-closure equations, we have three constraint equations in all four joint variables—we have $\mathbf{q} = (\theta_1, \phi_1, \phi_2, \phi_3)$ with actuated joint $\boldsymbol{\theta} = \theta_1$ and the passive joints $\boldsymbol{\phi} = (\phi_1, \phi_2, \phi_3)$. In this approach, for the four-bar mechanism, we have $n = 1$, $m = 3$, and $J = 4$.

The approach based on multiplication and equating 4×4 transformation matrices is quite general, but it has two main difficulties:

- Presence of multi-DOF joints, such as spherical (S) and Hooke (U) joints, in a loop.
- Obtaining independent loops in the presence of several loops.

The first difficulty can be overcome by representing a multi-DOF joint by two or more one-DOF joints and obtain an equivalent 4×4 transformation matrix for a multi-DOF joint. For example, an S joint can be considered to be three intersecting rotary (R) joints (similar to the wrist of a PUMA) and a C joint can be considered as a combination of a rotary (R) and a prismatic (P) joint.

The second difficulty is much harder to overcome since it is not straight forward to determine independent loops whenever there are more than three loops. For example, consider the well known Stewart–Gough platform shown in Fig. 4.2. The Stewart–Gough platform consists of a (moving) top platform and a (fixed) base and six legs connecting the top platform and base at points P_i, B_i, where $i = 1, \ldots, 6$, respectively. Each leg has universal (U) or a Hooke joint with two degrees of freedom at the base connection point, a prismatic (P) joint in middle, and a spherical (S) joint at the platform connection point. For this parallel manipulator, we have $\lambda = 6$, $N = 14$, $J = 18$, and $\sum_{i=1}^{J} F_i = 18 + 12 + 6 = 36$. Grübler's formula gives dof = 6 and the six prismatic (P) joints are considered to be actuated. With two joint variables for the U joint and the 3 joint variables for the S joint in each leg, we have 30 passive variables. We can have many different loops—for example, five of the form B_i-P_i—P_{i+1}-B_{i+1}—B_i, where $i = 1, 2, 3, 4, 5$; four of the form B_i-P_i—P_{i+2}-B_{i+2}—B_i, where $i = 1, 2, 3, 4$; and three of the form B_i-P_i—P_{i+3}-B_{i+3}—B_i, where $i = 1, 2, 3$. Each of the 12 loops can give rise to a loop-closure equation of the form of Eqn (4.2) which can have potentially six independent equations (three for position and three for orientation). It is not clear how to obtain the 30 independent equations for the 30 passive variables.

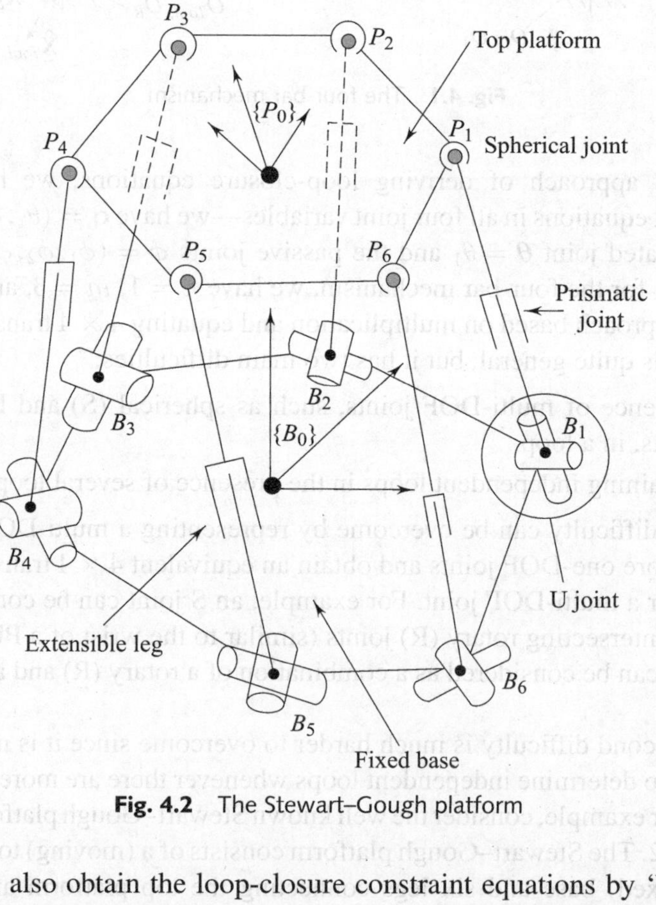

Fig. 4.2 The Stewart–Gough platform

We can also obtain the loop-closure constraint equations by 'breaking' a loop in a closed-loop mechanism or a parallel manipulator. Breaking the loops converts a closed-loop mechanism or a parallel manipulator into one or more serial manipulators which can be mathematically represented and analysed using tools of DH parameters and 4×4 homogeneous transformation matrices developed for serial manipulators in Chapter 3. Then we use the joint and other constraints presented in Section 2.5 to again 'close' the loop. This approach has advantages over the approach using the multiplication of 4×4 transformation matrices discussed earlier and is illustrated with the help of the four-bar mechanism discussed earlier.

One way to break the loop in the four-bar mechanism is to break it at the third rotary joint as shown in Fig. 4.3(a). Then we get a 2R planar serial manipulator and a 1R serial planar manipulator. The DH tables for these two manipulators are given in Example 2.4, Chapter 2. From the DH table, we can obtain transformation matrices ${}^{L}_{1}[T]$, ${}^{1}_{2}[T]$ for the 2R manipulator and ${}^{R}_{1}[T]$ for the 1R manipulator. Using the length l_2 for the length of the second link of the planar 2R manipulator and l_3 as the length of the link of the planar 1R manipulator, we can obtain ${}^{L}_{Tool}[T]$ and ${}^{R}_{Tool}[T]$. From the last column of ${}^{L}_{Tool}[T]$ we can extract the position vector ${}^{L}\mathbf{p}$ of the tip of the second link of the planar 2R serial manipulator [see Fig. 4.3(a)] (the point ${}^{L}\mathbf{p}$ is also a point on the third rotary joint axis). Since, it is planar manipulator, only the X and Y components are of interest and these are given by

$$
\begin{aligned}
x &= l_1 \cos\theta_1 + l_2 \cos(\theta_1 + \phi_2) \\
y &= l_1 \sin\theta_1 + l_2 \sin(\theta_1 + \phi_2)
\end{aligned}
\tag{4.4}
$$

Likewise, from the last column of ${}^{R}_{Tool}[T]$, we can extract the vector ${}^{R}\mathbf{p}$, and its X and Y components are given by

$$
\begin{aligned}
x &= l_3 \cos\phi_1 \\
y &= l_3 \sin\phi_1
\end{aligned}
\tag{4.5}
$$

where l_3 is the length of the link of the 1R manipulator [see Fig. 4.3(a)].

We can now apply the constraint imposed by the rotary joint (see Section 2.5.1). In this case, since the motion is planar, we have from the second of Eqns (2.41)

$$
\begin{aligned}
x &= l_1 \cos\theta_1 + l_2 \cos(\theta_1 + \phi_2) = l_0 + l_3 \cos\phi_1 \\
y &= l_1 \sin\theta_1 + l_2 \sin(\theta_1 + \phi_2) = l_3 \sin\phi_1
\end{aligned}
\tag{4.6}
$$

where l_0 is the distance along the X axis as shown in Fig. 4.3(a).

Equations (4.6) represent two equations in three joint variables $(\theta_1, \phi_1, \phi_2)$, out of which θ_1 is an actuated variable and the other two are passive. The variables $(\theta_1, \phi_1, \phi_2)$ determine the configuration of the planar four bars and thus $\mathbf{q} = (\theta_1, \phi_1, \phi_2)$ and the rotation at the third joint ϕ_3 is not an element of $\mathbf{q}$. Equations (4.6) are a possible set of constraint equations for

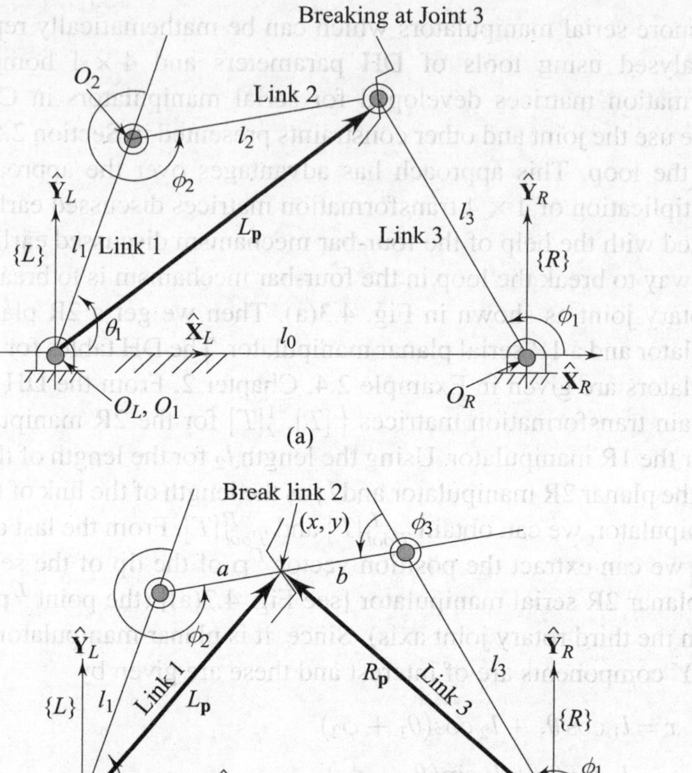

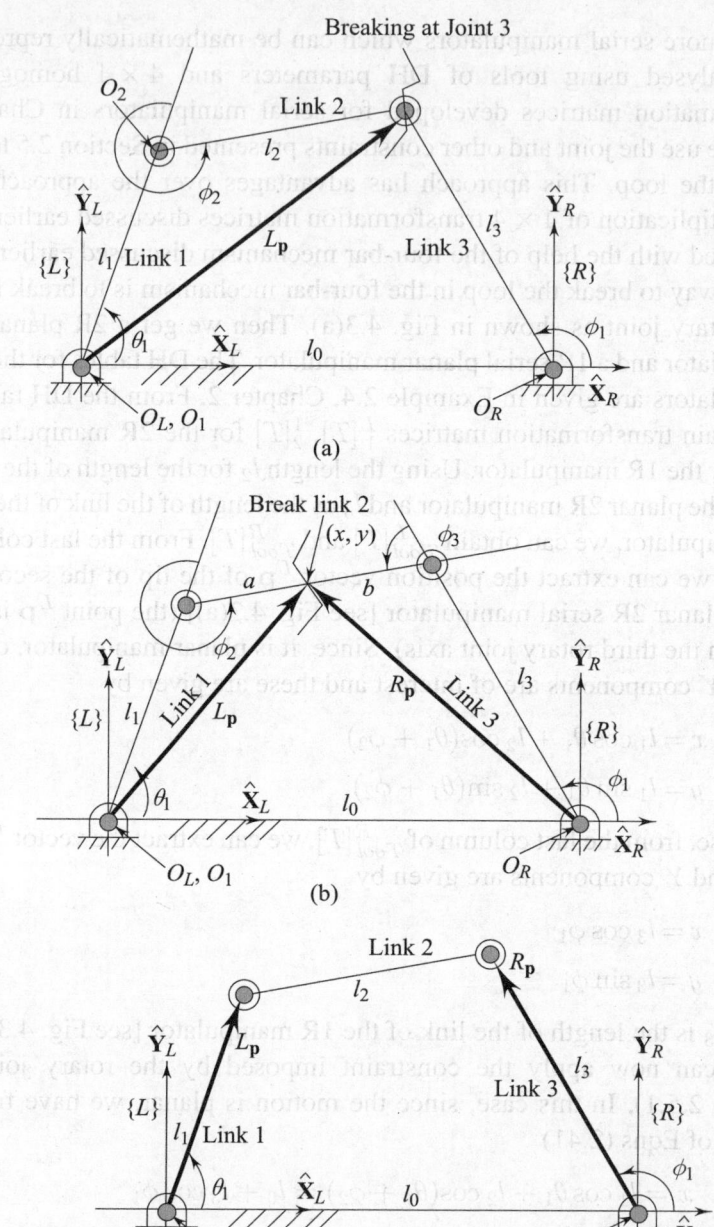

Fig. 4.3 The four-bar mechanism 'broken' in different ways

the planar four-bar mechanism. Comparing with Eqns (4.3), we can observe that we have a lesser number of equations and the dimension of $\mathbf{q}$ is 3 as compared to 4 earlier. It may be noted that the constraint equations in the first of Eqns (2.41), dealing with rotation matrices and the axis of the rotary joint, are not needed for a planar case.

Another way to break the loop is to break the second link as shown in Fig. 4.3(b). In this case, we get two 2R planar manipulators. We can obtain the X and Y coordinates of ${}^L\mathbf{p}$ as

$$
\begin{aligned}
x &= l_1 \cos\theta_1 + a\cos(\theta_1 + \phi_2) \\
y &= l_1 \sin\theta_1 + a\sin(\theta_1 + \phi_2)
\end{aligned}
\tag{4.7}
$$

where a is the distance as shown in Fig. 4.3(b).

Likewise, the X and Y components of the vector ${}^R\mathbf{p}$ are given by

$$
\begin{aligned}
x &= l_3 \cos\phi_1 + b\cos(\phi_1 + \phi_3) \\
y &= l_3 \sin\phi_1 + b\sin(\phi_1 + \phi_3)
\end{aligned}
\tag{4.8}
$$

where b is the distance as shown in Fig. 4.3(b), $l_2 = a + b$ and the angle ϕ_3 is as shown in Fig. 4.3(b). It may be noted that ϕ_3 used above and shown in Fig. 4.3(b) is defined in a different way than in Fig. 4.1.

We can now impose the constraint that the broken link is actually rigid and we get

$$
\begin{aligned}
x &= l_1 \cos\theta_1 + a\cos(\theta_1 + \phi_2) = l_0 + l_3\cos\phi_1 + b\cos(\phi_1 + \phi_3) \\
y &= l_1 \sin\theta_1 + a\sin(\theta_1 + \phi_2) = l_3\sin\phi_1 + b\sin(\phi_1 + \phi_3) \\
\theta_1 &+ \phi_2 = \phi_1 + \phi_3 + \pi
\end{aligned}
\tag{4.9}
$$

where the last equation is obtained by noting that the X axes of the coordinate frames fixed at the ends of the two equivalent planar 2R manipulators point in opposite directions. Equations (4.9) represent three equations in four joint variables, $(\theta_1, \phi_1, \phi_2, \phi_3)$, out of which only θ_1 is an actuated variable and the other three are passive. Equations (4.9) are essentially equivalent to Eqns (4.3) and (4.6) since ϕ_3 can be eliminated easily.

Finally, we can also derive a single constraint equation by obtaining the vectors locating the second and the third rotary joints from the two fixed ends, i.e., ${}^L\mathbf{p}$ and ${}^R\mathbf{p}$ as shown in Fig. 4.3(c). We can get

$$
\begin{aligned}
{}^L\mathbf{p} &= (l_1\cos\theta_1, l_1\sin\theta_1)^T \\
{}^R\mathbf{p} &= (l_3\cos\phi_1, l_3\sin\phi_1)^T
\end{aligned}
\tag{4.10}
$$

and enforce the constraint of constant length l_2 to obtain

$$\eta_1(\theta_1, \phi_1) = (l_1 \cos\theta_1 - l_0 - l_3 \cos\phi_1)^2 + (l_1 \sin\theta_1 - l_3 \sin\phi_1)^2 \\ -l_2^2 = 0 \tag{4.11}$$

It may be noted that in this case we have only one constraint equation[2] and $\mathbf{q}$ consists of only (θ_1, ϕ_1) with θ_1 as the actuated variable and ϕ_1 as the passive variable. When this equation is used for kinematic analysis, additional effort is required to obtain values of the other two joint variables.

In this text, we will use the concept of breaking the loop, obtaining relevant transformation matrices, and then imposing the joint or other constraint to close the loop. Our goal is to use the least possible number of passive joint variables (least possible m) in $\mathbf{q}$ for analysis and obtain the simplest and adequate number of equations, $\eta_i(\mathbf{q}) = 0$, where $i = 1, ..., m$. The m equations required for the m passive joint variables will be called the loop-closure constraint equations.

4.4 Direct Kinematics of Parallel Manipulators

Unlike the direct kinematics problem of a serial manipulator, not all the joint variables can be specified in the case of parallel manipulators and closed-loop mechanisms. Only the n actuated variables are specified, and m passive joint variables must be computed. In addition, unlike in a serial manipulator, there may not be a 'natural' end-effector or an output link in a closed-loop mechanisms or a parallel manipulator. For example, in the kinematic analysis of a four-bar mechanism, the output link is often the second link, called the *coupler*, and we are interested in the motion of a point on this coupler. However, for function generation, the third link is the chosen output link.

The direct kinematics problem for a parallel manipulator or a closed-loop mechanism consists of two parts. In the first part, the problem can be stated as: given all the link parameters, the geometry and the n actuated joint variables, determine the m passive joint variables. In the second part: obtain the position and orientation of a chosen end-effector or an output link once the n actuated variables and the m passive variables (or $\mathbf{q}$) are known. The first part, for most closed-loop mechanisms and parallel manipulators, has a complexity similar (often more) to the inverse kinematics problem in serial manipulators. It involves (a) determining the m loop-closure constraint equations and (b) the solution of non-linear equations for the m passive variables in terms of

[2] In the four-bar kinematics this is well known as Freudenstein's equation (see Uicker et al. 2003).

the given n actuated variables. As discussed in the preceding section, there are several ways to determine the most appropriate loop-closure constraint equations for a closed-loop mechanism or a parallel manipulator, and there are no general approaches available. Once the m loop-closure constraint equations are available, the goal is to obtain analytical expressions for the m joint variables and this can be done (at least attempted) by using elimination methods, such as Sylvester's dialytic elimination method, presented in Chapter 3. Once $\mathbf{q}$ is known, additional effort is required to compute the position and orientation of the chosen end-effector.

We illustrate the solution of the direct kinematics problem for three representative examples.

Example 4.1 The planar four-bar mechanism

The four-bar mechanism, shown in Fig. 4.1, is one of the simplest possible closed-loop mechanisms and has been studied extensively (see, for example, Uicker et al. 2003). In this example, we use the loop-closure equations given in Eqn (4.9) and obtain analytical expressions for the passive variables as a function of the actuated variable θ_1. We also obtain the equation of the curve traced by point ${}^L\mathbf{p}$ with coordinates (x, y) where the coupler is 'broken' [see Fig. 4.3(b)]. We proceed as follows.

Using the third equation in the first of two Eqns (4.9), we get

$$\begin{aligned} x - l_0 &= l_3 \cos\phi_1 - b\cos(\theta_1 + \phi_2) \\ y &= l_3 \sin\phi_1 - b\sin(\theta_1 + \phi_2) \end{aligned} \tag{4.12}$$

and denoting $\theta_1 + \phi_2$ by δ, squaring and adding, and after simplification, we get

$$A_1 \cos\delta + B_1 \sin\delta + C_1 = 0 \tag{4.13}$$

where

$$\begin{aligned} A_1 &= x - l_0, \quad B_1 = y \\ C_1 &= (1/2b)[(x - l_0)^2 + y^2 + b^2 - l_3^2] \end{aligned} \tag{4.14}$$

Similarly from the first part of two Eqns (4.9), we get

$$\begin{aligned} x &= l_1 \cos\theta_1 + a\cos(\theta_1 + \phi_2) \\ y &= l_1 \sin\theta_1 + a\sin(\theta_1 + \phi_2) \end{aligned} \tag{4.15}$$

which after squaring, adding, and after simplification gives

$$A_2 \cos\delta + B_2 \sin\delta + C_2 = 0 \tag{4.16}$$

where

$$A_2 = x, \quad B_2 = y$$
$$C_2 = (1/2a)[l_1^2 - a^2 - x^2 - y^2] \tag{4.17}$$

Equations (4.13) and (4.16) can be converted into a pair of quadratics by using the tangent half-angle substitutions [see Chapter 3, Eqns (3.14)]. Following Sylvester's dialytic elimination method (see Section 3.9 in Chapter 3), we can find the eliminant. In this case, the eliminant is given by

$$(A_1B_2 - A_2B_1)^2 = (A_1C_2 - A_2C_1)^2 + (B_1C_2 - B_2C_1)^2 \tag{4.18}$$

and

$$\delta = -2\tan^{-1}\left(\frac{A_1C_2 - A_2C_1}{(B_1C_2 - B_2C_1) + (A_1B_2 - A_2B_1)}\right) \tag{4.19}$$

Equation (4.18) is a function of the X and Y coordinates of ${}^L\mathbf{p}$ and the geometry of the four-bar mechanism. After some simplification, we get

$$\begin{aligned} 4a^2b^2{l_0}^2y^2 = {} & [b(x - l_0)(l_1^2 - a^2 - x^2 - y^2) - ax\{(x - l_0)^2 + y^2 \\ & + b^2 - l_3^2\}]^2 + y^2[b(l_1^2 - a^2 - x^2 - y^2) - a\{(x - l_0)^2 \\ & + y^2 + b^2 - l_3^2\}]^2 \end{aligned} \tag{4.20}$$

Equation (4.20) is known as the coupler curve.[3] It can be observed that the coupler curve is a sixth-degree curve since the highest power of x or y is 6 in the above equation.

The elimination procedure gives δ as a function of (x, y) and the geometry of the four-bar mechanism. Since θ_1 is given, we can obtain

$$\phi_2 = \delta - \theta_1 = -2\tan^{-1}\left(\frac{A_1C_2 - A_2C_1}{(B_1C_2 - B_2C_1) + (A_1B_2 - A_2B_1)}\right) - \theta_1 \tag{4.21}$$

The angle ϕ_1 can be obtained from Eqn (4.11). Simplifying Eqn (4.11), we can get an equation of the form

$$\begin{aligned} l_0^2 + l_1^2 + l_3^2 - l_2^2 - 2\,l_0l_1\cos\theta_1 &= \cos\phi_1(2l_1l_3\cos\theta_1 - 2l_0l_3) \\ &\quad + \sin\phi_1(2l_1l_3) \end{aligned} \tag{4.22}$$

which can be solved for ϕ_1 for a given θ_1.

Finally, ϕ_3 can be solved from the third equation in Eqns (4.9) and we can get

$$\phi_3 = \theta_1 + \phi_2 - \phi_1 - \pi \tag{4.23}$$

[3] The coupler curve for a four-bar mechanism is extensively studied in classical kinematics of mechanisms. For a more general form of the coupler curve and its interesting properties, see Chapter 6 of Hartenberg and Denavit (1964).

Numerical example

Using Grashof's criterion (see also Example 4.4), we choose link lengths so as ensure that the input link of the four-bar mechanism can rotate fully. We choose $l_0 = 5.0, l_1 = 1.0, l_2 = 3.0$, and $l_3 = 4.0$. Figure 4.4 gives a plot of ϕ_1 (in rad) with respect to the actuated variable θ_1 (in rad). Equation (4.22) gives two values of ϕ_1 for a given θ_1 and both the branches are shown in Fig. 4.4. Once ϕ_1 is known, the other two joint angles ϕ_2 and ϕ_3 can be obtained from Eqns (4.21) and (4.23) (see also exercise problem 4.1), respectively. Once ϕ_2 is known, we can also plot the curve traced by the coupler point by using Eqn 4.15. The curve traced for $a = 2.5$ and $b = 0.5$ and the link dimensions mentioned earlier is shown in Fig. 4.5. It may be noted that ϕ_2 can have two values for a given θ_1 and this results in two possible coupler curves. Both the coupler curves are shown in Fig. 4.5.

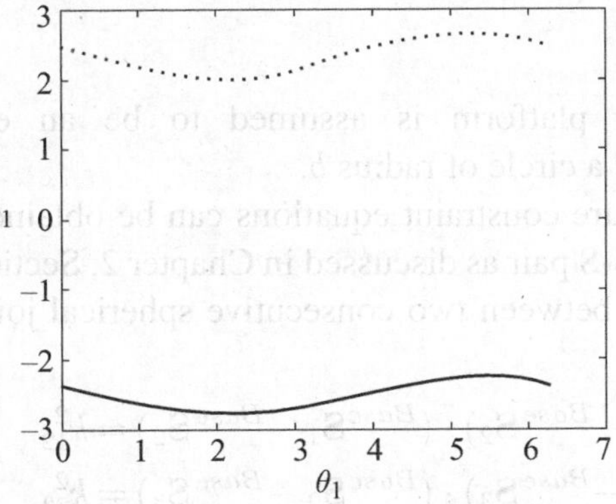

Fig. 4.4 Plot of ϕ_1 versus θ_1 for a four-bar mechanism

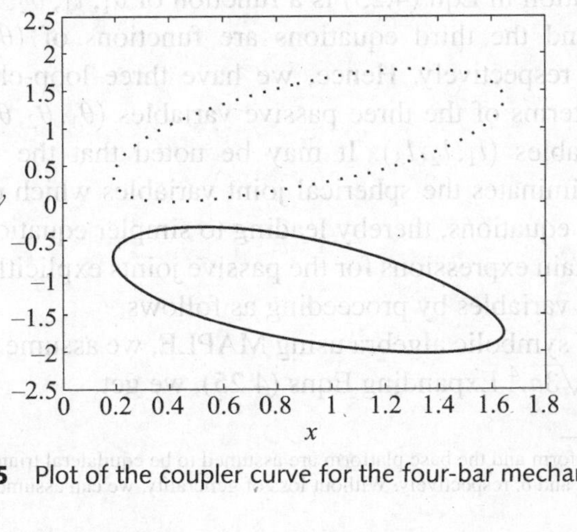

Fig. 4.5 Plot of the coupler curve for the four-bar mechanism

Example 4.2 A three-DOF parallel manipulator

In Chapter 2, Example 2.5, we presented the DH tables for the legs of a three-DOF parallel manipulator (see Fig. 2.22). In this example, we solve the direct kinematics problem for this parallel manipulator.

To obtain the loop-closure constraint equations for the three-DOF parallel manipulator, we start with the position vectors locating the spherical joints with respect to a $\{Base\}$. These are given by

$$
\begin{aligned}
{}^{Base}\mathbf{S}_1 &= (b - l_1\cos\theta_1, 0, l_1\sin\theta_1)^T \\
{}^{Base}\mathbf{S}_2 &= \left(-\frac{b}{2} + \frac{1}{2}l_2\cos\theta_2, \frac{\sqrt{3}}{2}b - \frac{\sqrt{3}}{2}l_2\cos\theta_2, l_2\sin\theta_2\right)^T \\
{}^{Base}\mathbf{S}_3 &= \left(-\frac{b}{2} + \frac{1}{2}l_3\cos\theta_3, -\frac{\sqrt{3}}{2}b + \frac{\sqrt{3}}{2}l_3\cos\theta_3, l_3\sin\theta_3\right)^T
\end{aligned}
\tag{4.24}
$$

where the base platform is assumed to be an equilateral triangle circumscribed by a circle of radius b.

The loop-closure constraint equations can be obtained by imposing the constraint of an S-S pair as discussed in Chapter 2, Section 2.5.6. Assuming that the distance between two consecutive spherical joints is k_{12}, k_{23}, and k_{31}, we can write

$$
\begin{aligned}
({}^{Base}\mathbf{S}_1 - {}^{Base}\mathbf{S}_2)\cdot({}^{Base}\mathbf{S}_1 - {}^{Base}\mathbf{S}_2) &= k_{12}^2 \\
({}^{Base}\mathbf{S}_2 - {}^{Base}\mathbf{S}_3)\cdot({}^{Base}\mathbf{S}_2 - {}^{Base}\mathbf{S}_3) &= k_{23}^2 \\
({}^{Base}\mathbf{S}_3 - {}^{Base}\mathbf{S}_1)\cdot({}^{Base}\mathbf{S}_3 - {}^{Base}\mathbf{S}_1) &= k_{31}^2
\end{aligned}
\tag{4.25}
$$

The first equation in Eqn (4.25) is a function of θ_1, l_1, θ_2, and l_2. Likewise the second and the third equations are functions of $(\theta_2, l_2, \theta_3, l_3)$, and $(\theta_1, l_1, \theta_3, l_3)$ respectively. Hence, we have three loop-closure constraint equations in terms of the three passive variables $(\theta_1, \theta_2, \theta_3)$ and the three actuated variables (l_1, l_2, l_3). It may be noted that the use of S-S pair constraints eliminates the spherical joint variables which do not appear in the constraint equations, thereby leading to simpler equations.

We can obtain expressions for the passive joints explicitly in terms of the actuated joint variables by proceeding as follows.

For ease of symbolic algebra using MAPLE, we assume $b = 1$ and $k_{12} = k_{23} = k_{31} = \sqrt{3}a$.[4] Expanding Eqns (4.25), we get

[4] The moving platform and the base platform are assumed to be equilateral triangles circumscribed by circles of radii a and b, respectively. Without loss of generality, we can assume that the dimension of

$$\begin{aligned}
\eta_1(\mathbf{q}) &= 3 - 3a^2 + l_1^2 + l_2^2 + l_1 l_2 c_1 c_2 - 2 l_1 l_2 s_1 s_2 - 3 l_1 c_1 - 3 l_2 c_2 = 0 \\
\eta_2(\mathbf{q}) &= 3 - 3a^2 + l_2^2 + l_3^2 + l_2 l_3 c_2 c_3 - 2 l_2 l_3 s_2 s_3 - 3 l_2 c_2 - 3 l_3 c_3 = 0 \qquad (4.26) \\
\eta_3(\mathbf{q}) &= 3 - 3a^2 + l_3^2 + l_1^2 + l_3 l_1 c_3 c_1 - 2 l_3 l_1 s_3 s_1 - 3 l_3 c_3 - 3 l_1 c_1 = 0
\end{aligned}$$

where $\mathbf{q}$ represents the configuration space of this three-DOF parallel manipulator and it consists of the actuated variables l_1, l_2, l_3 and the passive variables θ_1, θ_2, θ_3. It may be noted that the constraint equations are not functions of all the joint variables. For example, $\eta_1(\mathbf{q})$ is a function of only l_1, l_2, θ_1, and θ_2. This fact can be used to eliminate θ_1 from the first and third equations in Eqn (4.26) to obtain an equation in θ_2 and θ_3. This resulting equation can be used with the second equation in Eqn (4.26) to eliminate θ_2 to get a single equation in θ_3.

To eliminate θ_1 from the first and third equations in Eqn (4.26), we rewrite them as

$$A_i c_1 + B_i s_1 + C_i = 0, \qquad i = 1, 2 \qquad (4.27)$$

where

$$\begin{aligned}
A_1 &= l_1 l_2 c_2 - 3 l_1, \\
B_1 &= -2 l_1 l_2 s_2 \\
C_1 &= 3 - 3a^2 + l_1^2 + l_2^2 - 3 l_2 c_2 \\
A_2 &= l_1 l_3 c_3 - 3 l_1, \\
B_2 &= -2 l_1 l_3 s_3, \\
C_2 &= 3 - 3a^2 + l_1^2 + l_3^2 - 3 l_3 c_3 \qquad (4.28)
\end{aligned}$$

Using the tangent half-angle formulas for $\cos\theta_1$ and $\sin\theta_1$ in Eqn (4.27), we can get two quadratic equations in $x_1 = \tan(\theta_1/2)$, and using the elimination theory discussed in Section 3.9, we can obtain the eliminant as

$$\begin{aligned}
\eta_4(l_1, l_2, l_3, \theta_2, \theta_3) = (A_1 C_2 - A_2 C_1)^2 + (B_1 C_2 - B_2 C_1)^2 \\
-(A_1 B_2 - A_2 B_1)^2 = 0 \qquad (4.29)
\end{aligned}$$

where A_i, B_i, C_i $(i = 1, 2)$ are as given in Eqn (4.28).

In Eqn (4.29), the passive joint variables θ_2 and θ_3 occur in quadratic terms such as $(c_2 c_3)^2$, $(s_2 c_3)^2$, etc. together with the linear terms c_2, s_2, c_3, and s_3 (see Exercise 4.5). To eliminate θ_2, we use the tangent half-angle formulas for $\cos\theta_2$ and $\sin\theta_2$ in Eqn (4.29) and get a quartic equation in $x_2 = \tan(\theta_2/2)$ with the coefficients of the quartic being functions of θ_3, l_i $(i = 1, 2, 3)$ and constants. The second equation in Eqn (4.26) also yields a quadratic in x_2 with the coefficients as functions of θ_3, l_2, l_3 and constants. Following Sylvester's dialytic method, we can eliminate θ_2 between the quartic and

---

the moving platform is scaled by b.

the quadratic (see also Exercise 3.15) to get an eliminant containing powers of sine and cosine of θ_3. Finally, substituting tangent half-angle formulas for $\cos\theta_3$, $\sin\theta_3$ and its powers, and simplifying, we get an eighth-degree polynomial of the form

$$q_8(x_3^2)^8 + q_7(x_3^2)^7 + \cdots + q_1(x_3^2) + q_0 = 0 \tag{4.30}$$

where $x_3 = \tan(\theta_3/2)$.

The coefficients q_i ($i = 1, ..., 8$) are very large and only the first and the last one (obtained by MAPLE) are given below. They are obtained after extensive simplification and cancelling common factors.

$$q_8 = (p_0a^4 + p_1a^3 + p_2a^2 + p_3a + p_4)^2(p_0a^4 - p_1a^3 + p_2a^2 - p_3a + p_4)^2$$

$$q_0 = (r_0a^4 + r_1a^3 + r_2a^2 + r_3a + r_4)^2(r_0a^4 - r_1a^3 + r_2a^2 - r_3a + r_4)^2 \tag{4.31}$$

where $r_0 = p_0 = -9$, $r_1 = 12(l_3 - 3)$, $p_1 = 12(l_3 + 3)$, $r_2 = 3[l_1^2 + l_2^2 - l_3\,(l_3 - 10) - 15]$, $p_2 = 3[l_1^2 + l_2^2 - l_3\,(l_3 + 10) - 15]$, $r_3 = -2(l_3 - 3)\,(l_1^2 + l_2^2 + l_3^2 - 3)$, $p_3 = -2(l_3 + 3)\,(l_1^2 + l_2^2 + l_3^2 - 3)$, $r_4 = l_3^4 - 8l_3^3 + 3l_2^2 + 18l_3^2 - 2l_3(l_2^2 + 6) - l_1^2\,(l_2^2 + 2l_3 - 3)$, and $p_4 = l_3^4 + 8l_3^3 + 3l_2^2 + 18l_3^2 + 2l_3\,(l_2^2 + 6) + l_1^2\,(l_2^2 + 2l_3 - 3)$.

A natural output link for this three-DOF parallel manipulator is the moving platform. For the position of the moving platform, we choose the centroid. The position vector of the centroid with respect to the fixed $\{Base\}$ is given by

$$^{Base}\mathbf{p} = \frac{1}{3}(^{Base}\mathbf{S}_1 + {}^{Base}\mathbf{S}_2 + {}^{Base}\mathbf{S}_3) \tag{4.32}$$

where $^{Base}\mathbf{S}_i$ $(i = 1, 2, 3)$ are known from Eqns (4.24) once the passive joints can be solved for given actuated joint variables.

The orientation of the moving platform is also known once the position vectors $^{Base}\mathbf{S}_i$ $(i = 1, 2, 3)$ are known. The columns of the rotation matrix $^{Base}_{Top}[R]$ can be obtained as

$$^{Base}_{Top}[R] = \left[\frac{^{Base}\mathbf{S}_1 - {}^{Base}\mathbf{S}_2}{|^{Base}\mathbf{S}_1 - {}^{Base}\mathbf{S}_2|}\;\; \widehat{\mathbf{Y}} \;\; \frac{(^{Base}\mathbf{S}_1 - {}^{Base}\mathbf{S}_2) \times (^{Base}\mathbf{S}_1 - {}^{Base}\mathbf{S}_3)}{|(^{Base}\mathbf{S}_1 - {}^{Base}\mathbf{S}_2) \times (^{Base}\mathbf{S}_1 - {}^{Base}\mathbf{S}_3)|}\right] \tag{4.33}$$

where $\widehat{\mathbf{Y}}$ is obtained from the cross product of the third and first columns.

Numerical example

Since the polynomial given in Eqn (4.30) is eight degree in $\tan(\theta_3/2)^2$, it is not possible to obtain analytical expressions for the passive variables θ_1, θ_2, and θ_3. We can numerically solve for the passive variables for given values of the actuated variables, l_1, l_2, and l_3 using MATLAB. For $a = 1/2$, and for $l_1 = 2/3$, $l_2 = 3/5$ and $l_3 = 3/4$, we get two sets of values

$\theta_3 = 0.8111, 0.8028$ rad and two additional sets with a negative sign and same magnitude. For the positive values, we can compute the corresponding θ_2 as $0.4809, 0.2851$ rad and θ_1 as $0.7471, 0.7593$ rad, respectively. For the set $(0.7471, 0.4809, 0.8111)$, the position vector ${}^{Base}\mathbf{p}$ is obtained as $(0.0117, -0.0044, 0.4248)$ and the rotation matrix ${}^{Base}_{Top}[R]$ is given by

$$ {}^{Base}_{Top}[R] = \begin{pmatrix} 0.8602 & 0.5069 & -0.0564 \\ -0.4681 & 0.8285 & 0.3074 \\ 0.2026 & -0.2380 & 0.9499 \end{pmatrix} $$

It may be noted that the numbers given above are rounded off to 4 places of decimal.

Example 4.3 A six-DOF parallel manipulator

In Chapter 2, Example 2.6, we described a six-DOF manipulator (see Fig. 2.23). The DH table for a 'finger' was presented and the position vectors of the spherical joints with respect to a coordinate system $\{Base\}$ were obtained. These are reproduced below:

$$ {}^{Base}\mathbf{p}_1 = \begin{pmatrix} \cos\theta_1[l_{11} + l_{12}\cos\psi_1 + l_{13}\cos(\psi_1 + \phi_1)] \\ -d + \sin\theta_1[l_{11} + l_{12}\cos\psi_1 + l_{13}\cos(\psi_1 + \phi_1)] \\ h + l_{12}\sin\psi_1 + l_{13}\sin(\psi_1 + \phi_1) \end{pmatrix} \quad (4.34) $$

$$ {}^{Base}\mathbf{p}_2 = \begin{pmatrix} \cos\theta_2[l_{21} + l_{22}\cos\psi_2 + l_{23}\cos(\psi_2 + \phi_2)] \\ d + \sin\theta_2[l_{21} + l_{22}\cos\psi_2 + l_{23}\cos(\psi_2 + \phi_2)] \\ h + l_{22}\sin\psi_2 + l_{23}\sin(\psi_2 + \phi_2) \end{pmatrix} \quad (4.35) $$

$$ {}^{Base}\mathbf{p}_3 = [R(\widehat{\mathbf{Y}}, \gamma)] \begin{pmatrix} \cos\theta_3[l_{31} + l_{32}\cos\psi_3 + l_{33}\cos(\psi_3 + \phi_3)] \\ \sin\theta_3[l_{31} + l_{32}\cos\psi_3 + l_{33}\cos(\psi_3 + \phi_3)] \\ l_{32}\sin\psi_3 + l_{33}\sin(\psi_3 + \phi_3) \end{pmatrix} \quad (4.36) $$

In this example, we can again use the constraints imposed by a S-S pair, as discussed in Eqns (4.25), and we will get the three loop-closure constraint equations of the form

$$ \begin{aligned} |{}^{Base}\mathbf{p}_1 - {}^{Base}\mathbf{p}_2|^2 &= k_{12}^2 \\ |{}^{Base}\mathbf{p}_2 - {}^{Base}\mathbf{p}_3|^2 &= k_{23}^2 \\ |{}^{Base}\mathbf{p}_3 - {}^{Base}\mathbf{p}_1|^2 &= k_{31}^2 \end{aligned} \quad (4.37) $$

where k_{12}, k_{23}, and k_{31} are constants.

For the six-DOF manipulator, the first equation is a function of θ_1, ψ_1, ϕ_1 and θ_2, ψ_2, ϕ_2; the second equation is a function of θ_2, ψ_2, ϕ_2 and θ_3, ψ_3, ϕ_3; and, finally, the third equation is a function of θ_3, ψ_3, ϕ_3 and θ_1, ψ_1, ϕ_1. For the direct kinematics problem, we are given the six actuated variables, assumed to be $\theta_1, \psi_1, \theta_2, \psi_2, \theta_3$, and ψ_3, and our task is to find out the passive variables ϕ_1, ϕ_2, and ϕ_3. To obtain explicit expressions for the passive variables, we use Sylvester's dialytic elimination method.

From the first and the third equation in Eqn (4.37), we can eliminate ϕ_1 and get an equation containing the passive joint variables (ϕ_2, ϕ_3) in addition to the actuated variables. From this resulting equation and the second equation in Eqn (4.37), we can eliminate ϕ_2 to get a single equation containing only ϕ_3 and the actuated joint variables. In this case, the final eliminant is a sixteenth-degree polynomial in $\tan(\phi_3/2)$. The analytical expressions obtained using MAPLE for the coefficients for this sixteenth-degree polynomial are extremely large and in the following a numerical result is provided.

Numerical example

For the numerical example, we assume $d = 1/2$, $h = \sqrt{3}/2$, $l_{i1} = 1$, $l_{i2} = 1/2$, $l_{i3} = 1/4$ $(i = 1, 2, 3)$, $\gamma = \pi/4$ and $k_{12} = k_{23} = k_{13} = \sqrt{3}/2$. The actuated joint variables, all in radians, are chosen as $\theta_1 = 0.1$, $\psi_1 = -1.0$, $\theta_2 = 0.1$, $\psi_2 = -1.2$, $\theta_3 = 0.3$, and $\psi_3 = 1.0$. The sixteenth-degree polynomial is obtained as

$$\begin{aligned} 0.00012t_3^{16} &- 0.00182t_3^{15} + 0.01376t_3^{14} - 0.05230t_3^{13} + 0.13148t_3^{12} \\ &- 0.24391t_3^{11} + 0.35247t_3^{10} - 0.40965t_3^{9} + 0.38696t_3^{8} \\ &- 0.29811t_3^{7} + 0.18502t_3^{6} - 0.09104t_3^{5} + 0.03433t_3^{4} \\ &- 0.00968t_3^{3} + 0.00201t_3^{2} - 0.00037t_3 + 0.00006 = 0 \end{aligned}$$

where $t_3 = \tan(\phi_3/2)$.

The above sixteenth-degree polynomial can be solved numerically and we get two real values of ϕ_3 as $(0.8831, 1.8239)$ rad. The corresponding values of ϕ_1 and ϕ_2 are $(0.3679, 0.1146)$ rad and $(1.4548, 1.0448)$ rad, respectively.

To obtain the position of the gripped object, we can again obtain the expression for the centroid similar to Eqn (4.32). The position vector of the centroid is given by $(1.3768, 0.2624, 0.1401)$ for the first set of values of the passive joint variables. Likewise, for the orientation of the gripped object we can use the expressions in Eqn (4.33). The rotation matrix ${}^{Base}_{Object}[R]$ is given as

$${}^{Base}_{Object}[R] = \begin{pmatrix} 0.0306 & 0.2099 & -0.9773 \\ -0.9811 & 0.1806 & 0.0695 \\ 0.1910 & -0.9609 & 0.2004 \end{pmatrix}$$

4.5* Direct Kinematics of Stewart–Gough Platform

The direct kinematics problem for a Stewart–Gough platform, shown in Fig. 4.2, has been an object of intense study by several researchers (see, for example, Nanua et al. 1990, Dasgupta & Mruthyunjaya 2000, and Merlet 2001) for a long time. One of the reasons is the varied uses of a Stewart–Gough platform in flight simulators, sensors, positioning devices, machine tools, vibration isolation, and other applications requiring large structural stiffness (Merlet 2001). The other important reason is the level of difficulty—it is one of the hardest problems in parallel manipulator kinematics analogous to the inverse kinematics problem for a general 6R manipulator.

The direct kinematics problem may be stated as follows: given the displacements at the six prismatic joints, obtain the position and orientation of the moving platform. The aim is to obtain analytical expressions for the position and orientation variables as univariate polynomials. In this section, we present the formulation for the direct kinematics problem and outline the elimination steps required for arriving at this polynomial. Figure 4.2 shows a 6-6 Stewart platform with the Hooke joints in the base represented as two intersecting rotary (R) joints. To derive the direct kinematics equations, we first assign a coordinate systems and label the variables. The $\{B_0\}$ is chosen at a suitable point on the fixed base and the $\{P_0\}$ is chosen on the top platform as shown in Fig. 4.2. Figure 4.6 shows a typical leg of the Stewart platform. The translation vector ${}^{B_0}\mathbf{t}$ denotes the location of the origin of $\{P_0\}$ with respect to $\{B_0\}$ and the orientation of the top platform with respect to the fixed base is given by ${}^{B_0}_{P_0}[R]$. The vector ${}^{B_0}\mathbf{t}$ has three components $(t_x, t_y, t_z)^T$, and the rotation matrix ${}^{B_0}_{P_0}[R]$ can be written in terms of nine direction cosines r_{ij}, where $i, j = 1, 2, 3$. There are six relationships (see Chapter 2, Section 2.2.1) between the nine r_{ij}'s.

From Fig. 4.6, an arbitrary platform point P_i can be written in $\{B_0\}$ as

$$ {}^{B_0}\mathbf{p}_i = {}^{B_0}_{P_0}[R]\,{}^{P_0}\mathbf{p}_i + {}^{B_0}\mathbf{t} \tag{4.38} $$

where ${}^{P_0}\mathbf{p}_i$ is a constant vector which due to the choice of $\{P_0\}$ is given by $(p_{i_x}, p_{i_y}, 0)^T$.

Denoting the location of the corresponding base point B_i by the vector ${}^{B_0}\mathbf{B}_i$, we can obtain the leg vector ${}^{B_0}\mathbf{S}_i$ as

$$ {}^{B_0}\mathbf{S}_i = {}^{B_0}_{P_0}[R]\,{}^{P_0}\mathbf{p}_i + {}^{B_0}\mathbf{t} - {}^{B_0}\mathbf{b}_i \tag{4.39} $$

where ${}^{B_0}\mathbf{b}_i$ is a constant vector which due to the choice of $\{B_0\}$ is given by $(b_{i_x}, b_{i_y}, 0)^T$.

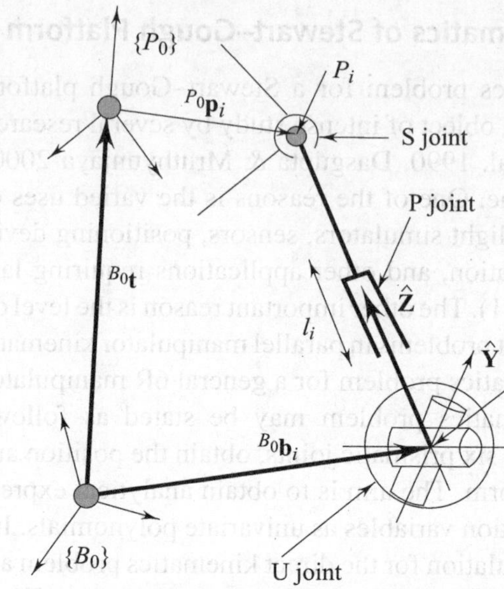

Fig. 4.6 A leg of a Stewart platform

The magnitude of the leg vector, denoted by l_i, can be obtained as

$$l_i^2 = (r_{11}p_{i_x} + r_{12}p_{i_y} + t_x - b_{i_x})^2 + (r_{21}p_{i_x} + r_{22}p_{i_y} + t_y - b_{i_y})^2 + (r_{31}p_{i_x} + r_{32}p_{i_y} + t_z - b_{i_z})^2 \quad (4.40)$$

which, using the constraints between the elements r_{ij}, can be written as

$$(t_x^2 + t_y^2 + t_z^2) + 2p_{i_x}(r_{11}t_x + r_{21}t_y + r_{31}t_z) + 2p_{i_y}(r_{12}t_x + r_{22}t_y + r_{32}t_z) - 2b_{i_x}(t_x + p_{i_x}r_{11} + p_{i_y}r_{12}) - 2b_{i_y}(t_y + p_{i_x}r_{21} + p_{i_y}r_{22}) + b_{i_x}^2 + b_{i_y}^2 + p_{i_x}^2 + p_{i_y}^2 - l_i^2 = 0 \quad (4.41)$$

For the six legs, $i = 1, ..., 6$, we get six equations of the type shown in Eqn. (4.41). In addition, we have three constraint equations for the relevant r_{ij}'s given by

$$\begin{aligned} r_{11}^2 + r_{21}^2 + r_{31}^2 &= 1 \\ r_{12}^2 + r_{22}^2 + r_{32}^2 &= 1 \\ r_{11}r_{12} + r_{21}r_{22} + r_{31}r_{32} &= 0 \end{aligned} \quad (4.42)$$

Equations (4.41) and (4.42) are nine quadratic equations in nine unknowns, t_x, t_y, t_z, r_{11}, r_{12}, r_{21}, r_{22}, r_{31}, and r_{32}. However, it can be noticed that all quadratic terms in Eqn. (4.41) enter as the square of the magnitude of the translation vector $(t_x^2 + t_y^2 + t_z^2)$, and the X and Y component of the vector

${}^{B_0}\mathbf{t}$, $(r_{11}t_x + r_{21}t_y + r_{31}t_z)$ and $(r_{12}t_x + r_{22}t_y + r_{32}t_z)$, respectively. This observation allows us to reduce the set of nine quadratics to six quadratic and three linear equations in the nine unknowns (Dasgupta & Mruthyunjaya 1994).

The set of six quadratic and three linear equations is the starting point for solving the direct kinematics problem of a general 6-6 Stewart platform manipulator, and the goal is to eliminate eight of these variables from the nine equations to arrive at a univariate polynomial in one unknown. This is one of the hardest problems in parallel manipulator kinematics and several researchers have worked on it. From the six quadratic and three linear equations, an upper bound of a 64th-degree polynomial was easily established using Bezout's theorem [see Section 3.9 and Semple & Roth (1949) listed at the end of Chapter 3]. Hence it was believed for some time that a general 6-6 Stewart platform could have up to 64 distinct configurations. Several researchers later estimated 40, 48, 54 configurations based on geometrical arguments, and it was finally proved that the direct kinematics of a general 6–6 Stewart–Gough platform leads to a univariate 40th-order polynomial (Raghavan 1993). Attempts are, however, still being made to obtain the simplest explicit expressions for the coefficients of this 40th-degree polynomial (Innocenti 2001).

We can make the following observations from the above formulation and Eqns (4.41) and (4.42).

- The third column of the rotation matrix ${}^{B_0}_{P_0}[R]$, i.e., the elements r_{13}, r_{23}, and r_{33}, does not appear in the equations. This is due to the choice of $\{B_0\}$ and $\{P_0\}$ and the resulting 0 in the Z coordinates of P_0 and B_0.
- The passive joint variables in the Hooke (U) joint do not appear as in the case of the previous three- and six-DOF examples (see Examples 4.2 and 4.3). This is a result of the fact that, with the prismatic joint locked, the distance between the centre of the spherical joint and the point of intersection of the two equivalent rotary joints in a Hooke joint will remain constant.[5] The formulation only involves task space variables, namely, the position and orientation of the top platform, and the translation at the prismatic (P) joints.

One consequence of the task space formulation discussed in this section is that we need to do additional work to obtain values of the passive joint variables in the Hooke joint (see Examples 4.6 and 4.7 and also Exercise 4.14). In addition, as we will see in Chapter 5, a singularity analysis in

[5] This is 'similar' to the S-S pair constraint discussed in Section 2.5.6.

terms of velocities is difficult for a Stewart platform with a task space formulation. To overcome these problems, we present a formulation of the direct kinematics of a general 6-6 Stewart platform in terms of configuration space variables in similar spirit to the three- and six-DOF examples discussed earlier.

We assume that the Hooke joint in leg i of the Stewart platform can be considered to be equivalent to two intersecting rotary joints. Further, we assume that the two rotations are like two successive Euler rotations (see Section 2.2.3), ϕ_i about $\widehat{\mathbf{Y}}_i$ and ψ_i about $\widehat{\mathbf{X}}_i$, respectively. Figure 4.6 shows the two rotary (R) joints and prismatic (P) joint in the leg of a Stewart platform. Each leg of the Stewart platform can be thus considered to be analogous to a R-R-P-S serial manipulator. The position vector of the point P_i, with respect to B_0, can be given as

$$\begin{aligned}{}^{B_0}\mathbf{p}_i &= {}^{B_0}\mathbf{b}_i + [R(\widehat{\mathbf{Z}}, \gamma_i)][R(\widehat{\mathbf{Y}}, \phi_i)][R(\widehat{\mathbf{X}}, \psi_i)](0, 0, l_i)^T \\ &= {}^{B_0}\mathbf{b}_i + l_i \begin{pmatrix} \cos\gamma_i \sin\phi_i \cos\psi_i + \sin\gamma_i \sin\psi_i \\ \sin\gamma_i \sin\phi_i \cos\psi_i - \cos\gamma_i \sin\psi_i \\ \cos\phi_i \cos\psi_i \end{pmatrix}\end{aligned} \tag{4.43}$$

where, as shown in Fig. 4.6, the vector ${}^{B_0}\mathbf{b}_i$ locates the origin of the $\{i\}$; γ_i is a constant determining the orientation of $\{i\}$, attached to the Hooke joint at the leg i, with respect to the coordinate system $\{B_0\}$; l_i is the translation of the prismatic (P) joint; and $[R(\widehat{\mathbf{Y}}, \phi_i)]$, $[R(\widehat{\mathbf{X}}, \psi_i)]$ denote the rotation matrices corresponding to rotations about the two intersecting rotary (R) joint axes, respectively. It can be seen that the vector ${}^{B_0}\mathbf{p}_i$ is a function of the two passive joint variables, ϕ_i and ψ_i, and the actuated joint variable l_i.

With the expression of the vector ${}^{B_0}\mathbf{p}_i$, we can now use the S-S pair constraints (see Section 2.5.6) to get six constraint equations of the form

$$\begin{aligned}\eta_1(\mathbf{q}) &= |{}^{B_0}\mathbf{p}_1 - {}^{B_0}\mathbf{p}_2|^2 - d_{12}^2 = 0 \\ \eta_2(\mathbf{q}) &= |{}^{B_0}\mathbf{p}_2 - {}^{B_0}\mathbf{p}_3|^2 - d_{23}^2 = 0 \\ \eta_3(\mathbf{q}) &= |{}^{B_0}\mathbf{p}_3 - {}^{B_0}\mathbf{p}_4|^2 - d_{34}^2 = 0 \\ \eta_4(\mathbf{q}) &= |{}^{B_0}\mathbf{p}_4 - {}^{B_0}\mathbf{p}_5|^2 - d_{45}^2 = 0 \\ \eta_5(\mathbf{q}) &= |{}^{B_0}\mathbf{p}_5 - {}^{B_0}\mathbf{p}_6|^2 - d_{56}^2 = 0 \\ \eta_6(\mathbf{q}) &= |{}^{B_0}\mathbf{p}_6 - {}^{B_0}\mathbf{p}_1|^2 - d_{61}^2 = 0\end{aligned} \tag{4.44}$$

In addition to the above six constraint equations, for all the six points, P_i (where $i = 1, ..., 6$) to lie on a plane, we have

$$\eta_7(\mathbf{q}) = |{}^{B_0}\mathbf{p}_1 - {}^{B_0}\mathbf{p}_3|^2 - d_{13}^2 = 0$$

$$\begin{aligned}
\eta_8(\mathbf{q}) &= |^{B_0}\mathbf{p}_1 - {}^{B_0}\mathbf{p}_4|^2 - d_{14}^2 = 0 \\
\eta_9(\mathbf{q}) &= |^{B_0}\mathbf{p}_1 - {}^{B_0}\mathbf{p}_5|^2 - d_{15}^2 = 0 \\
\eta_{10}(\mathbf{q}) &= (^{B_0}\mathbf{p}_1 - {}^{B_0}\mathbf{p}_3) \times (^{B_0}\mathbf{p}_1 - {}^{B_0}\mathbf{p}_4) \cdot (^{B_0}\mathbf{p}_1 - {}^{B_0}\mathbf{p}_2) = 0 \\
\eta_{11}(\mathbf{q}) &= (^{B_0}\mathbf{p}_1 - {}^{B_0}\mathbf{p}_4) \times (^{B_0}\mathbf{p}_1 - {}^{B_0}\mathbf{p}_5) \cdot (^{B_0}\mathbf{p}_1 - {}^{B_0}\mathbf{p}_3) = 0 \\
\eta_{12}(\mathbf{q}) &= (^{B_0}\mathbf{p}_1 - {}^{B_0}\mathbf{p}_5) \times (^{B_0}\mathbf{p}_1 - {}^{B_0}\mathbf{p}_6) \cdot (^{B_0}\mathbf{p}_1 - {}^{B_0}\mathbf{p}_4) = 0
\end{aligned} \tag{4.45}$$

where d_{ij} is the known distance between the spherical joints S_i and S_j on the top platform.

The twelve equations given in Eqns (4.44) and (4.45) are functions of the twelve passive variables ϕ_i, ψ_i (where $i = 1, ..., 6$), and the six actuated joint variables l_i (where $i = 1, ..., 6$). To solve the direct kinematics platform for the general 6-6 Stewart–Gough platform, we have to eliminate 11 passive variables from these 12 equations. It may be noted that any single equation does not contain all the 12 passive variables—for example, the first equation in Eqn (4.44) is a function of only ϕ_1, ψ_1, l_1, ϕ_2, ψ_2, and l_2. It may be further noted that the 12 equations are not unique and one can have other combinations.

It can be clearly seen that solving the direct kinematics problem with the 12 constraint equations is a much harder task than the task space formulation with six quadratic and three linear equations. For this reason the solution of the direct kinematics problem of a Stewart platform manipulator is never attempted in configuration space.

4.6 Mobility of Parallel Manipulators

In the analysis of serial manipulator, we defined the concept of a workspace as the positions and orientations in 3D space which could be achieved by the manipulator end-effector. For parallel manipulators and closed-loop mechanisms, more important than the notion of a workspace of a chosen end-effector is the notion of limits on the possible range of motion of the joints present in the closed-loop mechanism or the parallel manipulator. In particular, knowledge of the possible range of motion of the actuated joint is required for control of the parallel manipulator—if the range of motion of the actuated joint is limited, it would be unwise to attempt to drive an actuator past these limits. The limits on the range of motion for the joints in a parallel manipulator or a closed-loop mechanism define the concept of mobility. The mobility of a parallel manipulator or a closed-loop mechanism arises from its geometry, which in turn is captured by the loop closure or the constraint equations. Values of joint variables which do not satisfy the constraint equations lead to configurations where the closed-loop mechanism or a parallel manipulator cannot be assembled. In other words, the range of passive and actuated joint variables which satisfy the constraint equations determine the mobility of the parallel manipulator or the closed-loop mechanisms.

In serial manipulators, the existence of real solutions of the inverse kinematics problem determined the workspace. In contrast, in parallel manipulators, the existence of solutions of the direct kinematics problem determines the mobility. As shown earlier in the chapter, the direct kinematics problem involves elimination of $m-1$ passive joint variables from the m constraint equation to obtain a polynomial in one of the passive joint variables. The mobility of a parallel manipulator can be obtained by evaluating conditions of the existence of real solution(s) of the passive joint variable from this polynomial. As shown in the examples earlier, the eliminant polynomials are extremely complex and analytical expressions for the existence of real solutions are not possible except for the simplest four-bar mechanism,[6] and the range of motion of a joint can be obtained only by numerical simulation (see also Chapter 5 for discussion on singularities of parallel manipulators). In the following example, we derive the conditions for the actuated joint of a four-bar mechanism to have complete rotatability—in kinematics literature this is known as a crank and the conditions derived below form a part of the very well-known Grashof's criterion [see, for example, Williams and Reinholtz (1986) and Uicker et al. (2003)].

Example 4.4 The planar four-bar mechanism

To obtain the mobility of the actuated joint, we start with Eqn (4.11). After simplification, Eqn (4.11) can be written as

$$P \cos \phi_1 + Q \sin \phi_1 + R = 0 \tag{4.46}$$

where P, Q, and R are given by

$$\begin{aligned} P &= 2l_0 l_3 - 2l_1 l_3 c_1 \\ Q &= -2l_1 l_3 s_1 \\ R &= l_0^2 + l_1^2 + l_3^2 - l_2^2 - 2l_0 l_1 c_1 \end{aligned} \tag{4.47}$$

where l_0, l_1, l_2, and l_3 are the link lengths as shown in Fig. 4.1, and c_1, s_1 are the sine and cosine of the actuated joint angle θ_1, respectively.

---

[6] If the polynomial obtained after elimination is more than a quartic, analytic expressions for the existence of real roots are, in general, not possible.

Using tangent half-angle substitutions [see Chapter 3, Eqns (3.14)] we can reduce Eqn (4.46) to a quadratic in $\tan\phi_1/2$, and this can be solved to obtain

$$\phi_1 = 2\tan^{-1}\left(\frac{-Q \pm \sqrt{P^2+Q^2-R^2}}{R-P}\right) \tag{4.48}$$

For ϕ_1 to have real solutions $P^2+Q^2-R^2 \geq 0$ and in the limiting case of two ϕ_1's coinciding, we have $P^2+Q^2-R^2 = 0$. For this limiting case, after simplification, we get

$$c_1 = \frac{l_0^2 + l_1^2 - l_3^2 - l_2^2 \pm 2l_3l_2}{2l_0l_1} \tag{4.49}$$

This equation gives the bounds on θ_1 for the limiting case.

If the actuated joint θ_1 is to have full rotatability, i.e., $0 \leq \theta_1 \leq 2\pi$, then there *cannot* be a solution to Eqn (4.49). In this case, for Eqn (4.49) to have no solution (and θ_1 to have no bounds) we must have $c_1 > 1$ or $c_1 < -1$.[7] The condition $c_1 > 1$ and $c_1 < -1$ leads to

$$(l_0 - l_1)^2 > (l_3 - l_2)^2 \tag{4.50}$$

and

$$(l_0 + l_1) < (l_3 + l_2) \tag{4.51}$$

It can be verified that the other two conditions resulting from $c_1 > 1, c_1 < -1$ lead to $l_3 + l_2 + l_1 < l_0$ and $l_0 + l_1 + l_2 < l_3$, respectively, and for these conditions the four-bar mechanism cannot be assembled.

Equation (4.50) gives rise to four inequalities:

$$\begin{aligned} l_0 - l_1 &> l_3 - l_2 \\ l_0 - l_1 &> l_2 - l_3 \\ l_1 - l_0 &> l_3 - l_2 \\ l_1 - l_0 &> l_2 - l_3 \end{aligned} \tag{4.52}$$

and for the case of $l_1 < l_0$, we get

$$\begin{aligned} l_0 + l_2 &> l_1 + l_3 \\ l_0 + l_3 &> l_1 + l_2 \end{aligned} \tag{4.53}$$

[7] Alternately, we can convert Eqn (4.49) to a quadratic polynomial using the tangent half-angle substitution for θ_1 and then impose the condition that the roots of the quadratic are imaginary. It may be noted that to obtain full mobility of θ_1, we first insisted that ϕ_1 be real and then θ_1 be imaginary.

Equations (4.51) and (4.53) imply that l_0, l_2, and l_3 are all larger than l_1, i.e., l_1 is the shortest link and either of l_0, l_2, and l_3 must be the largest link. With this reasoning, Eqns (4.51) and (4.53) can be concisely written as $s + l < p + q$, where s, l are the shortest and largest links, respectively, and p, q are two other intermediate links.

Likewise, for the case of $l_1 > l_0$, we get

$$\begin{aligned} l_1 + l_2 &> l_0 + l_3 \\ l_1 + l_3 &> l_0 + l_2 \end{aligned} \tag{4.54}$$

and again using Eqns (4.51) and (4.54), we can show that l_0 is the shortest link in this case and either of l_1, l_2, and l_3 must be largest link. In this case, also, we can concisely represent Eqns (4.51) and (4.54) as $s + l < p + q$, where s, l are the shortest and largest links, respectively, and p, q are two other intermediate links.

In summary, for the joint θ_1 to rotate fully, we must have the condition that the sum of the shortest and the largest link must be less than the sum of the two intermediate links. Readers familiar with four-bar kinematics will recognize this as Grashof's criterion for the input link to be a crank. By first obtaining the condition for real θ_1 as opposed to real ϕ_1 and obtaining conditions for no real solutions for ϕ_1, we can also obtain the condition for the output link to rotate fully.

Example 4.5 A three-DOF parallel manipulator

In the example of the planar four-bar mechanism, due to its simplicity, we could obtain analytical expressions for mobility of its actuated joint. In the case of the three-DOF parallel manipulator, it is not possible to obtain analytical expressions from the analysis of the direct kinematics equations, since the order of the polynomial is eight degree in x_3^2 as shown in Eqn (4.30). In this case we have to resort to numerical computations.

To obtain the mobility of the actuated joints, we use Eqn (4.30). The three-DOF parallel manipulator will satisfy the constraint equations if the roots of the eight-degree polynomial are real and positive—if x_3^2 is negative, the angle θ_3 will be imaginary. By numerical search, we can obtain all possible (l_1, l_2, l_3) such that Eqn (4.30) yields at least one real and positive root. In Fig. 4.7, we show some points in the (l_1, l_2, l_3) space for which the parallel manipulator cannot be assembled and points where it can be assembled. The range of $l_i (i = 1, 2, 3)$ is assumed to be $[0.5\ 1.5]$ and a is chosen to be 0.5. In Fig. 4.7, the points marked as '$*$' are values of l_1, l_2, l_3 for which there are no real and positive values of x_3^2 and the manipulator cannot be assembled. The points marked with '$\cdot$' have at least one real and positive value of x_3^2

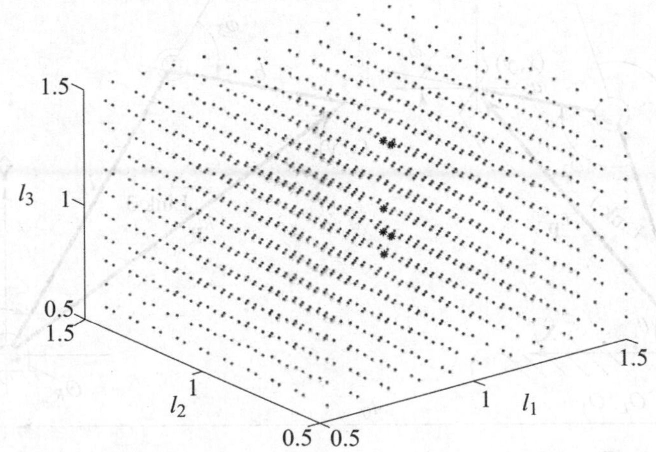

Fig. 4.7 Values of (l_1, l_2, l_3) for real θ_3 (marked by '·') and imaginary θ_3 (marked by '$*$')

and hence we can get real θ_3 for these l_i's. As with any numerical search, a more accurate picture of the mobility region will require finer search.

4.7 Inverse Kinematics of Parallel Manipulators

The inverse kinematics problem for parallel manipulators can be stated as follows: given the geometry, link parameters and the position and orientation of a chosen output link with respect to a fixed frame, find the passive and the actuated joint variables. The inverse kinematics problem for a parallel manipulator is simpler than the direct kinematics problem since the problem can be solved without worrying about the loops or the loop-closure constraint equations. The key idea is that we can 'break' the mechanism and obtain the joint angles by solving the inverse kinematics problems, in 'parallel', for the resulting serial manipulators. This concept is illustrated with the help of a simple planar four-bar mechanism discussed in Section 4.3.

Let us assume that the floating coupler is the chosen output link and we are given the position of a point ${}^L\mathbf{p}$ and the rotation matrix ${}^L_2[R]$ of this coupler link (see Fig. 4.8). Since the four-bar mechanism is planar, we are given x, y coordinates of a point on the coupler and the orientation of the coupler link denoted by angle ϕ. We assume that the lengths l_0, l_1, $l_2 = a + b$; a, b, and l_3 are known. Hence, we can write

$$
\begin{aligned}
x &= l_1 \cos\theta_1 + a\cos(\theta_1 + \phi_2) \\
y &= l_1 \sin\theta_1 + a\sin(\theta_1 + \phi_2)
\end{aligned}
\tag{4.55}
$$

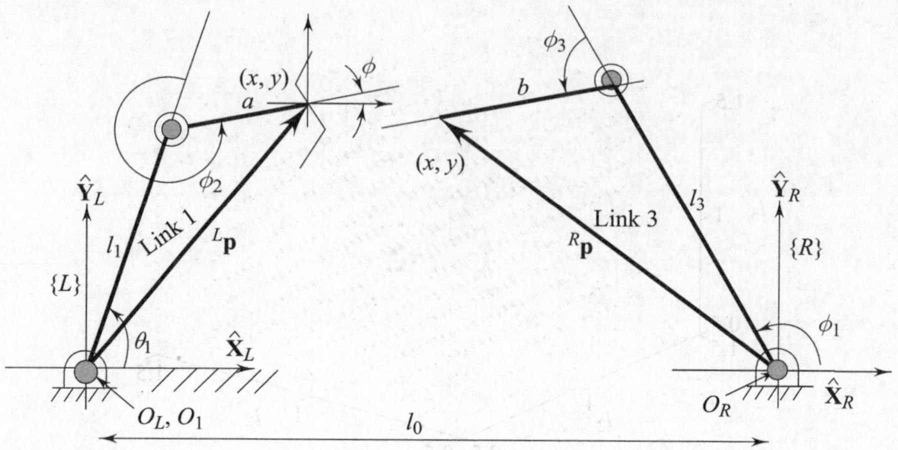

Fig. 4.8 Inverse kinematics of a four-bar mechanism

where x and y are given. In addition, the angle ϕ denoting the orientation of link 2 is given by

$$\phi = \theta_1 + \phi_2 - 2\pi \tag{4.56}$$

We can solve for θ_1 and ϕ_2 as

$$\begin{aligned} \theta_1 &= \text{Atan2}(y - a\sin\phi,\ x - a\cos\phi) \\ \phi_2 &= \phi - \theta_1 \end{aligned} \tag{4.57}$$

In a similar manner, considering the equations

$$\begin{aligned} x &= l_0 + l_3\cos\phi_1 + b\cos(\phi_1 + \phi_3) \\ y &= l_3\sin\phi_1 + b\sin(\phi_1 + \phi_3) \\ \phi &= \phi_1 + \phi_3 - \pi \end{aligned} \tag{4.58}$$

we can solve for ϕ_1 and ϕ_3.

It may be noted that ϕ obtained as $\theta_1 + \phi_2 - 2\pi$ and as $\phi_1 + \phi_3 - \pi$ must be same. In addition, one can intuitively see that the four-bar mechanism is a one-DOF mechanism and only one of (x, y, ϕ) can be independent—x and y are related through the sixth-degree polynomial given in Eqn (4.20) and ϕ must satisfy

$$x\cos\phi + y\sin\phi = (1/2a)(x^2 + y^2 - a^2 - l_1^2) \tag{4.59}$$

The constraints on the given position and orientation of the chosen output link, x, y, ϕ, are analogous to the case of the inverse kinematics of serial manipulators when $n < 6$ (see Section 3.7). The inverse kinematics of the four-bar mechanism can only be solved when the given position and orientation is consistent.

The concept of 'breaking' leads to serial manipulators and the most complex case is that of a 6R serial manipulator. Hence, the complexity of the inverse kinematics problem in parallel manipulators is less than or equal to the complexity of the inverse kinematics problem of a general 6R manipulator discussed in Chapter 3. It may be noted that if the process of 'breaking' results in one or more redundant serial manipulators, and it is known that the parallel manipulator has dof ≤ 6, then we must try a different approach of 'breaking' so as to get unique joint values. Now we will consider some examples using inverse kinematics.

Example 4.6 A six-DOF hybrid manipulator

For the inverse kinematic problem of the six-DOF hybrid manipulator shown in Fig. 2.23, we are given the position and orientation of the gripped object with respect to $\{Base\}$. Our task is to obtain the rotations at the nine joints in the three 'fingers'. As discussed in the example of the four-bar mechanism, we can solve the inverse kinematics problem in 'parallel'.

Figure 4.9 shows one 'finger' of the six-DOF parallel manipulator and the gripped object. We assume that the vector ${}^{Base}\mathbf{p}$ denoting a point on the gripped object (for example, the centroid) and the orientation of the gripped object, denoted by the rotation matrix ${}^{Base}_{Object}[R]$, are available. In the $\{Object\}$ coordinate system, the coordinate of the centre of the spherical joint, point S_1, is a known constant. Denoting this by ${}^{Object}\mathbf{S}_1$, we can get

$$ {}^{Base}\mathbf{S}_1 = {}^{Base}_{Object}[R]\,{}^{Object}\mathbf{S}_1 + {}^{Base}\mathbf{p}_{Object} \tag{4.60} $$

From Eqn (2.66), we can write

$$ (x, y, z)^T = \begin{pmatrix} \cos\theta_1[l_{11} + l_{12}\cos\psi_1 + l_{13}\cos(\psi_1 + \phi_1)] \\ -d + \sin\theta_1[l_{11} + l_{12}\cos\psi_1 + l_{13}\cos(\psi_1 + \phi_1)] \\ h + l_{12}\sin\psi_1 + l_{13}\sin(\psi_1 + \phi_1) \end{pmatrix} \tag{4.61} $$

where $(x, y, z)^T$ is the right-hand side of Eqn (4.60).

Equation (4.61) is similar to an RRR manipulator and the joint variables θ_1, ψ_1, and ϕ_1 can be obtained by elimination as discussed in Chapter 3. From the equations in Eqn (4.61), we can get

$$ x^2 + (y + d)^2 + (z - h)^2 = l_{11}^2 + l_{12}^2 + l_{13}^2 + 2l_{11}l_{12}\cos\psi_1 + 2l_{12}l_{13}\cos\phi_1 + 2l_{11}l_{13}\cos(\psi_1 + \phi_1) \tag{4.62} $$

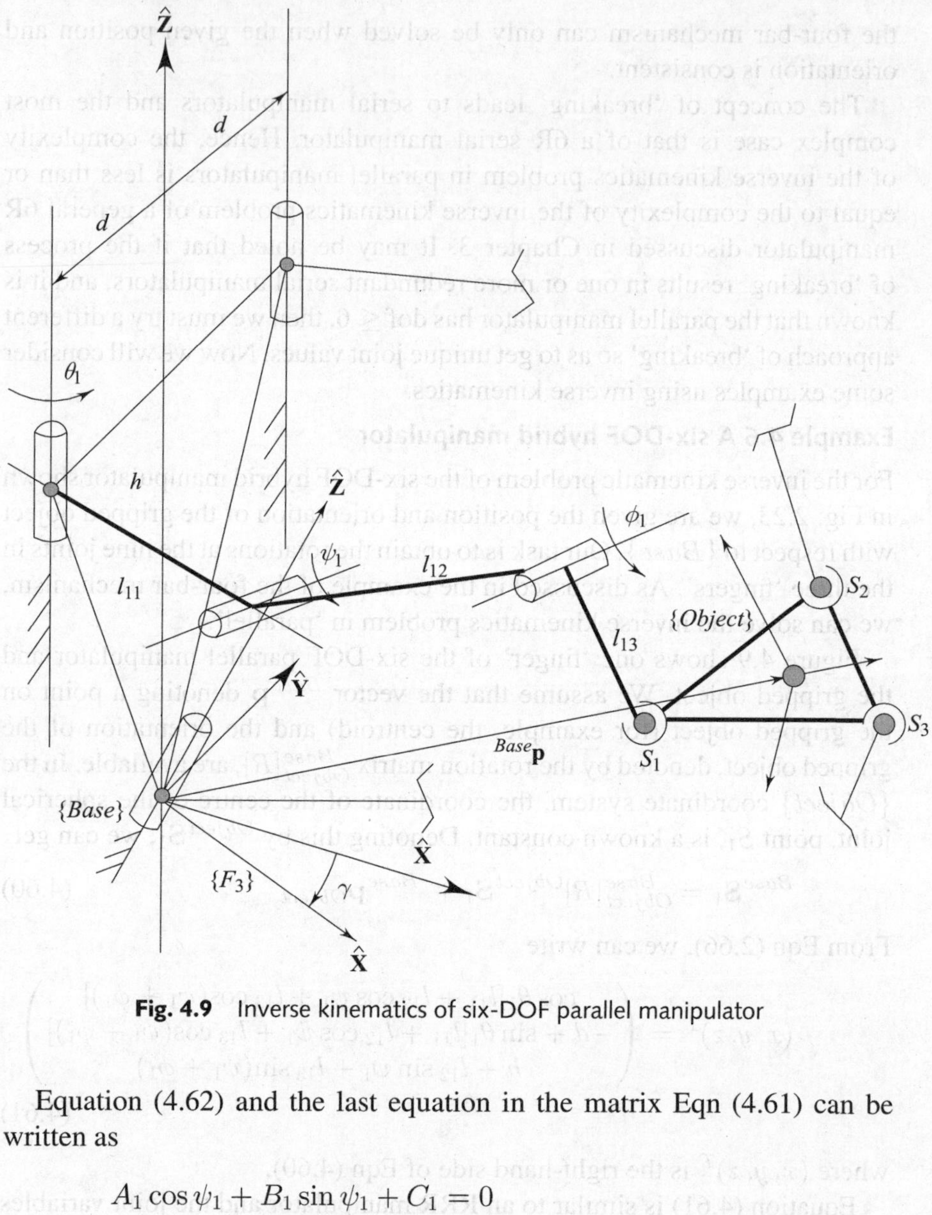

Fig. 4.9 Inverse kinematics of six-DOF parallel manipulator

Equation (4.62) and the last equation in the matrix Eqn (4.61) can be written as

$$
\begin{aligned}
&A_1 \cos\psi_1 + B_1 \sin\psi_1 + C_1 = 0 \\
&A_2 \cos\psi_1 + B_2 \sin\psi_1 + C_2 = 0 \qquad (4.63)
\end{aligned}
$$

where

$$
\begin{aligned}
A_1 &= 2l_{11}l_{12} + 2l_{11}l_{13}\cos\phi_1, \quad A_2 = l_{13}\sin\phi_1 \\
C_1 &= l_{11}^2 + l_{12}^2 + l_{13}^2 + 2l_{12}l_{13}\cos\phi_1 - x^2 - (y+d)^2 - (h-z)^2 \\
B_1 &= -2l_{11}l_{13}\sin\phi_1, B_2 = l_{12} + l_{13}\cos\phi_1, \quad C_2 = h - z \qquad (4.64)
\end{aligned}
$$

Following Sylvester's dialytic method, we can obtain the eliminant as

$$4l_{11}^2(l_{12}^2 + l_{13}^2 + 2l_{12}l_{13}\cos\phi_1) = C_1^2 + 4l_{11}^2(h - z)^2 \tag{4.65}$$

and on using the tangent half-angle formulas for $\cos\phi_1$ and $\sin\phi_1$, we get a quartic equation in

$$a_4x^4 + a_3x^3 + a_2x^2 + a_1x + a_0 = 0 \tag{4.66}$$

where $x = \tan(\phi_1/2)$ (see Exercise 4.13).

The elimination procedure also gives

$$\psi_1 = -2\tan^{-1}\left(\frac{A_1C_2 - A_2C_1}{(B_1C_2 - B_2C_1) + (A_1B_2 - A_2B_1)}\right) \tag{4.67}$$

and

$$\theta_1 = \text{Atan2}(y + d, x) \tag{4.68}$$

The joint variables θ_2, ψ_2, and ϕ_2 for the second 'finger' and the joint variables θ_3, ψ_3, and ϕ_3 for the third 'finger' can be obtained in parallel by obtaining ${}^{Base}\mathbf{S}_2$ and ${}^{Base}\mathbf{S}_3$ in a manner similar to Eqns (4.60) and (4.61). It may be noted that the passive variables ϕ_1, ϕ_2, ϕ_3 must also satisfy the constraint equations given by Eqn (4.37).

Example 4.7 The Stewart platform

The inverse kinematics of the six-DOF Stewart–Gough platform can also be solved by recognizing that we can again find the coordinates of the spherical joints with respect to the origin of $\{B_0\}$. As shown in Eqn (4.38), from a given orientation of the top platform, ${}^{B_0}_{P_0}[R]$, the constant vector locating the centre of the spherical joint P_1 with respect to the top platform ${}^{P_0}\mathbf{p}_1$ and the translation vector ${}^{B_0}\mathbf{t}$, we can obtain ${}^{B_0}\mathbf{p}_1$. Denoting the components of ${}^{B_0}\mathbf{p}_1$ by $(x, y, z)^T$, and by using Eqn (4.43), we can write

$$[R(\widehat{\mathbf{Z}}, \gamma_i)]^T[(x, y, z)^T - {}^{B_0}\mathbf{b}_1] = [R(\widehat{\mathbf{Y}}, \phi_i)][R(\widehat{\mathbf{X}}, \psi_i)](0, 0, l_i)^T$$

$$= l_1\begin{pmatrix}\sin\phi_1\cos\psi_1 \\ -\sin\psi_1 \\ \cos\phi_1\cos\psi_1\end{pmatrix} \tag{4.69}$$

Equation (4.69) is a set of three non-linear equations in three unknowns, l_1, ψ_1, ϕ_1, and in terms of knowns x, y, z, and γ_1. The unknowns can be solved as follows.

Squaring and adding the three equations in Eqn (4.69), we get

$$l_1 = \pm\sqrt{[(x, y, z)^T - {}^{B_0}\mathbf{b}_1]^2} \tag{4.70}$$

and to obtain ϕ_1 and ψ_1, we use the Atan2 function to get

$$\psi_1 = \text{Atan2}(-Y, \pm\sqrt{X^2+Z^2})$$
$$\phi_1 = \text{Atan2}(X/\cos\psi_1, Z/\cos\psi_1) \tag{4.71}$$

where X, Y, Z are the components of the vector $[R(\widehat{\mathbf{Z}},\gamma_i)]^T[(x,y,z)^T - {}^{B_0}\mathbf{b}_1]$ in the left-hand side of Eqn (4.69). It may be noted that the above procedure can be used for each of the legs, in 'parallel', and we can find the joint variables l_i, ψ_i, and ϕ_i for $i = 1, \ldots, 6$.

Exercises

4.1 From the Grashof criterion for a four-bar mechanism, discussed in Example 4.4 obtain link lengths such that the four-bar mechanism is a double-crank mechanism. Plot ϕ_1, ϕ_2, and ϕ_3 versus θ_1, and plot the coupler curve for a chosen double-crank configuration.

4.2 Figure 4.10 shows a two-DOF planar five-bar mechanism. Assume that θ_1 and θ_2 are actuated variables. Determine expressions for the passive variables ϕ_1, ϕ_2, and ϕ_3 in terms of the link lengths as shown in Fig. 4.10.

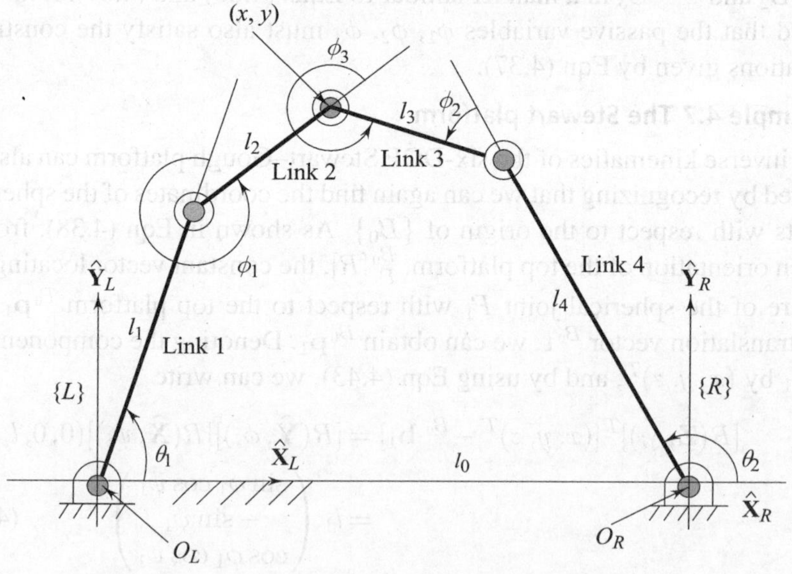

Fig. 4.10 The planar five-bar mechanism

4.3 For the five-bar mechanism, shown in Fig. 4.10, determine the conditions on the link lengths such that both the actuated links are cranks.

†4.4 For a set of chosen link lengths such that both the actuated links are cranks, plot the locus of point (x, y) shown in Fig. 4.10.

†4.5 Use MAPLE to derive and verify the constraint equations for the three-DOF parallel manipulator, Eqn (4.25), discussed in Example 4.2.

†4.6 Use MAPLE to derive and verify the eighth-degree polynomial obtained for the three-DOF parallel manipulator discussed in Example 4.2.

4.7 Figure 4.11 shows a spatial parallel manipulator. The joints at the connection points at the base and the top platform are Hooke (U) joints, and each leg has a 'U-P-U' configuration. Verify that the parallel manipulator has three degrees of freedom. Formulate and solve the direct kinematics problem for the U-P-U parallel manipulator shown in Fig. 4.11.

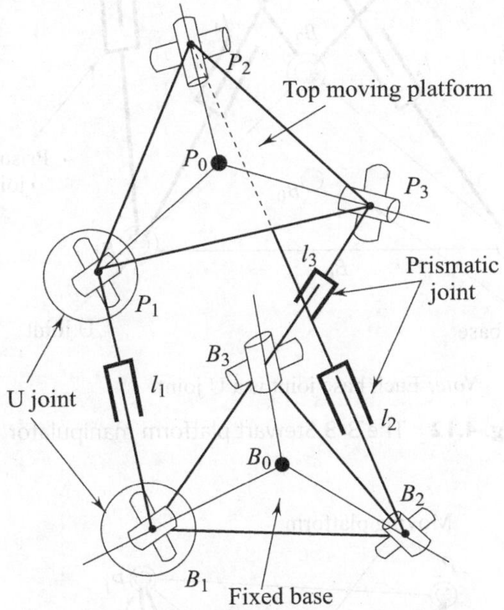

Fig. 4.11 The U-P-U parallel manipulator

4.8 Figure 4.12 shows a simplified Stewart platform configuration. This is known as the 3-3 Stewart platform configuration since there are only three connection points in the fixed base and the top platform. Each leg has a U-P-S configuration. Formulate and solve the direct kinematics problem for this manipulator.

4.9 Figure 4.13 shows another simplified Stewart platform configuration. This is known as the 6-3 Stewart platform configuration since there are six connection points in the fixed base and three on the top platform. Each leg has a U-P-S configuration. Formulate and solve the direct kinematics problem for this manipulator.

4.10 Figure 4.14 shows two planar 2R manipulators handling an object. We assume that the point of contact between the two manipulators and the object can be modelled as a rotary (R) joint. What is the degree of freedom of the resulting parallel manipulator? Formulate and solve the direct kinematics problem for this equivalent parallel manipulator.

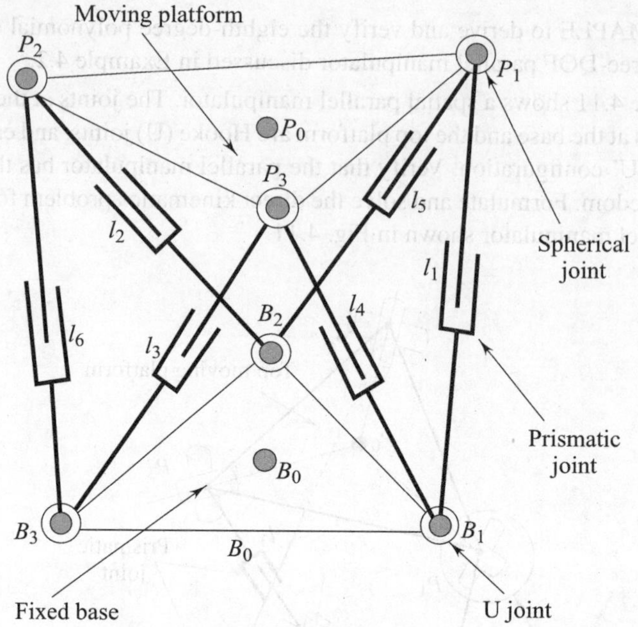

Note: Each base joint is a U joint

Fig. 4.12 The 3-3 Stewart platform manipulator

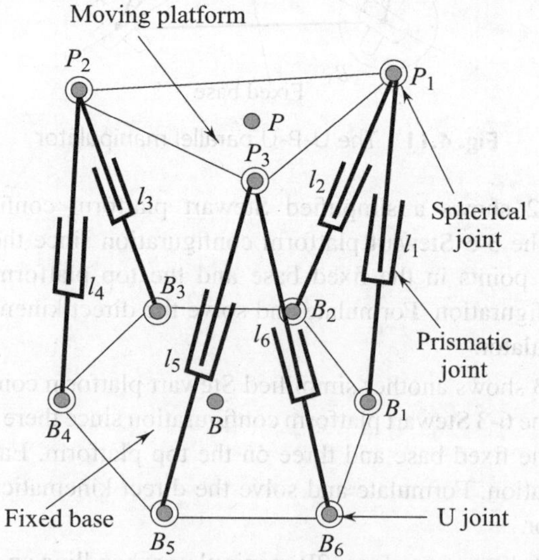

Note: Each base joint is an U joint

Fig. 4.13 The 6-3 Stewart platform manipulator

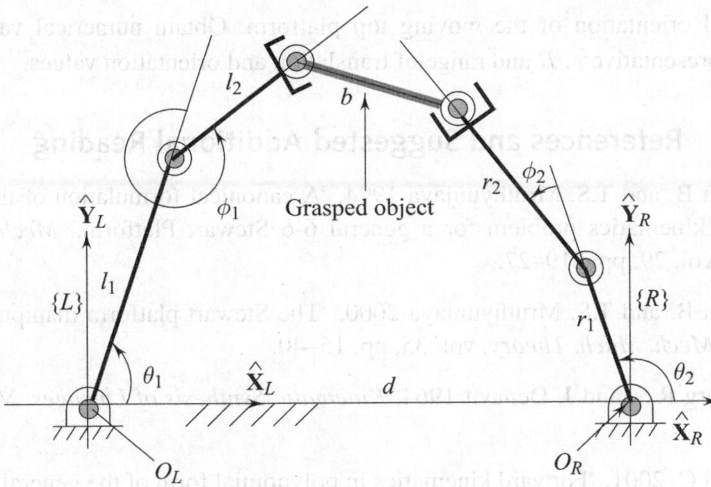

Fig. 4.14 Two 2R manipulators handling an object

†4.11 For a six-DOF parallel, as discussed in Example 4.3, assume $2d = 1$, $h = \sqrt{3}/2$, and $k_{12} = k_{23} = k_{31} = \sqrt{3}k$. Further assume that all the 'fingers' are identical and $l_{11} = 1$, $l_{12} = 1/2$, and $l_{13} = 1/4$. Write a MATLAB program to vary the six actuated joints, in convenient steps, and collect the coordinates (x, y, z) of the centroid of the gripped object for which the direct kinematics problem can be solved. Numerically obtain and plot the cloud of points (x, y, z) for $k = 1$.

†4.12 Intuitively, as the size of the 'gripped' object (or k) becomes large or very small with other dimensions remaining the same, the workspace of the gripped object or the possible range of (x, y, z) is expected to be small. Perform the numerical analysis in Exercise 4.11 for several values of k greater and less than 1. Discuss what happens to the 'volume' of the cloud of points as k increases or decreases from 1.

†4.13 Derive the coefficients of the quartic equation discussed in Example 4.6 for the six-DOF parallel manipulator using MAPLE.

†4.14 While designing a Stewart platform one of the important tasks is choosing the spherical joint. From a purely kinematic perspective, to choose a spherical joint, we need to know the possible range of rotations at the spherical joint. Assume each leg of a Stewart platform can be modelled as an R-R-R-P-R-R serial manipulator where the first three R joints intersect and model an S joint. Derive analytical expressions to obtain the rotations at a spherical joint given a translation vector and the orientation of the top platform. Assume that the moving top and fixed base platforms are regular hexagons circumscribed by circles of radii r and R, respectively.

†4.15 Write a MATLAB program to obtain the range of rotation at the spherical joint for given numerical values of r and R and a range for the translation

and orientation of the moving top platform. Obtain numerical values for representative r, R and range of translation and orientation values.

References and Suggested Additional Reading

Dasgupta B. and T.S. Mruthyunjaya 1994, 'A canonical formulation of the direct position kinematics problem for a general 6-6 Stewart Platform,' *Mech. Mach. Theory*, vol. 29, pp. 819–27.

Dasgupta B. and T.S. Mruthyunjaya 2000, 'The Stewart platform manipulator: A review,' *Mech. Mach. Theory*, vol. 35, pp. 15–40.

Hartenberg R.S. and J. Denavit 1964, *Kinematic Synthesis of Linkages*, McGraw-Hill.

Innocenti C. 2001, 'Forward kinematics in polynomial form of the general Stewart Platform,' *Trans. AMSE J. Mech. Des.*, vol. 123, pp. 254–60.

Merlet J.-P. 2001, *Parallel Robots*, Kluwer Academic, Dordrecht.

Nanua P., K.J. Waldron, and V. Murthy 1990, 'Direct kinematics solution of a Stewart Platform,' *IEEE Trans. Robotic. Autom.*, vol. 6, no. 4, pp. 438–43.

Raghavan M. 1993, 'The Stewart Platform of general geometry has 40 configurations,' *Trans. ASME J. Mech. Des.*, vol. 115, pp. 277–82.

Uicker J.J., G.R. Pennock, and J.E. Shigley 2003, *Theory of Machines and Mechanisms*, 3rd edn, Oxford University Press.

Williams, R.L. and C.F. Reinholtz 1986, 'Proof of Grashof's law using polynomial discriminants,' *Trans. ASME J. Mech. Transm. Autom. Des.*, vol. 108, pp. 562–64.

5

Velocity Analysis and Statics of Manipulators

5.1 Introduction

In Chapters 3 and 4, we dealt with the positional and orientational aspects of the links of serial and parallel manipulators. In this chapter, we discuss the change of position and orientation, with respect to time, of the links of a serial and parallel manipulator. The linear velocity of a point on a rigid body is defined as the time derivative of the position vector of that point. The angular velocity of a rigid body is defined in terms of the time derivative of the rotation matrix describing the orientation of the rigid body. We show that there are two possible (related) angular velocities. In serial manipulators, with single-DOF rotary and prismatic joints, we develop an iterative formula to obtain the linear and angular velocities of the links. We then present the important concept of a manipulator Jacobian. The Jacobian matrix relates the Cartesian linear and angular velocities of the end-effector of a manipulator to the joint rates. The singularities of the workspace of a serial manipulator are related to the vanishing of the determinant of the Jacobian matrix. At a singularity, a serial manipulator can lose one or more degrees of freedom. In parallel manipulators and closed-loop mechanisms, the time derivatives of the loop-closure constraint equations are used to define an equivalent Jacobian. It is shown that a parallel manipulator or closed-loop mechanism has different kinds of singularities where it can lose or gain one or more degrees of freedom. Finally, in this chapter, we consider the static equilibrium analysis of a serial or parallel manipulator as a structure with its joint locked. It is shown that the manipulator Jacobian relates the external (Cartesian) forces and moments acting on the end-effector to the joint torques and forces, and the loss and gain of degrees of freedom is related to the ability of a manipulator to withstand external forces and moments.

5.2 Linear and Angular Velocities of a Rigid Body

Consider a coordinate system $\{i\}$ attached to a rigid body at a point O_i as shown in Fig. 5.1. The point O_i can be described with respect to another fixed coordinate system $\{0\}$ by the position vector ${}^0\mathbf{O}_i$. The linear velocity of O_i with respect to $\{0\}$ is defined as

$$ {}^0\mathbf{V}_{O_i} \triangleq \frac{d}{dt}\,{}^0\mathbf{O}_i(t) = \lim_{\Delta t \to 0} \frac{{}^0\mathbf{O}_i(t+\Delta t) - {}^0\mathbf{O}_i(t)}{\Delta t} \tag{5.1} $$

where ${}^0\mathbf{O}_i(t)$ and ${}^0\mathbf{O}_i(t+\Delta t)$ are position vectors of O_i at two instants of time.

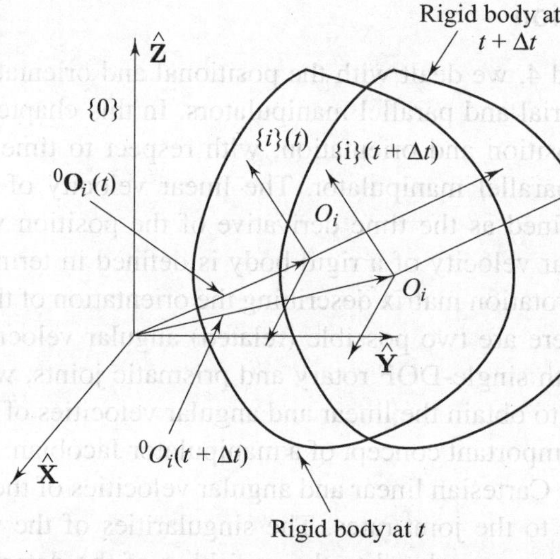

Fig. 5.1 Linear velocity of a rigid body

The linear velocity vector, as defined above, clearly depends on the coordinate system in which the two position vectors at t and $t + \Delta t$ are described—it is possible that the vector $\mathbf{O}_i$ does not change with t in a coordinate system and the limit is zero. In another coordinate system, the limit is nonzero. The linear velocity vector, like any other vector, can always be described with respect to a different coordinate system $\{j\}$ once it is available in a coordinate system. Using the rotation matrix ${}^j_0[R]$, we can write

$$ {}^j\left({}^0\mathbf{V}_{O_i}\right) = {}^j_0[R]\,{}^0\mathbf{V}_{O_i} \tag{5.2} $$

Hence, for the linear velocity vector, two coordinate systems are involved, namely, the coordinate system in which the subtraction and the limit is taken (or the differentiation is done) and the coordinate system in which it is described. If not explicitly mentioned, the differentiation will be done in a fixed reference system (such as $\{0\}$) and we will use the expressions in the form given in the right-hand side of Eqn (5.2) when we wish to express the velocity vector in a different coordinate system. Unless explicitly stated, the leading superscript in the velocity vector will always denote the coordinate system in which the differentiation is performed. If the leading superscript is dropped, then the coordinate system is some understood fixed reference coordinate system.

Unlike the linear velocity, the angular velocity of a rigid body cannot be obtained as a time derivative of the three independent quantities (such as Euler angles) representing the orientation of the rigid body. To derive the angular velocity of a rigid body from the rotation matrix ${}^0_i[R]$, we recall some of the properties of the rotation matrix from Chapter 2. For the rotation matrix ${}^0_i[R]$,[1] we can write

$${}^0_i[R]\,{}^0_i[R]^T = [U]$$

where $[U]$ is a 3×3 identity matrix.

The above equation is valid for all time t, and by differentiating with respect to t, we get

$${}^0_i[\dot{R}]\,{}^0_i[R]^T + {}^0_i[R]\,{}^0_i[\dot{R}]^T = [0]$$

where the derivative of the matrix implies the derivative of all the components of the matrix. The above equation can also be written as

$${}^0_i[\dot{R}]\,{}^0_i[R]^T + ({}^0_i[\dot{R}]\,{}^0_i[R]^T)^T = [0]$$

which implies that

$${}^0_i[\Omega]_R \triangleq {}^0_i[\dot{R}]\,{}^0_i[R]^T \tag{5.3}$$

is a skew-symmetric matrix, and we can write

$${}^0_i[\Omega]_R = \begin{pmatrix} 0 & -\omega_z^s & \omega_y^s \\ \omega_z^s & 0 & -\omega_x^s \\ -\omega_y^s & \omega_x^s & 0 \end{pmatrix} \tag{5.4}$$

[1] It may be recalled that the inverse of a rotation matrix is the same as its transpose.

The product of any skew-symmetric matrix with a vector $(p_x, p_y, p_z)^T$ in $\Re^3$ can be thought of as a cross product, and we can write

$$ {}^0_i[\Omega]_R(p_x, p_y, p_z)^T = \begin{pmatrix} \omega^s_y p_z - \omega^s_z p_y \\ \omega^s_z p_x - \omega^s_x p_z \\ \omega^s_x p_y - \omega^s_y p_x \end{pmatrix} = {}^0\boldsymbol{\omega}^s_i \times {}^0\mathbf{p} \tag{5.5} $$

The skew-symmetric matrix ${}^0_i[\Omega]_R$ is called the angular velocity matrix and the vector ${}^0\boldsymbol{\omega}^s_i$ is called the angular velocity vector of a rigid body $\{i\}$ with respect to $\{0\}$. As can be seen the angular velocity vector is not a straightforward differentiation as the linear velocity vector. This is more easily seen from the following example.

As shown in Chapter 2, the rotation matrix in terms of Z-Y-Z Euler angles is given as

$$ \begin{aligned} {}^A_B[R] &= \begin{pmatrix} c_\alpha & -s_\alpha & 0 \\ s_\alpha & c_\alpha & 0 \\ 0 & 0 & 1 \end{pmatrix} \begin{pmatrix} c_\beta & 0 & s_\beta \\ 0 & 1 & 0 \\ -s_\beta & 0 & c_\beta \end{pmatrix} \begin{pmatrix} c_\gamma & -s_\gamma & 0 \\ s_\gamma & c_\gamma & 0 \\ 0 & 0 & 1 \end{pmatrix} \\ &= \begin{pmatrix} c_\alpha c_\beta c_\gamma - s_\alpha s_\gamma & -c_\alpha c_\beta s_\gamma - s_\alpha c_\gamma & c_\alpha s_\beta \\ s_\alpha c_\beta c_\gamma + c_\alpha s_\gamma & -s_\alpha c_\beta s_\gamma + c_\alpha c_\gamma & s_\alpha s_\beta \\ -s_\beta c_\gamma & s_\beta s_\gamma & c_\beta \end{pmatrix} \end{aligned} \tag{5.6} $$

Using Eqns (5.3) and (5.4), we can write the three components of the angular velocity vector in terms of the three Euler angles and their derivatives as

$$ \begin{aligned} \omega^s_x &= \dot{\gamma}\cos\alpha\sin\beta - \dot{\beta}\sin\alpha \\ \omega^s_y &= \dot{\gamma}\sin\alpha\sin\beta + \dot{\beta}\cos\alpha \\ \omega^s_z &= \dot{\gamma}\cos\beta + \dot{\alpha} \end{aligned} \tag{5.7} $$

The angular velocity matrix obtained from ${}^0_i[\dot{R}]\,{}^0_i[R]^T$ is called the right invariant since we started with the identity ${}^0_i[R]\,{}^0_i[R]^T = [U]$. The angular velocity vector obtained this way is also called a space-fixed angular velocity and is indicated by the superscript s.

We can also start with the identity ${}^0_i[R]^T\,{}^0_i[R] = [U]$. Proceeding exactly as above, we will get a skew-symmetric matrix:

$$ {}^0_i[\Omega]_L \triangleq {}^0_i[R]^T\,{}^0_i[\dot{R}] \tag{5.8} $$

which can be written as

$$ {}^0_i[\Omega]_L = \begin{pmatrix} 0 & -\omega^b_z & \omega^b_y \\ \omega^b_z & 0 & -\omega^b_x \\ -\omega^b_y & \omega^b_x & 0 \end{pmatrix} \tag{5.9} $$

and we can define an angular velocity vector ${}^0\boldsymbol{\omega}_i^b$ from the three components $(\omega_x^b, \omega_y^b, \omega_z^b)$. For the Z-Y-Z rotation matrix given in Eqn (5.6), the three components are

$$\begin{aligned}
\omega_x^b &= -\dot{\alpha}\cos\gamma\sin\beta + \dot{\beta}\sin\gamma \\
\omega_y^b &= \dot{\alpha}\sin\beta\sin\gamma + \dot{\beta}\cos\gamma \\
\omega_z^b &= \dot{\alpha}\cos\beta + \dot{\gamma}
\end{aligned} \tag{5.10}$$

The matrix ${}^0_i[\Omega]_L$ is called the left-invariant angular velocity matrix. The vector ${}^0\boldsymbol{\omega}_i^b$ is called the body-fixed angular velocity vector of a rigid body $\{i\}$ with respect to $\{0\}$ and is indicated by the superscript b. The two skew-symmetric matrices are related, like any tensor, as

$${}^0_i[\Omega]_R = {}^0_i[R]\,{}^0_i[\Omega]_L {}^0_i[R]^T \tag{5.11}$$

and the two angular velocities are related as

$${}^0\boldsymbol{\omega}_i^s = {}^0_i[R]{}^0\boldsymbol{\omega}_i^b \tag{5.12}$$

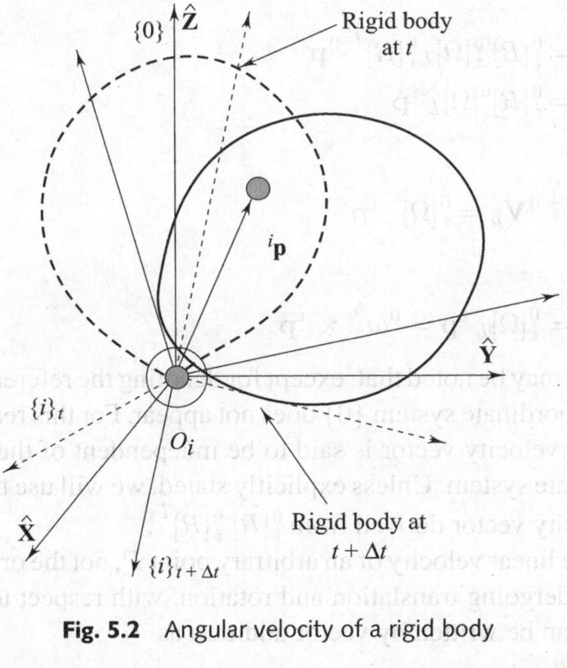

Fig. 5.2 Angular velocity of a rigid body

To get more insight into the two forms of the angular velocity matrices and the vectors, we consider a pure rotation of a rigid body as shown in Fig. 5.2—the vectors ${}^0\mathbf{O}_i(t)$ and ${}^0\mathbf{O}_i(t + \Delta t)$ are coincident and only the elements of the rotation matrix ${}^i_0[R]$ change with time. Consider a point P in the rigid body located by the vector ${}^i\mathbf{p}$ and fixed in $\{i\}$. We can write

$$ {}^0\mathbf{p} = {}^0_i[R]{}^i\mathbf{p} $$

and since the point P is fixed in $\{i\}$, we get

$$ {}^0\dot{\mathbf{p}} \triangleq {}^0\mathbf{V}_p = {}^0_i[\dot{R}]{}^i\mathbf{p} $$

and noting that ${}^0_i[R]^{-1} = {}^0_i[R]^T$, we can write

$$ \begin{aligned} {}^0\mathbf{V}_p &= {}^0_i[\dot{R}]\,{}^0_i[R]^T\,{}^0\mathbf{p} \\ &= {}^0_i[\Omega]_R\,{}^0\mathbf{p} \\ &= {}^0\boldsymbol{\omega}^s_i \times {}^0\mathbf{p} \end{aligned} \tag{5.13} $$

It may be noted that the coordinate system $\{i\}$ does not appear (except in denoting that we are computing the angular velocity vector of rigid body $\{i\}$) and for this reason the space-fixed angular velocity vector is said to be independent of the choice of the body coordinate system. We can also substitute the expression for ${}^0_i[\Omega]_R$ from Eqn (5.11) and rewrite Eqn (5.13) as

$$ \begin{aligned} {}^0\mathbf{V}_p &= {}^0_i[R]\,{}^0_i[\Omega]_L{}^0_i[R]^T\,{}^0\mathbf{p} \\ &= {}^0_i[R]{}^0_i[\Omega]_L\,{}^i\mathbf{p} \end{aligned} $$

and get

$$ {}^0_i[R]^{-1}\,{}^0\mathbf{V}_p = {}^0_i[\Omega]_L\,{}^i\mathbf{p} $$

which yields

$$ {}^i\mathbf{V}_p = {}^0_i[\Omega]_L\,{}^i\mathbf{p} = {}^0\boldsymbol{\omega}^b_i \times {}^i\mathbf{p} \tag{5.14} $$

In this case, it may be noted that, except for denoting the reference coordinate system, the coordinate system $\{0\}$ does not appear. For this reason the body-fixed angular velocity vector is said to be independent of the choice of the fixed coordinate system. Unless explicitly stated, we will use the space-fixed angular velocity vector derived from ${}^0_i[\dot{R}]\,{}^0_i[R]^T$.

Finally, the linear velocity of an arbitrary point P, not the origin, on a rigid body $\{i\}$, undergoing translation and rotation, with respect to a coordinate system $\{0\}$ can be written by vector addition as

$$ {}^0\mathbf{V}_p = {}^0\mathbf{V}_{O_i} + {}^0\boldsymbol{\omega}_i \times {}^0_i[R]{}^i\mathbf{p} + {}^0_i[R]{}^i\mathbf{V}_p \tag{5.15} $$

where the last term appears if the point P is not fixed in $\{i\}$ and the vector ${}^0\boldsymbol{\omega}_i$ is the space-fixed angular velocity vector.

5.3 Linear and Angular Velocities of Links in Serial Manipulators

The linear and angular velocity vectors follow the rules of vector algebra and, together with Eqn (5.15), we can find the linear and angular velocities of links of a manipulator. For serial manipulators with rotary or prismatic joints, it is particularly easy to obtain the linear and angular velocities of link $\{i\}$ in terms of the linear and angular velocities of a link $\{i-1\}$ and the type of joint connecting the two links. To start with, we assume that the two links are connected by a rotary joint as shown in Fig. 2.9. We start with Eqn (2.39) and taking the derivative with respect to time, we can write

$$ {}_i^0[\dot{R}]\, {}_i^0[R]^T = \frac{d}{dt}\left\{ {}_{i-1}^{0}[R]\ {}_{i}^{i-1}[R(\widehat{\mathbf{k}},\theta_i)]\right\}\left\{ {}_{i}^{i-1}[R(\widehat{\mathbf{k}},\theta_i)]^T\ {}_{i-1}^{0}[R]^T\right\} \tag{5.16}$$

Using the definition of the right-invariant angular velocity matrix [Eqn (5.3)], we can write

$$ {}_i^0[\Omega]_R = {}_{i-1}^{0}[\Omega]_R + {}_{i-1}^{0}[R]\left\{ {}_{i}^{i-1}[\dot{R}(\widehat{\mathbf{k}},\theta_i)]\, {}_{i}^{i-1}[R(\widehat{\mathbf{k}},\theta_i)]^T\right\} {}_{i-1}^{0}[R]^T \tag{5.17}$$

To simplify the term inside the brackets {} in the above equation, we make use of the result (see Exercise 2.6)

$$ {}_{i}^{i-1}[R(\widehat{\mathbf{k}},\theta_i)] = e^{({}_{i}^{i-1}[\mathcal{K}]\theta_i)}$$

where the matrix ${}_{i}^{i-1}[\mathcal{K}]$ is the skew-symmetric form of the rotation axis vector $\widehat{\mathbf{k}}$ and θ_i is the rotation at the rotary joint. It may be noted that the rotation axis is fixed in $\{i-1\}$ and $\{i\}$ and we can write

$$ \frac{d}{dt} e^{({}_{i}^{i-1}[\mathcal{K}]\theta_i)} = {}_{i}^{i-1}[\mathcal{K}]\dot{\theta}_i\, e^{({}_{i}^{i-1}[\mathcal{K}]\theta_i)} \tag{5.18}$$

Using the result from Eqn (5.18) in Eqn (5.17) and the properties of a rotation matrix, we can get

$$\begin{aligned} {}_i^0[\Omega]_R &= {}_{i-1}^{0}[\Omega]_R + {}_{i-1}^{0}[R]\ {}_{i}^{i-1}[\mathcal{K}]\ {}_{i-1}^{0}[R]^T\dot{\theta}_i \\ &= {}_{i-1}^{0}[\Omega]_R + {}_{i}^{0}[\mathcal{K}]\dot{\theta}_i \end{aligned} \tag{5.19}$$

Equation (5.19) can be re-written in vector form as

$$ {}^0\boldsymbol{\omega}_i = {}^0\boldsymbol{\omega}_{i-1} + {}^0\,\widehat{\mathbf{k}}_i\dot{\theta}_i \tag{5.20}$$

where ${}^0\boldsymbol{\omega}_{(\cdot)}$ is the space-fixed angular velocity vector.

In serial manipulators, the joint axis is always chosen to be the Z axis. We can pre-multiply both sides of Eqn (5.20) by ${}_0^i[R]$ and after simplification get

$$ {}^i\boldsymbol{\omega}_i = {}_{i-1}^{i}[R]^{i-1}\boldsymbol{\omega}_{i-1} + \dot{\theta}_i(0\ 0\ 1)^T \tag{5.21}$$

where ${}^i\boldsymbol{\omega}_i$ denotes ${}_0^i[R]^0\boldsymbol{\omega}_i$ and ${}^{i-1}\boldsymbol{\omega}_{i-1}$ denotes ${}_0^{i-1}[R]^0\boldsymbol{\omega}_{i-1}$.

To obtain the linear velocity of the origin of $\{i\}$ in terms of linear and angular velocities of $\{i-1\}$, we start from Eqn (2.40) which, for two consecutive links a serial manipulator, can be written as

$$ {}^0\mathbf{O}_i = {}^0\mathbf{O}_{i-1} + {}^{0}_{i-1}[R]^{i-1}\mathbf{O}_i \tag{5.22} $$

since the origins O_i, O_{i-1} are on the joint axis i and $i-1$, respectively (see also Fig. 2.9). Taking the derivative of both sides of Eqn (5.22) with respect to time t and noting that ${}^{i-1}\mathbf{O}_i$ is constant in $\{i-1\}$, we can write

$$ {}^0\mathbf{V}_{O_i} = {}^0\mathbf{V}_{O_{i-1}} + {}^0\boldsymbol{\omega}_{i-1} \times {}^{0}_{i-1}[R]^{i-1}\mathbf{O}_i \tag{5.23} $$

and similar to the angular velocity vector, we can rewrite Eqn (5.23) as

$$ {}^i\mathbf{V}_i = {}^{i}_{i-1}[R]({}^{i-1}\mathbf{V}_{i-1} + {}^{i-1}\boldsymbol{\omega}_{i-1} \times {}^{i-1}\mathbf{O}_i) \tag{5.24} $$

where ${}^i\mathbf{V}_i$ and ${}^{i-1}\mathbf{V}_{i-1}$ denote ${}^{i}_{0}[R]^0\mathbf{V}_i$ and ${}^{i-1}_{0}[R]^0\mathbf{V}_{i-1}$, respectively, and represent linear velocities of the origins of $\{i-1\}$ and $\{i\}$, respectively.

If the joint i is prismatic, then we can start with the constraint Eqn (2.42). Proceeding in a manner similar to the rotary joint case discussed above, we can write

$$ \begin{aligned} {}^{i}_{i-1}[R]^{i-1}\boldsymbol{\omega}_i &\stackrel{\Delta}{=} {}^i\boldsymbol{\omega}_i = {}^{i}_{i-1}[R]^{i-1}\boldsymbol{\omega}_{i-1} \\ {}^{i}_{i-1}[R]^{i-1}\mathbf{V}_i &\stackrel{\Delta}{=} {}^i\mathbf{V}_i = {}^{i}_{i-1}[R]({}^{i-1}\mathbf{V}_{i-1} + {}^{i-1}\boldsymbol{\omega}_{i-1} \times {}^{i-1}\mathbf{O}_i) \\ &\quad + \dot{d}_i(0\;0\;1)^T \end{aligned} \tag{5.25} $$

Equations (5.21), (5.24), and (5.25) can be used to obtain, iteratively, the linear and angular velocities of any link in a serial manipulator containing rotary (R) or prismatic (P) joints. To start the iteration, we need to know ${}^0\boldsymbol{\omega}_0$ and ${}^0\mathbf{V}_0$. If the manipulator is fixed to a base, i.e., if $\{0\}$ is fixed, then both ${}^0\boldsymbol{\omega}_0$ and ${}^0\mathbf{V}_0$ are $\mathbf{0}$. If they are moving, for example, if the manipulator is mounted on a moving platform, then they will not be zero and appropriate values need to be used. In addition, once the linear and angular velocities are obtained from the above iterative formulae, they can be written in any other coordinate system by multiplication with appropriate rotation matrices.

Example 5.1 The planar 3R manipulator

Figure 2.17 shows the planar 3R manipulator described before. We use the above iterative formulas to obtain the linear and angular velocities of each link.

Since $\{0\}$ is fixed we have ${}^0\boldsymbol{\omega}_0 = \mathbf{0}$ and ${}^0\mathbf{V}_0 = \mathbf{0}$. From Eqns (5.21) and (5.24), we have

$i = 1$

$$^1\boldsymbol{\omega}_1 = (0 \;\; 0 \;\; \dot{\theta}_1)^T$$
$$^1\mathbf{V}_1 = \mathbf{0}$$

$i = 2$

$$^2\boldsymbol{\omega}_2 = (0 \;\; 0 \;\; \dot{\theta}_1 + \dot{\theta}_2)^T$$
$$^2\mathbf{V}_2 = \begin{pmatrix} c_2 & s_2 & 0 \\ -s_2 & c_2 & 0 \\ 0 & 0 & 1 \end{pmatrix} \begin{pmatrix} 0 \\ l_1\dot{\theta}_1 \\ 0 \end{pmatrix} = \begin{pmatrix} l_1 s_2 \dot{\theta}_1 \\ l_1 c_2 \dot{\theta}_1 \\ 0 \end{pmatrix}$$

$i = 3$

$$^3\boldsymbol{\omega}_3 = (0 \; 0 \; \dot{\theta}_1 + \dot{\theta}_2 + \dot{\theta}_3)^T$$
$$^3\mathbf{V}_3 = \begin{pmatrix} (l_1 s_{23} + l_2 s_3)\dot{\theta}_1 + l_2 s_3 \dot{\theta}_2 \\ (l_1 c_{23} + l_2 c_3)\dot{\theta}_1 + l_2 c_3 \dot{\theta}_2 \\ 0 \end{pmatrix}$$

We can also calculate the linear and angular velocities for the tool coordinate system as shown in Fig. 2.17.
$i =$ Tool

$$^{Tool}\boldsymbol{\omega}_{Tool} = (0 \; 0 \; \dot{\theta}_1 + \dot{\theta}_2 + \dot{\theta}_3)^T$$
$$^{Tool}\mathbf{V}_{Tool} = \begin{pmatrix} (l_1 s_{23} + l_2 s_3)\dot{\theta}_1 + l_2 s_3 \dot{\theta}_2 \\ (l_1 c_{23} + l_2 c_3 + l_3)\dot{\theta}_1 + (l_2 c_3 + l_3)\dot{\theta}_2 + l_3 \dot{\theta}_3 \\ 0 \end{pmatrix}$$

The linear and angular velocities of the tool can be easily found in the fixed coordinate system $\{0\}$ by pre-multiplying with the rotation matrix ${}^0_3[R]$. We get

$$^0\boldsymbol{\omega}_{Tool} = (0 \; 0 \; \dot{\theta}_1 + \dot{\theta}_2 + \dot{\theta}_3)^T \tag{5.26}$$

and

$$^0\mathbf{V}_{Tool} = \begin{pmatrix} -l_1 s_1 \dot{\theta}_1 - l_2 s_{12}(\dot{\theta}_1 + \dot{\theta}_2) - l_3 s_{123}(\dot{\theta}_1 + \dot{\theta}_2 + \dot{\theta}_3) \\ l_1 c_1 \dot{\theta}_1 + l_2 c_{12}(\dot{\theta}_1 + \dot{\theta}_2) + l_3 c_{123}(\dot{\theta}_1 + \dot{\theta}_2 + \dot{\theta}_3) \\ 0 \end{pmatrix} \tag{5.27}$$

5.4 Serial Manipulator Jacobian

Equations (5.26) and (5.27) can be written in a compact form as

$$
{}^0\mathcal{V}_{Tool} \triangleq \begin{pmatrix} {}^0\mathbf{V}_{Tool} \\ -- \\ {}^0\omega_{Tool} \end{pmatrix}
= \begin{bmatrix} -l_1 s_1 - l_2 s_{12} - l_3 s_{123} & -l_2 s_{12} - l_3 s_{123} & -l_3 s_{123} \\ l_1 c_1 + l_2 c_{12} + l_3 c_{123} & l_2 c_{12} + l_3 c_{123} & l_3 c_{123} \\ 0 & 0 & 0 \\ -- & -- & -- \\ 0 & 0 & 0 \\ 0 & 0 & 0 \\ 1 & 1 & 1 \end{bmatrix} \begin{pmatrix} \dot{\theta}_1 \\ \dot{\theta}_2 \\ \dot{\theta}_3 \end{pmatrix} \quad (5.28)
$$

where ${}^0\mathcal{V}_{Tool}$ is a 6×1 entity containing the linear and angular velocities of $\{Tool\}$. It must be noted that ${}^0\mathcal{V}_{Tool}$ is *not* a 6×1 vector in the linear algebra sense. This is because it contains two different quantities, the linear velocity and the angular velocity, which have different units. It may be, however, recalled that the quantity $({}^0\omega_{Tool};\, {}^0\mathbf{V}_{Tool})$ is termed as a twist (see Section 2.8) in theoretical kinematics. We will use '–' or ';' to separate the linear and angular velocities to remind us that ${}^0\mathcal{V}_{Tool}$ or $({}^0\mathbf{V}_{Tool};{}^0\omega_{Tool})^T$ is not a vector.

The matrix in the square brackets relates the linear and angular velocities of the tool with the joint velocities. This matrix denoted by ${}^{0}_{Tool}[J(\Theta)]$ is called the Jacobian matrix for the planar 3R manipulator. It may be noted that the Jacobian matrix is for the end-effector or the $\{Tool\}$ coordinate system, and this is indicated by the subscript *Tool*. In addition, since the linear and angular velocities are described in a fixed coordinate system $\{0\}$, the leading superscript denotes that the Jacobian matrix is described in $\{0\}$.

The manipulator Jacobian matrix is not a proper matrix in the linear algebra sense. This is because the first and the last three rows represent two different things. The top three rows are related to the linear velocity and have units of length, whereas the bottom three rows are related to the angular velocity and have no units. Hence one needs to be very careful with matrix operations using the manipulator Jacobian matrix. For example, it makes no sense to obtain a *condition number*[2] of this matrix. The choice of length units, from say meters to millimeters, can change the condition number. The Jacobian matrix is best viewed as a map which takes joint rates to Cartesian linear and angular velocities of an end-effector and, to remind us that the

---

[2] The condition number of a matrix is the ratio of the absolute value of the largest to the smallest eigenvalues.

Jacobian matrix is not a proper matrix, we will separate the top and bottom halves of a Jacobian matrix by '–'.

The Jacobian matrix can be derived for any serial manipulator with rotary and prismatic joints simply by computing the linear and angular velocity vectors and rewriting them as a matrix equation as done for the planar 3R manipulator (see Exercise 5.5). More generally, one can define a Jacobian for any differentiable vector function. Conceptually, we can think of the robot kinematics as a differentiable vector function $\mathcal{X} = \mathbf{\Psi}(\mathbf{\Theta})$, where $\mathbf{\Theta} = (\theta_1, \theta_2, \ldots, \theta_n)$ denotes the n joint variables. The three position and three orientation variables of the end-effector are denoted by $\mathcal{X}$[3]. The manipulator Jacobian matrix is the matrix of first partial derivatives of $\mathbf{\Psi}$ with respect to the joint variables and the ith column of $[J(\mathbf{\Theta})]$ is the partial derivatives of $\mathbf{\Psi}$ with respect to θ_i. We can write

$$[J(\mathbf{\Theta})] = \left[\frac{\partial \mathbf{\Psi}}{\partial \theta_1} \frac{\partial \mathbf{\Psi}}{\partial \theta_2} \cdots \frac{\partial \mathbf{\Psi}}{\partial \theta_n}\right]$$

The Jacobian matrix plays a very important role in serial manipulator kinematics. We list some of the main features of ${}^{0}_{Tool}[J(\mathbf{\Theta})]$.

1. The elements of the Jacobian matrix are non-linear functions of the joint variables $\mathbf{\Theta}$. When the manipulator is in motion, it is a time-varying matrix. However, at any instant of time or at a given configuration ($\mathbf{\Theta}$ known), it gives the linear and angular velocities of the end-effector or the $\{Tool\}$ coordinate system for known joint velocities. The relationship between the linear and angular velocities of the end-effector and the joint velocities, at a given $\mathbf{\Theta}$, is linear.
2. The Jacobian matrix can be obtained for any link by computing the linear and angular velocities up to that link and then writing the linear angular velocities in the matrix form. In a serial manipulator, it is most useful to obtain the Jacobian for the end-effector.
3. The Jacobian matrix is always with respect to a coordinate system—the same coordinate system in which the linear and angular velocities are obtained. Unless explicitly stated, it is with respect to the fixed $\{0\}$ coordinate system. The Jacobian matrix can always be written in any other coordinate system (see Exercise 5.6) by using appropriate rotation matrices.
4. The Jacobian matrix is usually non-square and of dimension $m \times n$, where m is the number of coordinates required to specify the motion

[3] For example, $\mathcal{X}$ denotes the three Cartesian position variables $(x,\ y,\ z)$ and the three Euler angles $(\alpha,\ \beta, \gamma)$.

in Cartesian space and n is the number of actuated joints. If the motion is in $\Re^3$, then $m = 6$ (three for position and three for orientation), and if the motion is in a plane, $m = 3$.

5. If ${}^{0}_{Tool}[J(\Theta)]$ is square, i.e., $m = n$, and if the determinant $\det({}^{0}_{Tool}[J(\Theta)]) \neq 0$, then we can invert ${}^{0}_{Tool}[J(\Theta)]$ [see Section 5.6 for an analysis of singularities of a serial manipulator when $\det({}^{0}_{Tool}[J(\Theta)]) = 0$]. We get

$$\dot{\Theta} = {}^{0}_{Tool}[J(\Theta)]^{-1}\, {}^{0}\mathcal{V}_{Tool} \tag{5.29}$$

The above relationship can be used to obtain the joint velocities required to achieve a desired linear and angular velocities of the $\{Tool\}$ coordinate system. The above relationship is analogous to the inverse problem in serial manipulator kinematics discussed in Chapter 3.

The Jacobian matrix can also be visualized in a more geometric manner and this is illustrated with the help of a planar 2R manipulator shown in Fig. 5.3.

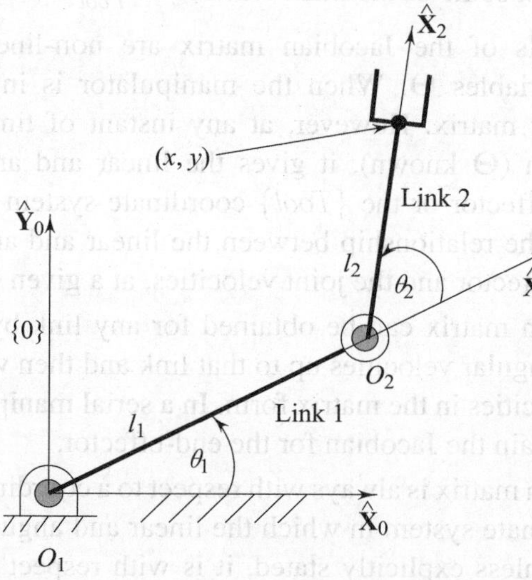

Fig. 5.3 A planar 2R manipulator

The velocity of the point (x, y) shown in Fig. 5.3 is given by

$$\mathbf{V} \triangleq \begin{pmatrix} \dot{x} \\ \dot{y} \end{pmatrix} = \begin{bmatrix} -l_1 s_1 - l_2 s_{12} & -l_2 s_{12} \\ l_1 c_1 + l_2 c_{12} & l_2 c_{12} \end{bmatrix} \begin{pmatrix} \dot{\theta}_1 \\ \dot{\theta}_2 \end{pmatrix} \tag{5.30}$$

where $\mathbf{V}$ denotes the linear velocity vector, l_1, l_2 are the link lengths, $\dot{\theta}_1$, $\dot{\theta}_2$ are the joint rates at the two rotary joints, and the matrix inside square brackets is the Jacobian matrix for the planar 2R manipulator described in $\{0\}$.

The dot product of the linear velocity vector with itself can be written as

$$\mathbf{V}^2 \triangleq \mathbf{V} \cdot \mathbf{V} = g_{11}\dot{\theta}_1^2 + 2g_{12}\dot{\theta}_1\dot{\theta}_2 + g_{22}\dot{\theta}_2^2 \tag{5.31}$$

where $g_{ij}(i, j = 1, 2)$ are the elements of a matrix $[\,g\,]$, which is the same as $[J(\mathbf{\Theta})]^T[J(\mathbf{\Theta})]$. For the planar 2R manipulator the g_{ij}'s are

$$\begin{aligned} g_{11} &= l_1^2 + l_2^2 + 2l_1l_2c_2 \\ g_{12} &= g_{21} = l_2^2 + l_1l_2c_2 \\ g_{22} &= l_2^2 \end{aligned} \tag{5.32}$$

It may be noted that the elements g_{ij}'s could be, in general, functions of all the joint variables $\mathbf{\Theta}$. However, for the planar 2R example, g_{11} and g_{12} are functions of θ_2 alone and g_{22} is a constant.

The maximum and minimum of $\mathbf{V}^2$ subject to a constraint, $\dot{\theta}_1^2 + \dot{\theta}_2^2 = 1$, can be obtained by solving $\partial \mathbf{V}^{*2}/\partial\dot{\theta}_i = 0$ $(i = 1, 2)$, where

$$\mathbf{V}^{*2} = g_{11}\dot{\theta}_1^2 + 2g_{12}\dot{\theta}_1\dot{\theta}_2 + g_{22}\dot{\theta}_2^2 - \lambda(\dot{\theta}_1^2 + \dot{\theta}_2^2 - 1)$$

Performing the partial differentiation reduces to an eigenvalue problem

$$[\,g\,]\dot{\mathbf{\Theta}} - \lambda\dot{\mathbf{\Theta}} = 0 \tag{5.33}$$

and the eigenvalues are given by

$$\lambda_{1,2} = (1/2)\{(g_{11} + g_{22}) \pm [(g_{11} + g_{22})^2 - 4(g_{11}g_{22} - g_{12}^2)]^{1/2}\} \tag{5.34}$$

Since $[\,g\,]$ is real, symmetric, and positive definite (see Exercise 5.7), the eigenvalues are always real and positive, and assuming $\lambda_1 > \lambda_2$, the maximum and minimum values of $|\mathbf{V}|$ are

$$|\mathbf{V}|_{\text{max}} = \sqrt{\lambda_1}, \quad |\mathbf{V}|_{\text{min}} = \sqrt{\lambda_2} \tag{5.35}$$

In the case of a square Jacobian matrix, the eigenvalue of $[J(\mathbf{\Theta})]$ is the square-root of the eigenvalues of $[J(\mathbf{\Theta})]^T[J(\mathbf{\Theta})]$ or $[\,g\,]$ (see Strang 1976) and, hence, the maximum and minimum $|\mathbf{V}|$ for the planar 2R manipulator are the eigenvalues of $[J(\mathbf{\Theta})]$. It may be noted that if a constraint $\dot{\theta}_1^2 + \dot{\theta}_2^2 = k^2$ is used, then the maximum and minimum $|\mathbf{V}|$ are scaled by k.

From the relationship $\mathbf{V} = [J(\mathbf{\Theta})]\dot{\mathbf{\Theta}}$, we can also write (dropping the dependence of $\mathbf{\Theta}$ for convenience)

$$[J]^T\mathbf{V} = [\,g\,]\dot{\mathbf{\Theta}}$$

and for non-singular $[\,g\,]$, we can get

$$\mathbf{V}^T([J][\,g\,]^{-1})([J][\,g\,]^{-1})^T\mathbf{V} = \dot{\mathbf{\Theta}}^T\dot{\mathbf{\Theta}}$$

For a planar 2R manipulator, the matrix $([J][g]^{-1})([J][g]^{-1})^T$ is symmetric and of rank 2. Hence if $\dot{\mathbf{\Theta}}^T\dot{\mathbf{\Theta}} = 1$, the above equation reduces to $(\dot{x},\dot{y})^T([J][\,g\,]^{-1})([J][\,g\,]^{-1})^T(\dot{x},\dot{y}) = 1$. From linear algebra, we know that an expression of the form $\mathbf{x}^T[A]\mathbf{x} = 1$, with $[A]$ symmetric and non-singular, describes an ellipse. Hence, we conclude that the tip of the linear velocity vector traces an ellipse and the semi-major and semi-minor axes of the ellipse are $\sqrt{\lambda_1}$ and $\sqrt{\lambda_2}$, respectively. It may be noted that the size of the ellipse will be scaled by k if a constraint $\dot{\mathbf{\Theta}}^T\dot{\mathbf{\Theta}} = k^2$ is used, but the shape of the ellipse does not change with k.

For the planar 2R manipulator, the eigenvalues of $[\,g\,]$ are only functions of θ_2 and hence the shape and size of the ellipse will be different for different values of θ_2. For the planar 2R manipulator, we show the ellipse traced by the tip of the linear velocity vector in Fig. 5.4 at a point in the workspace of the 2R manipulator.

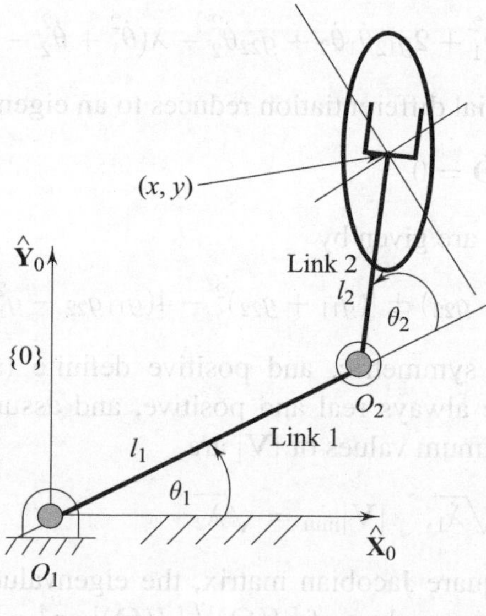

Fig. 5.4 Velocity ellipse for a planar 2R manipulator

The shape of the velocity ellipse indicates which directions are 'easier' to move for given joint rates—the magnitude of the linear velocity vector is larger along the major axis of the ellipse and hence it is easier to move along the major axis as compared to the minor axis. If the ellipse reduces to a circle, then it is equally easy to move in all directions. All points in the workspace, where the ellipse is a circle, are called isotropic. This concept was first developed by Salisbury (1982). In general, a manipulator is said to be in an isotropic configuration if the eigenvalues of $[J(\Theta)]$ (or $[\ g\ ]$) are equal or the condition number of $[J(\Theta)]$ of $[\ g\ ]$ is unity. For the planar 2R manipulator of Fig. 5.3, the eigenvalues of $[\ g\ ]$ are given in Eqn (5.34) and they can be equal only if

$$g_{11} = g_{22} \quad \text{and} \quad g_{12} = 0 \tag{5.36}$$

From the expressions of g_{ij}'s given in Eqn (5.32), the above conditions imply that

$$l_1^2 + 2l_1l_2c_2 = 0 \quad \text{and} \quad l_2^2 + l_1l_2c_2 = 0$$

and this is only possible if

$$l_1 = \sqrt{2}l_2, \quad c_2 = -\frac{1}{\sqrt{2}} \tag{5.37}$$

Equation (5.37) implies that a planar 2R manipulator can posses isotropic configurations only if the link lengths have a ratio of $\sqrt{2}$, and for these link dimensions, the second joint should be at an angle of 135°. Since θ_1 can take any value between $[0, 2\pi]$, all the isotropic configurations lie on a circle.

The velocity ellipse can also degenerate to a line and, in this case, the tip of the manipulator can have velocity only along one direction. A manipulator in such a configuration is said to be in a singular configuration and this is discussed in detail in Section 5.6.

The above idea of a velocity ellipse can be easily extended to spatial motion. For spatial manipulators with two degrees of freedom, the locus of the end-effector position traces a surface in $\Re^3$. In this case, the tip of the linear velocity vector lies in the tangent plane at any point on the surface, and the velocity ellipse lies on this tangent plane. If the manipulator has three degrees of freedom and the motion is in $\Re^3$, the Jacobian matrix maps a unit sphere in $\dot{\Theta}$ space to a velocity ellipsoid in $\Re^3$. The shape and size of the linear velocity ellipsoid can again be obtained from the eigenvalues of $[\ g\ ]$ which will now be a 3×3 matrix.

The geometric picture of an ellipse traced by the tip of the linear velocity vector matrix can also be easily extended to the angular velocity vector. We can, in a similar manner, obtain the eigenvalues of $[\ g\ ]$ obtained from the Jacobian associated with the angular velocity vector and obtain an angular velocity ellipse or an ellipsoid. The extension of the geometric picture to a 6×6 Jacobian matrix, for a six-DOF manipulator, containing the linear and angular velocity vectors does not make much sense due to the fact that the Jacobian matrix is not a proper matrix in the mathematical sense. As mentioned earlier, the top half is associated with the linear velocity of the end-effector and the bottom half is associated with the angular velocity of the end-effector and units of both are different. By choosing length units as meters or millimeters, the eigenvalues can be made different, and it makes no sense to equate two different quantities and define quantities such as isotropic configurations. Extension of the concept of velocity ellipse or ellipsoid requires advanced mathematical concepts of dual vectors, lines, and screws. These are beyond the scope of this text. The interested reader is referred to the paper by Ghosal and Ravani (1998) and Bandyopadhyay and Ghosal (2004b) and the references contained in them.

The geometric concepts of velocity ellipse or ellipsoid can also be applied to parallel or hybrid manipulators. However, this requires the concept of parallel manipulator Jacobian which is discussed in the following section.

5.5 Parallel Manipulator Jacobians

For parallel manipulators and closed-loop mechanisms, obtaining the Jacobian matrix is more complicated. The main difficulty is the presence of passive joints and the loop-closure constraint equations which may not contain all the joint variables—in deriving the loop-closure constraint equations for the three-DOF parallel manipulator and the six-DOF hybrid parallel manipulator in Example 4.1, the joint variables associated with the spherical (S) joints do not appear. In addition, the algorithm for obtaining the linear and angular velocities of a link in a serial manipulator in terms of the linear and angular velocities of the previous link, as described in Section 5.3, cannot be directly applied, since, due to the presence of one or more loops, we can reach a link in more ways than one.

To obtain the angular velocity of a chosen end-effector link, we use the fact that the angular velocity can be obtained directly from the rotation matrix as shown in Section 5.2. The rotation matrix ${}^{0}_{Tool}[R]$ of the chosen end-effector can be found if the position vector of three points, not lying on a straight line, on the end-effector are known (see Exercise 2.1). Once the rotation matrix is known, we can compute the space-fixed angular velocity[4] from $({}^{0}_{Tool}[\dot{R}])\ {}^{0}_{Tool}[R]^T$. The linear velocity of the end-effector can be obtained as a derivative, with respect to time, of any of the three position vectors. However, in many platform type parallel manipulators, the typical point of interest is the centroid of the platform. The centroid, as shown in Example 4.2, is the arithmetic mean of the three position vectors. Hence, the

[4] In this textbook we will often equate the angular velocity to the skew-symmetric matrix $[\dot{R}][R]^T$ [see Eqn (5.38)]. This is in the sense that the angular velocity vector can be *extracted* from the skew-symmetric matrix.

angular and linear velocities can be obtained as

$$ {}^0\boldsymbol{\omega}_{Tool} = \frac{d}{dt}({}^0_{Tool}[R]){}^0_{Tool}[R]^T = {}^0_{Tool}[J_{\boldsymbol{\omega}}(\mathbf{q})]\,\dot{\mathbf{q}} \tag{5.38} $$

$$ {}^0\mathbf{V}_{Tool} = \frac{1}{3}({}^0\dot{\mathbf{p}}_1 + {}^0\dot{\mathbf{p}}_2 + {}^0\dot{\mathbf{p}}_3) = {}^0_{Tool}[J_{\mathbf{V}}(\mathbf{q})]\,\dot{\mathbf{q}} $$

where ${}^0_{Tool}[J_{\boldsymbol{\omega}}(\mathbf{q})]$ and ${}^0_{Tool}[J_{\mathbf{V}}(\mathbf{q})]$ are the Jacobian matrices corresponding to the linear and angular velocities of the end-effector $\{Tool\}$ and $\dot{\mathbf{q}}$ is a vector of time derivatives of the configuration variables $\mathbf{q}$.

In general, the linear and the angular velocity vectors are a function of all the configuration variables $\mathbf{q}$ and their derivatives with respect to time, $\dot{\mathbf{q}}$. However, in parallel manipulators and closed-loop mechanisms, only the n of the $\mathbf{q}$'s, namely the θ_i's are actuated and the m ϕ_i's are passive, and it is desirable to obtain the linear and angular velocities only in terms of the $\dot{\theta}_i$'s. To eliminate the derivatives of the passive variables, we make use of the loop-closure constraint equations. For a parallel manipulator with m passive variables, we have m constraint equations of the form $\eta_i(q_1, \ldots, q_{n+m}) = 0 (i = 1, \ldots, m)$ (see Chapter 4, Section 4.3). We can rewrite them as a vector equation

$$ \boldsymbol{\eta}(\mathbf{q}) = \boldsymbol{\eta}(\boldsymbol{\theta}, \boldsymbol{\phi}) = \mathbf{0} \tag{5.39} $$

Differentiating Eqn (5.39) with respect to t, and rearranging, we can write

$$ [K(\mathbf{q})]\dot{\boldsymbol{\theta}} + [K^*(\mathbf{q})]\dot{\boldsymbol{\phi}} = \mathbf{0} \tag{5.40} $$

where the columns of the $m \times n$ matrix $[K(\mathbf{q})]$ are the partial derivatives of $\boldsymbol{\eta}(\mathbf{q})$ with respect to the actuated variables $\theta_i (i = 1, \ldots, n)$ and the columns of $m \times m$ matrix $[K^*(\mathbf{q})]$ are the partial derivatives of $\boldsymbol{\eta}(\mathbf{q})$ with respect to the passive variables $\phi_i (i = 1, \ldots, m)$. It may be noted that $[K^*(\mathbf{q})]$ is always an $m \times m$ square matrix and, in general, the matrices $[K(\mathbf{q})]$ and $[K^*(\mathbf{q})]$, as indicated, are functions of all the $n + m$ configuration variables $\mathbf{q} = (\boldsymbol{\theta}, \boldsymbol{\phi})$. For convenience, we drop the dependence on $\mathbf{q}$ in the rest of the discussion.

If $\det([K^*]) \neq 0$, i.e., the matrix $[K^*]$ is not singular (see Section 5.6 for an analysis of singularities of parallel manipulators), we can solve for $\dot{\boldsymbol{\phi}}$ from Eqn (5.40) and write

$$ \dot{\boldsymbol{\phi}} = -[K^*]^{-1}[K]\dot{\boldsymbol{\theta}} \tag{5.41} $$

Similar to the partitioning in Eqn (5.40), we can also rewrite Eqn (5.38) as

$$ {}^0\boldsymbol{\omega}_{Tool} = [J_{\boldsymbol{\omega}}]\dot{\boldsymbol{\theta}} + [J^*_{\boldsymbol{\omega}}]\dot{\boldsymbol{\phi}} \tag{5.42} $$

$$ {}^0\mathbf{V}_{Tool} = [J_{\mathbf{V}}]\dot{\boldsymbol{\theta}} + [J^*_{\mathbf{V}}]\dot{\boldsymbol{\phi}} $$

where the superscript 0, subscript *Tool*, and the functional dependence of the matrices on **q** are dropped for convenience.

We can now substitute $\dot{\phi}$ from Eqn (5.41) in Eqn (5.42) and get

$$
\begin{aligned}
{}^0\omega_{Tool} &= ([J_\omega] - [J^*_\omega][K^*]^{-1}[K])\dot{\boldsymbol{\theta}} \\
{}^0\mathbf{V}_{Tool} &= ([J_V] - [J^*_V][K^*]^{-1}[K])\dot{\boldsymbol{\theta}}
\end{aligned}
\tag{5.43}
$$

For parallel manipulators and closed-loop mechanisms, we can define an equivalent $[J_\omega]_{\text{eq}}$ and an equivalent $[J_V]_{\text{eq}}$ as

$$[J_V]_{\text{eq}} \triangleq [J_V] - [J^*_V][K^*]^{-1}[K] \tag{5.44}$$

and

$$[J_\omega]_{\text{eq}} \triangleq [J_\omega] - [J^*_\omega][K^*]^{-1}[K] \tag{5.45}$$

and we can write

$${}^0\mathcal{V}_{Tool} \triangleq \begin{pmatrix} {}^0\mathbf{V}_{Tool} \\ -- \\ {}^0\omega_{Tool} \end{pmatrix} = {}^0_{Tool}[J_{\text{eq}}]\dot{\boldsymbol{\theta}} \tag{5.46}$$

where the $6 \times n$ matrix, ${}^0_{Tool}[J_{\text{eq}}]$, is constructed by placing the $3 \times n$ rows from $[J_\omega]_{\text{eq}}$ below the $3 \times n$ rows of $[J_V]_{\text{eq}}$.

For parallel manipulators and closed-loop mechanisms, the matrix ${}^0_{Tool}[J_{\text{eq}}]$ plays the same role as the Jacobian matrix in serial manipulators—at a known configuration **q** and for given actuated joint rates $\dot{\boldsymbol{\theta}}$, we can use the relation in Eqn (5.46) to obtain the linear and angular velocities of the chosen end-effector $\{Tool\}$. The equivalent linear velocity Jacobian can also be used to obtain a $[\,g_{\text{eq}}\,]$ for parallel manipulators. Dropping the superscripts and subscripts for convenience, we obtain

$$[\,g_{\text{eq}}\,] = ([J_V] - [J_V{}^*][K^*]^{-1}[K])^T([J_V] - [J_V{}^*][K^*]^{-1}[K]) \tag{5.47}$$

The matrix $[\,g_{\text{eq}}\,]$ is symmetric and positive definite.[5] We can again state that for a constraint of the form $\dot{\boldsymbol{\theta}}^T\dot{\boldsymbol{\theta}} = k^2$, the tip of the linear velocity vector lies on an ellipse or an ellipsoid. The shape and size of the ellipse or the ellipsoid are again determined by the eigenvalues of the matrix $[\,g_{\text{eq}}\,]$. Similar to the treatment of the linear velocity vector, we can also obtain a $[\,g_{\text{eq}}\,]$ for the angular velocity vector by using $[J_\omega]_{\text{eq}}$ defined as $[J_\omega] - [J^*_\omega][K^*]^{-1}[K]$.

It can be, however, seen from Eqn (5.47) that the computation of the ellipse or ellipsoid is much more complicated. Firstly, the equivalent Jacobian's are functions of all the configuration variables **q** and, to compute the ellipse

[5] The matrix $[\,g_{\text{eq}}\,]$ is clearly symmetric since it is of the form $[A]^T[A]$. It is also positive definite provided that $\det([K^*]) \neq 0$ and $([J] - [J^*][K^*]^{-1}[K])$ is non-singular

or ellipsoids, we have to perform direct kinematics and solve the passive variables for given actuated variables. Secondly, the matrix $[g_{\text{eq}}]$ cannot be obtained if $[K^*]$ is singular. We discuss the singularities of parallel manipulators in Section 5.6. In the rest of the section, we give two examples of obtaining the various Jacobian matrices for parallel manipulators and plot of the velocity ellipse for a three-DOF parallel manipulator.

Example 5.2 The planar four-bar mechanism

To derive the matrices $[K]$ and $[K^*]$, we start with the constraint equations of a four-bar mechanism given in Chapter 4, Eqn (4.6). These are reproduced below:

$$\eta_1(\mathbf{q}) \stackrel{\Delta}{=} l_1 \cos\theta_1 + l_2 \cos(\theta_1 + \phi_2) - l_0 - l_3 \cos\phi_1 = 0$$

$$\eta_2(\mathbf{q}) \stackrel{\Delta}{=} l_1 \sin\theta_1 + l_2 \sin(\theta_1 + \phi_2) - l_3 \sin\phi_1 = 0 \tag{5.48}$$

In these equations, θ_1 is the actuated joint variable and (ϕ_1, ϕ_2) are the passive joint variable. Differentiating these equations with respect to time, we get, after re-arranging,

$$\begin{pmatrix} -l_1 \sin\theta_1 - l_2 \sin(\theta_1 + \phi_2) \\ l_1 \cos\theta_1 + l_2 \cos(\theta_1 + \phi_2) \end{pmatrix} \dot{\theta}_1 + \begin{pmatrix} l_3 \sin\phi_1 & -l_2 \sin(\theta_1 + \phi_2) \\ -l_3 \cos\phi_1 & l_2 \cos(\theta_1 + \phi_2) \end{pmatrix} \begin{pmatrix} \dot{\phi}_1 \\ \dot{\phi}_2 \end{pmatrix} = \mathbf{0}$$

and we get

$$[K] = \begin{pmatrix} -l_1 \sin\theta_1 & - l_2 \sin(\theta_1 + \phi_2) \\ l_1 \cos\theta_1 & + l_2 \cos(\theta_1 + \phi_2) \end{pmatrix},$$

$$[K^*] = \begin{bmatrix} l_3 \sin\phi_1 & -l_2 \sin(\theta_1 + \phi_2) \\ -l_3 \cos\phi_1 & l_2 \cos(\theta_1 + \phi_2) \end{bmatrix} \tag{5.49}$$

As expected, the matrix $[K^*]$ is a square 2×2 matrix and is a function of actuated and passive joint variables.

Example 5.3 A three-DOF parallel manipulator

In Eqn (4.26), the three loop-closure equations were presented. To derive expressions for $[K]$ and $[K^*]$, we perform the derivative of $\eta_i(\mathbf{q}) (i = 1, 2, 3)$ with respect to time. The matrix $[K]$ is associated with the derivative of the actuated variables $\mathbf{l}$ and is given by

$$[K] = \begin{bmatrix} 2l_1 - 3c_1 + l_2 c_1 c_2 - 2l_2 s_1 s_2 & 2l_2 - 3c_2 + l_1 c_1 c_2 - 2l_1 s_1 s_2 & 0 \\ 0 & 2l_2 - 3c_2 + l_3 c_2 c_3 - 2l_3 s_2 s_3 & 2l_3 - 3c_3 + l_2 c_2 c_3 - 2l_2 s_2 s_3 \\ 2l_1 - 3c_1 + l_3 c_1 c_3 - 2l_3 s_1 s_3 & 0 & 2l_3 - 3c_3 + l_1 c_1 c_3 - 2l_1 s_1 s_3 \end{bmatrix} \tag{5.50}$$

The matrix $[K^*]$ is associated with the derivative of the passive joint variables $\dot{\boldsymbol{\theta}}$ and is given by

$$[K^*] = \begin{bmatrix} 3l_1s_1 - l_1l_2s_1c_2 & 3l_2s_2 - l_1l_2c_1s_2 & 0 \\ -2l_1l_2c_1s_2 & -2l_1l_2s_1c_2 & \\ 0 & 3l_2s_2 - l_2l_3s_2c_3 & 3l_3s_3 - l_2l_3c_2s_3 \\ & -2l_2l_3c_2s_3 & -2l_2l_3s_2c_3 \\ 3l_1s_1 - l_1l_3s_1c_3 & 0 & 3l_3s_3 - l_1l_3c_1s_3 \\ -2l_1l_3c_1s_3 & & -2l_1l_3s_1c_3 \end{bmatrix} \tag{5.51}$$

The position vector of the centroid is given in Eqn (4.32). To obtain $[J_{\mathrm{V}}]$ and $[J_{\mathrm{V}}{}^*]$, we differentiate Eqn (4.32) with respect to time and collect the coefficients of the passive and actuated joint rates. We get

$$[J_{\mathrm{V}}] = (1/3) \begin{bmatrix} -c_1 & (1/2)c_2 & (1/2)c_3 \\ 0 & (-\sqrt{3}/2)c_2 & (\sqrt{3}/2)c_3 \\ s_1 & s_2 & s_3 \end{bmatrix} \tag{5.52}$$

and

$$[J_{\mathrm{V}}{}^*] = (1/3) \begin{bmatrix} l_1s_1 & -(1/2)l_2s_2 & (-1/2)l_3s_3 \\ 0 & (\sqrt{3}/2)l_2s_2 & (-\sqrt{3}/2)l_3s_3 \\ l_1c_1 & l_2c_2 & l_3c_3 \end{bmatrix} \tag{5.53}$$

To obtain $[J_{\boldsymbol{\omega}}]$ and $[J_{\boldsymbol{\omega}}{}^*]$, we first compute the space-fixed angular velocity vector from ${}^{Base}_{Top}[\dot{R}]\ {}^{Base}_{Top}[R]^T$ as discussed in Section 5.2. The rotation matrix ${}^{Base}_{Top}[R]$ can be obtained once the position vector of the three spherical joints is known with respect to $\{Base\}$ and is given in Eqn (4.33). Using MAPLE one can, in principle, compute the analytical expressions for the components of the angular velocity vector and from it extract analytical expressions for $[J_{\boldsymbol{\omega}}]$ and $[J_{\boldsymbol{\omega}}{}^*]$ as has been done for $[J_{\mathrm{V}}]$ and $[J_{\mathrm{V}}{}^*]$. The analytical expressions are, however, extremely large and are not presented in this text. It is much simpler to obtain numerical values at a given **q**, and for $l_1 = 2/3$, $l_2 = 3/5$, $l_3 = 3/4$ and corresponding $\theta_1 = 0.7593$, $\theta_2 = 0.2851$, $\theta_3 = 0.8028$ rad, we get

$$[J_{\boldsymbol{\omega}}] = \begin{pmatrix} 0.0644 & -0.2801 & -0.9548 \\ -1.1519 & 0.1361 & 0.4815 \\ -0.1953 & 0.5339 & -0.3803 \end{pmatrix},$$

$$[J_{\boldsymbol{\omega}}{}^*] = \begin{pmatrix} -0.0307 & 0.6686 & -0.3398 \\ -0.3961 & 0.4256 & 0.1713 \\ 0.3069 & 0.0308 & -0.1353 \end{pmatrix}$$

The analytical expressions for $[J\text{v}]_{\text{eq}}$ and $[J\omega]_{\text{eq}}$ are even more harder to obtain as it involves obtaining the inverse of $[K^*]$ as given in Eqns (5.44) and (5.45) and we again give a numerical example for the values of joint variables chosen above. We get

$$[J\text{v}]_{\text{eq}} = \begin{pmatrix} -0.2313 & 0.5372 & 0.0114 \\ 0.0722 & -0.6758 & 0.1951 \\ 1.1765 & -1.6830 & 0.9223 \end{pmatrix},$$

$$[J\omega]_{\text{eq}} = \begin{pmatrix} 2.1409 & -6.4331 & 0.4665 \\ 0.0072 & -4.1216 & 1.6048 \\ 0.1565 & 0.4570 & -0.3285 \end{pmatrix}$$

For the numerical example, we assume that all sides of the top platform are equal and it is inscribed in a circle of radius 0.5, i.e., $a = 1/2$ (see Chapter 4, Example 4.2). For $(l_1, l_2, l_3) = (0.5, 1.0, 2.0)$ m, the solution of the direct kinematics problem gives the values of the passive variables $(\theta_1, \theta_2, \theta_3)$ as $(0.4, 0.7535, 0.2402)$ rad. The tip of the linear velocity vector of the centroid of the top platform lies on the ellipsoid shown in Fig. 5.5 as three sectional views and a 3D plot. The maximum, intermediate, and minimum velocities along the principal axes of the ellipsoid are given by $0.3724, 0.3162, 0.2031$ m/sec, respectively. The directions

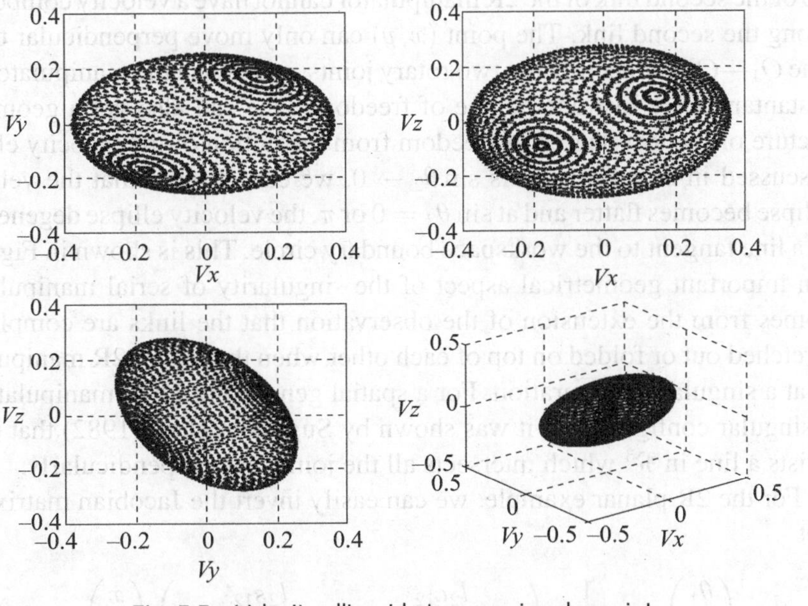

Fig. 5.5 Velocity ellipsoid at a non-singular point

of the corresponding principal axes are $(0.9921, -0.0394, 0.1187)^T$, $(0.1166, 0.6338, -0.7646)^T$ and $(-0.0452, 0.7724, 0.6335)^T$, respectively.

5.6 Singularities of Serial and Parallel Manipulators

If $\det({}^0_{Tool}[J(\Theta)]) = 0$, for some value(s) of the joint variables Θ, then the serial manipulator is said to be in a singular configuration. At a singular configuration, the Jacobian matrix of a serial manipulator 'loses' rank and the inverse does not exist. At such a configuration, we cannot obtain the joint rates $\dot{\Theta}$ for given linear and angular velocities of the end-effector or $\{Tool\}$. In these configurations, the serial manipulator loses one or more degrees of freedom, and there are some directions in which the manipulator cannot move. This fact can be seen clearly from the example of the planar 2R manipulator discussed earlier. For the 2R manipulator example, the relationship between the linear velocity of the point (x, y) and the joint rates $(\dot{\theta}_1, \dot{\theta}_2)$ is given in Eqn (5.30). The Jacobian matrix is given by

$$
{}^0_{Tool}[J(\Theta)] = \begin{bmatrix} -l_1 s_1 - l_2 s_{12} & -l_2 s_{12} \\ l_1 c_1 + l_2 c_{12} & l_2 c_{12} \end{bmatrix} \tag{5.54}
$$

For this example, the condition $\det({}^0_{Tool}[J(\Theta)]) = 0$ reduces to $\sin\theta_2 = 0$. This implies that θ_2 is 0 or π radians, i.e., the second link is stretched completely or folded on top of the first link. Clearly, in this configuration the tip of the second link of the 2R manipulator cannot have a velocity component along the second link. The point (x, y) can only move perpendicular to the line O_1—O_2 connecting the two rotary joints, and hence the manipulator has instantaneously lost one degree of freedom. One can also get a geometric picture of the lost degree of freedom from the notion of the velocity ellipse discussed in Section 5.4. As $\sin\theta_2 \to 0$, we can observe that the velocity ellipse becomes flatter and at $\sin\theta_2 = 0$ or π, the velocity ellipse degenerates to a line tangent to the workspace boundary circle. This is shown in Fig. 5.6. An important geometrical aspect of the singularity of serial manipulators comes from the extension of the observation that the links are completely stretched out or folded on top of each other when the planar 2R manipulator is at a singular configuration. For a spatial general six-DOF manipulator at a singular configuration, it was shown by Sugimoto et al. (1982) that there exists a line in $\Re^3$ which intersects all the joint axes perpendicularly.

For the 2R planar example, we can easily invert the Jacobian matrix and get

$$
\begin{pmatrix} \dot{\theta}_1 \\ \dot{\theta}_2 \end{pmatrix} = \frac{1}{l_1 l_2 s_2} \begin{pmatrix} l_2 c_{12} & l_2 s_{12} \\ -l_1 c_1 - l_2 c_{12} & -l_1 s_1 - l_2 s_{12} \end{pmatrix} \begin{pmatrix} \dot{x} \\ \dot{y} \end{pmatrix} \tag{5.55}
$$

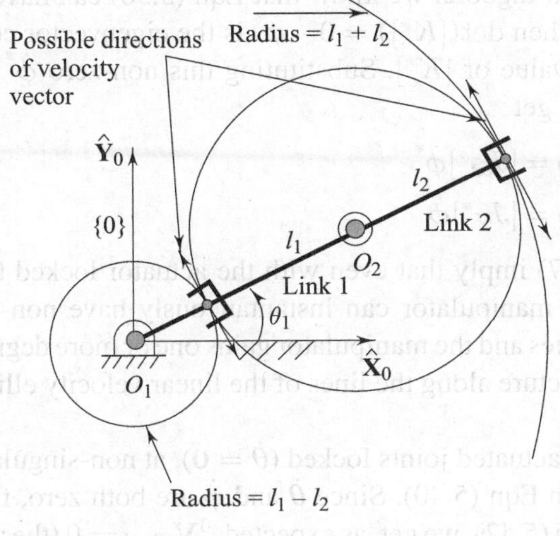

Fig. 5.6 Singular configurations for a planar 2R manipulator

From Eqn (5.55), one can observe that s_2 is in the denominator and as $\theta_2 \rightarrow 0$ or π, the left-hand side $(\dot{\theta}_1, \dot{\theta}_2)^T \rightarrow \infty$. This effect is not limited to the simple planar 2R manipulator discussed here and occurs in all serial manipulators (see Wang & Waldron 1987). Prior knowledge of the singular configurations in a serial manipulator is extremely important since, if the manipulator motion is such that $\det({}^0_{Tool}[J(\mathbf{\Theta})])$ is close to zero, the joint velocities tend to become large and this could cause problems for the servo controller of the robot.

Similar to the serial manipulators, when the equivalent Jacobian loses rank, a parallel manipulator is in a singular configuration and the actuated joint rates $\dot{\mathbf{\Theta}} \rightarrow \infty$ at a singular configuration. Similar to the serial manipulator, when $\det({}^0_{Tool}[J_{\text{eq}}]) = 0$, the end-effector of a parallel manipulator loses one or more degrees of freedom. Again, similar to a serial manipulator, we can visualize the singularity as a degeneracy of the velocity ellipse or the ellipsoid.

Unlike serial manipulators, parallel manipulators can exhibit another kind of singularity. This can be seen from Eqn (5.40). Consider all actuated joints as locked, i.e., $\dot{\boldsymbol{\theta}} = \mathbf{0}$. From Eqn (5.40), if $\det([K^*]) \neq 0$, we get $\dot{\boldsymbol{\phi}} = \mathbf{0}$. This implies that all the passive joints are also locked and the parallel manipulator becomes a structure. However, consider the case when $\dot{\boldsymbol{\theta}} = \mathbf{0}$ and $\det([K^*]) = 0$. Equation (5.40) now represents a homogenous equation

$$[K^*(\mathbf{q})]\dot{\boldsymbol{\phi}} = \mathbf{0} \tag{5.56}$$

and, from linear algebra, we know that Eqn (5.56) can have a non-zero $\dot{\boldsymbol{\phi}}^*$ as a solution when $\det([K^*]) = 0$ – $\dot{\boldsymbol{\phi}}^*$ is the eigenvector corresponding to the zero eigenvalue of $[K^*]$. Substituting this non-zero $\dot{\boldsymbol{\phi}}^*$ (and $\dot{\boldsymbol{\theta}} = \mathbf{0}$) in Eqn (5.42), we get

$$
\begin{aligned}
{}^0\boldsymbol{\omega}_{Tool} &= [J_{\boldsymbol{\omega}}{}^*]\dot{\boldsymbol{\phi}}^* \\
{}^0\mathbf{V}_{Tool} &= [J_{\mathbf{V}}{}^*]\dot{\boldsymbol{\phi}}^*
\end{aligned}
\tag{5.57}
$$

Equations (5.57) imply that even with the actuator locked the end-effector of the parallel manipulator can instantaneously have non-zero linear and angular velocities and the manipulator gains one or more degrees of freedom. A geometric picture along the lines of the linear velocity ellipse or ellipsoid is as follows.

With all the actuated joints locked ($\dot{\boldsymbol{\theta}} = \mathbf{0}$), at non-singular positions, we get $\dot{\boldsymbol{\phi}} = \mathbf{0}$ from Eqn (5.40). Since $\dot{\boldsymbol{\theta}}$ and $\dot{\boldsymbol{\phi}}$ are both zero, from the second equation in Eqn (5.42), we get, as expected, ${}^0\mathbf{V}_{Tool} = 0$ (the angular velocity ${}^0\boldsymbol{\omega}_{Tool}$ is also zero). Hence at a non-singular position with actuated joints locked, we can think of the tip of the velocity vector tracing an ellipsoid of zero size. At a singularity, the matrix $[K^*]$ loses rank. If the rank is $(m - 1)$ then we can extract the eigenvector of $[K^*]$ corresponding to the zero eigenvalue of $[K^*]$. Let the eigenvector corresponding to the zero eigenvalue be $\boldsymbol{\phi}_1$. Since $C_1\boldsymbol{\phi}$ is also an eigenvector, with C_1 any scaling constant, from the second equation in Eqn (5.42), we get

$$
{}^0\mathbf{V}_{Tool} = C_1[J_{\mathbf{V}}{}^*]\dot{\boldsymbol{\phi}}_1 \tag{5.58}
$$

and there can be motion along the direction of $[J_{\mathbf{V}}{}^*]\boldsymbol{\phi}_1$. In this case, we can think of the zero velocity ellipsoid 'growing' into a line. If the rank of the matrix $[K^*]$ is $(m - 2)$, then with a similar reasoning we can get

$$
{}^0\mathbf{V}_{Tool} = C_1[J_{\mathbf{V}}{}^*]\dot{\boldsymbol{\phi}}_1 + C_2[J_{\mathbf{V}}{}^*]\dot{\boldsymbol{\phi}}_2 \tag{5.59}
$$

where $\boldsymbol{\phi}_1$, $\boldsymbol{\phi}_2$ are the two eigenvectors corresponding to the two zero eigenvalues of $[K^*]$ and C_1, C_2 are the two scaling constants. If we normalize C_i $(i = 1, 2)$ to be between -1 and $+1$ (or $C_1^2 + C_2^2 = 1$), then the tip of the velocity vector traces an ellipse.[6] If the rank of $[K^*]$ is $(m - 3)$, then the tip of the velocity vector will lie on an ellipsoid. If the rank is less than $(m - 3)$, then we have a situation similar to the redundant serial manipulator.

The singularity associated with the gain of one or more degrees of freedom occurs in all parallel and hybrid manipulators. It has been pointed out by Hunt

[6] The quantities C_1 and C_2 are similar to $\dot{\theta}_1$ and $\dot{\theta}_2$, and $C_1^2 + C_2^2 = 1$ is similar to the constraint $\dot{\theta}_1^2 + \dot{\theta}_2^2 = 1$ used for the planar 2R example in Section 5.4. Hence, by following the reasoning in Section 5.4, we can prove that the tip of ${}^0\mathbf{V}_{Tool}$ for a parallel manipulator lies on an ellipse.

(1991) that a fully parallel six-DOF manipulator (where the end-effector is connected directly to the fixed base by a single actuated joint as in a Stewart platform manipulator) can only exhibit the gain kind of singularity whereas a hybrid parallel manipulator (such as the six-DOF three-fingered hand discussed in Chapters 2 and 4) can exhibit both loss and gain kinds of singularity. The singularity analysis of parallel manipulators, especially for important ones such as a Stewart platform, is still an active topic of research. The interested reader is referred to some of the important references, which are by no means exhaustive, listed at the end of the chapter (Hunt 1986, Litvin et al. 1990, Merlet 1991, Gosselin and Angeles 1990, Zlatanov 1995, Park & Kim 1999). In Section 5.9, we will look at the relationship of gain singularity with the notion of resistance to externally applied forces and moments on a parallel manipulator. We conclude this discussion with some of the more advanced concepts in the singularity analysis of parallel manipulators and closed-loop mechanisms.

- At a configuration where $\det([K^*]) = 0$, as discussed above, a parallel manipulator gains one or more degrees of freedom instantaneously. It is possible under a special choice of link dimensions and geometry, a parallel manipulator gains one or more degrees of freedom over a *finite* range of motion of the configuration variables. A simple example is that of a two-DOF five-bar mechanism shown in Fig. 4.10. For the condition $l_2 = l_3$, and $\theta_1, \theta_2, l_0, l_1$, and l_4 such that the fixed link, link 1, and link 4 form a triangle as shown in Fig. 5.7, links 2 and 3 can rotate from 0 to 2π about the point O_2. With θ_1 and θ_2 locked, the five-bar mechanism was expected to be a structure. However, we can see that it has gained finite motion.
  The finite motion at a gain singularity is related to the partial derivative of $\det([K^*])$, with respect to one or more passive joint variables, being zero in addition to $\det([K^*]) = 0$. The interested reader is referred to Bandyopadhyay and Ghosal (2004a) for details.
- In closed-loop mechanisms and parallel manipulators, one comes across another phenomenon of *dwell* of a link. Dwell of a passive link refers to a situation when the link is at rest for instantaneous or finite motion of the actuators. From Eqn (5.41), if $\det([K^*])$ is non-singular, then we get a non-zero $\dot{\phi}$. A dwell will occur if a passive joint variable is not influenced by an actuated joint variable, i.e., $\partial\phi_i/\partial\theta_j = 0$ for a particular i and j. It can be seen that the (i, j) element of $(-[K^*]^{-1}[K])$ is $\partial\phi_i/\partial\theta_j$. Since $\partial\phi_i/\partial\theta_j$ $(j = 1, \ldots, m)$ gives the ith row of $(-[K^*]^{-1}[K])$, the passive link associated with ϕ_i will instantaneously dwell if the ith row of $(-[K^*]^{-1}[K])$ is null. For finite

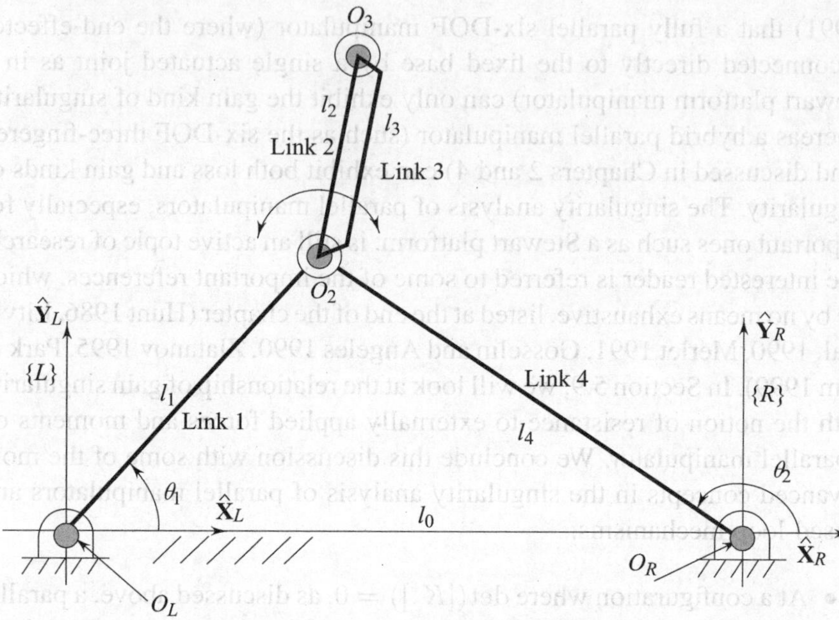

Fig. 5.7 Finite motion at gain singularity

dwell, the above condition is maintained over a finite span of motion of the input joints and this can only happen for special link lengths and geometry. The interested reader is referred to Bandyopadhyay and Ghosal (2004a) for details.

Now we present examples of singularities in two parallel manipulators.

Example 5.4 The planar four-bar mechanism

The $[K^*]$ matrix is given in Eqn (5.49). The condition $\det([K^*]) = 0$ gives

$$l_2 l_3 \sin(\theta_1 + \phi_2 - \phi_1) = 0$$

This implies that

$$\theta_1 + \phi_2 - \phi_1 = n\pi, \; n = 0, 1, 2, \cdots \tag{5.60}$$

To understand what Eqn (5.60) means, we go back to Eqn (4.3). Substituting the result from Eqn (5.60) in third equation in Eqn (4.3), we can get

$$\phi_3 = 3\pi - n\pi$$

and from Fig. 4.1, we can conclude that links 2 and 3 are parallel when $\det([K^*]) = 0$. Due to the definition of ϕ_2 and ϕ_3 (see Fig. 4.1), the value of n must be 1 (or $\phi_3 = 2\pi$ or 0) and the configuration of the four-bar mechanism for $\det([K^*]) = 0$ is shown in Fig. 5.8. It may be mentioned that for a four-bar mechanism to achieve the configuration shown in Fig. 5.8,

$$\cos\theta_1 = \frac{l_0^2 + l_1^2 - (l_2 + l_3)^2}{2l_1 l_0}$$

and since $-1 \leq \cos\theta_1 \leq 1$, we will get $(l_0 + l_1) \geq (l_2 + l_3)$.

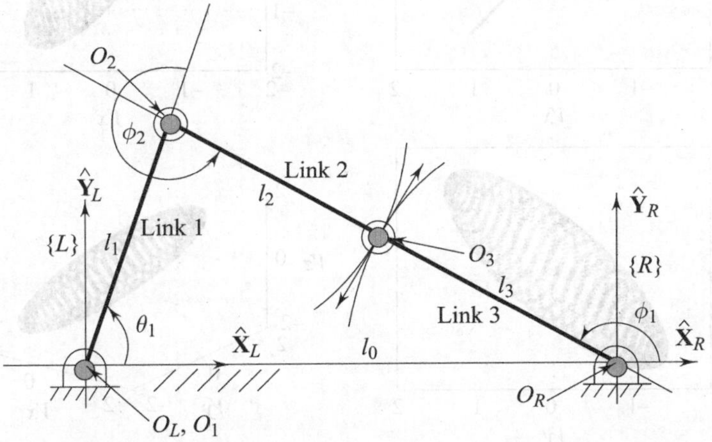

Fig. 5.8 Singular configuration for a planar four-bar mechanism

The fact that the four-bar mechanism instantaneously gains a degree of freedom at the configuration shown in Fig. 5.8 is seen from the following reasoning: with the actuated joint locked, the point O_2 is also fixed. The two links, link 2 and link 3, are in a straight line and the point O_3 can have instantaneous velocity along the common tangent to the two circles as shown in Fig. 5.8.

Example 5.5 A three-DOF parallel manipulator

For the three-DOF parallel manipulator, if $\det([J_V]_{eq}) = 0$, the linear velocity ellipsoid described by the centroid of the top platform degenerates to an ellipse [see Ghosal and Ravani (2001) for more details]. For (l_1, l_2, l_3) given by $(0.5, 1.0, 1.9710)$ m and the corresponding passive variables $(\theta_1, \theta_2, \theta_3)$ given by $(1.1691, 0.4781, 0.2355)$ rad, it can be shown that $\det([J_V]_{eq}) = 0$. The linear velocity ellipse at this configuration is shown in sectional and 3D views in Fig. 5.9. It may be noted that this is not in contradiction to the statement by Hunt (1991) that a fully parallel manipulator is not capable of losing degrees of freedom. This is due to the fact that this is not a six-DOF manipulator and in addition, we are only considering the linear velocity vector.

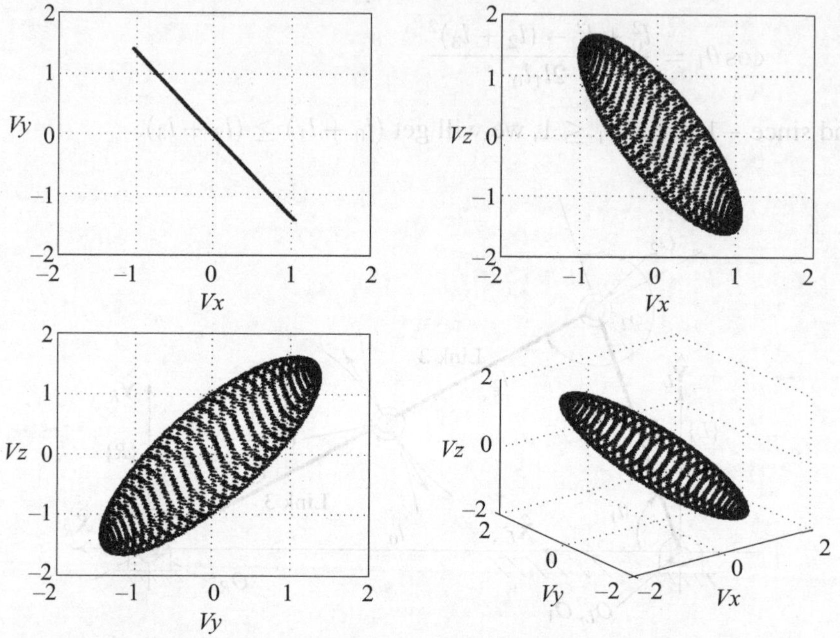

Fig. 5.9 Linear velocity ellipse at a loss singular point

The three-DOF parallel manipulator will gain one or more degrees of freedom when $\det([K^*]) = 0$ and this can be expanded as

$$\begin{aligned}\det([K^*]) = {} & (3l_1s_1 - l_1l_2s_1c_2 - 2l_1l_2c_1s_2) \times (3l_2s_2 - l_2l_3s_2c_3 \\ & -2l_2l_3c_2s_3) \times (3l_3s_3 - l_1l_3c_1s_3 - 2l_1l_3s_1c_3) \\ & +(3l_1s_1 - l_1l_3s_1c_3 - 2l_1l_3c_1s_3) \times (3l_2s_2 - l_1l_2c_1s_2 \\ & -2l_1l_2s_1c_2) \times (3l_3s_3 - l_2l_3c_2s_3 - 2l_2l_3s_2c_3) = 0 \end{aligned} \quad (5.61)$$

Equation (5.61) is a function of all the passive and active joint variables and, again together with the three loop closure equations, Eqn (4.26), represent a set of four equations in six variables. Thus the singularities resulting in a gain of one or more degrees of freedom lie on a 2D surface. It is very difficult to get analytical expressions for this surface and we present numerical results [see Basu and Ghosal (1997) and Ghosal and Ravani (2001) for details].

At the values of leg lengths (l_1, l_2, l_3) given by $(0.575, 0.483, 0.544)$ m, respectively, and the corresponding passive variables $(\theta_1, \theta_2, \theta_3)$ given by $(-0.3441, -0.0138, 0.2320)$ rad, $\det[K^*]$ is found to be very close to zero. The eigenvalues of $[K^*]$ are approximately -0.5565, 0, and 0.4509, respectively, and the three eigenvectors corresponding to the three eigenvalues are $(-0.8098, 0.3571, -0.4656)^T$, $(-0.3109, -0.8743, -0.3727)^T$, and $(-0.0877, -0.4781, -0.8739)^T$. Hence, at this point, the mechanism gains one degree of freedom and the velocity of the centroid, with all actuated joints locked, is given as

$$ {}^0\mathbf{V}_{Tool} = \begin{pmatrix} -0.0647 \\ 0 \\ 0.1804 \end{pmatrix} \dot{\theta}_1 + \begin{pmatrix} 0.0011 \\ -0.0019 \\ 0.1610 \end{pmatrix} \dot{\theta}_2 + \begin{pmatrix} -0.0208 \\ -0.0361 \\ 0.1763 \end{pmatrix} \dot{\theta}_3 \quad (5.62) $$

where $(\dot{\theta}_1, \dot{\theta}_2, \dot{\theta}_3)^T$ is the eigenvector $\alpha \times (-0.3109, -0.8743, -0.3727)^T$ with α arbitrary. It is clear that the velocity vector lies along a straight line and the mechanism has gained instantaneously a degree of freedom at this singular point. Figure 5.10 shows the straight line traced by the tip of the linear velocity vector of the centroid at the singular point. It also shows the three sectional view and a 3D plot.

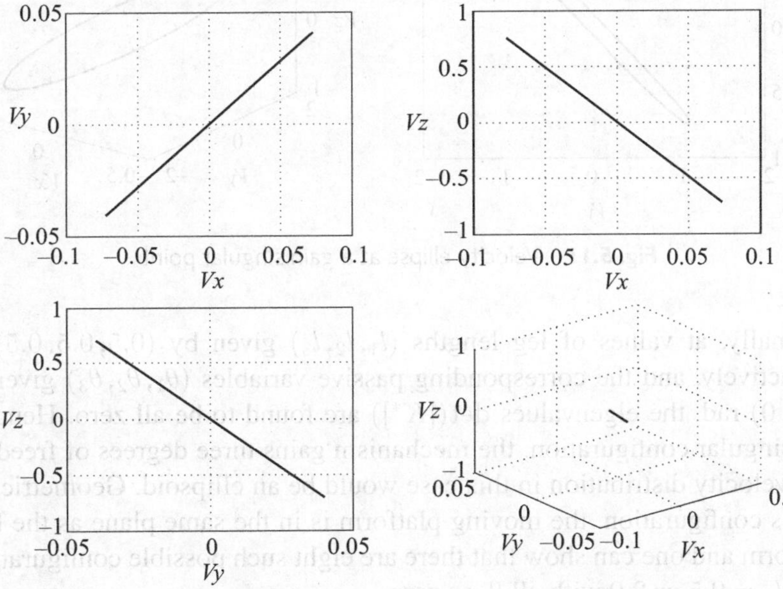

Fig. 5.10 Velocity at a gain singular point

At the values of leg lengths (l_1, l_2, l_3) given by $(1.9363, 2.9998, 1.9363)$ m, respectively, and the corresponding passive variables $(\theta_1, \theta_2, \theta_3)$ given by $(1.3096, 0.9817, 1.3096)$ rad, $\det[K^*]$ is also found to be very close to zero. The eigenvalues of $[K^*]$ are approximately 0, 0, 3.9680, respectively. Hence, at this configuration, the mechanism gains two degrees of freedom. The velocity distribution, in this case an ellipse, is shown in Fig. 5.11. The singularities corresponding to gain of two degrees of freedom lie on a curve in $\Re^3$.

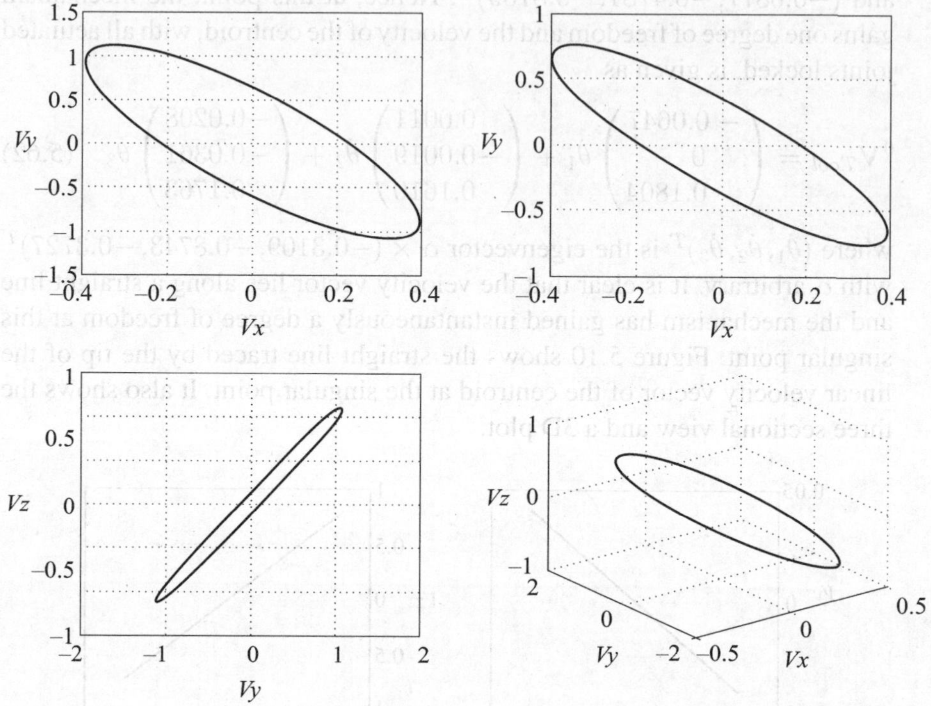

Fig. 5.11 Velocity ellipse at a gain singular point

Finally, at values of leg lengths (l_1, l_2, l_3) given by $(0.5, 0.5, 0.5)$ m, respectively, and the corresponding passive variables $(\theta_1, \theta_2, \theta_3)$ given by $(0, 0, 0)$ rad, the eigenvalues $\det([K^*])$ are found to be all zero. Hence at this singular configuration, the mechanism gains three degrees of freedom. The velocity distribution in this case would be an ellipsoid. Geometrically, at this configuration, the moving platform is in the same plane as the base platform and one can show that there are eight such possible configurations with $l_i = 0.5$ or 2.0 with all θ_i as zero.

5.7 Statics of Serial Manipulators

When the joints of a serial manipulator are locked, the manipulator can be viewed as a structure. It is of interest, from a design viewpoint, to obtain static forces and moments acting on the links of manipulator structure when the end-effector is subjected to external forces and moments. The external forces and moments on the end-effector can arise if the manipulator end-effector is pushing some object or carrying a payload. It is also of interest to obtain the joint forces or torques which can maintain the static equilibrium of the system. To obtain the forces and moments acting on a link or the joint forces and moments required to maintain equilibrium, we use the concept of a free-body diagram and use the equations of static equilibrium.

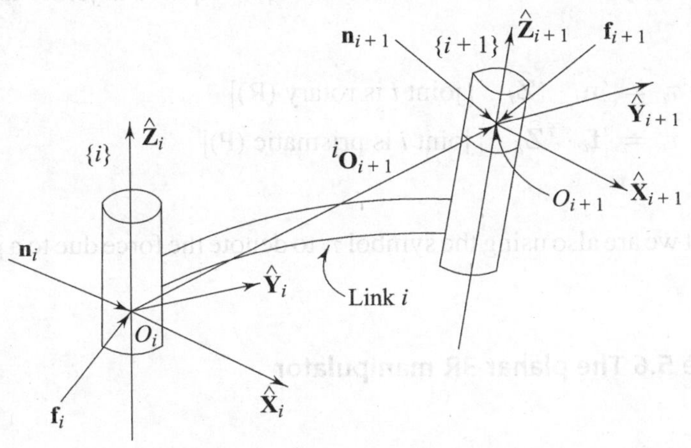

Fig. 5.12 Free-body diagram of a link

Figure 5.12 shows two intermediate rotary (R) joints and a link of a manipulator. We use the symbols $\mathbf{f}_i$ and $\mathbf{n}_i$ to denote the forces and moments, respectively, exerted on link $\{i\}$ by link $\{i-1\}$. For static equilibrium, the summation of forces and moments acting on link $\{i\}$ must be zero. Hence we have

$$^i\mathbf{f}_i - ^i\mathbf{f}_{i+1} = 0$$

The force $\mathbf{f}_{i+1}$ is the force on link $\{i+1\}$ exerted by link $\{i\}$, and hence the force on link $\{i\}$ exerted by link $\{i+1\}$ will be equal and of opposite sign. The leading superscript i signifies that the vectors are described in $\{i\}$. Summing the moments on link $\{i\}$ about its origin O_i, we get

$$^i\mathbf{n}_i \; - \; ^i\mathbf{n}_{i+1} \; - \; ^i\mathbf{O}_{i+1} \; \times \; ^i\mathbf{f}_{i+1} = 0$$

where $^i\mathbf{O}_{i+1}$ is the vector from O_i to O_{i+1}.

The above two equations can be re-arranged and written as

$$\begin{aligned} ^i\mathbf{f}_i &= {}^{\;\;i}_{i+1}[R] \; ^{i+1}\mathbf{f}_{i+1} \\ ^i\mathbf{n}_i &= {}^{\;\;i}_{i+1}[R] \; ^{i+1}\mathbf{n}_{i+1} + \; ^i\mathbf{O}_{i+1} \; \times \; ^i\mathbf{f}_i \end{aligned} \tag{5.63}$$

Equations (5.63) can be used to compute forces and moments when the forces and moments on the end-effector, $i = n$, are known. If the manipulator is not in contact with the environment or not carrying a payload, then ${}^{n+1}\mathbf{f}_{n+1} = {}^{n+1}\mathbf{n}_{n+1} = 0$.

To answer the question, what forces and torques need to be applied at the joints to keep the manipulator in static equilibrium, we recognize that the joints can apply forces or torques only along the $\hat{Z}$ axis. All other components are resisted by the structure. Hence, the torque required at joint i is given by

$$\begin{aligned} \tau_i &= {}^i\mathbf{n}_i \cdot {}^i\widehat{\mathbf{Z}}_i \quad \text{[joint } i \text{ is rotary (R)]} \\ \tau_i &= {}^i\mathbf{f}_i \cdot {}^i\widehat{\mathbf{Z}}_i \quad \text{[joint } i \text{ is prismatic (P)]} \end{aligned} \tag{5.64}$$

Note that we are also using the symbol τ_i to denote the force due to a prismatic joint.

Example 5.6 The planar 3R manipulator

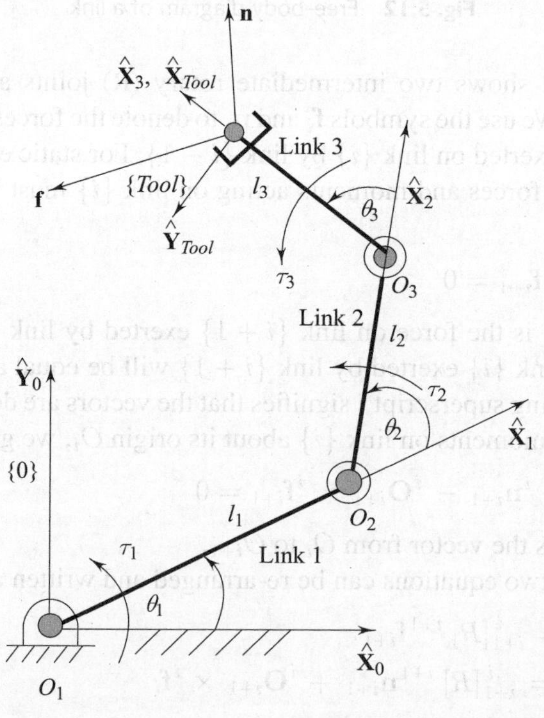

Fig. 5.13 A 3R manipulator applying force and moment

The three-link planar manipulator of Fig. 5.13 is applying a force and moment on the environment. These are given by

$$\begin{aligned} {}^0\mathbf{f}_{Tool} &= (f_x,\ f_y,\ 0)^T \\ {}^0\mathbf{n}_{Tool} &= (0,\ 0,\ n_z)^T \end{aligned}$$

To apply the algorithm, it is easier to obtain the forces ${}^{Tool}\mathbf{f}_{Tool}$ and ${}^{Tool}\mathbf{n}_{Tool}$. Denoting these by $(f'_x,\ f'_y,\ 0)^T$ and $(0,\ 0,\ n'_z)^T$, we get

$$\begin{pmatrix} f'_x \\ f'_y \\ 0 \end{pmatrix} = \begin{bmatrix} c_{123} & s_{123} & 0 \\ -s_{123} & c_{123} & 0 \\ 0 & 0 & 1 \end{bmatrix} \begin{pmatrix} f_x \\ f_y \\ 0 \end{pmatrix} \tag{5.65}$$

and

$$(0,\ 0,\ n'_z)^T = (0,\ 0,\ n_z)^T \tag{5.66}$$

Equation (5.66) results from the fact that the rotation matrix ${}^0_1[R]$ has (0, 0, 1) in the last column and row.

We now apply the iterative formulas given in Eqn (5.63) from the last link going towards the base.

$i = 3$

$$\begin{aligned} {}^3\mathbf{f}_3 &= (f'_x,\ f'_y,\ 0)^T \\ {}^3\mathbf{n}_3 &= (0, 0, n'_z + l_3 f'_y)^T \end{aligned}$$

$i = 2$

$$\begin{aligned} {}^2\mathbf{f}_2 &= (c_3 f'_x - s_3 f'_y,\ s_3 f'_x + c_3 f'_y,\ 0)^T \\ {}^2\mathbf{n}_2 &= (0,\ 0,\ n'_z + l_2(s_3 f'_x + c_3 f'_y) + l_3 f'_y)^T \end{aligned}$$

$i = 1$

$$\begin{aligned} {}^1\mathbf{f}_1 &= (c_{23} f'_x - s_{23} f'_y,\ s_{23} f'_x + c_{23} f'_y,\ 0)^T \\ {}^1\mathbf{n}_1 &= (0,\ 0,\ n'_z + l_1(s_{23} f'_x + c_{23} f'_y) + l_2(s_3 f'_x + c_3 f'_y) + l_3 f'_y)^T \end{aligned}$$

We can now calculate the torques required at the joints to keep the manipulator in equilibrium from Eqn (5.64). We have

$$\begin{aligned} \tau_1 &= {}^1\mathbf{n}_1 \cdot {}^1\hat{\mathbf{Z}}_1 = n'_z + f'_x(l_1 s_{23} + l_2 s_3) + f'_y(l_1 c_{23} + l_2 c_3 + l_3) \\ \tau_2 &= {}^2\mathbf{n}_2 \cdot {}^2\hat{\mathbf{Z}}_2 = n'_z + f'_x l_2 s_3 + f'_y(l_2 c_3 + l_3) \\ \tau_3 &= {}^3\mathbf{n}_3 \cdot {}^3\hat{\mathbf{Z}}_3 = n'_z + f'_y l_3 \end{aligned} \tag{5.67}$$

Using Eqns (5.65) and (5.66), Eqn (5.67) can be re-arranged as

$$\begin{pmatrix} \tau_1 \\ \tau_2 \\ \tau_3 \end{pmatrix} = \begin{bmatrix} -l_1 s_1 - l_2 s_{12} - l_3 s_{123} & l_1 c_1 + l_2 c_{12} + l_3 c_{123} & 0 & 0 & 0 & 1 \\ -l_2 s_{12} - l_3 s_{123} & l_2 c_{12} + l_3 c_{123} & 0 & 0 & 0 & 1 \\ -l_2 s_{123} & l_3 c_{123} & 0 & 0 & 0 & 1 \end{bmatrix} \begin{pmatrix} f_x \\ f_y \\ 0 \\ 0 \\ 0 \\ n_z \end{pmatrix} \tag{5.68}$$

It can be seen that the term in the square bracket is the transpose of the Jacobian matrix in Eqn (5.28). In fact this observation is true for any serial manipulator and to see why this is not a coincidence, we invoke the principal of virtual work.

As we had done in the case of linear and angular velocities, we denote the forces and moments acting on the end-effector by

$${}^0\mathcal{F}_{Tool} \triangleq \begin{pmatrix} {}^0\mathbf{f}_{Tool} \\ -- \\ {}^0\mathbf{n}_{Tool} \end{pmatrix} = (f_x\ f_y\ f_z;\ n_x\ n_y\ n_z)^T \tag{5.69}$$

It may be noted that ${}^0\mathcal{F}_{Tool}$ is not a 6×1 vector since forces and moments have different units. The quantity ${}^0\mathcal{F}_{Tool}$ is called a wrench in theoretical kinematics, and a wrench can be thought of as screw with a magnitude which has units of force. We will use '–' or ';' to separate the forces and moments.

The virtual work done by ${}^0\mathcal{F}_{Tool}$ for an infinitesimal Cartesian displacement of the end-effector ${}^0\delta\mathcal{X}_{Tool}$ is given by ${}^0\mathcal{F}_{Tool} \cdot {}^0\,\delta\mathcal{X}_{Tool}$.[7] This must be equal to the work done by the joint torques or forces given by $\boldsymbol{\tau} \cdot \delta\boldsymbol{\Theta}$, where $\boldsymbol{\tau}$ is the vector of joint torques or forces. Hence we have

$${}^0\mathcal{F}_{Tool} \cdot {}^0\delta\mathcal{X}_{Tool} \triangleq {}^0\mathbf{f}_{Tool} \cdot \delta\mathbf{x} + {}^0\mathbf{n}_{Tool} \cdot \delta\boldsymbol{\theta} = \boldsymbol{\tau} \cdot \delta\boldsymbol{\Theta}$$

However, from the definition of Jacobian, we have

$${}^0\delta\mathcal{X}_{Tool} = {}^0_{Tool}[J(\boldsymbol{\Theta})]\delta\boldsymbol{\Theta}$$

and hence we have

$${}^0\mathcal{F}_{Tool} \cdot {}^0_{Tool}[J(\boldsymbol{\Theta})]\delta\boldsymbol{\Theta} = \boldsymbol{\tau} \cdot \delta\boldsymbol{\Theta}$$

The above equations hold true for all $\delta\boldsymbol{\Theta}$, and hence we have

$$\boldsymbol{\tau} = {}^0_{Tool}[J(\boldsymbol{\Theta})]^T\, {}^0\mathcal{F}_{Tool} \tag{5.70}$$

[7] The quantity ${}^0\delta\mathcal{X}_{Tool}$ is not a 6×1 vector since it is composed of infinitesimal change in position and orientation of the end-effector. The infinitesimal change in position and orientation could be written in a form $(\delta\mathbf{x}; \delta\boldsymbol{\theta})^T$, where $\delta\mathbf{x} = (\delta x,\ \delta y,\ \delta z)^T$ denote the infinitesimal change in the position coordinates and $\delta\boldsymbol{\theta} = (\delta\alpha,\ \delta\beta,\ \delta\gamma)^T$ represent infinitesimal change in three Euler angles. One can also use some other representation of orientation to represent infinitesimal change in orientation. See also Section 2.8 for the notion of a *twist*.

It may be noted that if the forces and moments are acting at some other point (and not at the end-effector), or the forces and moments are given in some other coordinate system, then the Jacobian matrix is for that point or has to be described in that coordinate system. It may also be noted that if the Jacobian matrix is singular, then the end-effector cannot exert static forces as desired. This topic is discussed in detail in Section 5.9.

Equation (5.70) is a very useful relationship since it allows us to convert Cartesian forces and moments into joint torques. A similar relationship is derived for parallel manipulators and this is discussed in Section 5.8. The relationship between external forces and/or moments and joint torques will be used in Cartesian and force control schemes in later chapters.

5.8 Statics of Parallel Manipulators

For serial manipulators, the forces and torques at the actuated joints required to maintain static equilibrium are given by Eqn (5.70). This relationship is true for parallel manipulators also as the derivation of Eqn (5.70), from the principle of virtual work, is equally applicable to parallel manipulators. However, for parallel manipulators we should use

$$\boldsymbol{\tau} = {}^{0}_{Tool}[J_{\text{eq}}]^{T}\, {}^{0}\mathcal{F}_{Tool} \tag{5.71}$$

where ${}^{0}\mathcal{F}_{Tool}$ is the 6×1 vector of forces and moments applied or resisted by a chosen end-effector, ${}^{0}_{Tool}[J_{\text{eq}}]$ denotes the equivalent Jacobian which is a function of all the configuration variables $\mathbf{q}$, and $\boldsymbol{\tau}$ is taken to be the vector of forces or torques applied at the actuated joint alone—at the passive joints, there are no actuators. The equivalent Jacobian is given in Eqn (5.46), and as pointed out in Section 5.5, the equivalent Jacobian for a parallel manipulator requires obtaining $[K^*]^{-1}$ [see Eqns (5.44) and (5.45)]. This makes it very difficult to obtain closed-form expressions, except for very simple parallel manipulators. In addition, in parallel manipulators, we are often more interested in the force and moment that can be applied or resisted by them. Although we can obtain the force and moment exerted by an end-effector of a parallel manipulator from

$${}^{0}\mathcal{F}_{Tool} = {}^{0}_{Tool}[J(\mathbf{q})_{\text{eq}}]^{-T}\, \boldsymbol{\tau} \tag{5.72}$$

computing ${}^{0}_{Tool}[J(\mathbf{q})_{\text{eq}}]^{-T}$ is even harder.[8] To overcome these difficulties, we obtain a force transformation matrix (Merlet 2000, Agrawal & Roth 1992, Dasgupta & Mruthyunjaya 2000) which directly relates ${}^{0}\mathcal{F}_{Tool}$ to the

[8] We use the symbol $[(\cdot)]^{-T}$ to denote $([(\cdot)]^{T})^{-1}$ in this text.

actuated joint forces or torques. The procedure is illustrated with the help of a Stewart platform manipulator schematically shown in Fig. 4.2. As mentioned in Chapter 4, Section 4.5, a general 6—6 Stewart platform manipulator consists of a moving platform and a fixed base which are connected by six legs. A moving platform point P_i is connected to a base point B_i by a leg containing a passive spherical joint at one end, an actuated prismatic joint in the middle, and a passive universal or a Hooke joint at the other end. A typical leg of a Stewart platform manipulator is shown in Chapter 4, Fig. 4.6. The vector along the leg, denoted by ${}^{B_0}\mathbf{S}_i$, is given in Eqn (4.39) and is reproduced here for convenience.

$$ {}^{B_0}\mathbf{S}_i = {}^{B_0}_{P_0}[R]\,{}^{P_0}\mathbf{p}_i + {}^{B_0}\mathbf{t} - {}^{B_0}\mathbf{b}_i \tag{5.73} $$

We can define a unit vector along ${}^{B_0}\mathbf{S}_i$ as

$$ {}^{B_0}\mathbf{s}_i = \frac{{}^{B_0}\mathbf{S}_i}{l_i} \tag{5.74} $$

where the magnitude of the leg vector is l_i.

Denoting the force exerted by the actuated prismatic joint by f_i, the corresponding force vector is $f_i^{B_0}\mathbf{s}_i$. The line of action of the force vector is located by ${}^{B_0}\mathbf{b}_i$ and hence the corresponding moment vector is $f_i({}^{B_0}\mathbf{b}_i \times {}^{B_0}\mathbf{s}_i)$. Denoting the external force and moment acting at a point, instantaneously coincident with the origin of $\{B_0\}$ and on the moving platform, by ${}^0\mathcal{F}_{Tool}$, we get

$$ {}^{B_0}\mathcal{F}_{Tool} \triangleq \begin{pmatrix} {}^{B_0}\mathbf{F}_{Tool} \\ \hline {}^{B_0}\mathbf{M}_{Tool} \end{pmatrix} = \begin{bmatrix} \sum_{i=1}^{6} {}^{B_0}\mathbf{s}_i f_i \\ \hline \sum_{i=1}^{6} ({}^{B_0}\mathbf{b}_i \times {}^{B_0}\mathbf{s}_i) f_i \end{bmatrix} \tag{5.75} $$

where ${}^{B_0}\mathbf{F}_{Tool}$, ${}^{B_0}\mathbf{M}_{Tool}$ are the 3×1 force and moment vectors, respectively, acting on the chosen $\{Tool\}$. It may be noted that the vectors are described in a fixed coordinate system $\{B_0\}$.

Equation (5.75) can be written in a matrix form

$$ {}^{B_0}\mathcal{F}_{Tool} = {}^{B_0}_{Tool}[\,H\,]\mathbf{f} \tag{5.76} $$

where the force transformation matrix ${}^{B_0}_{Tool}[\,H\,]$ is given by

$$ {}^{B_0}_{Tool}[\,H\,] = \begin{bmatrix} {}^{B_0}\mathbf{s}_1 & {}^{B_0}\mathbf{s}_2 & \cdots & {}^{B_0}\mathbf{s}_6 \\ \hline {}^{B_0}\mathbf{b}_1 \times {}^{B_0}\mathbf{s}_1 & {}^{B_0}\mathbf{b}_2 \times {}^{B_0}\mathbf{s}_2 & \cdots & {}^{B_0}\mathbf{b}_6 \times {}^{B_0}\mathbf{s}_6 \end{bmatrix} \tag{5.77} $$

and $\mathbf{f}$ is the vector of forces applied at the prismatic joints $(f_1, f_2, \ldots, f_6)^T$. It may be noted that, like the manipulator Jacobian matrix, the force transformation matrix is not a proper matrix in the linear algebra sense.

The top and bottom halves of ${}^{B_0}_{Tool}[\,H\,]$ represent two different quantities and have different units.

Comparing Eqn (5.76) with Eqn (5.72), we can see that ${}^{B_0}_{Tool}[\,H\,]$ is indeed the same as ${}^{0}_{Tool}[J(\mathbf{q})_{\text{eq}}]^{-T}$—given the joint forces, we can obtain the forces acting on the moving platform by ${}^{0}_{Tool}[J(\mathbf{q})_{\text{eq}}]^{-T}$. The advantage of using ${}^{0}_{Tool}[\,H\,]$ in a Stewart platform is that we do not need to compute the inverse of $[K^*]$ for the equivalent Jacobian and the inverse of the equivalent Jacobian. As mentioned in Eqn (4.45), there are 12 loop-closure constraint equations for a Stewart platform and obtaining the inverse of $[K^*]$, a 12×12 matrix, would be almost impossible analytically and extremely difficult numerically.

The force transformation matrix is easily obtained for any purely parallel manipulator (see Exercise 5.14). If the actuated joint in the leg of the purely parallel manipulator is a rotary joint, then the force transformation matrix will be of the form

$$ {}^{B_0}_{Tool}[\,H\,] = \begin{bmatrix} {}^{B_0}\mathbf{b}_1 \times {}^{B_0}\mathbf{s}_1 & {}^{B_0}\mathbf{b}_2 \times {}^{B_0}\mathbf{s}_2 & \dots & {}^{B_0}\mathbf{b}_n \times {}^{B_0}\mathbf{s}_n \\ \hline {}^{B_0}\mathbf{s}_1 & {}^{B_0}\mathbf{s}_2 & \dots & {}^{B_0}\mathbf{s}_n \end{bmatrix} \tag{5.78} $$

for n legs connecting a moving platform to a fixed base.

For a parallel manipulator with more than one actuated joint in a leg (as in the six-DOF example discussed in Example 2.6), more effort is required to obtain $[\,H\,]$. One approach would be to consider each leg, up to the connection point on the moving platform or the end-effector, as a serial manipulator and then obtain, by following the algorithm described in Section 5.7, the relationship between the forces or torques at the actuated joints and the force exerted by the end of each leg on the end-effector. Once all the forces and moments exerted on the end-effector are known, we can use equations of static equilibrium to obtain the relationship between the actuated forces or torques in each leg and the moving end-effector. Alternately, we can use

$$ {}^{B_0}_{Tool}[\,H\,] = {}^{0}_{Tool}[J(\mathbf{q})_{\text{eq}}]^{-T} \tag{5.79} $$

if obtaining and inverting the equivalent Jacobian is not very hard.

5.9 Singularity in Force Domain

If the force transformation matrix can be inverted, then we can obtain the actuated joint forces or torques given the external force and moment applied on or resisted by the end-effector. We can get

$$ \mathbf{f} = {}^{B_0}_{Tool}[\,H\,]^{-1}\, {}^{B_0}\mathcal{F}_{Tool} \tag{5.80} $$

If the force transformation matrix is singular, i.e., $\det([\,H\,]) = 0$, then we cannot obtain the actuated joint forces or torques for a given externally applied $\mathcal{F}_{Tool}$. Similar to the situation when the velocity Jacobian is singular, the actuated joint forces or torques $\mathbf{f}$ tend to infinity. We can get a clearer understanding of this phenomenon by constructing a force ellipsoid similar to the linear velocity ellipsoid in the case of the serial manipulators. We use the example of the Stewart platform discussed earlier to illustrate the force ellipsoid.

From Eqn (5.75) the external force $\mathbf{F}$ can be written as

$$\mathbf{F} = [H_{\mathbf{F}}]\mathbf{f} = \begin{bmatrix} \mathbf{s}_1 & \mathbf{s}_2 & \mathbf{s}_3 & \mathbf{s}_4 & \mathbf{s}_5 & \mathbf{s}_6 \end{bmatrix} \mathbf{f} \tag{5.81}$$

The square of the magnitude of $\mathbf{F}$ can be obtained by taking the dot product with itself, and denoting $[H_{\mathbf{F}}]^T[H_{\mathbf{F}}]$ by $[g_{\mathbf{F}}]$, we can write

$$\mathbf{F}^T\mathbf{F} = \mathbf{f}^T[g_{\mathbf{F}}]\mathbf{f} \tag{5.82}$$

The maximum, intermediate, and minimum values of $\mathbf{F}^T\mathbf{F}$ subject to a constraint of the form $\mathbf{f}^T\mathbf{f} = 1$ are the eigenvalues of $[g_{\mathbf{F}}]$ and since the rank of $[g_{\mathbf{F}}]$ is 3 ($[H_{\mathbf{F}}]$ has at most rank 3), similar to the linear velocity ellipsoid, the tip of the force vector $\mathbf{F}$ lies on an ellipsoid in $\Re^3$. If the rank of $[g_{\mathbf{F}}]$ drops to 2, the force ellipsoid shrinks to an ellipse and the Stewart platform manipulator cannot apply a force normal to the plane of the ellipse. If the rank of $[g_{\mathbf{F}}]$ drops to 1, the Stewart platform cannot apply any force in a plane. If for any parallel manipulator the rank of $[g_{\mathbf{F}}]$ is zero, it cannot apply any external force. As an example, we consider a Stewart platform with the fixed base and the moving platform as regular hexagons of same size. In addition, we consider the configuration of all legs parallel to the vertical. In this case the $[\,H\,]$ matrix is given by

$$[H] = \begin{pmatrix} 0 & 0 & 0 & 0 & 0 & 0 \\ 0 & 0 & 0 & 0 & 0 & 0 \\ 1 & 1 & 1 & 1 & 1 & 1 \\ b_{1_y} & b_{2_y} & b_{3_y} & b_{4_y} & b_{5_y} & b_{6_y} \\ -b_{1_x} & -b_{2_x} & -b_{3_x} & -b_{4_x} & -b_{5_x} & -b_{6_x} \\ 0 & 0 & 0 & 0 & 0 & 0 \end{pmatrix} \tag{5.83}$$

where the vector $\mathbf{b}_i$ locating the base points is given by $(b_{i_x}, b_{i_y}, 0)^T$, where $i = 1, 2, ..., 6$.

In this case, since three rows are null, $\det([\,H\,])$ is clearly zero and the Stewart platform is in a singular configuration. In addition, $[H_{\mathbf{F}}]$ has $(0, 0, 1)$ in all its columns and clearly the matrix $[g_{\mathbf{F}}]$ has rank 1. In this case the tip of the force vector $\mathbf{F}$ can only lie along a line. For this example, one can easily visualize that only external force along the vertical direction can be resisted and all forces in the horizontal plane cannot be resisted. Hence, the Stewart platform in this configuration has singularity along F_x and F_y.

Similar to the force ellipsoid, we can also consider the moment vector $\mathbf{M}$ part of the $[\,H\,]$ matrix to obtain directions along which the Stewart platform cannot apply moments. In the example considered above, the bottom half of the $[\,H\,]$ is also of rank 2, and one can intuitively see that moments along the vertical axis, M_z, cannot be resisted.

The above intuitive analysis is not possible to extend to general Stewart platforms in arbitrary configurations. To obtain singular directions of a general Stewart platform, we can obtain eigenvectors of $[g_{\mathbf{F}}]$ corresponding to zero eigenvalues and then map these to the $[\mathbf{F};\mathbf{M}]$ by $[\,H\,]$. These eigenvectors are the singular directions of force and moments. It may be mentioned that the eigensystem analysis can be done analytically since it involves the solution of at most cubic equations. Using this approach we can obtain configurations of the Stewart platform which cannot apply various components of forces or moments (Ranganath et al. 2004).

If the prismatic joints of a Stewart platform are locked, it becomes a structure and the Stewart platform can resist forces and moments along any direction. However, if the force transformation matrix is singular, the Stewart platform, with its joints locked, cannot resist external applied force or moment along certain directions, and an infinitely small force or moment applied to the Stewart platform along certain directions will result in instantaneous motion. To understand the relationship between $\det([\,H\,]) = 0$ and instantaneous gain of one or more degrees of freedom, we again consider the example of simple planar four-bar mechanism.

With the actuated joint locked, the point O_2 in a four-bar mechanism will also become fixed as shown in Fig. 5.14(a). For a given θ_1, l_0, and l_1, the length d opposite to θ_1 is known. In Fig. 5.14(b), we draw the planar truss structure determined by link 2, link 3, and the side O_2—O_R which is now considered to be fixed. It may be noted that the angles α_1 and α_2 can be computed in terms of θ_1, ϕ_1, and ϕ_2.

We now consider a force $\mathbf{F}$, with components F_x, F_y, acting at an angle β (measured with respect to a line parallel to O_2—O_R) at point O_3. We denote the axial forces[9] along the links O_2—O_3 and O_3—O_R of the planar truss by T_1 and T_2, respectively. By using a free-body diagram we can obtain

$$\begin{pmatrix} F_x \\ F_y \end{pmatrix} = \begin{bmatrix} \cos\alpha_1 & -\cos\alpha_2 \\ \sin\alpha_1 & \sin\alpha_2 \end{bmatrix} \begin{pmatrix} T_1 \\ T_2 \end{pmatrix} \tag{5.84}$$

[9] The planar truss has rotary joints and hence the forces along the members can only be axial.

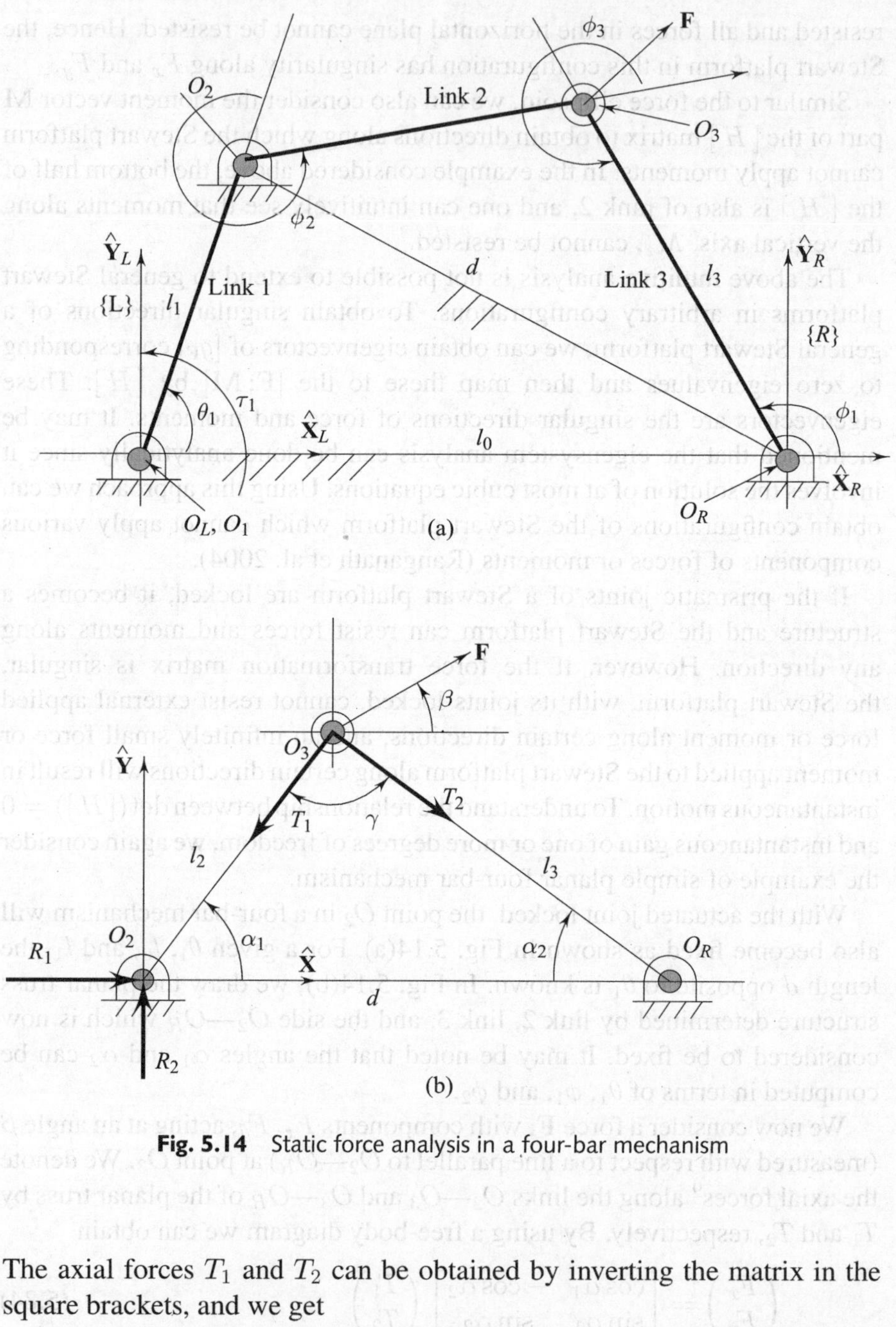

Fig. 5.14 Static force analysis in a four-bar mechanism

The axial forces T_1 and T_2 can be obtained by inverting the matrix in the square brackets, and we get

$$\begin{pmatrix} T_1 \\ T_2 \end{pmatrix} = \frac{1}{\sin(\alpha_1 + \alpha_2)} \begin{bmatrix} \sin\alpha_2 & \cos\alpha_2 \\ -\sin\alpha_1 & \cos\alpha_1 \end{bmatrix} \begin{pmatrix} F_x \\ F_y \end{pmatrix} \tag{5.85}$$

From T_1, we can now obtain the reactions R_1 and R_2 at the joint O_2 [see Fig. 5.14(b)] and these are given by

$$\begin{pmatrix} R_1 \\ R_2 \end{pmatrix} = \frac{1}{\sin(\alpha_1 + \alpha_2)} \begin{bmatrix} \cos\alpha_1 \sin\alpha_2 & \cos\alpha_1 \cos\alpha_2 \\ \sin\alpha_1 \sin\alpha_2 & \sin\alpha_1 \cos\alpha_2 \end{bmatrix} \begin{pmatrix} F_x \\ F_y \end{pmatrix} \quad (5.86)$$

and the torque required in joint 1, τ_1, to keep the four-bar mechanism in equilibrium is given by

$$\begin{aligned} \tau_1 &= R_1 l_1 s_1 - R_2 l_1 c_1 \\ &= \frac{1}{\sin(\alpha_1 + \alpha_2)} [l_1 s_1 \quad - l_1 c_1] \begin{bmatrix} \cos\alpha_1 \sin\alpha_2 & \cos\alpha_1 \cos\alpha_2 \\ \sin\alpha_1 \sin\alpha_2 & \sin\alpha_1 \cos\alpha_2 \end{bmatrix} \begin{pmatrix} F_x \\ F_y \end{pmatrix} \end{aligned} \quad (5.87)$$

Equation (5.87) is similar to Eqn (5.80) and the matrix in the square brackets is the inverse of the force transformation matrix, namely $[\ H\ ]^{-1}$, for the planar four-bar mechanism—the force transformation matrix, $[\ H\ ]$, in this case is the 2×2 matrix inside the square brackets in Eqn (5.84). The matrix $[\ H\ ]^{-1}$ does not exist if $\sin(\alpha_1 + \alpha_2) = 0$ or the angle γ [see Fig. 5.14(b)] is π radians. This in turn implies that ϕ_3 in the four-bar mechanism is 2π, and, geometrically, the links 2 and 3 of the four-bar mechanism are aligned.

With $\alpha_1 + \alpha_2$ as zero (or γ as π), the axial force $T_1 \to \infty$ except if $\mathbf{F}$ is along link 2 or link 3 (or if $\beta = 0$)—if β is zero and for γ as π, the axial force T_1 has a finite value of $|\mathbf{F}|/2$. The singular direction for $\mathbf{F}$ corresponds to $\beta = \pi/2$ with links 2 and 3 aligned. From Eqns (5.86) and (5.87) any force applied along the singular direction will give rise to infinite reactions R_1 and R_2, and infinite actuated joint torque τ_1 will be required to maintain static equilibrium.

In Example 5.4, we had shown that when links 2 and 3 of the four-bar mechanism were aligned, $\det[K^*]$ was zero and the four-bar mechanism gained a degree of freedom instantaneously. Comparing this result with the above analysis, we can see that $\det[K^*] = 0$ is equivalent to $\det[\ H\ ] = 0$ at least for the four-bar mechanism. The direction of the gained degree of freedom was shown in Fig. 5.8, and, again from the above analysis, we can see that the direction of gained degree of freedom is the same as the singular direction of force. We can conclude that, at least for the four-bar mechanism, the loss of rank of $[K^*]$ is equivalent to the loss of rank of $[\ H\ ]$. This result can be proved for general parallel manipulators from the principle of virtual work, and is left as an exercise problem for the readers.

5.10 Resolution of Redundancy at Velocity Level

In Section 5.4, it was mentioned that for a square Jacobian matrix we can obtain the joint rates for given linear and angular velocities of the end-effector [see Eqn (5.29)]. In redundant manipulators (see Chapter 3, Section 3.8) the Jacobian matrix is not square—the number of joint variables is more than six for spatial motion (or more than three for planar motion). For such redundant manipulators, to obtain $\dot{\Theta}$, a common approach is to use the pseudo-inverse of the Jacobian matrix (Strang 1976). This technique is discussed in this section.

We consider the relationship

$$^0\mathcal{V}_{Tool} = {}^0_{Tool}[J(\Theta)]\dot{\Theta} \tag{5.88}$$

with the dimension of $^0\mathcal{V}_{Tool}$ as m, the dimension of $\dot{\Theta}$ as n, and the Jacobian matrix is $m \times n$ with $n > m$. Physically, we have more number of joints than what is required to achieve the motion in the task space. For example, in a planar 3R manipulator if we are interested only in the linear velocity $(\dot{x}, \dot{y})^T$ of the end-effector, then the manipulator is redundant since there are three joint rates $(\dot{\theta}_1, \dot{\theta}_2, \dot{\theta}_3)^T$.

The pseudo-inverse of an $m \times n$ $(n > m)$ matrix $[J(\Theta)]$ is defined as

$$[J(\Theta)]^{\#} = [J(\Theta)]^T([J(\Theta)][J(\Theta)]^T)^{-1} \tag{5.89}$$

and we can get

$$\dot{\Theta} = [J(\Theta)]^{\#0}\mathcal{V}_{Tool} + ([U] - [J(\Theta)]^{\#}[J(\Theta)])\dot{\mathcal{W}} \tag{5.90}$$

where $[U]$ denotes the identity matrix and $([U] - [J(\Theta)]^{\#}[J(\Theta)])\dot{\mathcal{W}}$ is an arbitrary vector from the null space of $[J(\Theta)]$. The pseudo-inverse solution, without the null-space term, has the attractive property of minimizing joint rates—one can show (see Strang 1976) that the computed $\dot{\Theta}$ for a given $^0\mathcal{V}_{Tool}$ minimizes $\dot{\Theta}^T\dot{\Theta}$. The null-space term has been used by several researchers to avoid obstacles, maximize a manipulability index defined by $\det([[J(\Theta)][J(\Theta)]^T]^{1/2})$, and to avoid joint limits. These are discussed in detail in the textbook by Nakamura (1991) and the references contained in this textbook. We, however, make a few observations regarding the resolution of redundancy using the pseudo-inverse of the Jacobian matrix.

1. The pseudo-inverse based schemes are purely local and numerical, and, in general, it is not possible to obtain any analytical results. For example, although we know that the joint rates are minimized at each position, it is not clear what is the minimum in a trajectory in the workspace.
2. The pseudo-inverse schemes operate at the velocity level and not at the position and orientation level.
3. Intuitively, we expect that the local (or global) properties at a position and orientation in $\Re^3$ (or along a trajectory in $\Re^3$) will be different if reached (or traversed) by a redundant manipulator as compared to

a non-redundant manipulator. The pseudo-inverse scheme, due to its numerical nature, does not yield insights into these differences.

An approach, not using the pseudo-inverse, is presented in Ghosal and Roth (1988). In this approach, the authors make use of the concept of velocity ellipse (ellipsoid) and use redundancy to obtain isotropic conditions in a planar redundant 3R manipulator. The redundant joint rate is computed in a manner so that the end-effector can achieve isotropy over a large region of the workspace of the redundant 3R manipulator, and, unlike the example of the planar 2R discussed at the end of Section 5.6, the isotropy is not limited to one circle in the workspace and does not occur only for special ratios of link lengths. The isotropy treatment is restricted to linear velocity and can be extended to angular velocity. However, the basic nature of the Jacobian matrix restricts its extension to a combination of linear and angular velocities.

Exercises

5.1 In the development of the angular velocity vector, we defined a space-fixed and a body-fixed angular velocity vector. Can you think of any example where the body-fixed angular velocity would be put to use?

5.2 Derive the space-fixed and body-fixed angular velocity matrices and vectors for a rotation matrix obtained from a X-Y-Z Euler angle rotations of a rigid body.

5.3 Given a rotation matrix ${}^{A}_{B}[R]$ in terms of the axis and angle, i.e., as $[R(\widehat{\mathbf{k}}, \phi)]$, obtain the space-fixed and body-fixed angular velocity vector components in terms of $\widehat{\mathbf{k}}$ and ϕ and their derivatives with respect to time.

5.4 Derive the recursive equations, Eqn (5.25), for a serial manipulator with two consecutive links connected by prismatic (P) joints.

†5.5 Using MAPLE derive the expressions for the elements of the Jacobian matrix of a PUMA 560 robot described in Example 2.2.

5.6 Obtain the Jacobian matrix for the planar 3R manipulator in $\{Tool\}$. What is the general rule for transforming a $6 \times n$ Jacobain matrix between two coordinate systems?

5.7 Prove that $[g]$ is positive definite. Obtain the eigenvectors for $[g]$ corresponding to the eigenvalues λ_1 and λ_2 given in Eqn (5.34). What do the eigenvectors mean?

5.8 For the five-bar mechanism shown in Fig. 4.10, obtain $[K]$ and $[K^*]$. Describe geometrically, similar to the planar four-bar example, what happens when $\det([K^*]) = 0$.

†5.9 Verify using MAPLE the expressions for the elements of $[J]$, $[J^*]$, $[K]$, and $[K^*]$ obtained for the three-DOF parallel manipulator discussed in Example 5.3.

5.10 Sketch the eight configurations where all the eigenvalues of $\det([K^*])$ are zeros for the three-DOF parallel manipulator discussed in Example 5.5.

†5.11 Write a MATLAB program to search and obtain a configuration of the three-DOF manipulator where the linear velocity ellipsoid degenerates to an ellipse.

†5.12 Write a MATLAB program to search and obtain a configuration where the three-DOF manipulator gains a degree of freedom.

†5.13 Using MAPLE obtain the expressions for the elements of $[J]$, $[J^*]$, $[K]$, and $[K^*]$ for the six-DOF parallel manipulator discussed in Examples 2.6 and 4.3.

†5.14 Obtain the force transformation matrix for the three-DOF parallel manipulator discussed in Example 5.3.

†5.15 Obtain ${}^{0}_{Tool}[\,H\,]$ for the six-DOF parallel manipulator discussed in Examples 2.6 and 4.3.

5.16 Using the principle of virtual work, show that $\det[K^*] = 0$ implies $\det[\,H\,] = 0$ and vice versa for an general parallel manipulator or a closed-loop mechanism.

†5.17 For the redundant planar 3R manipulator, discussed in Chapter 3, Section 3.8, numerically obtain and plot the values of $\dot{\theta}_1$, $\dot{\theta}_2$, and $\dot{\theta}_3$ for a straight line trajectory along a line parallel to the $\widehat{\mathbf{Y}}$ axis. Assume $l_1 = 5$, $l_2 = 3$, $l_3 = 1$ m and assume that the tip of the manipulator moves at a constant speed of 1 m/sec.

5.18 For a six-DOF serial manipulator in a singular configuration, it is known that all the joint axes intersect a line perpendicularly. For the PUMA 560 discussed in Example 2.2 obtain the equation of this line.

5.19 Sketch the three-DOF parallel manipulator of Example 5.2 when it gains one degree of freedom. What general result can you guess about the configurations when this manipulator gains a degree of freedom.

5.20 For the six-DOF parallel manipulator discussed in Examples 2.6 and 4.3, obtain and sketch a configuration where it will lose one degree of freedom.

References and Suggested Additional Reading

Agrawal, S.K. and B. Roth 1992, 'Statics of in-parallel manipulator systems', *Trans. ASME, J. Mech. Eng. Des.*, vol. 114, pp. 564–68.

Bandyopadhyay, S. and A. Ghosal 2004a, 'Analysis of configuration space singularities of closed loop mechanisms and parallel manipulators', *Mech. Mach. Theory*, vol. 39, pp. 519–44.

Bandyopadhyay, S. and A. Ghosal 2004b, 'Analytical determination of principal twists in serial, parallel and hybrid manipulators using dual vectors and matrices', *Mech. Mach. Theory*, vol. 39, pp. 1289–1305.

Basu, D. and A. Ghosal 1997 'Singularity analysis of platform-type multi-loop spatial mechanism', *Mech. Mach. Theory*, vol. 32, pp. 375–89.

Dasgupta, B. and T.S. Mruthyunjaya 1998, 'Force redundancy in parallel manipulators: theoretical and practical issues', *Mech. Mach. Theory*, vol. 33, pp. 727–42.

Ghosal, A. and B. Ravani 1998, 'A dual ellipse is a cylindroid', *Proc. ASME DETC'98*, Atlanta, USA, September 1998.

Ghosal, A. and B. Roth 1988, 'A new approach for kinematic resolution of redundancy', *Int. J. Robotic. Res.*, vol. 7, pp. 22–35.

Ghosal, A. and B. Ravani 2001, 'Differential geometric analysis of singularities of point trajectories of serial and parallel manipulators', *Trans. ASME J. Mech. Design*, vol. 123, pp. 80–89.

Gosselin, C. and J. Angeles 1990, 'Singularity analysis of closed loop kinematic chains', *IEEE Trans. Robotic. Autom.*, vol. 6, pp. 281–90.

Hunt, K.H. 1986, 'Special configurations of robot arms via screw theory, Part 1. The Jacobian and its matrix cofactors', *Robotica*, vol. 4, pp. 171–79.

Hunt, K.H., A.E. Samuel, and P.R. McAree 1991, 'Special configurations of multi-finger multi-freedom gripper—A kinematic study', *Int. J. Robotic. Res.*, vol. 10, pp. 123–34.

Litvin, F.L., Y. Zhang, V. Parenti Castelli, and C. Innocenti 1990, 'Singularities, configurations and displacement functions for manipulators', *Int. J. Robotic. Res.*, vol. 5, pp. 52–65.

Merlet, J.-P. 1991, 'Singularity configurations of parallel manipulators and Grassman geometry', *Int. J. Robotic. Res.*, vol. 10, no. 2, pp. 123–34.

Merlet, J.-P. 2001, *Parallel Robots*, Kluwer Academic, Dordrecht.

Nakamura, Y. 1991, *Advanced Robotics: Redundancy and Optimization*, Addison-Wesley,

Park, F.C. and J. W. Kim 1999, 'Singularity analysis of closed loop kinematic chains', *Trans. ASME, J. Mech. Eng. Des.*, vol. 121, no. 1, pp. 32–38.

Ranganath, R., P.S. Nair, T.S. Mruthyunjaya, and A. Ghosal 2004, 'A force-torque sensor based on a Stewart platform in a near singular configuration', *Mech. Mach. Theory*, vol. 39, pp. 971–98.

Salisbury, J.K. 1982, *Kinematics and force analysis of articulated hands*, PhD Thesis, Stanford University.

Strang, G. 1976, *Linear Algebra and its Application*, Academic Press.

Sugimoto, K., J. Duffy, and K.H. Hunt, 1982, 'Special configurations of spatial mechanisms and robot arms', *Mech. Mach. Theory*, vol. 17, pp. 119–32.

Wang, S.L. and K.J. Waldron 1987, 'A study of the singular configurations of serial manipulators', *Trans. ASME J. Mech. Transm. Autom. Des.*, vol. 109, pp. 14–20.

Zlatanov, D., R.G. Fenton, and B. Benhabib 1995, 'A unifying framework for classification and interpretation of mechanism singularities', *Trans. ASME J. Mech. Eng. Des.*, vol. 117, no. 4, pp. 566–72.

6 Dynamics of Manipulators

6.1 Introduction

In previous chapters we discussed the kinematics and statics of serial and parallel manipulators where we did not introduce the causes of motion. In this chapter, we consider the forces required to cause the motion and discuss methods to derive equations of motion of serial and parallel manipulators. The main assumption we make in discussing dynamics of manipulators is that the links and joints are rigid. This is never true since the links are always flexible to some extent, even though most industrial robots are built quite solidly. In addition, due to transmission elements, such as gears, there is always some flexibility at the joints. Flexibility of links and joints introduces significant complexity in the kinematics and dynamics of manipulators, and this has been a subject of active research in the robotics community. In Chapter 9, we will discuss modelling and simulation of flexible-link manipulators.

There are several ways to derive the equations of motion of a manipulator. The two well-known methods are the Newton–Euler and Lagrangian formulations.[1] The Newton–Euler formulation involves determining the linear and angular accelerations of each link of a manipulator and the use of the well-known concept of a free-body diagram of a link wherein all the forces and moments acting on the link are described. The final step is to use the familiar Newton's law and Euler's equation to each free-body (link) and using the constraints at the joints. The Lagrangian formulation is based on computing the scalar kinetic and potential energy of each link of a manipulator in terms of generalized coordinates and their derivatives with respect to time. The next step is to determine a scalar function, called the Lagrangian, and finally to take partial and ordinary derivatives of the Lagrangian with respect to generalized coordinates, their derivatives and time. Clearly, a given manipulator gives the same equations of motion by all methods.

[1] One can also use Kane's method to derive dynamic equations of motion, for details see Kane (1983).

The Lagrangian formulation can be easily automated in the sense that the explicit symbolic equations of motion can be derived using symbolic manipulation software such as MATHEMATICA or MAPLE.[2] The Lagrangian formulation can be directly applied to serial manipulators. For parallel manipulators and closed-loop mechanisms with loop-closure constraint equations, we use the concept of Lagrange multipliers which are related to the constraint forces. It may be mentioned that there are several commercially available software packages, such as ADAMS (2002) which can be readily used for numerical simulations of a wide range of mechanical systems including manipulators. However, any deeper analysis, such as the nature or importance of the terms in equations of motion, is only possible with a symbolic derivation of the equations of motion.

There are two main problems in manipulator dynamics. The direct problem deals with the motion of a manipulator given the externally applied forces and moments. This problem is related to the simulation of the manipulator and involves the solution of ordinary differential equations or differential algebraic equations with given initial conditions. The inverse problem deals with obtaining joint torques or forces given the motion of the manipulator as a function of time and requires algebraic and trigonometric computations. The inverse problem is useful for sizing of actuators in a robot and for model-based control (the topic of control is discussed in detail in Chapter 8). One of the main focusses of past researchers in the field of manipulator and multi-body dynamics has been the issue of computational efficiency for inverse and direct problems. This issue has now found renewed interest in the field of computational biology, wherein algorithms first developed by robot researchers are being used to simulate the folding of proteins[3] (Klepeis et al. 2002).

With N denoting the number of links, the first Lagrangian formulation based algorithm for the inverse dynamics problem of a serial manipulator was shown to be of $\mathcal{O}(N^4)$ (Uicker 1967, Hollerbach 1983) whereas the recursive Newton–Euler algorithm for a serial manipulator was known to be $\mathcal{O}(N)$ (Orin et al. 1979, Luh et al. 1980).[4] Very soon the Lagrangian

---

[2] The Newton–Euler and Kane's methods can and have been automated and there exist commercially available software packages based on these methods.

[3] Protein molecules consist of a large number of carbon and other atoms subjected to intermolecular forces and arranged in a serial chain often with closed loops. Starting from an initial configuration, the protein molecule changes shape or folds to a unique configuration under the action of forces from the environment and other intermolecular forces. Since the chains are very large, of the order of 1000 'links', fast recursive algorithms are required to simulate the motion of these chains.

[4] In the complexity analysis of algorithms for manipulator dynamics, typically only the number of multiplications and divisions are counted. An $\mathcal{O}(N)$ algorithm implies that the number of multiplication or divisions increases linearly with N, and although the constant coefficient is often not mentioned, it may be large and hence important.

formulation was shown to be equivalent to the Newton–Euler algorithm, i.e., $\mathcal{O}(N)$ (Silver 1982). For the direct problem, Walker and Orin (1982) used the techniques of the inverse problem to derive an $\mathcal{O}(N^3)$ algorithm. An $\mathcal{O}(N)$ algorithm for manipulators with spherical joints was first developed by Armstrong (1979), and later Featherstone (1983, 1987) developed an $\mathcal{O}(N)$ algorithm for general serial manipulators with rotary and prismatic joints.

In the case of parallel manipulators, with closed loops, there is the additional complexity of passive joints and constraints arising from the non-linear, holonomic loop-closure equations. It is generally very difficult to eliminate these passive joint variables, and algorithms, typically, use both passive and actuated joint variables to develop the equations of motion. Once the equations of motion, in terms of actuated and passive joints, are derived, they need to be solved together with the loop-closure constraint equations. Instead of ordinary differential equations (ODEs) in the case of serial manipulators, for parallel manipulators and closed-loop mechanisms, we need to solve a set of ordinary differential equations and algebraic equations also known as DAE's.

In this chapter, we use the Lagrangian formulation to obtain equations of motion of serial and parallel manipulators. In Section 6.7 of the chapter, we briefly present and discuss recursive algorithms for serial and parallel manipulators. We start our discussion with the notion of inertia of rigid bodies and links of a manipulator.

6.2 Inertia of a Link

Unlike a point mass, for a rigid body, we often talk of the inertia of the rigid body. The inertia of a rigid body, in three-dimensional space, is described by the so-called inertia tensor. Figure 6.1 shows an arbitrary rigid body along with a coordinate system $\{0\}$ fixed to the rigid body. The inertia tensor of the rigid body, defined with respect to the coordinate system $\{0\}$, is given by

$$ {}^0[I] = \begin{bmatrix} I_{xx} & I_{xy} & I_{xz} \\ I_{xy} & I_{yy} & I_{yz} \\ I_{xz} & I_{yz} & I_{zz} \end{bmatrix} \tag{6.1} $$

The elements of the tensor are given by

$$ I_{xx} = \int_V (y^2 + z^2)\, \rho\, dV, \quad I_{xy} = -\int_V xy\, \rho\, dV, $$

$$ I_{xz} = -\int_V xz\, \rho\, dV, \qquad I_{yy} = \int_V (x^2 + z^2)\, \rho\, dV, $$

$$I_{yz} = -\int_V yz\,\rho\,dV, \quad I_{zz} = \int_V (x^2 + y^2)\,\rho\,dV \tag{6.2}$$

where the rigid body is composed of differential volume elements dV with density ρ. A representative volume element located at ${}^0\mathbf{p} = (x,\, y,\, z)^T$ is shown in Fig. 6.1, and the mass m of the rigid body is given by $\int_V \rho dV$.

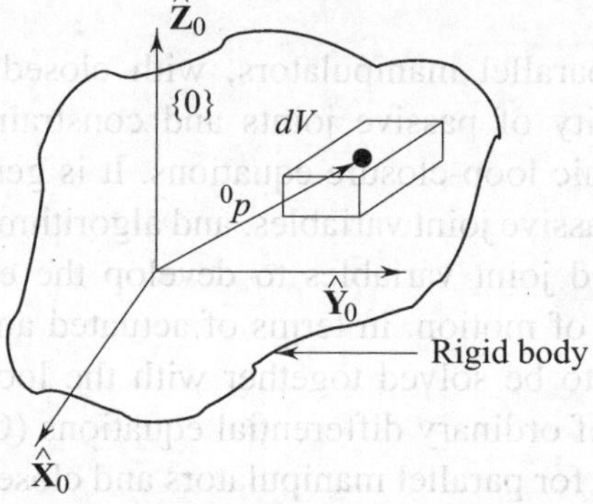

Fig. 6.1 Inertia of a rigid body

The inertia tensor is positive definite and symmetric. Hence the eigenvalues of the 3×3 matrix ${}^0[I]$ are real and positive. The three eigenvalues are called the principal moments of inertias and the associated eigenvectors are called the principal axes.[5] The inertia tensor in any other rotated coordinate system $\{A\}$, with its origin coincident with $\{0\}$, can be obtained as

$${}^A[I] = {}^A_0[R]\,{}^0[I]\,{}^A_0[R]^T \tag{6.3}$$

It can be shown that a rotation matrix whose columns are the eigenvectors associated with the eigenvalues can be used to daigonalize the matrix ${}^0[I]$ [see any textbook on linear algebra, such as by Strang (1976), for similarity transformations].

To obtain the elements of the inertia tensor for a link i, we will fix the coordinate system, $\{C_i\}$, at the centre of mass of the link. The orientation of the $\{C_i\}$ coordinate system is parallel to the link coordinate system $\{i\}$ (see the DH convention in Chapter 2). To obtain the inertia tensor of a geometrically complex robot link, we can divide the link of the robot into simple shapes whose inertias are known from standard tables. Then we make use of the well-known parallel axis theorem.

[5] If the rigid body possesses symmetry with two or more of the eigenvalues as equal, the associated eigenvectors are arbitrary—for example, for a cylinder, any two perpendicular directions in the circular cross section are the principal directions.

6.3 The Lagrangian Formulation

In the Lagrangian formulation, we compute the kinetic and potential energies of each of the links of a manipulator. The kinetic energy of link i with mass m_i and inertia ${}^0[I]_i$ is given as

$$\mathrm{KE}_i = \frac{1}{2}\, m_i\, {}^0\mathbf{V}_{C_i} \cdot {}^0\mathbf{V}_{C_i} + \frac{1}{2}\, {}^0\boldsymbol{\omega}_i \cdot {}^0[I]_i\, {}^0\boldsymbol{\omega}_i \tag{6.4}$$

where the first term is due to the linear velocity of the link's centre of mass and the second term is due to the angular velocity of the link. The quantities ${}^0\mathbf{V}_{C_i}$ and ${}^0\boldsymbol{\omega}_i$ are the linear and angular velocities of the centre of mass and link $\{i\}$, respectively, with respect to a fixed coordinate system $\{0\}$. The linear and angular velocities with respect to $\{0\}$ can be expressed in $\{i\}$ (see Section 5.2), and we can write

$$\begin{aligned} {}^0\mathbf{V}_{C_i} &= {}^0_i[R]\, {}^i\mathbf{V}_{C_i} \\ {}^0\boldsymbol{\omega}_i &= {}^0_i[R]\, {}^i\boldsymbol{\omega}_i \end{aligned} \tag{6.5}$$

and substituting Eqns (6.5) in Eqn (6.4), we get

$$\mathrm{KE}_i = \frac{1}{2}\, m_i\, {}^i\mathbf{V}_{C_i} \cdot {}^i\mathbf{V}_{C_i} + \frac{1}{2}\, {}^i\boldsymbol{\omega}_i \cdot {}^{C_i}[I]_i\, {}^i\boldsymbol{\omega}_i \tag{6.6}$$

where we have used the fact that ${}^i_0[R]^T\, {}^i_0[R]$ is the identity matrix, the link coordinate system $\{i\}$ is parallel to $\{C_i\}$. We used Eqn (6.3) for transforming the inertia matrix in $\{C_i\}$ coordinate system.

The expression for ${}^i\boldsymbol{\omega}_i$, in terms of ${}^{i-1}\boldsymbol{\omega}_{i-1}$ for serial manipulators with rotary and prismatic joints, was derived in Section 5.3 and we restate them here for convenience. We can write

$$\begin{aligned} {}^i\boldsymbol{\omega}_i &= {}^{i}_{i-1}[R]\, {}^{i-1}\boldsymbol{\omega}_{i-1} + \dot{\theta}_i (0\ 0\ 1)^T && \text{(joint } i \text{ is rotary)} \\ &= {}^{i}_{i-1}[R]\, {}^{i-1}\boldsymbol{\omega}_{i-1} && \text{(joint } i \text{ is prismatic)} \end{aligned} \tag{6.7}$$

These equations can be used in an iterative manner, from the fixed base ($i = 0$), to obtain the angular velocity of all the links of the serial manipulator.

Similarly, the expression for the linear velocity of the centre of mass of link $\{i\}$ can be obtained from Eqn (5.24), and we can write

$${}^i\mathbf{V}_{C_i} = {}^i\mathbf{V}_i + {}^i\boldsymbol{\omega}_i \times {}^i\mathbf{p}_{C_i} \tag{6.8}$$

where the position vector ${}^i\mathbf{p}_{C_i}$ locates the centre of mass with respect to the origin O_i of the link coordinate system $\{i\}$. Again, the above equation can be used to iteratively obtain the linear velocity of the centre of mass starting from the fixed base.

In the case of parallel manipulators, it is easier to use the derivative formulas as discussed in Section 5.5. We restate the formulas for the angular velocity and the linear velocity of the centre of mass of a link $\{i\}$ in terms of the time derivatives of the rotation matrix and the position vector, respectively. We can write

$$
\begin{aligned}
{}^{0}\boldsymbol{\omega}_i &= {}^{0}_{i}[\dot{R}]\,{}^{0}_{i}[R]^T \\
{}^{0}\mathbf{V}_{C_i} &= \frac{d}{dt}({}^{0}\mathbf{p}_{C_i})
\end{aligned}
\tag{6.9}
$$

where ${}^{0}\mathbf{p}_{C_i}$ is the position vector of the centre of mass of link i. In closed-loop mechanisms or parallel manipulators, we can also judiciously use the iterative formulas for serial portions. It may be noted that the angular and linear velocities will be in terms of the actuated and the passive joint variables and their derivatives, i.e., functions of $\mathbf{q}$ and $\dot{\mathbf{q}}$.

The potential energy[6] of link i is defined as

$$
\text{PE}_i = -m_i {}^{0}\mathbf{g} \cdot {}^{0}\mathbf{p}_{C_i} \tag{6.10}
$$

where ${}^{0}\mathbf{g}$ is the gravity vector (of magnitude 9.81 m/sec^2), and ${}^{0}\mathbf{p}_{C_i}$ is the location of the centre of mass of link i from the zero or reference potential energy surface. Since we will take derivatives to derive the equations of the manipulator, the value of the reference potential energy does not matter.

Once the kinetic and potential energies of the individual links are known, we define a scalar, called the Lagrangian, of the system. The Lagrangian is a function of the generalized coordinates $\mathbf{q}$ and its derivative $\dot{\mathbf{q}}$ and is given as

$$
\mathcal{L}(\mathbf{q}, \dot{\mathbf{q}}) = \sum_{i}^{N} (\text{KE}_i - \text{PE}_i) \tag{6.11}
$$

where KE_i and PE_i are the kinetic and potential energies of a link as derived in Eqns (6.4) and (6.10), respectively, and N is the total number of links (excluding the fixed link). In serial manipulators, with single-DOF joints (θ_i or d_i for rotary or prismatic joints) the dimension n of the vector of generalized coordinates $\mathbf{q}$ is the same as N.

Once the Lagrangian $\mathcal{L}(\mathbf{q}, \dot{\mathbf{q}})$ of a mechanical system is known by using techniques from calculus of variation (see Goldstein 1980), we can obtain

[6] In this development we have assumed that potential energy is due to gravity alone. If springs or other energy-storage devices are present, we can appropriately modify the expression of the potential energy of the link. For example, if torsional springs are present at joint i, then we add a term of the form $\frac{1}{2}k_i\theta_i^2$ (see Example 6.5) in the expression for the potential energy.

the equations of motion. For serial manipulators with no loop closure or other constraints, the equations of motion are given by

$$\frac{d}{dt}\left(\frac{\partial \mathcal{L}}{\partial \dot{q}_i}\right) - \frac{\partial \mathcal{L}}{\partial q_i} = Q_i, \quad i = 1, 2, \ldots, n \tag{6.12}$$

where Q_i's are the externally applied generalized forces. In case the joint torques or forces are the only externally applied generalized forces, we get

$$Q_i = \tau_i, \quad i = 1, \ldots, n \tag{6.13}$$

where the symbol τ_i is used to denote a torque in the case of a rotary joint or a force in the case of a prismatic joint.

Once the derivatives with respect to q_i, $\dot{q}_i$, and t are done, the equations of motion of a serial manipulator can be written in a matrix form as

$$[\mathbf{M}(\mathbf{q})]\ddot{\mathbf{q}} + [\mathbf{C}(\mathbf{q}, \dot{\mathbf{q}})]\dot{\mathbf{q}} + \mathbf{G}(\mathbf{q}) = \boldsymbol{\tau} \tag{6.14}$$

where $[\mathbf{M}(\mathbf{q})]$ is the $n \times n$ mass matrix, $[\mathbf{C}(\mathbf{q}, \dot{\mathbf{q}})]$ is the $n \times n$ matrix with $[\mathbf{C}(\mathbf{q}, \dot{\mathbf{q}})]\dot{\mathbf{q}}$ representing an $n \times 1$ vector of centripetal and Coriolis terms, [7] $\mathbf{G}(\mathbf{q})$ is the $n \times 1$ vector containing the gravity terms, and $\boldsymbol{\tau}$ is $n \times 1$ vector of joint torques or forces. The mass matrix is always positive definite and symmetric, and in particular the total kinetic energy of a serial manipulator is given by

$$\mathrm{KE} = \frac{1}{2}\dot{\mathbf{q}}^T[\mathbf{M}(\mathbf{q})]\dot{\mathbf{q}} \tag{6.15}$$

and from physical arguments $\mathrm{KE} \geq 0$ for $|\dot{\mathbf{q}}| \neq \mathbf{0}$ and zero only when $|\dot{\mathbf{q}}| = \mathbf{0}$. The components of the $n \times n$ matrix, $[\mathbf{C}(\mathbf{q}, \dot{\mathbf{q}})]$, containing the centripetal and Coriolis terms can be obtained from the mass matrix as

$$C_{ij} = \frac{1}{2}\sum_{k=1}^{n}\left(\frac{\partial M_{ij}}{\partial q_k} + \frac{\partial M_{ik}}{\partial q_j} - \frac{\partial M_{kj}}{\partial q_i}\right)\dot{q}_k \tag{6.16}$$

Finally the gravity terms can be obtained from the expression of the potential energy as

$$G_i = \frac{\partial(\mathrm{PE})}{\partial q_i} \tag{6.17}$$

In parallel manipulators and closed-loop mechanisms, we have m loop-closure (holonomic) constraint equations of the form

$$\eta_i(\mathbf{q}) = 0, \quad i = 1, 2, \ldots, m \tag{6.18}$$

[7] In this text we will interchangeably use $[\mathbf{C}(\mathbf{q}, \dot{\mathbf{q}})]\dot{\mathbf{q}}$ and $\mathbf{C}(\mathbf{q}, \dot{\mathbf{q}})$ to denote the centripetal and Coriolis terms.

where with the n actuated joint variables $\boldsymbol{\theta}$, we also have m passive joint variables ϕ making the dimension of $\mathbf{q}$ as $n+m$ (see Section 4.4). To obtain the equations of motion, we use the concept of *Lagrange multipliers* (Goldstein 1980, Haug 1989) and write

$$\bar{\mathcal{L}}(\mathbf{q},\dot{\mathbf{q}}) = \mathcal{L}(\mathbf{q},\dot{\mathbf{q}}) - \sum_{j=1}^{m} \lambda_j \eta_j(\mathbf{q}) \tag{6.19}$$

where the $m\lambda_j$'s are the introduced Lagrange multipliers which need to be determined.

Using calculus of variation and noting that the constraints $\eta_j(\mathbf{q}) = 0$ are only functions of $\mathbf{q}$ (or holonomic), equations of motion can be written as

$$\frac{d}{dt}\left(\frac{\partial \bar{\mathcal{L}}}{\partial \dot{q}_i}\right) - \frac{\partial \bar{\mathcal{L}}}{\partial q_i} = \tau_i + \sum_{j=1}^{m} \lambda_j \frac{\partial \eta_j(\mathbf{q})}{\partial q_i},$$
$$i = 1, 2, \ldots, n+m \tag{6.20}$$

and, in a matrix form, we can write

$$[\mathbf{M}(\mathbf{q})]\ddot{\mathbf{q}} + [\mathbf{C}(\mathbf{q},\dot{\mathbf{q}})]\dot{\mathbf{q}} + \mathbf{G}(\mathbf{q}) = \boldsymbol{\tau} + [\boldsymbol{\Psi}(\mathbf{q})]^{\mathrm{T}}\boldsymbol{\lambda} \tag{6.21}$$

where $\boldsymbol{\lambda}$ is the $m \times 1$ vector of Lagrange multipliers, the constraint matrix $[\boldsymbol{\Psi}(\mathbf{q})]$ is obtained from the partial derivatives of m constraint equations, given in Eqn (6.18), with respect to q_i, $\mathbf{q}$ is the $(n+m)\times 1$ vector of generalized coordinates, and all other terms have the same meaning as those for serial manipulators. It may be noted that matrix $[\boldsymbol{\Psi}]$ is a concatenation of the matrices $[K]$ and $[K^*]$ discussed in Section 5.5.

To determine the Lagrange multipliers, we twice differentiate the m constraint equations with respect to t and get

$$[\boldsymbol{\Psi}(\mathbf{q})]\ddot{\mathbf{q}} + [\dot{\boldsymbol{\Psi}}(\mathbf{q})]\dot{\mathbf{q}} = \mathbf{0} \tag{6.22}$$

where $[\dot{\boldsymbol{\Psi}}]$ is a $m \times (n+m)$ matrix containing the time derivatives of each of the elements of $[\boldsymbol{\Psi}]$.

Since the mass matrix is always invertible, we can write

$$\ddot{\mathbf{q}} = [\mathbf{M}]^{-1}(\boldsymbol{\tau} - [\mathbf{C}]\dot{\mathbf{q}} - \mathbf{G}) + [\mathbf{M}]^{-1}[\boldsymbol{\Psi}]^{\mathrm{T}}\boldsymbol{\lambda} \tag{6.23}$$

where we dropped the functional dependence of the terms on $\mathbf{q}$ and $\dot{\mathbf{q}}$ for convenience. Substituting $\ddot{\mathbf{q}}$ in Eqn (6.22), we get after rearranging

$$\boldsymbol{\lambda} = -([\boldsymbol{\Psi}][\mathbf{M}]^{-1}[\boldsymbol{\Psi}]^{\mathrm{T}})^{-1}\{[\dot{\boldsymbol{\Psi}}]\dot{\mathbf{q}} + [\boldsymbol{\Psi}][\mathbf{M}]^{-1}(\boldsymbol{\tau} - [\mathbf{C}]\dot{\mathbf{q}} - \mathbf{G})\} \tag{6.24}$$

Finally, $\boldsymbol{\lambda}$ can be substituted back into Eqn (6.21) to get the equations of motion as

$$[\mathbf{M}]\ddot{\mathbf{q}} = \mathbf{f} - [\boldsymbol{\Psi}]^T([\boldsymbol{\Psi}][\mathbf{M}]^{-1}[\boldsymbol{\Psi}]^T)^{-1}\{[\boldsymbol{\Psi}][\mathbf{M}]^{-1}\mathbf{f} + [\dot{\boldsymbol{\Psi}}]\dot{\mathbf{q}}\} \quad (6.25)$$

where $\mathbf{f}$ denotes $(\boldsymbol{\tau} - [\mathbf{C}]\dot{\mathbf{q}} - \mathbf{G})$.

It may be noted that the mass matrix $[\mathbf{M}]$ is a $(n+m)\times(n+m)$ positive definite and symmetric matrix and the centripetal/Coriolis terms and the gravity terms are $(n+m)\times 1$ vectors. The quantity $[\boldsymbol{\Psi}(\mathbf{q})]^T\boldsymbol{\lambda}$ contains the constraint forces and moments associated with the loop-closure constraints (see also Chapter 5). Although the constraint forces and moments do not do any work (see Exercise 6.3), the constraint forces and moments are often required for a mechanical design of the joints and links, and many software packages, such as ADAMS (2002), compute and provide the constraint forces and moments.

The above procedure can be adapted in the case of multi-body systems subjected to non-holonomic, non-integrable constraints and constraints containing explicit functions of time. In such a situation, we start with constraints given in the so-called Pfaffian form[8]

$$\boldsymbol{\Phi}(t) + [\boldsymbol{\Psi}(\mathbf{q})]\dot{\mathbf{q}} = \mathbf{0} \quad (6.26)$$

Differentiating the constraint Eqn (6.26), we get

$$[\boldsymbol{\Psi}]\ddot{\mathbf{q}} + [\dot{\boldsymbol{\Psi}}]\dot{\mathbf{q}} + \dot{\boldsymbol{\Phi}}(t) = \mathbf{0} \quad (6.27)$$

where the dependence of $[\boldsymbol{\Psi}]$ on $\mathbf{q}$ is dropped for convenience.

In the case of non-integrable non-holonomic constraints, it is not possible to write a Lagrangian as in Eqn (6.19) using the Lagrange multipliers. However, the constraint forces and moments arising out of the non-holonomic constraints can still be accommodated in the equations of motion on the right-hand side, and we can obtain the equations of motion as

$$\begin{aligned}[\mathbf{M}]\ddot{\mathbf{q}} = \mathbf{f} - [\boldsymbol{\Psi}]^T([\boldsymbol{\Psi}][\mathbf{M}]^{-1}[\boldsymbol{\Psi}]^T)^{-1}\{[\boldsymbol{\Psi}][\mathbf{M}]^{-1}\mathbf{f}\\ +\dot{\boldsymbol{\Phi}}(t) + [\dot{\boldsymbol{\Psi}}]\dot{\mathbf{q}}\}\end{aligned} \quad (6.28)$$

and $\boldsymbol{\lambda}$ is given by

$$\begin{aligned}\boldsymbol{\lambda} = -([\boldsymbol{\Psi}][\mathbf{M}]^{-1}[\boldsymbol{\Psi}]^T)^{-1}\{\dot{\boldsymbol{\Phi}}(t) + [\dot{\boldsymbol{\Psi}}]\dot{\mathbf{q}}\\ +[\boldsymbol{\Psi}][\mathbf{M}]^{-1}(\boldsymbol{\tau} - [\mathbf{C}]\dot{\mathbf{q}} - \mathbf{G})\}\end{aligned} \quad (6.29)$$

The equations of motion, Eqn (6.14), (6.25), or (6.28) does not contain effects of friction, flexibility, backlash, and other unmodelled dynamics.

---

[8] We use non-holonomic constraints for modelling and simulation of a wheeled mobile robot in Chapter 10.

To take into account friction, equations of motion are modified by adding a dissipative term $\mathbf{F}(\mathbf{q}, \dot{\mathbf{q}})$ to the right-hand side—we assume that the friction, in general, may be a function of generalized coordinates and its derivative with respect to time. It may be noted that this term is very much different from the Coriolis and centripetal terms which do not do work. The equations of motion for a serial manipulator can then be written as

$$\boldsymbol{\tau} = [\mathbf{M}(\mathbf{q})]\ddot{\mathbf{q}} + \mathbf{C}(\mathbf{q}, \dot{\mathbf{q}}) + \mathbf{G}(\mathbf{q}) + \mathbf{F}(\mathbf{q}, \dot{\mathbf{q}}) \tag{6.30}$$

and similar equations of motion of a parallel manipulator or a closed-loop mechanism can be derived by adding $\mathbf{F}(\mathbf{q}, \dot{\mathbf{q}})$ at the appropriate place.

Example 6.1: Equations of motion of a planar 2R manipulator

As mentioned earlier, the Lagrangian formulation is quite 'mechanical' in nature. The evaluation of the kinetic and potential energies, and obtaining the partial and ordinary derivatives, can be easily automated using a symbolic software tool such as MATHEMATICA or MAPLE. The mass matrix, the centripetal and Coriolis terms, and the gravity term can be extracted from expressions of the kinetic energy and the potential energy [see Eqns (6.15) to (6.17)]. In this example, we provide a trace of the steps used to obtain the symbolic equation of a planar 2R manipulator using MAPLE.

Figure 6.2 shows a two-link, planar 2R manipulator. The gravity vector is along the negative ${}^0\widehat{\mathbf{Y}}_0$ axis. The geometrical and inertial properties of each link i are given by the set (m_i, l_i, r_i, I_i) denoting its mass, length, location

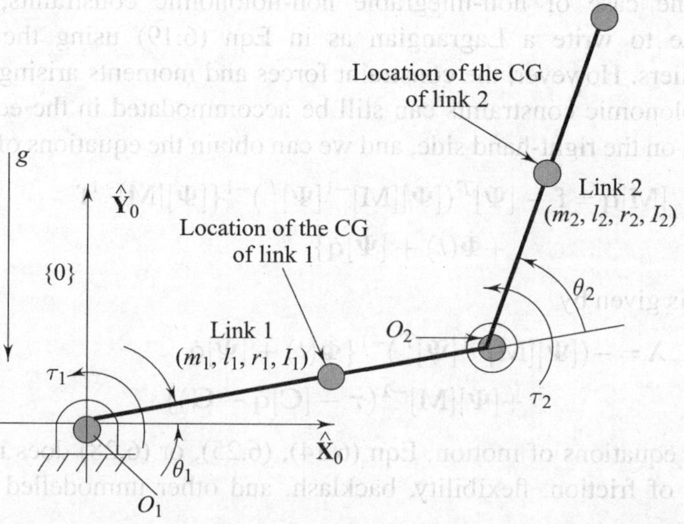

Fig. 6.2 A 2R manipulator

of the centre of mass (assumed to be along the $\widehat{\mathbf{X}}_i$), and the I_{zz} component of the inertia matrix, respectively. Since the motion is in the X-Y plane, only the I_{zz} component of the inertia matrix of each link is relevant in this example. The generalized coordinates in this example are the joint variables θ_1 and θ_2 and the joint torques are denoted by τ_1 and τ_2.

We use the expressions for linear and angular velocity of links 1 and 2 derived in Example 5.1 and then derive the expressions for the total kinetic and potential energies of the 2R manipulator. The total kinetic energy of the links is given by

$$\mathrm{KE} = \frac{1}{2}m_1(r_1\dot{\theta}_1)^2 + \frac{1}{2}I_1\dot{\theta}_1^2 + \frac{1}{2}I_2(\dot{\theta}_1+\dot{\theta}_2)^2 + \frac{1}{2}m_2\Big[l_1^2\dot{\theta}_1^2 + r_2^2(\dot{\theta}_1+\dot{\theta}_2)^2 + 2l_1r_2c_2\dot{\theta}_1(\dot{\theta}_1+\dot{\theta}_2)\Big] \tag{6.31}$$

where the first two terms are for link 1 and the second two terms are for link 2, respectively. The total potential energy is given by

$$\mathrm{PE} = m_1 g r_1 s_1 + m_2 g(l_1 s_1 + r_2 s_{12}) \tag{6.32}$$

We begin by giving as input to MAPLE the expression for the total kinetic and potential energies.[9] It may be noted that the inputs to MAPLE start with a sign > and have ASCII text fonts whereas the MAPLE outputs are in standard math fonts.

```
>  restart;
>  KE:=  1/2*m[1]*(r[1]*dtheta[1])^2
>  + 1/2* I[1]*dtheta[1]^2
>  + 1/2*I[2]*(dtheta[1] + dtheta[2])^2
>  + 1/2*m[2]*(l[1]^2*dtheta[1]^2
>  + r[2]^2*(dtheta[1]+dtheta[2])^2
>  +2*l[1]*r[2]*cos(theta[2])*dtheta[1]*(dtheta[1]
>  + dtheta[2])  ); #0
```

$$\mathrm{KE} := \frac{1}{2}\,m_1\,r_1^2\,dtheta_1^2 + \frac{1}{2}\,I_1\,dtheta_1^2 + \frac{1}{2}\,I_2\,(dtheta_1 + dtheta_2)^2 + \frac{1}{2}m_2(l_1^2\,dtheta_1^2 + r_2^2\,(dtheta_1 + dtheta_2)^2 + 2\,l_1\,r_2\cos(\theta_2)\,dtheta_1\,(dtheta_1 + dtheta_2))$$

```
>  PE:= m[1]*g*r[1]*sin(theta[1])
>  + m[2]*g*(l[1]*sin(theta[1])
>  + r[2]*sin(theta[1]+theta[2]));
```

[9] We can also start with expressions for linear and angular velocities derived in Example 5.1 and let MAPLE compute the kinetic and potential energies of each link and then the total kinetic and potential energy. However, for the sake of brevity, we skip these steps in this example.

$$\text{PE} := m_1\, g\, r_1 \sin(\theta_1) + m_2\, g\, (l_1 \sin(\theta_1) + r_2 \sin(\theta_1 + \theta_2))$$

where dtheta[1] and dtheta[2] are used as symbols for $\dot{\theta}_1$ and $\dot{\theta}_2$, respectively.

With the total kinetic and potential, we now compute the Lagrangian.

```
>   L:=KE-PE;
```

$$L := \frac{1}{2}\, m_1\, r_1^2\, dtheta_1^2 + \frac{1}{2}\, I_1\, dtheta_1^2 + \frac{1}{2}\, I_2\, (dtheta_1 + dtheta_2)^2$$
$$+ \frac{1}{2} m_2 (l_1^2\, dtheta_1^2 + r_2^2\, ((dtheta_1 + dtheta_2)^2$$
$$+ 2\, l_1\, r_2 \cos(\theta_2)\, dtheta_1\, (dtheta_1 + dtheta_2))$$
$$-\, m_1\, g\, r_1 \sin(\theta_1) - m_2\, g\, (l_1 \sin(\theta_1) + r_2 \sin(\theta_1 + \theta_2))$$

To make MAPLE understand that θ_1, θ_2, dtheta[1], and dtheta[2] are also functions of time t, we declare the variable, symb_time in the next step. This variable is used later for substitution.

```
>   symb_time:=
>   {theta[1]=theta[1](t),dtheta[1]=diff(theta[1](t),
>   t),theta[2]=theta[2]
>   (t),dtheta[2]=diff(theta[2](t),t)};
```

$$symb\_time := \{\theta_1 = \theta_1(t),\, dtheta_1 = \tfrac{d}{dt}\, \theta_1(t),\, \theta_2 = \theta_2(t),\, dtheta_2 = \tfrac{d}{dt}\, \theta_2(t)\}$$

We start with the partial differentiation of the Lagrangian with respect to θ_1 and θ_2.

```
>   diff(L,theta[1]);
```

$$-m_1\, g\, r_1 \cos(\theta_1) - m_2\, g\, (l_1 \cos(\theta_1) + r_2 \cos(\theta_1 + \theta_2))$$

```
>   diff(L,theta[2]);
```

$$-m_2\, l_1\, r_2 \sin(\theta_2)\, dtheta_1\, (dtheta_1 + dtheta_2) - m_2\, g\, r_2 \cos(\theta_1 + \theta_2)$$

In the next step, we perform the partial differentiation of the Lagrangian with respect to $\dot{\theta}_1$ and $\dot{\theta}_2$, simplify and collect various terms. The terms after partial differentiation with respect to $\dot{\theta}_1$ and $\dot{\theta}_2$ are denoted by temp_1 and temp_2, respectively.

```
>   temp_1:=collect(simplify(subs(symb_time,diff
>   (L,dtheta[1]))),
>    {m[1],m[2]});
```

$$temp\_1 := m_1\, {r_1}^2\, \%1 +$$
$$({l_1}^2\, \%1 + l_1\, r_2 \cos(\theta_2(t))\, (\tfrac{d}{dt}\, \theta_2(t)) + {r_2}^2\, \%1 + {r_2}^2\, (\tfrac{d}{dt}\, \theta_2(t))$$
$$+\, 2\, l_1\, r_2 \cos(\theta_2(t))\, \%1) m_2 + I_1\, \%1 + I_2\, \%1 + I_2\, (\tfrac{d}{dt}\, \theta_2(t))$$
$$\%1 := \tfrac{d}{dt}\, \theta_1(t)$$

The use of the substitution symb_time in the above step can be noticed. In addition, it may be noted that %1 is used to denote $(d/dt)\theta_1(t)$ in the above expressions. This is to fit the full expressions in a single line and does not appear in the actual MAPLE output.

```
> temp_2:=collect(simplify(subs(symb_time,
> diff(L,dtheta[2]))),
> {m[1],m[2]});
```

$$temp\_2 := (r_2{}^2\,(\tfrac{d}{dt}\,\theta_1(t)) + r_2{}^2\,(\tfrac{d}{dt}\,\theta_2(t)) + l_1\,r_2\cos(\theta_2(t))\,(\tfrac{d}{dt}\,\theta_1(t)))\,m_2 + I_2\,(\tfrac{d}{dt}\,\theta_1(t)) + I_2\,(\tfrac{d}{dt}\,\theta_2(t))$$

Next we differentiate temp_1 and temp_2 with respect to t:

```
> diff(temp_1,t);
```

$$m_1\,r_1{}^2\,\%1 + (l_1{}^2\,\%1 - l_1\,r_2\sin(\theta_2(t))\,(\tfrac{d}{dt}\,\theta_2(t))^2 + l_1\,r_2\cos(\theta_2(t))\,(\tfrac{d^2}{dt^2}\,\theta_2(t)) + r_2{}^2\,\%1 + r_2{}^2\,(\tfrac{d^2}{dt^2}\,\theta_2(t)) - 2\,l_1\,r_2\sin(\theta_2(t))\,(\tfrac{d}{dt}\,\theta_2(t))\,(\tfrac{d}{dt}\,\theta_1(t)) + 2\,l_1\,r_2\cos(\theta_2(t))\,\%1)m_2 + I_1\,\%1 + I_2\,\%1 + I_2\,(\tfrac{d^2}{dt^2}\,\theta_2(t))$$

$$\%1 := \tfrac{d^2}{dt^2}\,\theta_1(t)$$

Note The use of %1, to denote $(d^2/dt^2)\theta_1(t)$ in the above expressions, is to fit the full expressions in a single line and does not appear in the actual MAPLE output.

```
> diff(temp_2,t);
```

$$(r_2{}^2\,(\tfrac{d^2}{dt^2}\,\theta_1(t)) + r_2{}^2\,(\tfrac{d^2}{dt^2}\,\theta_2(t)) - l_1\,r_2\sin(\theta_2(t))\,(\tfrac{d}{dt}\,\theta_2(t))\,(\tfrac{d}{dt}\,\theta_1(t)) + l_1\,r_2\cos(\theta_2(t))\,(\tfrac{d^2}{dt^2}\,\theta_1(t)))m_2 + I_2\,(\tfrac{d^2}{dt^2}\,\theta_1(t)) + I_2\,(\tfrac{d^2}{dt^2}\,\theta_2(t))$$

Finally, we get the two equations of motion by performing the following steps in MAPLE.

```
> eqn[1]:= collect(simplify(diff(subs(symb_time,
> diff(L,dtheta[1])),t)
> - subs(symb_time,diff(L,theta[1]))
> - tau[1],trig),{diff(theta[1](t),t,t),
> diff(theta[2](t),t,t)});
> eqn[2]:= collect(simplify(diff(subs
> (symb_time,diff(L,dtheta[2])),t)
> - subs(symb_time,diff(L,theta[2]))
> -tau[2],trig),
> {diff(theta[1](t),t,t),
> diff(theta[2](t),t,t)});
```

The MAPLE output of the last two inputs can be rearranged by taking τ_1 and τ_2 to the right-hand side, and rewriting the resulting equation with the symbols $\ddot{\theta}_i$, $\dot{\theta}_i$, θ_i, $i = 1, 2$. We get

$$\begin{aligned}
\tau_1 = & \ddot{\theta}_1(I_1 + I_2 + m_2 l_1^2 + m_1 r_1^2 + m_2 r_2^2 + 2 m_2 l_1 r_2 c_2) \\
& + \ddot{\theta}_2(I_2 + m_2 r_2^2 + m_2 l_1 r_2 c_2) - m_2 l_1 r_2 s_2 (2\dot{\theta}_1 + \dot{\theta}_2)\dot{\theta}_2 \\
& + m_2 g(l_1 c_1 + r_2 c_{12}) + m_1 g r_1 c_1 \\
\tau_2 = & \ddot{\theta}_1(I_2 + m_2 r_2^2 + m_2 l_1 r_2 c_2) + \ddot{\theta}_2(I_2 + m_2 r_2^2) \\
& + m_2 l_1 r_2 s_2 \dot{\theta}_1^2 + m_2 r_2 g c_{12}
\end{aligned} \tag{6.33}$$

The above two second-order ordinary differential equations can be written in the standard matrix form as

$$\begin{pmatrix} \tau_1 \\ \tau_2 \end{pmatrix} = \begin{bmatrix} I_1 + I_2 + m_2 l_1^2 + m_1 r_1^2 + m_2 r_2^2 + 2 m_2 l_1 r_2 c_2 & I_2 + m_2 r_2^2 + m_2 l_1 r_2 c_2 \\ I_2 + m_2 r_2^2 + m_2 l_1 r_2 c_2 & I_2 + m_2 r_2^2 \end{bmatrix} \begin{pmatrix} \ddot{\theta}_1 \\ \ddot{\theta}_2 \end{pmatrix}$$
$$+ \begin{pmatrix} -m_2 l_1 r_2 s_2 (2\dot{\theta}_1 + \dot{\theta}_2)\dot{\theta}_2 \\ m_2 l_1 r_2 s_2 \dot{\theta}_1^2 \end{pmatrix} + \begin{pmatrix} m_2 g(l_1 c_1 + r_2 c_{12}) + m_1 g r_1 c_1 \\ m_2 r_2 g c_{12} \end{pmatrix} \tag{6.34}$$

where the 2×2 matrix is called the mass matrix, the 2×1 vector containing the quadratic terms $\dot{\theta}_1\dot{\theta}_2$ etc. is called the centripetal/Coriolis term, and the 2×1 vector containing g is called the gravity term. The matrix Eqn (6.34) or the two ODEs in Eqn (6.33) represent the closed-form dynamic equations of motion of a planar 2R manipulator.

Example 6.2: Equations of motion of a planar four-bar mechanism

Figure 6.3 shows the planar four-bar mechanism. The geometry and the inertial parameters of each of the three links are given by the set (m_i, l_i, r_i, I_i) for $i = 1, 2, 3$. Again, since the motion is in a plane, only the I_{zz} component of the inertia matrix of each link is relevant.

```
> restart;
```

In this example it is useful to start MAPLE with the in-built package for linear algebra where symbolic matrix manipulation can be performed.

```
> with(LinearAlgebra):
```

In the case of planar four-bar mechanism, the total kinetic and the potential energy can be obtained by using the notion of 'breaking' the mechanisms at joint 3 (see Section 4.3), and treating the four-bar mechanism as a planar 2R manipulator with joint variables (θ_1, ϕ_2) and a single link manipulator with the joint variable ϕ_1.

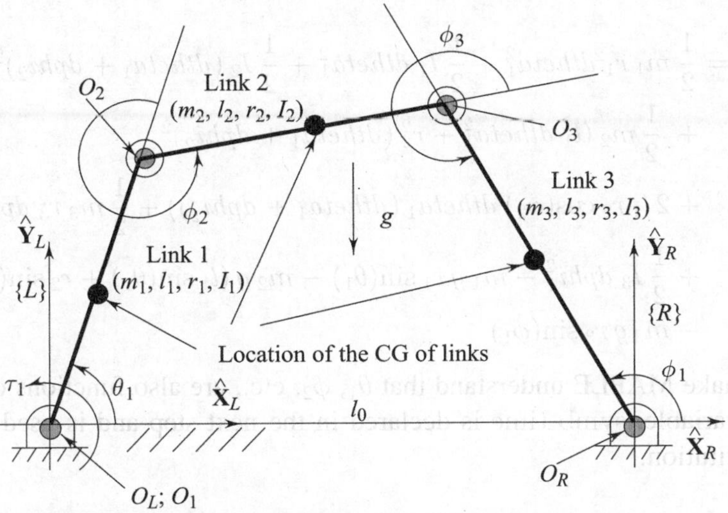

Fig. 6.3 A planar four-bar mechanism

```
>   KE:=  1/2*m[1]*(r[1]*dtheta[1])^2
>   + 1/2* I[1]*dtheta[1]^2
>   + 1/2*I[2]*(dtheta[1] + dphi[2])^2
>   + 1/2*m[2]*(l[1]^2*dtheta[1]^2
>   + r[2]^2*(dtheta[1]+dphi[2])^2
>   +2*l[1]*r[2]*cos(phi[2])*dtheta[1]*(dtheta[1]
>   + dphi[2])) + m[3]*(r[3]*dphi[1])^2/2
>   + 1/2*I[3]*dphi[1]^2;
```

$$
\begin{aligned}
KE := &\frac{1}{2}\, m_1\, r_1^2\, dtheta_1^2 + \frac{1}{2}\, I_1\, dtheta_1^2 \\
&+ \frac{1}{2}\, I_2\, (dtheta_1 + dphi_2)^2 + \frac{1}{2}\, m_2\, (l_1^2\, dtheta_1^2 + r_2^2\, (dtheta_1 \\
&+ dphi_2)^2 + 2\, l_1\, r_2 \cos(\phi_2)\, dtheta_1\, (dtheta_1 + dphi_2)) \\
&+ \frac{1}{2}\, m_3\, r_3^2\, dphi_1^2 + \frac{1}{2}\, I_3\, dphi_1^2
\end{aligned}
$$

```
>   PE:= m[1]*g*r[1]*sin(theta[1])
>   + m[2]*g*(l[1]*sin(theta[1])
>   + r[2]*sin(theta[1]+phi[2]))
>   + m[3]*g*r[3]*sin(phi[1]);
```

$$
\begin{aligned}
PE := &\, m_1\, g\, r_1 \sin(\theta_1) + m_2\, g\, (l_1 \sin(\theta_1) + r_2 \sin(\theta_1 + \phi_2)) \\
&+ m_3\, g\, r_3 \sin(\phi_1)
\end{aligned}
$$

From the total kinetic and potential energies, we compute the Lagrangian.

```
>   L:=KE-PE;
```

$$L := \frac{1}{2} m_1 r_1{}^2 dtheta_1^2 + \frac{1}{2} I_1\, dtheta_1^2 + \frac{1}{2} I_2\,(dtheta_1 + dphi_2)^2$$
$$+ \frac{1}{2} m_2\,(l_1^2\, dtheta_1^2 + r_2^2\,(dtheta_1 + dphi_2)^2$$
$$+ 2\, l_1\, r_2 \cos(\phi_2)\, dtheta_1\,(dtheta_1 + dphi_2)) + \frac{1}{2} m_3\, r_3^2\, dphi_1^2$$
$$+ \frac{1}{2} I_3\, dphi_1^2 - m_1\, g\, r_1 \sin(\theta_1) - m_2\, g\,(l_1 \sin(\theta_1) + r_2 \sin(\theta_1 + \phi_2))$$
$$- m_3\, g\, r_3 \sin(\phi_1)$$

To make MAPLE understand that θ_1, ϕ_2, etc., are also functions of time t, the variable, symb_time is declared in the next step and is used later for substitution.

```
>   symb_time:=
>   {theta[1]=theta[1](t),dtheta[1]=
>   diff(theta[1](t),t),phi[2]=phi[2](t),
>   dphi[2]=diff(phi[2](t),t),phi[1]=phi[1](t),
>   dphi[1]=diff(phi[1](t),t)
>   };
```

$$symb\_time := \{$$
$$\theta_1 = \theta_1(t),\ dtheta_1 = \tfrac{d}{dt}\,\theta_1(t),\ \phi_2 = \phi_2(t),$$
$$dphi_2 = \tfrac{d}{dt}\,\phi_2(t),\ \phi_1 = \phi_1(t),\ dphi_1 = \tfrac{d}{dt}\,\phi_1(t)\}$$

The loop-closure constraint equations, corresponding to 'breaking' the mechanisms at joint 3 were derived in Section 4.3, Eqn (4.6), and are given as the MAPLE input.

```
>   constr[1]:= l[1]*cos(theta[1]) +
>   l[2]*cos(theta[1]+phi[2])-l[3]*cos(phi[1])
>   -l[0];
```

$$constr_1 := l_1 \cos(\theta_1) + l_2 \cos(\theta_1 + \phi_2) - l_3 \cos(\phi_1) - l_0$$

```
>   constr[2]:= l[1]*sin(theta[1]) +
>   l[2]*sin(theta[1]+phi[2])-l[3]*sin(phi[1]) ;
```

$$constr_2 := l_1 \sin(\theta_1) + l_2 \sin(\theta_1 + \phi_2) - l_3 \sin(\phi_1)$$

The equations of motion are obtained by performing the partial differentiation and differentiation with respect to time and by simplifying the resulting expressions and collecting the various terms. This is done in the next set of MAPLE inputs.

```
> eqn[1]:= collect(simplify(diff(subs
> (symb_time,diff(L,dtheta[1])),t)
> - subs(symb_time,diff(L,theta[1]))
> -tau[1],trig),[diff(theta[1](t),t,t),d
> iff(phi[2](t),t,t),diff(theta[1](t),t),
> diff(phi[2](t),t),g,lambda[1],l
> ambda[2]], distributed);
```

$$\begin{aligned} eqn_1 := &-\tau_1 + (m_2\, r_2 \cos(\theta_1(t) + \phi_2(t)) + m_1\, r_1 \cos(\theta_1(t)) \\ &+ m_2\, l_1 \cos(\theta_1(t)))\, g + (I_2 + m_2\, {r_2}^2 + I_1 \\ &+ m_2\, {l_1}^2 + 2\, m_2\, l_1\, r_2 \cos(\phi_2(t)) + m_1\, {r_1}^2)\, (\tfrac{d^2}{dt^2}\, \theta_1(t)) \\ &+ (I_2 + m_2\, {r_2}^2 + m_2\, l_1\, r_2 \cos(\phi_2(t)))\, (\tfrac{d^2}{dt^2}\, \phi_2(t)) \\ &- m_2\, l_1\, r_2 \sin(\phi_2(t))\, (\tfrac{d}{dt}\, \phi_2(t))^2 \\ &- 2\, m_2\, l_1\, r_2 \sin(\phi_2(t))\, (\tfrac{d}{dt}\, \phi_2(t))\, (\tfrac{d}{dt}\, \theta_1(t)) \end{aligned}$$

```
> eqn[2]:=collect(simplify(diff(subs
> (symb_time,diff(L,dphi[2])),t)
> -subs(symb_time,diff(L,phi[2]))
> -tau[2],trig),[diff(theta[1](t),t,t),
> diff(phi[2](t),t,t),diff(theta[1](t),t),
> diff(phi[2](t),t),g,lambda[1],lambda[2]],
> distributed);
```

$$\begin{aligned} eqn_2 := &-\tau_2 + m_2\, g\, r_2 \cos(\theta_1(t) + \phi_2(t)) + (I_2 + m_2\, {r_2}^2 \\ &+ m_2\, l_1\, r_2 \cos(\phi_2(t)))\, (\tfrac{d^2}{dt^2}\, \theta_1(t)) + (I_2 + m_2\, {r_2}^2)\, (\tfrac{d^2}{dt^2}\, \phi_2(t)) \\ &+ m_2\, l_1\, r_2 \sin(\phi_2(t))\, (\tfrac{d}{dt}\, \theta_1(t))^2 \end{aligned}$$

```
> eqn[3]:= collect(simplify(diff(subs
> (symb_time,diff(L,dphi[1])),t)
> -subs(symb_time,diff(L,phi[1]))
> -tau[3],trig),[diff(phi[1](t),t,t),
> diff(phi[1](t),t),g], distributed);
```

$$eqn_3 := -\tau_3 + m_3\, g\, r_3 \cos(\phi_1(t)) + (m_3\, {r_3}^2 + I_3)\, (\tfrac{d^2}{dt^2}\, \phi_1(t))$$

The mass matrix is obtained by collecting the coefficients of the second derivatives of the generalized coordinates. We also use the substitution time_symb to remove the explicit dependence of t on θ_1, ϕ_1, and ϕ_2 in the expressions of the elements of the mass matrix.

```
>mass_matrix:=
> Matrix(3,3,[[coeff(eqn[1],diff(theta[1](t),t,
> t)),coeff(eqn[1],diff(phi[2](t),t,t)),
> coeff(eqn[1],diff(phi[1](t),t,t))],
> [coeff(eqn[2],diff(theta[1](t),t,t)),
```

```
> coeff(eqn[2],diff(phi[2](t),t,t)),
> coeff(eqn[2],diff(phi[1](t),t,t))],
> [coeff(eqn[3],diff(theta[1](t),t,t)),
> coeff(eqn[3],diff(phi[2](t),t,t)),
> coeff(eqn[3],diff(phi[1](t),t,t))]]);
> time_symb:={theta[1](t)=theta[1],
> phi[2](t)=phi[2],
> phi[1](t)=phi[1]}:
```

$$time\_symb := \{\theta_1(t) = \theta_1,\ \phi_2(t) = \phi_2,\ \phi_1(t) = \phi_1\}$$

```
> M:=subs(time_symb,mass_matrix):
```

$$M := \begin{bmatrix} I_2 + m_2\,r_2^2 + I_1 + m_2\,l_1^2 + 2\,m_2\,l_1\,r_2\cos(\phi_2) + m_1\,r_1^2,\ \ I_2 + m_2\,r_2^2 + m_2\,l_1\,r_2\cos(\phi_2),\ \ 0 \\ I_2 + m_2\,r_2^2 + m_2\,l_1\,r_2\cos(\phi_2),\ \ I_2 + m_2\,r_2^2,\ \ 0 \\ 0,\ \ 0,\ \ m_3\,r_3^2 + I_3 \end{bmatrix}$$

The Coriolis/centripetal terms can be obtained by collecting the coefficients of the quadratic terms $\dot{\theta}_1\dot{\phi}_1$, etc., and the gravity terms can be obtained by collecting the terms containing only θ_1, etc. We can also use Eqn (6.16) to obtain the Coriolis/centripetal terms from the elements of the mass matrix, and Eqn (6.17) to obtain the gravity terms from the expression of the potential energy. It is easier to perform partial differentiation in Eqn (6.16) by defining generalized coordinates (q_1, q_2, q_3) to denote $(\theta_1, \phi_2, \phi_1)$.

```
> time_gen_cor:={theta[1](t)=q[1],
> phi[2](t)=q[2],phi[1](t)=q[3]}:
```

$$time\_gen\_cor := \{\theta_1(t) = q_1,\ \phi_2(t) = q_2,\ \phi_1(t) = q_3\}$$

```
> M:= subs(time_gen_cor,mass_matrix):
```

Denoting the Coriolis/centripetal terms as a 3×3 matrix, as in Eqn (6.16), we get

```
> C:=Matrix(3,3):

> for i from 1 to 3 do    for j from 1 to 3 do
> C[i,j]:=0;  for k from 1
> to 3 do  C[i,j]:= C[i,j]
> + 1/2*(diff(M[i,j],q[k]) + diff(M[i,k],q[j])
> - diff(M[k,j],q[i]))*diff(q[k](t),t);
> end do;  end do;  end do;
```

The generalized coordinates in the Coriolis/centripetal term can be converted back to θ_1, ϕ_2, ϕ_1 and their derivatives, and skipping a few steps, we can get

$$C := \begin{bmatrix} -m_2\,l_1\,r_2\sin(\phi_2)\,(\frac{d}{dt}\,\phi_2) & \begin{matrix}-m_2\,l_1\,r_2\sin(\phi_2)\,(\frac{d}{dt}\,\theta_1) \\ -m_2\,l_1\,r_2\sin(\phi_2)\,(\frac{d}{dt}\,\phi_2)\end{matrix} & 0 \\ m_2\,l_1\,r_2\sin(\phi_2)\,(\frac{d}{dt}\,\theta_1) & 0 & 0 \\ 0 & 0 & 0 \end{bmatrix}$$

The gravity vector is obtained as

```
> G:= Vector(3,[ diff(PE,theta[1]),
> diff(PE,phi[2]), diff(PE,phi[1])]);
```

$$G := \begin{bmatrix} m_1\, g\, r_1 \cos(\theta_1) + m_2\, g\, (l_1 \cos(\theta_1) + r_2 \cos(\theta_1 + \phi_2)) \\ m_2\, g\, r_2 \cos(\theta_1 + \phi_2) \\ m_3\, g\, r_3 \cos(\phi_1) \end{bmatrix}$$

The two constraint equations can be differentiated with respect to θ_1, ϕ_2, and ϕ_3, and we can obtain the $\boldsymbol{\Psi}$ matrix required for the equations of motion of the four-bar mechanism [see Eqn (6.21].

```
> psi:=
> Matrix(2,3,[[diff(constr[1],theta[1]),
> diff(constr[1],phi[2]),diff(cons
> tr[1],phi[1])],[diff(constr[2],theta[1]),
> diff(constr[2],phi[2]),diff(c
> onstr[2],phi[1])]]);
```

$$\psi := \begin{bmatrix} -l_1 \sin(\theta_1) - l_2 \sin(\%1) & -l_2 \sin(\%1) & l_3 \sin(\phi_1) \\ l_1 \cos(\theta_1) + l_2 \cos(\%1) & l_2 \cos(\%1) & -l_3 \cos(\phi_1) \end{bmatrix}$$

$$\%1 := \theta_1 + \phi_2$$

By using the constraint matrix, and as shown in Eqns (6.22)–(6.25), we can eliminate the Lagrange multipliers and obtain a set of three second-order, ordinary differential equations of motion for the planar four-bar mechanism.

6.4 Dynamic Equations in Cartesian Space

The dynamic equations, derived in previous sections, are called the *joint space* equations of motion since they involve the joint variables. It is often of interest to rewrite the equations in terms of Cartesian variables and Cartesian forces and moments acting on the end-effector. The equations of motion in the Cartesian space were first developed by Khatib (1987) for motion and force control, and can be symbolically written as

$$\mathcal{F} = [M_{\mathcal{X}}(\mathbf{q})]\ddot{\mathcal{X}} + \mathbf{C}_{\mathcal{X}}(\mathbf{q}, \dot{\mathbf{q}}) + \mathbf{G}_{\mathcal{X}}(\mathbf{q}) \qquad (6.35)$$

where $\mathcal{F}$ is a 6×1 entity of forces and moments acting on the end-effector, and $\mathcal{X}$ is an appropriate 6×1 entity representing the position and orientation of the end-effector. The term $[M_{\mathcal{X}}(\mathbf{q})]$ is analogous to the mass matrix, the term $\mathbf{C}_{\mathcal{X}}(\mathbf{q}, \dot{\mathbf{q}})$ is analogous to the Coriolis and centripetal term, and the term $\mathbf{G}_{\mathcal{X}}(\mathbf{q})$ is analogous to the gravity term.

We can derive relationships between the joint space terms and the Cartesian space terms by making use of the inverse and transpose of the

Jacobian. These are given as

$$\begin{aligned}
\boldsymbol{\tau} &= [J(\mathbf{q})]^T \mathcal{F} \\
[M_{\mathcal{X}}(\mathbf{q})] &= [J(\mathbf{q})]^{-T}[\mathbf{M}(\mathbf{q})][J(\mathbf{q})]^{-1} \\
\mathbf{C}_{\mathcal{X}}(\mathbf{q},\dot{\mathbf{q}}) &= [J(\mathbf{q})]^{-\mathrm{T}}(\mathbf{C}(\mathbf{q},\dot{\mathbf{q}}) - [\mathbf{M}(\mathbf{q})][J(\mathbf{q})]^{-1}[\dot{J}(\mathbf{q})]\dot{\mathbf{q}}) \\
\mathbf{G}_{\mathcal{X}}(\mathbf{q}) &= [J(\mathbf{q})]^{-T}\mathbf{G}(\mathbf{q})
\end{aligned} \tag{6.36}$$

where $[J(\mathbf{q})]^{-T}$ denotes the inverse of $[J(\mathbf{q})]^T$, and $[J(\mathbf{q})]$, $\mathcal{F}$, and $\mathcal{X}$ have to be written in the same coordinate system.

In Eqn (6.36), at least conceptually, $\mathbf{q}$ and $\dot{\mathbf{q}}$ can be replaced with the Cartesian variables by using the inverse kinematics and inverse Jacobian functions of the manipulator. However, the inverse kinematics of serial manipulators and direct kinematics of parallel manipulators cannot often be solved in closed form and almost always give rise to multiple solutions. Even if closed-form expressions are available, multiple solutions will lead to various cases of the formulas given in Eqn (6.36). Fortunately, in the most common applications of the Cartesian equations of motion (see Chapter 8, Section 8.8 on Cartesian model-based control schemes), replacement of the joint variables with Cartesian variables is not required and we leave Eqn (6.36) in the above form for convenience.

6.5 Inverse Dynamics of Manipulators

The inverse dynamics problem of serial manipulator is stated as: Given the kinematic and inertial parameters of a manipulator and a trajectory of the manipulator as a function of time, find the required joint torques. This problem is straight forward since it involves substituting $\mathbf{q}(t)$, $\dot{\mathbf{q}}(t)$, and $\ddot{\mathbf{q}}(t)$ in the right-hand side of the equations of motion and evaluating the left-hand side, $\boldsymbol{\tau}(t)$. The computation of $\boldsymbol{\tau}(t)$, as we will see in a later chapter, is required for model-based control. The maximum value of τ_i for representative trajectories also helps in the sizing and choice of the actuators.

Example 6.3: Inverse dynamics of planar 2R manipulator

Figure 6.2 shows the familiar 2R manipulator. The mass and inertial properties are assumed to be as follows.

Link	Length (m)	Mass (kg)	CG (m)	Inertia (kg m^2)
1	1.0	12.456	0.773	1.042
2	1.0	12.456	0.583	1.042

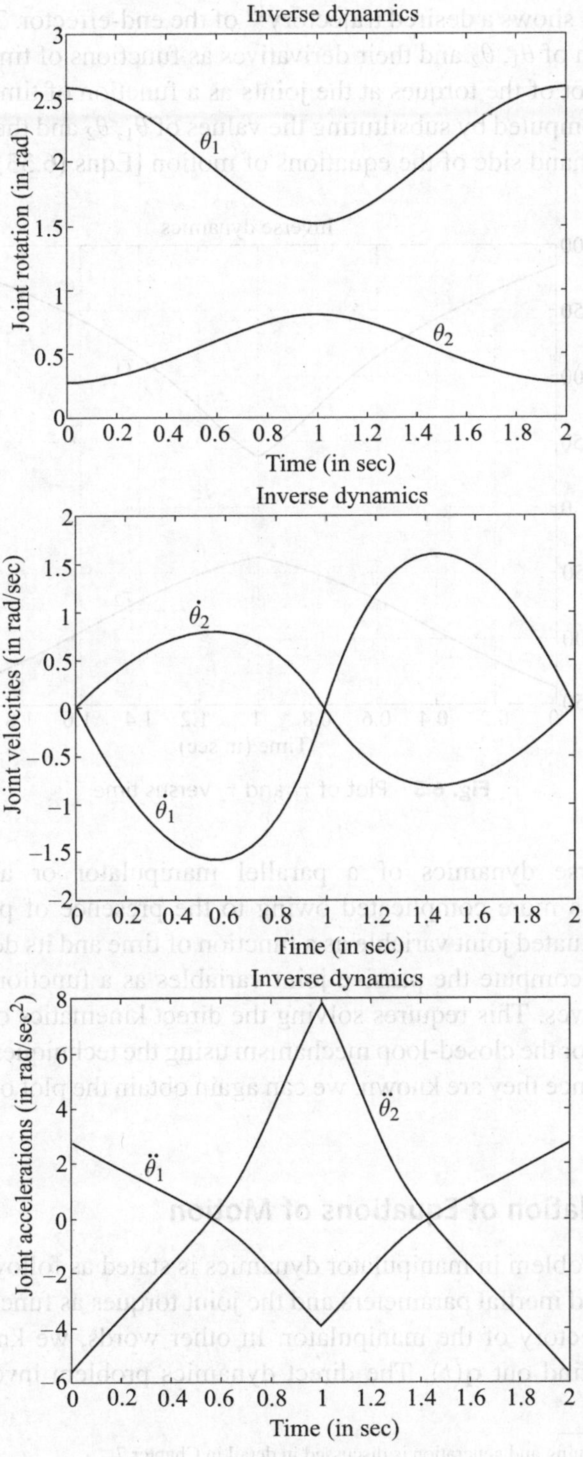

Fig. 6.4 Plot of desired trajectories

Figure 6.4 shows a desired trajectory[10] of the end-effector. The trajectory is in the form of θ_1, θ_2 and their derivatives as functions of time. Figure 6.5 shows the plot of the torques at the joints as a function of time. The torque values are computed by substituting the values of θ_1, θ_2 and their derivatives on the right-hand side of the equations of motion [Eqns (6.33)].

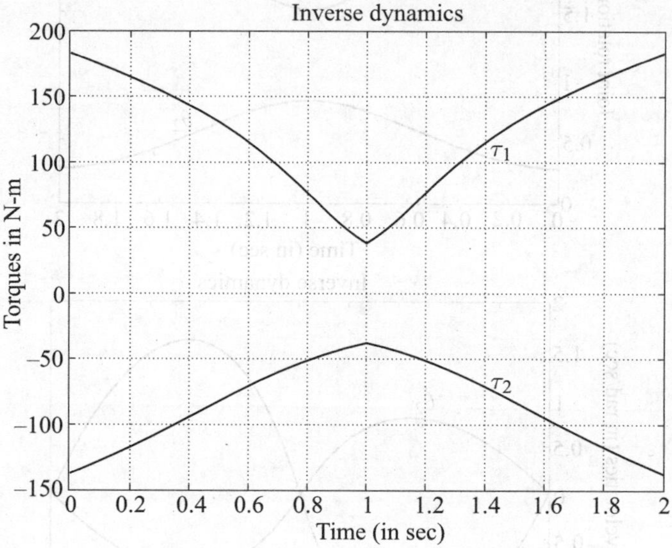

Fig. 6.5 Plot of τ_1 and τ_2 versus time

The inverse dynamics of a parallel manipulator or a closed-loop mechanism is more complicated owing to the presence of passive joints. Given the actuated joint variable as a function of time and its derivatives, we first need to compute the passive joint variables as a function of time and their derivatives. This requires solving the direct kinematics of the parallel manipulator or the closed-loop mechanism using the techniques discussed in Chapter 4. Once they are known, we can again obtain the plot of the actuated torque.

6.6 Simulation of Equations of Motion

The direct problem in manipulator dynamics is stated as follows: Given the kinematic and inertial parameters and the joint torques as functions of time, find the trajectory of the manipulator. In other words, we know $\boldsymbol{\tau}(t)$ and we wish to find out $\mathbf{q}(t)$. The direct dynamics problem involves solving

[10] Trajectory planning and generation is discussed in detail in Chapter 7.

equations of motion. Equations of motion, as shown earlier, are highly nonlinear and cannot be solved (integrated) analytically except in the simplest cases. Hence, obtaining $\mathbf{q}(t)$ for a given $\boldsymbol{\tau}(t)$ involves numerical integration. We will use in-built routines in MATLAB, such as ODE45, for numerical integration of the ODEs. Most of these routines require the equations of motion in a state space form together with initial conditions, $\mathbf{q}(0)$ and $\dot{\mathbf{q}}(\mathbf{0})$. We present below one way to rewrite the dynamic equations of motion of a manipulator in the state space form.

For serial manipulators, since the mass matrix $[\mathbf{M}(\mathbf{q})]$ is invertible, we can write

$$\ddot{\mathbf{q}} = [\mathbf{M}(\mathbf{q})]^{-1} [\boldsymbol{\tau} - \mathbf{C}(\mathbf{q}, \dot{\mathbf{q}}) - \mathbf{G}(\mathbf{q}) - \mathbf{F}(\mathbf{q}, \dot{\mathbf{q}})] \tag{6.37}$$

We define a $2n \times 1$ vector $\mathbf{X}$, where the first n are the joint variables, q_i $(i = 1, 2, \ldots, n)$ and the last n are the joint velocities $\dot{\mathbf{q}}_i$ $(i = 1, 2, \ldots, n)$. Hence we can write

$$\begin{aligned}
\dot{X}_1 &= X_{n+1} \\
\dot{X}_2 &= X_{n+2} \\
&\vdots \\
\dot{X}_n &= X_{2n} \\
\begin{pmatrix} \dot{X}_{n+1} \\ \vdots \\ \dot{X}_{2n} \end{pmatrix} &= [\mathbf{M}(\mathbf{X})]^{-1} [\boldsymbol{\tau} - \mathbf{C}(\mathbf{X}) - \mathbf{G}(\mathbf{X}) - \mathbf{F}(\mathbf{X})]
\end{aligned} \tag{6.38}$$

The above equation can be written in the compact state space, first-order form as

$$\dot{\mathbf{X}} = \mathbf{g}(\mathbf{X}, \boldsymbol{\tau}) \tag{6.39}$$

with initial conditions $\mathbf{X}(0)$.

In the case of parallel manipulators and closed-loop mechanisms, we use Eqn (6.25) and obtain the $2(n+m)$ first-order state equations as

$$\begin{aligned}
\dot{X}_1 &= X_{n+m+1} \\
\dot{X}_2 &= X_{n+m+2} \\
&\vdots \\
\dot{X}_{n+m} &= X_{2(n+m)}
\end{aligned} \tag{6.40}$$

$$\begin{pmatrix} \dot{X}_{n+m+1} \\ \vdots \\ \dot{X}_{2(n+m)} \end{pmatrix} = [\mathbf{M}]^{-1}(\mathbf{f} - [\boldsymbol{\Psi}]^T([\boldsymbol{\Psi}][\mathbf{M}]^{-1}[\boldsymbol{\Psi}]^T)^{-1}\{[\boldsymbol{\Psi}][\mathbf{M}]^{-1}\mathbf{f} \\ +[\dot{\boldsymbol{\Psi}}](X_1 \cdots X_{n+m})^T\})$$

where $\mathbf{f}$ denotes $(\boldsymbol{\tau} - [\mathbf{C}](X_{n+m+1} \cdots X_{2(n+m)})^T - \mathbf{G})$ and the dependence of the terms on $\mathbf{X}$ is dropped for convenience.

Although the state equations, as given in Eqn (6.40), appear to be ODEs, they must also satisfy the loop-closure constraints given in Eqn (6.18) and its first derivative with respect to time. This is known in literature as a system of differential-algebraic equations or DAEs. One of the problems in the numerical solution of DAEs is the stiff nature of equations. To solve stiff equations, we may need to use stiff solvers in MATLAB such as ODE15S or ODE23S, although for simple problems such as the four-bar mechanism discussed in Example 6.5 the non-stiff solver ODE45 is good enough. Stiff solvers are, usually, much slower than non-stiff solvers.

Another problem in the numerical solution of equations of motion for parallel manipulators and closed-loop mechanisms is the use of the second derivative of the constraint equations to solve for the Lagrange multipliers and the derivation of the equations of motion [see Eqns (6.22)–(6.25)]. Any small numerical error in acceleration will accumulate and errors in the position and velocity may increase uncontrollably. To keep errors small, there is need for stabilization. We present one scheme, called *Baumgarte stabilization* (Baumgarte 1983), which has been extensively used for controlling errors.

We assume the general case of m holonomic constraints, which are functions of $\mathbf{q}$ and t, given by

$$\eta_i(\mathbf{q}, t) = 0, \qquad i = 1, 2, \ldots, m \tag{6.41}$$

For Baumgarte stabilization, the second derivative form of the constraints, namely Eqn (6.27), is replaced by

$$([\boldsymbol{\Psi}]\ddot{\mathbf{q}} + [\dot{\boldsymbol{\Psi}}]\dot{\mathbf{q}} + \dot{\boldsymbol{\Phi}}(t)) + 2\alpha(\boldsymbol{\Phi}(t) + [\boldsymbol{\Psi}(\mathbf{q})]\dot{\mathbf{q}}) \\ +\beta^2\boldsymbol{\eta}(\mathbf{q}, t) = \mathbf{0} \tag{6.42}$$

where α and β are *constants* and $[\boldsymbol{\Psi}]$ contains the partial derivatives of $\boldsymbol{\eta}(\mathbf{q})$ with respect to $\mathbf{q}$, and the matrix $[\dot{\boldsymbol{\Psi}}]$ contains the time derivatives of each of the elements of $[\boldsymbol{\Psi}]$. It may be noted that for loop-closure equations in parallel manipulators and closed-loop mechanisms, $\boldsymbol{\eta}$ has no explicit dependence on time, for such cases $\boldsymbol{\Phi}(t)$ and its derivative are zero. To obtain the equation of motion, we obtain an expression for $\ddot{\mathbf{q}}$ as in Eqn (6.23), solve for $\boldsymbol{\lambda}$, and obtain an equation of motion similar to Eqn (6.25). Of course, since Eqn (6.42) is now being used, the right-hand side of the equation of motion will contain α and β and will be more complicated.

The motivation behind the use of Eqn (6.42) is its 'similarity' to the unforced equation of motion of a single mass–spring–damper system,[11] and the known fact that for a mass–spring–damper system with appropriate damping and natural frequency, the vibrations die down asymptotically. It is expected that, with the proper choice of α and β, the numerical errors in position and velocity would die down quickly as in an unforced mass–spring–damper system. This is, however, easier said than done since, unlike a single mass–spring–damper system, the equations of motion are very non-linear and complex, and there is no systematic way, other than by trial and error, to choose appropriate values of α and β.

Example 6.4: Simulation of a planar 2R manipulator

Figure 6.2 shows the familiar 2R manipulator. We choose the same kinematic and inertial parameters as for the inverse dynamics simulations. The equations of motion as given in Eqn (6.33) were first written in the state space form as discussed in the preceding section. The right-hand side is assumed to be zero, i.e., there is no external torque acting on the system. The manipulator tip is lifted such that θ_1 is $-90°$ and θ_2 is 45° and the system is allowed to move under gravity. The links of the manipulator oscillate. Figure 6.6 shows

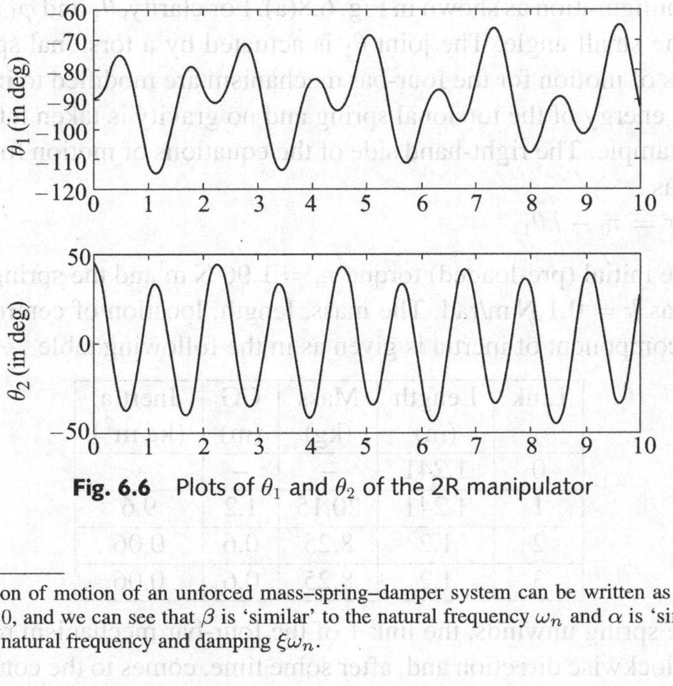

Fig. 6.6 Plots of θ_1 and θ_2 of the 2R manipulator

---

[11] The equation of motion of an unforced mass–spring–damper system can be written as $\ddot{x} + 2\xi\omega_n\dot{x} + {\omega_n}^2 x = 0$, and we can see that β is 'similar' to the natural frequency ω_n and α is 'similar' to the product of natural frequency and damping $\xi\omega_n$.

the plots of θ_1 and θ_2 as functions of time. Figure 6.7 shows the motion of the tip of the manipulator. One can see that the path traced by the tip is quite complicated.

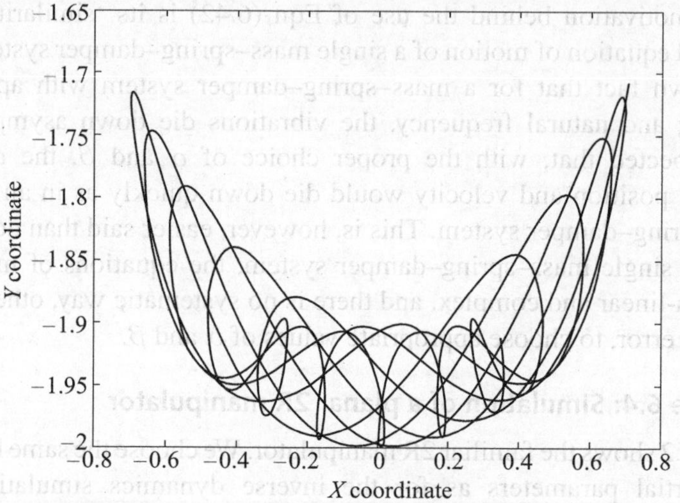

Fig. 6.7 Plot of tip motion in Cartesian space

Example 6.5: Simulation of a planar four-bar mechanism

Figure 6.8 shows a four-bar mechanism which is initially in a completely folded configuration as shown in Fig. 6.8(a). For clarity, θ_1 and ϕ_1 are shown with some small angle. The joint θ_1 is actuated by a torsional spring. The equations of motion for the four-bar mechanism are modified to include the potential energy of the torsional spring and no gravity is taken into account in this example. The right-hand side of the equations of motion for this case is given as

$$\tau = \tau_0 - k\theta_1$$

where the initial (pre-loaded) torque $\tau_0 = 1.96$ N m and the spring constant is given as $k = 0.1$ N m/rad. The mass, length, location of centre of mass, and I_{zz} component of inertia is given as in the following table.

Link	Length (m)	Mass (kg)	CG (m)	Inertia (kg m^2)
0	1.241	–	–	–
1	1.241	20.15	1.2	9.6
2	1.2	8.25	0.6	0.06
3	1.2	8.25	0.6	0.06

As the spring unwinds, the link 1 of the four-bar mechanism rotates in a counterclockwise direction and, after some time, comes to the configuration

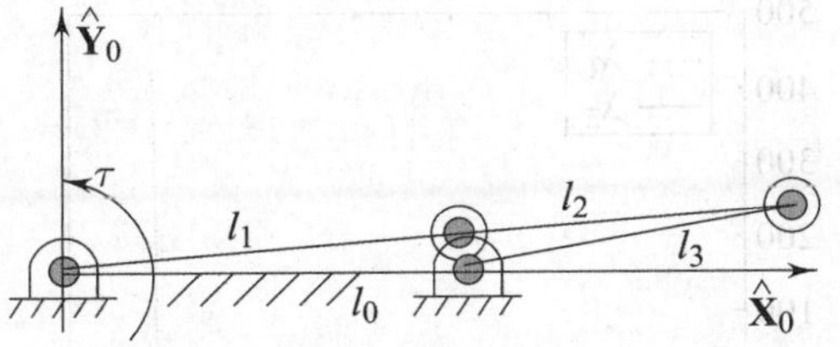

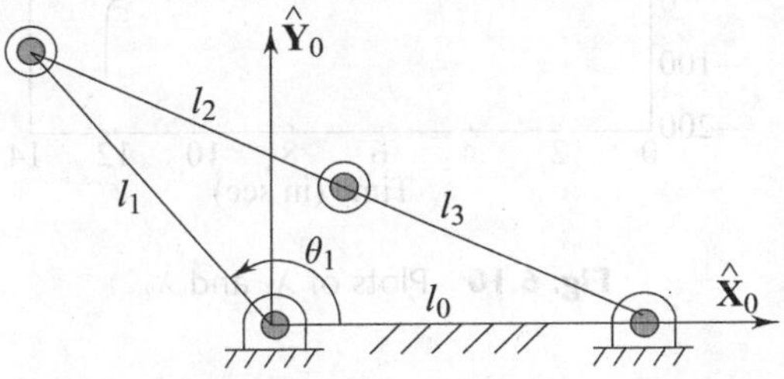

Fig. 6.8 A four-bar mechanism in two configurations

shown in Fig. 6.8(b). To simulate the motion, we use the equations of motion of a four-bar mechanism given in Example 6.2 with τ as given above. Figure 6.9 shows the plots of θ_1, ϕ_2, and ϕ_1 as functions of time. Figure 6.10 shows the behaviour of the Lagrange multipliers as a function of time.

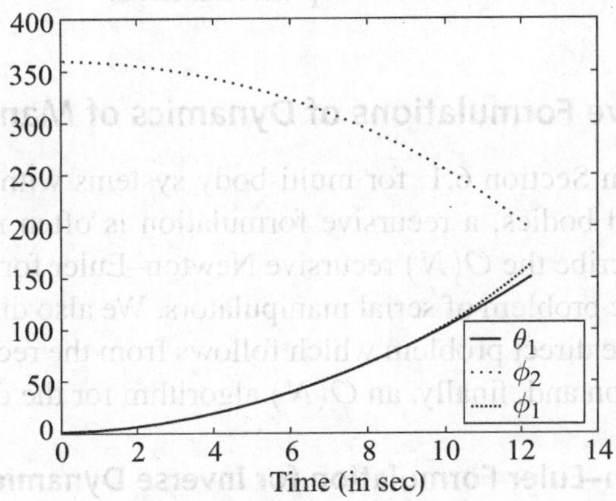

Fig. 6.9 Plots of θ_1, ϕ_2, and ϕ_1 of the four-bar mechanism

For simulation, we have assumed that all joints have zero velocity at $t = 0$. To avoid the singularity at the folded initial configuration, we have assumed $\theta_1 = 0.01$ rad at $t = 0$. The corresponding values of ϕ_1 and ϕ_2 are obtained from direct kinematics of the four-bar mechanism (see Example 4.1) Of

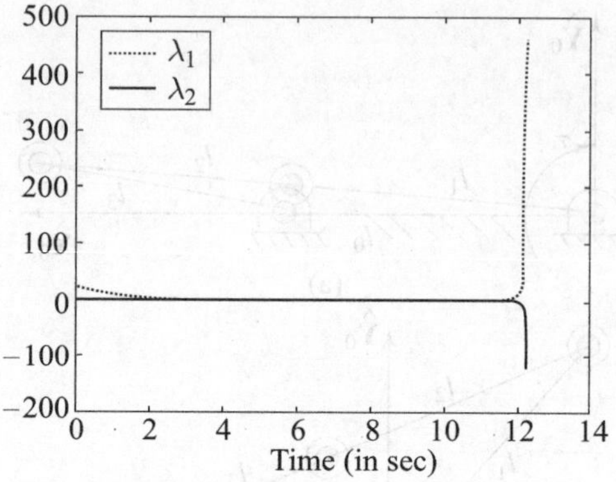

Fig. 6.10 Plots of λ_1 and λ_2

the two possible sets of values of ϕ_1 and ϕ_2, we choose the appropriate set given by $\phi_1 = 0.0102$ rad and $\phi_2 = 6.2698$ rad. Simulation is stopped at 12.25 sec with θ_1 reaching approximately 150°. At this configuration, the four-bar mechanism is very close to a singular configuration (see Chapter 5) — at $\theta_1 = 150.4°$ the four-bar mechanism becomes singular. One of the consequences of singularity, seen from the behaviour of the Lagrange multipliers in Fig. 6.10, is that the Lagrange multipliers tend to infinity and simulation cannot be performed further. This phenomenon is seen in all parallel manipulators and closed-loop mechanisms.

6.7 Recursive Formulations of Dynamics of Manipulators

As mentioned in Section 6.1, for multi-body systems with a large number of links or rigid bodies, a recursive formulation is often required. In this section, we describe the $\mathcal{O}(N)$ recursive Newton–Euler formulation for the inverse dynamic problem of serial manipulators. We also discuss an $\mathcal{O}(N^3)$ algorithm for the direct problem which follows from the recursive Newton–Euler formulation and, finally, an $\mathcal{O}(N)$ algorithm for the direct problem.

6.7.1 Newton–Euler Formulation for Inverse Dynamics

Newton–Euler equations require the computation of linear and angular accelerations of links. The linear acceleration of an arbitrary point $\mathbf{p}$ on the rigid body $\{i\}$ with respect to the coordinate system $\{0\}$ can be obtained by differentiating the velocity relation given in Eqn (5.15). We can write

$$^0\dot{\mathbf{V}}_p = \frac{d}{dt}(^0\mathbf{V}_{O_i} + {}^0\boldsymbol{\omega}_i \times {}^0_i[R]\,{}^i\mathbf{p} + {}^0_i[R]\,{}^i\mathbf{V}_p) \tag{6.43}$$

and by differentiating term by term and using the relationship

$$\frac{d}{dt}\left({}^0_i[R]\,{}^i\mathbf{p}\right) = {}^0_i[R]\,{}^i\mathbf{V}_p + {}^0\boldsymbol{\omega}_i \times {}^0_i[R]\,{}^i\mathbf{p} \tag{6.44}$$

we get

$$\begin{aligned} {}^0\dot{\mathbf{V}}_p = {}^0\dot{\mathbf{V}}_{O_i} + {}^0_i[R]\,{}^i\dot{\mathbf{V}}_p + 2\,{}^0\boldsymbol{\omega}_i \times {}^0_i[R]\,{}^i\mathbf{V}_p + {}^0\dot{\boldsymbol{\omega}}_i \times {}^0_i[R]\,{}^i\mathbf{p} \\ + {}^0\boldsymbol{\omega}_i \times ({}^0\boldsymbol{\omega}_i \times {}^0_i[R]\,{}^i\mathbf{p}) \end{aligned} \tag{6.45}$$

If ${}^i\mathbf{p}$ is a constant, then ${}^i\mathbf{V}_p = {}^i\dot{\mathbf{V}}_p = 0$ and the above equation can be simplified.

Equation (6.45) can be readily extended to obtain the linear acceleration of link $i+1$ in terms of the velocity and acceleration of the previous link i. When joint $i+1$ is rotary, we get

$${}^{i+1}\dot{\mathbf{V}}_{i+1} = {}^{i+1}_i[R][{}^i\dot{\mathbf{V}}_i + {}^i\dot{\boldsymbol{\omega}}_i \times {}^i\mathbf{p}_{i+1} + {}^i\boldsymbol{\omega}_i \times ({}^i\boldsymbol{\omega}_i \times {}^i\mathbf{p}_{i+1})] \tag{6.46}$$

If joint $i+1$ is prismatic, we get

$$\begin{aligned} {}^{i+1}\dot{\mathbf{V}}_{i+1} = {}^{i+1}_i[R][{}^i\dot{\mathbf{V}}_i + {}^i\dot{\boldsymbol{\omega}}_i \times {}^i\mathbf{p}_{i+1} + {}^i\boldsymbol{\omega}_i \times ({}^i\boldsymbol{\omega}_i \times {}^i\mathbf{p}_{i+1})] \\ +2\,{}^{i+1}\boldsymbol{\omega}_{i+1} \times \dot{d}_{i+1}\,{}^{i+1}\widehat{\mathbf{Z}}_{i+1} + \ddot{d}_{i+1}\,{}^{i+1}\widehat{\mathbf{Z}}_{i+1} \end{aligned} \tag{6.47}$$

When joint $i+1$ is rotary, the angular acceleration of link $i+1$ can be written as

$${}^{i+1}\dot{\boldsymbol{\omega}}_{i+1} = {}^{i+1}_i[R]\,{}^i\dot{\boldsymbol{\omega}}_i + {}^{i+1}_i[R]\,{}^i\boldsymbol{\omega}_i \times \dot{\theta}_{i+1}\,{}^{i+1}\widehat{\mathbf{Z}}_{i+1} + \ddot{\theta}_{i+1}\,{}^{i+1}\widehat{\mathbf{Z}}_{i+1} \tag{6.48}$$

When joint $i+1$ is prismatic, we get

$${}^{i+1}\dot{\boldsymbol{\omega}}_{i+1} = {}^{i+1}_i[R]\,{}^i\dot{\boldsymbol{\omega}}_i \tag{6.49}$$

The acceleration of the centre of mass of link i can be written as

$${}^i\dot{\mathbf{V}}_{C_i} = {}^i\dot{\mathbf{V}}_i + {}^i\dot{\boldsymbol{\omega}}_i \times {}^i\mathbf{p}_{C_i} + {}^i\boldsymbol{\omega}_i \times ({}^i\boldsymbol{\omega}_i \times {}^i\mathbf{p}_{C_i}) \tag{6.50}$$

where ${}^i\mathbf{p}_{C_i}$ is the position vector of the centre of mass of link i with respect to origin O_i.

In the Newton–Euler formulation, we use the well-known Newton's and Euler's equations for each link of the manipulator. Newton's and Euler's equations relate the external forces $\mathbf{F}$ and moments $\mathbf{M}$ acting on a rigid body to accelerations and velocities of the rigid body, and are given by

$$\begin{aligned} \mathbf{F} &= m_i^0\dot{\mathbf{V}}_{C_i} \\ \mathbf{N} &= {}^{C_i}[I_i]\,{}^0\dot{\boldsymbol{\omega}}_i + {}^0\boldsymbol{\omega}_i \times {}^{C_i}[I_i]\,{}^0\boldsymbol{\omega}_i \end{aligned} \tag{6.51}$$

where ${}^0\dot{\mathbf{V}}_{C_i}$, ${}^0\dot{\boldsymbol{\omega}}_i$ and ${}^0\boldsymbol{\omega}_i$, are the linear acceleration, angular acceleration, and angular velocity of the rigid body $\{i\}$ in a fixed or inertial coordinate system $\{0\}$, respectively. In the above equations the subscript or the superscript C_i refers to the centre of mass of the rigid body $\{i\}$, and m_i, ${}^{C_i}[I_i]$ denote the mass and the inertia matrix (with respect to coordinate system $\{C_i\}$) of the rigid body i, respectively. It may be recalled that for the links of a manipulator, $\{C_i\}$ is located at the centre of mass of link $\{i\}$ and $\{C_i\}$ is parallel to link coordinate system $\{i\}$.

With the recursive equations to obtain linear and angular acceleration, the recursive equations for the linear and angular velocities given in Section 5.2 and Eqn (6.51), we can now present the recursive Newton–Euler formulation for serial manipulators.

Outward Iterations for Velocities and Accelerations

We use the formulas derived in Section 5.3 for links 1 through $N-1$. We assume that all joints are rotary. If any joint is not rotary but prismatic, we have to make appropriate changes.

$\mathbf{i : 0 \rightarrow N - 1}$

$$\begin{aligned}
{}^{i+1}\boldsymbol{\omega}_{i+1} &= {}^{i+1}_{i}[R]\,{}^{i}\boldsymbol{\omega}_i + \dot{\theta}_{i+1}(0\;0\;1)^T \\
{}^{i+1}\dot{\boldsymbol{\omega}}_{i+1} &= {}^{i+1}_{i}[R]\,{}^{i}\dot{\boldsymbol{\omega}}_i + {}^{i+1}_{i}[R]{}^{i}\boldsymbol{\omega}_i \times \dot{\theta}_{i+1}(0\;0\;1)^T + \ddot{\theta}_{i+1}(0\;0\;1)^T \\
{}^{i+1}\dot{\mathbf{V}}_{i+1} &= {}^{i+1}_{i}[R][{}^{i}\dot{\mathbf{V}}_i + {}^{i}\dot{\boldsymbol{\omega}}_i \times {}^{i}\mathbf{p}_{i+1} \\
&\quad + {}^{i}\boldsymbol{\omega}_i \times ({}^{i}\boldsymbol{\omega}_i \times {}^{i}\mathbf{p}_{i+1})] \\
{}^{i+1}\dot{\mathbf{V}}_{C_{i+1}} &= {}^{i+1}\dot{\mathbf{V}}_{i+1} + {}^{i+1}\dot{\boldsymbol{\omega}}_{i+1} \times {}^{i+1}\mathbf{p}_{C_{i+1}} \\
&\quad + {}^{i+1}\boldsymbol{\omega}_{i+1} \times ({}^{i+1}\boldsymbol{\omega}_{i+1} \times {}^{i+1}\mathbf{p}_{C_{i+1}})
\end{aligned} \tag{6.52}$$

Once the linear acceleration of the centre of mass of the links and angular velocity and acceleration are known, by using Newton's and Euler's equations we can calculate the forces and moments acting on each link as

$$\begin{aligned}
{}^{i+1}\mathbf{F}_{i+1} &= m_{i+1}\,{}^{i+1}\dot{\mathbf{V}}_{C_{i+1}} \\
{}^{i+1}\mathbf{N}_{i+1} &= {}^{C_{i+1}}[I]_{i+1}\,{}^{i+1}\dot{\boldsymbol{\omega}}_{i+1} + {}^{i+1}\boldsymbol{\omega}_{i+1} \times {}^{C_{i+1}}[I]_{i+1}\,{}^{i+1}\boldsymbol{\omega}_{i+1}
\end{aligned} \tag{6.53}$$

Inward Iterations for Joint Torques

To compute the torques acting on the joints, we recall that the joints can apply torques (or forces for prismatic joints) only along their axis. Using free-body diagrams for each link (similar to what we had done during statics of manipulators), we can perform an inward iteration to obtain the joint torques. We have to use the expressions for ${}^{i}\mathbf{F}_i$ and ${}^{i}\mathbf{N}_i$ calculated during the outward iterations.

$\mathbf{i : N \to 1}$

$$
\begin{aligned}
{}^i\mathbf{f}_i &= {}^{i}_{i+1}[R]\,{}^{i+1}\mathbf{f}_{i+1} + {}^i\mathbf{F}_i \\
{}^i\mathbf{n}_i &= {}^{i}_{i+1}[R]\,{}^{i+1}\mathbf{n}_{i+1} + {}^i\mathbf{p}_{i+1} \times {}^{i}_{i+1}[R]^{i+1}\mathbf{f}_{i+1} \\
&\quad + {}^i\mathbf{p}_{C_i} \times {}^i\mathbf{F}_i + {}^i\mathbf{N}_i \qquad (6.54) \\
\tau_i &= {}^i\mathbf{n}_i \cdot {}^i\hat{\mathbf{Z}}i
\end{aligned}
$$

The above iterative algorithm does not include gravity. To include the effect of gravity, we simply have to set ${}^0\dot{\mathbf{V}}_0 = \mathbf{g}$, where $\mathbf{g}$ is the gravity vector. This is equivalent to saying that the fixed link (or base) is accelerating upwards with $1.0g$ acceleration.

6.7.2* Algorithms for Forward Dynamics

To recollect, the forward dynamics or the direct problem for a serial manipulator is to obtain $\ddot{\mathbf{q}}$, $\dot{\mathbf{q}}$, and $\mathbf{q}$ as functions of time t given the geometry and inertia parameters of each link and the external forces acting on the serial manipulator. The Newton–Euler formulation in Section 6.7.1 can be also used to obtain the equations of motion of a serial manipulator given in Eqn (6.14), and hence it forms the basis of a simple intuitive algorithm for the forward dynamics. We rewrite the equations of motion as

$$[\mathbf{M}(\mathbf{q})]\ddot{\mathbf{q}} = \boldsymbol{\tau} - [\mathbf{C}(\mathbf{q}, \dot{\mathbf{q}})]\dot{\mathbf{q}} - \mathbf{G}(\mathbf{q}) \qquad (6.55)$$

and solve for $\ddot{\mathbf{q}}$. It may be noted that Eqn (6.55) is a set of *linear* equations in $\ddot{\mathbf{q}}$ once the right-hand side and $[\mathbf{M}(\mathbf{q})]$ is known. We can compute the right-hand side by the $\mathcal{O}(N)$ Newton–Euler algorithm for inverse dynamics. The computational complexity of the $N \times N$ mass matrix is $\mathcal{O}(N^2)$ and the set of linear equations in Eqn (6.55) can be solved by Gaussian elimination, which has a complexity of $\mathcal{O}(N^3)$. This algorithm by Walker and Orin (1982) is very efficient since the coefficient of the N^3 portion is very small and this algorithm works best for typical serial manipulators with $N \leq 6$.

For large N, as in the simulation of protein folding and other computational biology problems, the $\mathcal{O}(N^3)$ algorithm is not suitable. Several researchers have developed $\mathcal{O}(N^2)$ and $\mathcal{O}(N)$ algorithms [see Jain (1991) and Featherstone and Orin (2000) for a review on algorithms for manipulator dynamics]. We present an $\mathcal{O}(N)$ algorithm developed by Featherstone (1983, 1987) known as *articulated-body algorithm*.

We start our discussion with a way to combine Newton's and Euler's equations of motion for a single rigid body. For a single rigid body of mass m and inertia tensor about the mass centre, ${}^C[I]$, subjected to a force $\mathbf{F}$ and a moment $\mathbf{N}_C$ acting at the mass centre, C, the equations of motion can be written as

$$\mathbf{F} = m\dot{\mathbf{V}}_C$$
$$\mathbf{N}_C = {}^C[I]\dot{\boldsymbol{\omega}} \tag{6.56}$$

where the cross-product term does not appear since all quantities are referred to the centre of mass. Equation (6.56) can be combined as

$$\mathcal{F}_C = \begin{bmatrix} m\,[U] & [0] \\ [0] & {}^C[I] \end{bmatrix} \mathcal{A}_C \tag{6.57}$$

where $[U]$ is the 3×3 identity matrix, the 6×1 entity $\mathcal{F}_C$ is the combined force and moment [see also Eqn (5.69)] acting on the rigid body and is defined as

$$\mathcal{F}_C \triangleq \begin{pmatrix} \mathbf{F} \\ --- \\ \mathbf{N}_C \end{pmatrix} \tag{6.58}$$

and the 6×1 quantity $\mathcal{A}_C$ is defined as

$$\mathcal{A}_C \triangleq \begin{pmatrix} \dot{\mathbf{V}}_C \\ --- \\ \dot{\boldsymbol{\omega}} \end{pmatrix} \tag{6.59}$$

It may be noted that the matrix in the square brackets in Eqn (6.57) is a 6×6 positive definite and symmetric matrix, and the 6×1 entity $\mathcal{A}_C$ is the derivative of $\mathcal{V}$ discussed in Section 5.4 and Eqn (5.28).

Instead of the centre of mass, if Newton's and Euler's equations are written about an arbitrary point O, with respect to which the centre of mass is given by the vector $\mathbf{r}_C$, Eqn (6.56) modifies to

$$\mathbf{F} = m[U](\dot{\mathbf{V}}_O - \mathbf{r}_C \times \dot{\boldsymbol{\omega}})$$
$$\mathbf{N}_O = {}^C[I]\dot{\boldsymbol{\omega}} + \mathbf{r}_C \times \mathbf{F} \tag{6.60}$$

and these two equations can be combined as

$$\mathcal{F}_O = [\mathcal{I}]\mathcal{A}_O \tag{6.61}$$

where the matrix in the square brackets is a 6×6 *equivalent* 'inertia' matrix containing the inertia tensor ${}^C[I]$, m, $[U]$, and a 3×3 skew-symmetric matrix $[r_C]$ given by

$$[r_C] = m \begin{bmatrix} 0 & -r_{C_z} & r_{C_y} \\ r_{C_z} & 0 & -r_{C_x} \\ -r_{C_y} & r_{C_x} & 0 \end{bmatrix} [U] \tag{6.62}$$

It may be mentioned that $[r_C]$ is the first moment of inertia about O.

Instead of a single rigid body if we consider a rigid body i as part of a serial manipulator with N rigid bodies connected by N joints and generalized forces Q_i's (can be a torque or a force) acting along each of the joint axis i, then Eqn (6.61) cannot be used. In such a case, we have to take into account, for the rigid body i, the effects of the links $i+1$ through N and the generalized forces Q_{i+1} through Q_N acting along the joint axis $i+1$ through N. The key goal in the articulated-body algorithm (ABA) is to obtain, for an arbitrary link i in a serial manipulator, an equation of the form

$$\mathcal{F}_i = [\mathcal{I}]_i^A \; \mathcal{A}_i + \mathcal{P}_i^A \tag{6.63}$$

where the 6×6 matrix $[\mathcal{I}]_i^A$ is called the *articulated-body inertia* (ABI) and the 6×1 quantity $\mathcal{P}_i^A$ is called the 'bias' term, which contains the effect of the generalized actuated forces and the effect of the links after link i. It may be noted that $[\mathcal{I}]_i^A$, in general, is not the 'equivalent' $[\mathcal{I}]$ as shown in Eqn (6.61) since the effect of the other links must be taken into account. In addition, we wish to obtain $[\mathcal{I}]_i^A$ and $\mathcal{P}_i^A$ in $\mathcal{O}(N)$ steps. This can be done by using the following formulas (Featherstone 1983, 1987):

$$\begin{aligned}
[\mathcal{I}]_i^A &= [\mathcal{I}]_i + [\mathcal{I}]_{i+1}^A - \frac{[\mathcal{I}]_{i+1}^A \, \mathcal{S}_{i+1} \, \mathcal{S}_{i+1}^T \, [\mathcal{I}]_{i+1}^A}{\mathcal{S}_{i+1}^T [\mathcal{I}]_{i+1}^A \mathcal{S}_{i+1}}, \; [\mathcal{I}]_N^A = [\mathcal{I}]_N \\
\mathcal{P}_i^A &= \mathcal{P}_{i+1}^A + \frac{[\mathcal{I}]_{i+1}^A \; \mathcal{S}_{i+1}(Q_{i+1} - \mathcal{S}_{i+1}^T \; \mathcal{P}_{i+1}^A)}{\mathcal{S}_{i+1}^T \; [\mathcal{I}]_{i+1}^A \; \mathcal{S}_{i+1}}, \; \mathcal{P}_N^A = \mathbf{0}
\end{aligned} \tag{6.64}$$

where $\mathcal{S}_{i+1}$ is a 6×1 entity representing the $(i+1)^{\text{th}}$ joint axis.[12]

Since the articulated-body inertia for the end-effector (last link), $[\mathcal{I}]_N^A$, and the bias term for the end-effector, $\mathcal{P}_N^A$, are known to be $[\mathcal{I}]_N$ and $\mathbf{0}$, respectively, we can start from $i = N-1$ and compute the articulated-body inertia and the bias term for each link up to $i = 1$. Featherstone (1983, 1987) discusses in detail that these computations can be done in $\mathcal{O}(N)$ steps.

Once $[\mathcal{I}]_i^A$ and $\mathcal{P}_i^A$ are known for each link, we can obtain the acceleration $\ddot{q}_i$ at each joint by the following steps.
The acceleration $\mathcal{A}_i$ can be related to $\mathcal{A}_{i-1}$ by

$$\mathcal{A}_i = \mathcal{A}_{i-1} + \mathcal{S}_i \ddot{q}_i, \; \mathcal{A}_0 = \mathbf{0} \tag{6.65}$$

[12] A joint axis is represented by a line $\mathcal{L}$, which in turn is represented by a pair of vectors $(\mathbf{Q}; \mathbf{Q}_0)$ with $\mathbf{Q} \cdot \mathbf{Q}_0 = 0$ as discussed in Section 2.8. The inner product (also called the reciprocal product) of two 6×1 entities, $(\mathbf{Q}_1 \,; \mathbf{Q}_{01})$ and $(\mathbf{Q}_2 \,; \mathbf{Q}_{02})$, is defined to be $(\mathbf{Q}_1 \cdot \mathbf{Q}_{02} + \mathbf{Q}_2 \cdot \mathbf{Q}_{01})$ and $\mathcal{S}_{i+1}^T \, {}^0\mathcal{P}_{i+1}^A$ is to be computed as the inner product of two 6×1 entities. See numerator of Eqn (2.75) in Section 2.8 for another use of the inner product of two 6×1 entities. See also Section 2.5 for the representation of various kinds of joints.

The generalized force Q_i is the component of $\mathcal{F}_i$ along $\mathcal{S}_i$, and we get

$$\mathcal{S}_i^T \mathcal{F}_i = Q_i \tag{6.66}$$

Using Eqn (6.63) and after simplification, we get

$$\ddot{q}_i = \frac{Q_i - \mathcal{S}_i^T\, [\mathcal{I}]_i^A\, \mathcal{A}_{i-1} - \mathcal{S}_i^T\, \mathcal{P}_i^A}{\mathcal{S}_i^T\, [\mathcal{I}]_i^A\, \mathcal{S}_i} \tag{6.67}$$

Since the acceleration of the fixed base is known, $\mathcal{A}_0 = \mathbf{0}$, we can compute the $\ddot{q}_1$ from Eqn (6.67) once the articulated-body inertia and the bias term are known for $i = 1$ and for a given actuated generalized force Q_1. In general, we iterate from $i = 1$ to N and obtain all the $\ddot{q}$'s in the serial manipulator. The computation of joint accelerations with the above formulas can be done in $\mathcal{O}(N)$ steps (Featherstone 1983, 1987).

6.7.3* Recursive Algorithms for Parallel Manipulators

The recursive algorithms of the preceding sub-section cannot be applied for parallel manipulators and closed-loop mechanisms due to the presence of passive joints and holonomic loop-closure constraints relating the passive and active joint variables. It is impractical and difficult to eliminate the passive joint variables using the loop-closure constraint equations. Most of the algorithms make use of the concept of Lagrange multipliers as discussed in Section 6.3. Equations of motion for a parallel manipulator or a closed-loop mechanisms given in Eqn (6.21) and the second derivative of the constraint equations given in Eqn (6.22) can be written in a matrix form as

$$\begin{bmatrix} [\mathbf{M}] & [\mathbf{\Psi}]^T \\ [\mathbf{\Psi}] & [\mathbf{0}] \end{bmatrix} \begin{pmatrix} \ddot{\mathbf{q}} \\ -\boldsymbol{\lambda} \end{pmatrix} = \begin{pmatrix} \boldsymbol{\tau} - [\mathbf{C}]\dot{\mathbf{q}} - \mathbf{G} \\ -[\dot{\mathbf{\Psi}}]\dot{\mathbf{q}} \end{pmatrix} \tag{6.68}$$

where the dependence of the terms on $\mathbf{q}$ and $\dot{\mathbf{q}}$ is dropped for convenience.

Equation (6.68) represents $n + m$ equations in n $\ddot{q}_i$'s and m λ_i's. Hence an intuitively simple scheme would be to solve for $\boldsymbol{\lambda}$ and $\ddot{\mathbf{q}}$ directly. This could be done in an $\mathcal{O}((n+m)^3)$ Gaussian elimination algorithm.

We can also solve for $\boldsymbol{\lambda}$, following the steps shown in Eqns (6.22) and (6.23) and write

$$([\mathbf{\Psi}][\mathbf{M}]^{-1}[\mathbf{\Psi}]^T)\boldsymbol{\lambda} = -[\dot{\mathbf{\Psi}}]\dot{\mathbf{q}} - [\mathbf{\Psi}][\mathbf{M}]^{-1}(\boldsymbol{\tau} - [\mathbf{C}]\dot{\mathbf{q}} - \mathbf{G}) \tag{6.69}$$

Once all the terms, such as $[\mathbf{\Psi}]$, $[\mathbf{M}]$, etc., are formed, Eq. (6.69) can be solved by an $\mathcal{O}(m^3)$ Gaussian elimination algorithm since there are only m linear equations in the m Lagrange multipliers. To form the terms $[\mathbf{M}]$, $[\mathbf{C}]\dot{\mathbf{q}}$, and $\mathbf{G}$, an $\mathcal{O}(n)$ inverse dynamics algorithm can be used, and the terms with $[\mathbf{\Psi}]$ would require an $\mathcal{O}(m^2)$ algorithm. Hence the total complexity of obtaining $\boldsymbol{\lambda}$ is $\mathcal{O}(nm^2 + m^3)$. Once $\boldsymbol{\lambda}$ is known, the parallel manipulator or the closed-loop mechanism is equivalent to a serial manipulator with an extra forcing term on the right-hand side [see Eqn (6.21)] and we can use an $\mathcal{O}(n)$ algorithm to obtain $\ddot{\mathbf{q}}$.

Formulation of efficient algorithms for parallel manipulators and large multi-body systems with loops is still a subject of research. There exist $\mathcal{O}(n + m^3)$ algorithms, with m being the number of closed-loops (see, for example, Lubich et al. 1992), and an iterative $\mathcal{O}(n)$ algorithm (Ascher & Lin 1999) which needs to be used k times for numerical convergence with k claimed to be 'independent' of m. The interested reader is referred to the literature cited above for details on these algorithms.

Exercises

6.1 In addition to the joint torques or forces, the end-effector of a serial manipulator is subjected to an external force ${}^0\mathbf{f}_{Tool}$ and a moment ${}^0\mathbf{n}_{Tool}$. What are the generalized forces Q_i $(i = 1, \ldots, n)$ in this case?

6.2 Derive the expression for C_{ij} given in Eqn (6.16).

6.3 Show that the constraint force given by $[\boldsymbol{\Psi}(\mathbf{q})]^T \boldsymbol{\lambda}$ does not do any work.

†6.4 If the planar four-bar mechanism is not broken at joint 3 (corresponding to ϕ_3) but at joint 4 (corresponding to ϕ_1), then we get a planar 3R manipulator with different constraint equations. Derive the symbolic equations of motion for this case and comment on the differences in the mass matrix, Coriolis/centripetal term, and the gravity term obtained this way with the ones obtained in Example 6.2.

†6.5 Obtain symbolic equations of motion of a planar five-bar mechanisms shown in Exercise 4.2 in Chapter 4.

†6.6 Obtain symbolic equations of motion of the spatial RRR manipulator shown in Exercise 3.5 in Chapter 3.

†6.7 Obtain symbolic equations of motion of three-DOF parallel manipulator shown in Fig. 2.21 discussed in Examples 2.5 and 4.2.

6.8 Derive expressions for $[M_{\mathcal{X}}(\mathbf{q})]$, $\mathbf{C}_{\mathcal{X}}(\mathbf{q}, \dot{\mathbf{q}})$, and $\mathbf{G}_{\mathcal{X}}(\mathbf{q})$ given in Eqn (6.36).

†6.9 For the planar 2R manipulator discussed in Example 6.3 assume that the trajectories of the joints are

$$\theta_1(t) = \frac{\pi}{2} \sin\left(\frac{\pi}{20} t\right)$$

$$\theta_2(t) = \frac{\pi}{4} \sin\left(\frac{\pi}{20} t\right)$$

Plot $\tau_1(t)$ and $\tau_2(t)$. What and where are the highest τ_1 and τ_2?

†6.10 The actuated joint of the four-bar mechanism discussed in Example 6.5 oscillates between 0° and 90° with θ_1 given as

$$\theta_1(t) = \frac{\pi}{2}\sin\left(\frac{\pi}{20}t\right)$$

What and where is the highest torque τ_1? Assume that the spring is not present for this problem and the joint is actuated by a motor.

†6.11 For the planar 2R manipulator discussed in Example 6.4, assume that there is viscous damping at the joints given by $0.01\dot{\theta}_1$ and $0.01\dot{\theta}_2$. Simulate the equations of motion with the viscous damping included and plot θ_1 and θ_2 as functions of time.

†6.12 Simulate the equations of motion obtained for the planar four-bar mechanisms from Exercise 6.4 using the same numerical values as in Example 6.5. Verify that the plot of θ_1 with respect to time t is the same. Comment on the differences between the plots of Lagrange multipliers, if any.

†6.13 The recursive Newton–Euler algorithm can be automated in MAPLE to yield the equations of motion for a serial manipulator. Write a MAPLE program to obtain the equations of motion for the planar 2R manipulator using the recursive Newton–Euler Algorithm.

6.14 Obtain the modified recursive Newton–Euler algorithm for inverse dynamics which takes into account prismatic joints in a serial manipulator.

6.15 The recursive Newton–Euler algorithm is known to be $\mathcal{O}(N)$. In addition to the order of an algorithm, it is also important to know what is the coefficient multiplying N. Obtain from the recent literature what is the constant term multiplying N in a recursive Newton–Euler algorithm for serial manipulators.

References and Suggested Additional Reading

ADAMS user manual, Mechanical Dynamics Inc., 2002.

Armstrong, W.W. 1979, 'Recursive solution to the equations of motion of an link manipulator', *Proceedings of the 5th World Congress on Theory of Machines and Mechanisms*, Montreal, pp. 1343–46.

Ascher, U. and P. Lin 1999, 'Sequential regularization methods for simulating mechanical systems with many closed loops', *SIAM J. Sci. Comput.*, vol. 21, pp. 1244–62.

Baumgarte, J.W. 1983, 'A new method of stabilization for holonomic constraints', *Trans. ASME, J. Appl. Mech.*, vol. 50, pp. 869–70.

Featherstone, R. 1983, 'The calculation of robot dynamics using articulated-body inertias', *Int. J. Robotic. Res.*, vol. 2, pp. 13–30.

Featherstone, R. 1987, *Robot Dynamics Algorithms*, Kluwer Academic Publishers.

Featherstone, R. and D. Orin 2000, 'Robot dynamics: Equations and algorithms', *Proceeding of IEEE Conference on Robotics and Automation*, San Francisco, pp. 826–34.

Goldstein, H. 1980, *Classical Mechanics*, Addison-Wesley.

Haug, E.J. 1989, *Computer-Aided Kinematics and Dynamics of Mechanical Systems: Basic Methods*, vol. 1, Allyn and Bacon.

Hollerbach, J.M. 1983, 'A recursive Lagrangian formulation of manipulator dynamics and a comparative study of dynamic formulation complexity', in *Robot Motion: Planning and Control*, M. Brady et al. (eds), MIT Press, Cambridge, MA.

Jain, A. 1991, 'Unified formulation of dynamics for serial rigid multi-body systems', *AIAA J. Guid. Control Dynam.*, vol. 14, pp. 531–42.

Kane, T.R. and D.A. Levinson 1983, 'The use of Kane's dynamical equations in robotics', *Int. J. Robotic. Res.*, vol. 2, pp. 3–21.

Khatib, O. 1987, 'A unified approach to motion and force control of robot manipulators: The operational space formulations', *IEEE Trans. Robot. Autom.*, vol. 3, pp. 43–53.

Klepeis, J.L., H.D. Schafroth, K.M. Westerberg, and C. A. Floudas 2002, 'Deterministic global optimization and ab initio approaches for the structure prediction of polypeptides, dynamics of protein folding, and protein-protein interactions', in *Computational Methods for Protein Folding: Advances in Chemical Physics*, vol. 120, John Wiley, New York.

Lubich, Ch, U. Nowak, U. Pöhle, and Ch. Engstler 1992, 'MEXX–Numerical software for the integration of constrained mechanical multi-body systems', Preprint SC 92-12, *Zuse Zentrum für Informationtechnik Berin* (ZIB), Germany.

Luh, J.Y.S., M.W. Walker, and R.P.C. Paul 1980, 'On-line computational scheme for mechanical manipulators', *Trans. ASME J. Dyn. Syst.*, vol. 102, pp. 69–76.

Orin, D.E., R.B. McGhee, M. Vukobratovic, and G. Hartoch 1979, 'Kinematic and kinetic analysis of open-chain linkages utilizing Newton-Euler Methods', *Math. Biosc.*, vol. 43, pp. 107–30.

Silver, W.M. 1982, 'On the equivalence of Lagrangian and Newton-Euler dynamics of manipulators', *Int. J. Robotic. Res.*, vol. 1, pp. 60–70.

Strang, G. 1976, *Linear Algebra and its Application*, Academic Press.

Uicker, J.J. 1967, 'Dynamic force analysis of spatial linkages', *Trans. ASME J. Appl. Mech.*, vol. 34, pp. 418–24.

Walker, M.W. and D.E. Orin 1982, 'Efficient dynamic computer simulation of robotic mechanisms', *Trans. ASME J. Dyn. Syst.*, vol. 104, pp. 205–11.

7

Trajectory Planning and Generation

7.1 Introduction

In this chapter, we deal with algorithms to plan and generate trajectories which describe the desired motion of a robot. By *trajectory* we mean the time history of position, velocity, and acceleration of either *actuated* joints or the end-effector of the robot. As we will see in Chapter 8, trajectory planning and generation is essential for robot control and for simulation.

The main issues in trajectory planning and generation are (a) ease and flexibility of planning, (b) efficient representation of the trajectory in a computer, and (c) generation of the desired robot trajectory at run time. By ease and flexibility of planning, we mean how the trajectory can be specified by the robot operator. There are two main ways in which the robot operator typically specifies the trajectory. The robot operator may want to specify the desired time history of a single or a group of joints, or the robot operator may simply state the desired time history of the end-effector or the tool of the manipulator. The first is called *joint space scheme* and the second is called *Cartesian space scheme*. Typically the planned and generated trajectory should be a *smooth* function of time in the sense that derivatives with respect to time, up to a desired order, must exist and be smooth. To what order the derivatives must be smooth is determined primarily by the operating characteristics of a robot. Most modern robots, however, require joint trajectories which are at least C^2, where the second derivative or the joint acceleration is continuous. Joint trajectories less than C^2 may lead to vibration and excessive wear in the gears and other mechanisms at the joint, and this in turn may result in loss of accuracy and repeatability of the robot over time.

Most often the robot operator specifies the initial (or a start) and the final (or goal) positions of a joint or the end-effector. Often the robot operator may specify the trajectory in terms of a start and goal, point and one or more intermediate or via points. Sometimes the desired velocities and/or accelerations at the start, goal, and via points may also be specified. In all such cases the robot operator should not have to specify complicated functions of time, which ensures smooth trajectories. The robot control computer and the trajectory planning algorithm should generate the exact path, velocity, and acceleration profiles between the start, intermediate, and end points.

By representation of the trajectory we mean how the computer internally stores the planned trajectory. Since issues of memory and storage costs are involved, the planned trajectory should be stored in an efficient manner. By generation, we mean how the desired trajectory is generated during the operation of the robot. Typically, the desired trajectory points are computed at a certain rate dictated by the digital controller. In typical robots this rate varies between 50 and 200 Hz. Hence, computations of trajectory points must be done in an efficient manner to minimize the number of floating point operations. With the advent of fast processors and cheap memory, the issues of representation and generation are not as critical as these were in the early history of robotics.

In this chapter, we will discuss all the above issues for joint space and Cartesian space schemes for robots. We start our discussion with joint space schemes.

7.2 Joint Space Schemes

Consider the task of actuating the ith joint of a robot from $\theta_i(t_0)$ to $\theta_i(t_f)$, where t_0 denotes the initial or start time and t_f denotes the end or final time, in a smooth manner. Fundamental to this task is the choice of an appropriate curve, $\theta_i(t)$, between the given initial and final values. Intuitively, there are many possible smooth function curves between $\theta_i(t_0)$ and $\theta_i(t_f)$, and, in fact there are infinitely many possible smooth trajectories between the initial and the final θ_i. Choosing a smooth curve between two given points is also known as *interpolation*, and this has been extensively studied in the field of geometric modelling and CAD [see, for example, Bezier curves, B-splines, NURBS in Mortenson's book (1985) or any other textbook on geometric modelling and CAD]. However, we restrict ourselves to simple polynomial curves since, as we see in this chapter, they are sufficient for the purpose of trajectory planning in robot manipulators.

The simplest possible polynomial curve between $\theta_i(t_0)$ and $\theta_i(t_f)$ is a straight line given by

$$\theta_i(t) = \frac{\theta_i(t_f) - \theta_i(t_0)}{t_f - t_0}(t - t_f) + \theta_i(t_f) \tag{7.1}$$

One can very easily determine $\theta_i(t)$ even if several intermediate or via points, as shown in Fig. 7.1, are specified—we will get a curve which is piece-wise linear. The problem with straight lines or piece-wise straight lines is apparent if we try to obtain the joint velocity and acceleration determined from the first and second derivatives of $\theta_i(t)$ with respect to time. The second plot in Fig. 7.1 shows the joint velocity, and it can be seen that it changes sign or magnitude in steps between two segments. The plot of acceleration will be even worse since we will get spikes at the instants when the velocity changes sign or magnitude and remains zero in a segment.

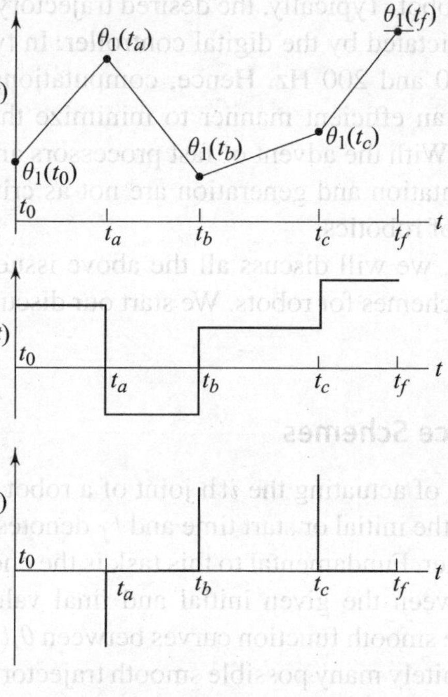

Fig. 7.1 Piece-wise linear joint trajectory

The simplest polynomial that ensures a continuous velocity and acceleration between given initial and final points is a cubic. A cubic is described by the equation

$$\theta_i(t) = a_0 + a_1 t + a_2 t^2 + a_3 t^3 \tag{7.2}$$

where a_0, a_1, a_2, and a_3 are four constant coefficients. To obtain the four coefficients, we specify that at the start or initial time, $t = t_0$, and at the final or end time, $t = t_f$. We have

$$\begin{aligned} \theta_i(t_0) &= \theta_i(0) \\ \theta_i(t_f) &= \theta_i(f) \end{aligned} \tag{7.3}$$

and the velocities at the initial and final times are

$$\begin{aligned}\dot{\theta}_i(t_0) &= \dot{\theta}_i(0)\\ \dot{\theta}_i(t_f) &= \dot{\theta}_i(f)\end{aligned} \tag{7.4}$$

The velocity and acceleration as a function of time can be derived by differentiating $\theta_i(t)$ in Eqn (7.2), and we can write

$$\begin{aligned}\dot{\theta}_i(t) &= a_1 + 2a_2 t + 3a_3 t^2\\ \ddot{\theta}_i(t) &= 2a_2 + 6a_3 t\end{aligned} \tag{7.5}$$

Without any loss of generality we can assume that $t_0 = 0$, and by using Eqns (7.3) and (7.4) in Eqns (7.2) and (7.5), we get

$$\begin{aligned}\theta_i(0) &= a_0\\ \theta_i(f) &= a_0 + a_1 t_f + a_2 t_f^2 + a_3 t_f^3\\ \dot{\theta}_i(0) &= a_1\\ \dot{\theta}_i(f) &= a_1 + 2a_2 t_f + 3a_3 t_f^2\end{aligned} \tag{7.6}$$

Solving the above four equations for the four unknown coefficients of the cubic, we get

$$\begin{aligned}a_0 &= \theta_i(0)\\ a_1 &= \dot{\theta}_i(0)\\ a_2 &= \frac{3}{t_f^2}[\theta_i(f) - \theta_i(0)] - \frac{2}{t_f}\dot{\theta}_i(0) - \frac{1}{t_f}\dot{\theta}_i(f)\\ a_3 &= -\frac{2}{t_f^3}[\theta_i(f) - \theta_i(0)] + \frac{1}{t_f^2}[\dot{\theta}_i(0) + \dot{\theta}_i(f)]\end{aligned} \tag{7.7}$$

The above equations give us the coefficients of a cubic polynomial connecting an arbitrary initial and final θ_i with an arbitrary initial and final $\dot{\theta}_i$.

Example 7.1: A cubic trajectory

The first joint of a 3R robot is to rotate from 30° to 60°. The initial and final angular velocities are +10 deg/sec, and −30 deg/sec. The motion is to be completed in 3.0 sec. On substitution in Eqn (7.7), we get

$$a_0 = 30, \quad a_1 = 10, \quad a_2 = 13.3333, \quad a_3 = -4.4444$$

Hence, the expressions for $\theta_1(t)$ and its derivatives are

$$\theta_1(t) = 30 + 10t + 13.3333t^2 - 4.4444t^3$$

$$\dot{\theta}_1(t) = 10 + 26.6666t - 13.3332t^2$$
$$\ddot{\theta}_1(t) = 26.6666 - 26.6664t$$

Figure 7.2 shows the plots of $\theta_1(t)$, $\dot{\theta}_1(t)$, and $\ddot{\theta}_1(t)$ as a function of time t. As expected the constraints on the initial and final positions and velocity are satisfied. We can also note that the plot of joint rotation, θ_1, as a function of t is a cubic, the plot of joint rate, $\dot{\theta}_i$, as a function of t is a quadratic, and the plot of joint acceleration as a function of t is a straight line.

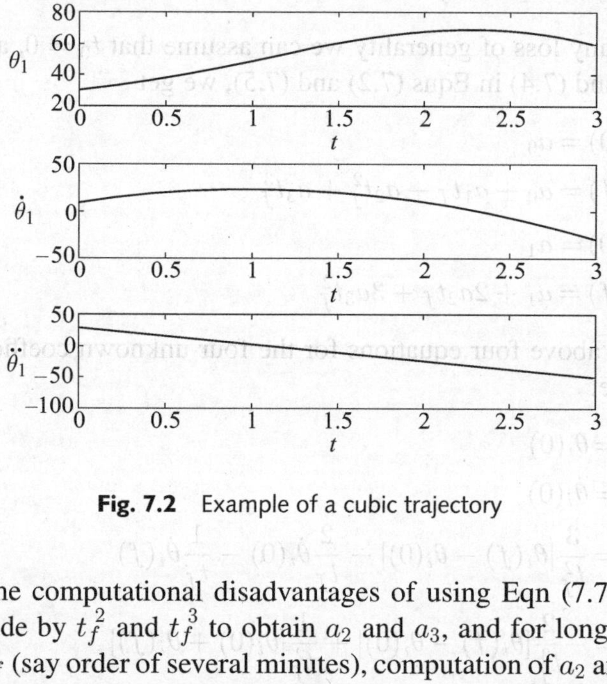

Fig. 7.2 Example of a cubic trajectory

One of the computational disadvantages of using Eqn (7.7) is that we need to divide by t_f^2 and t_f^3 to obtain a_2 and a_3, and for long trajectories with large t_f (say order of several minutes), computation of a_2 and a_3 can be error prone. To avoid this difficulty, we borrow the technique of scaling t as prevalent in the field of geometric modelling (Mortenson 1985). We define

$$u = t/t_f \tag{7.8}$$

and hence $u \in [0, 1]$. In terms of u, the cubic in Eqn (7.2) can be written as

$$\theta_i(u) = a_0 + a_1 u + a_2 u^2 + a_3 u^3 \tag{7.9}$$

and Eqn (7.6) becomes

$$\theta_i(0) = a_0, \qquad \theta_i(1) = a_0 + a_1 + a_2 + a_3$$
$$\theta_i'(0) = a_1, \qquad \theta_i'(1) = a_1 + 2a_2 + 3a_3$$

where θ_i' denotes the derivative of θ_i with respect to u, and the coefficients a_0, a_1, a_2, and a_3 are more simply given as

$$\begin{aligned}
a_0 &= \theta_i(0) \\
a_1 &= \theta_i'(0) \\
a_2 &= -3\theta_i(0) + 3\theta_i(1) - 2\theta_i'(0) - \theta_i'(1) \\
a_3 &= 2\theta_i(0) - 2\theta_i(1) + \theta_i'(0) + \theta_i'(1)
\end{aligned} \tag{7.10}$$

We can also rewrite $\theta_i(u)$, $\theta_i'(u)$, and $\theta_i''(u)$ as

$$\begin{aligned}
\theta_i(u) &= (2u^3 - 3u^2 + 1)\theta_i(0) + (-2u^3 + 3u^2)\theta_i(1) \\
&\quad + (u^3 - 2u^2 + u)\theta_i'(0) + (u^3 - u^2)\theta_i'(1) \\
\theta_i'(u) &= (6u^2 - 6u)\theta_i(0) + (-6u^2 + 6u)\theta_i(1) \\
&\quad + (3u^2 - 4u + 1)\theta_i'(0) + (3u^2 - 2u)\theta_i'(1) \\
\theta_i''(u) &= (12u - 6)\theta_i(0) + (-12u + 6)\theta_i(1) \\
&\quad + (6u - 4)\theta_i'(0) + (6u - 2)\theta_i'(1)
\end{aligned} \tag{7.11}$$

and using the above equation, we can generate $\theta_i(u)$, $\theta_i'(u)$, and $\theta_i''(u)$ for $u \in [0, 1]$. To obtain $\theta_i(t)$ and its derivatives, we can go back to t using the relationship between u and t given in Eqn (7.8).

From Fig. 7.2 and also from the last equation in Eqn (7.11), we note that there is no control over the initial and final acceleration. This is expected of a cubic where we can satisfy *at most* four constraints. In the above development, we have specified the joint position and velocity at the initial and final times. We can, of course, take another combination such as joint position, velocity, and acceleration at $u = 0$ and only joint position at $u = 1$ (see Exercise 7.1). However, in this combination, we will have no control over the joint velocity and acceleration at $u = 1$.

If we wish to specify the position, velocity, *and* acceleration at *both* initial and final times, then we have to use a higher-order polynomial. A fifth degree or a quintic polynomial with six coefficients is smooth up to the fourth derivative and one can specify the position, velocity, and acceleration at both ends (see Exercise 7.2). One can use other functions, such as $\sin(t)$, which are infinitely smooth. However, as can be seen in going from cubic to quintic, higher-order polynomials require computation of a larger number of coefficients and more computations to obtain the position, velocity, and acceleration. Hence, from the point of view of the efficient representation and generation of trajectories for the real time control of robots, one rarely goes beyond cubic polynomials. Only in a few research robots quintic polynomials have been tried for trajectory planning and generation. In this text we will restrict ourselves to cubic polynomials.

As can be seen from Eqns (7.2) and (7.5), we need to store only four parameters in a computer for each cubic. The amount of computation is also very low. By writing the cubic in a nested form

$$\theta_i(u) = a_0 + u(a_1 + u(a_2 + a_3 u)) \tag{7.12}$$

we can obtain the ith joint rotation, at a given u (or t), with three multiplications and three additions. Likewise for velocity, we need to perform two extra multiplications and two extra additions, and to compute acceleration, we need to perform one additional multiplication and addition.

To plan trajectories for a robot with n actuated joints, we simply have to use Eqns (7.10) n number of times. For coordinated motions, it must be ensured that the initial and final points and the specified velocities are reached at the same time. If via points are also specified (trajectory planning with via points is discussed in the following section), then in addition to the initial and final positions and velocities, we have to ensure that the timings at the via points are also correct.

7.3 Joint Space Schemes With Via Points

Often the robot programmer wishes to specify paths which include one or more intermediate via points in addition to the initial and final points. There are two possibilities in such situations:

1. The velocity at the k via point(s) is specified.
2. The velocity at the k via point(s) is not specified.

The first case is fairly simple. We will simply plan the trajectories for the $k+1$ segments[1] as $k+1$ cubics. We obtain the values of a_{0i}, a_{1i}, a_{2i}, and a_{3i} ($i = 1, 2, ..., k+1$) for each of the $k+1$ segments by using Eqns (7.10). Continuity in position and velocity is automatically ensured. However, continuity in acceleration is not maintained. This is clearly seen in the example that follows.

Example 7.2: A cubic trajectory with a via point

The first joint of a 3R robot is to rotate from 30° to 60° in 3 sec. It is also required that the joint is at 55° at $t = 2$ sec. The initial and final angular velocities are +10 deg/sec and −30 deg/sec. The velocity at the intermediate via point is −10 deg/sec.

[1] If k via points are given, then we have $k+1$ segments.

We obtain the trajectory in two segments from $t = 0$ to $t = 2$ sec and from $t = 0$ to $t = 1$ sec. Using the parameter t directly, from Eqn (7.7), we get for the first segment:

$$a_{01} = 30, \quad a_{11} = 10, \quad a_{21} = 13.75, \quad a_{31} = -6.25$$

For the second segment, we get

$$a_{02} = 55, \quad a_{12} = -10, \quad a_{22} = 65, \quad a_{32} = -50$$

Hence the expressions for $\theta_1(t)$ and its derivatives for both the segments are

$$\theta_1(t) = 30 + 10t + 13.75t^2 - 6.25t^3, \qquad 0 \le t \le 2$$

$$\theta_1(t) = 55 - 10t + 65t^2 - 50t^3, \qquad 2 \le t \le 3$$

$$\dot{\theta}_1(t) = 10 + 27.5t - 18.75t^2, \qquad 0 \le t \le 2$$

$$\dot{\theta}_1(t) = -10 + 130t - 150t^2, \qquad 2 \le t \le 3$$

$$\ddot{\theta}_1(t) = 27.5 - 37.5t, \qquad 0 \le t \le 2$$

$$\ddot{\theta}_1(t) = 130 - 300t, \qquad 2 \le t \le 3$$

Figure 7.3 shows the plots of $\theta_1(t)$, $\dot{\theta}_1(t)$, and $\ddot{\theta}_1(t)$ as a function of time. As expected, the position and velocity profiles are smooth, and the acceleration profile is discontinuous.

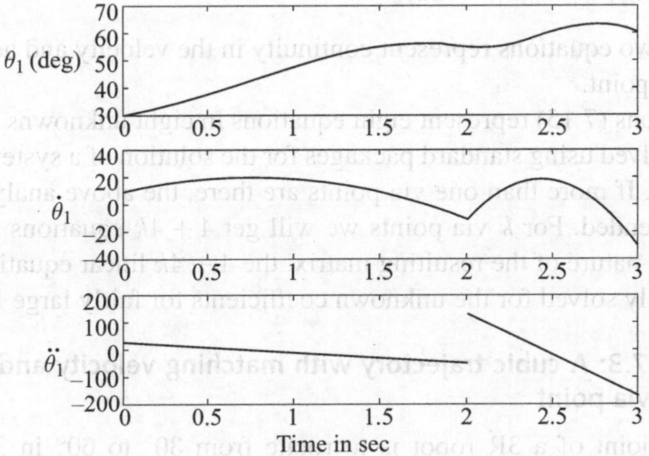

Fig. 7.3 A cubic trajectory with via point

In the second case, when the velocity at the via points is not specified, we have free choices, and one way to utilise the available freedom is to compute the joint velocities such that the accelerations at the via points are continuous. In order to maintain continuity in acceleration, we assume initial and final points to be $\theta_i(0)$ and $\theta_i(f)$, the via point $\theta_i(v)$, and the initial and final velocities $\dot{\theta}_i(0)$ and $\dot{\theta}_i(f)$. Each cubic is over an interval $t = 0$ to $t = t_{f_i}$ with $i = 1, 2$. The first cubic is assumed as

$$\theta_i(t) = a_{01} + a_{11}t + a_{21}t^2 + a_{31}t^3 \tag{7.13}$$

and the second cubic is assumed as

$$\theta_i(t) = a_{02} + a_{12}t + a_{22}t^2 + a_{32}t^3 \tag{7.14}$$

Substituting the given initial, final, via point, and the initial and final velocities, we get

$$\begin{aligned}
\theta_i(0) &= a_{01} \\
\dot{\theta}_i(0) &= a_{11} \\
\theta_i(v) &= a_{01} + a_{11}t_{f_1} + a_{21}t_{f_1}^2 + a_{31}t_{f_1}^3 \\
\theta_i(v) &= a_{02} \\
\theta_i(f) &= a_{02} + a_{12}t_{f_2} + a_{22}t_{f_2}^2 + a_{32}t_{f_2}^3 \\
\dot{\theta}_i(f) &= a_{12} + 2a_{22}t_{f_2} + 3a_{32}t_{f_2}^2 \\
a_{11} + 2a_{21}t_{f_1} &+ 3a_{31}t_{f_1}^2 = a_{12} \\
2a_{21} + 6a_{31}t_{f_1} &= 2a_{22}
\end{aligned} \tag{7.15}$$

The last two equations represent continuity in the velocity and acceleration at the via point.

Equations (7.15) represent eight equations in eight unknowns and can be readily solved using standard packages for the solution of a system of linear equations. If more than one via points are there, the above analysis can be easily extended. For k via points we will get $4 + 4k$ equations and due to the sparse nature of the resulting matrix, the $4 + 4k$ linear equations can be numerically solved for the unknown coefficients for fairly large k.

Example 7.3: A cubic trajectory with matching velocity and acceleration at a via point

The first joint of a 3R robot is to rotate from 30° to 60° in 3 sec. It is also required that the joint is at 55° at $t = 2$ sec. The initial and final angular velocities are +10 deg/sec and −30 deg/sec, and the velocity at the intermediate via point is not specified. We wish to plan a trajectory that ensures continuity of the velocity and acceleration at the via point.

From Eqns (7.15), we get

$$a_{01} = 30, \quad a_{11} = 10, \quad a_{21} = -1.0417, \quad a_{31} = 1.1458$$

For the second segment, we get

$$a_{02} = 55, \quad a_{12} = 19.5833, \quad a_{22} = 5.8333, \quad a_{32} = -20.4167$$

Figure 7.4 shows the plots of $\theta_1(t)$, $\dot{\theta}_1(t)$, and $\ddot{\theta}_1(t)$ obtained using the coefficients of the two segments given above. As can be seen, the velocity and accelerations are continuous at the via point.

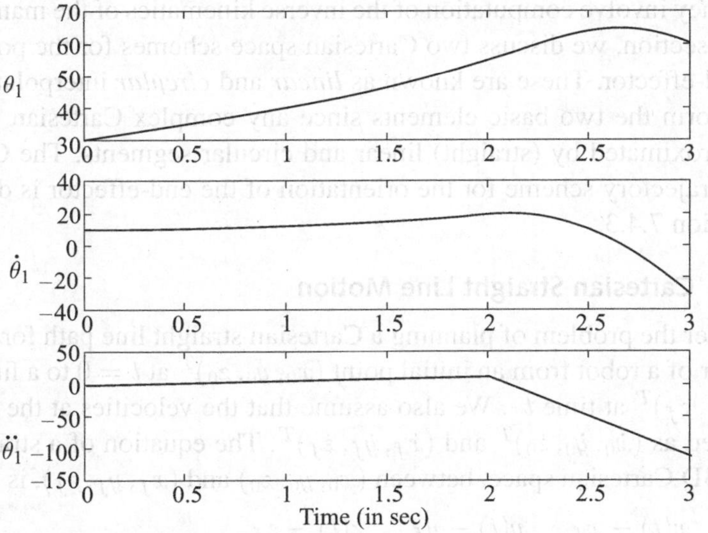

Fig. 7.4 A cubic trajectory with continuous acceleration

There are several other schemes to plan trajectories using higher- and lower-order polynomials. A well known scheme uses straight line segments, and to ensure continuity of velocity at the ends there is a parabolic blend region (Craig 1989). This scheme does not ensure continuity of acceleration in a segment and belongs to a class of functions which are only C^1 continuous. This scheme has been in extensive use for the control of actuators in various applications since very few coefficients need to be stored and generation in run time is very simple. However, with the advent of faster processors and large memory, one should be able to use the superior cubic-based trajectory planning schemes.

7.4 Cartesian Space Schemes

Joint space schemes are very useful if a single joint or a group of joints has to be moved between two positions. It becomes very difficult (or impossible) to visualise the required motion at all the joints to ensure a particular motion of the end-effector or the tool. It is much more easy and natural for the robot operator to specify a trajectory in the Cartesian space. For example, to plan a welding task, it is much more easier to specify the path of the welding torch (typically a straight line with a constant velocity along the path) with the required orientation of the torch. Hence, all trajectory planning systems provide for specifying the motion of the end-effector in the Cartesian space. Cartesian space schemes are, however, fairly computationally expensive since they involve computation of the inverse kinematics of the manipulator. In this section, we discuss two Cartesian space schemes for the position of the end-effector. These are known as *linear* and *circular* interpolation, and these form the two basic elements since any complex Cartesian path can be approximated by (straight) linear and circular segments. The Cartesian space trajectory scheme for the orientation of the end-effector is discussed in Section 7.4.3.

7.4.1 Cartesian Straight Line Motion

Consider the problem of planning a Cartesian straight line path for the end-effector of a robot from an initial point $(x_0, y_0, z_0)^T$ at $t = 0$ to a final point $(x_f, y_f, z_f)^T$ at time t_f. We also assume that the velocities at the ends are specified as $(\dot{x}_0, \dot{y}_0, \dot{z}_0)^T$ and $(\dot{x}_f, \dot{y}_f, \dot{z}_f)^T$. The equation of a straight line in the 3D Cartesian space, between (x_0, y_0, z_0) and (x_f, y_f, z_f), is given by

$$\frac{x(t) - x_f}{x_f - x_0} = \frac{y(t) - y_f}{y_f - y_0} = \frac{z(t) - z_f}{z_f - z_0} \tag{7.16}$$

or we can write

$$\begin{aligned} y(t) &= \left(\frac{y_f - y_0}{x_f - x_0}\right)(x(t) - x_f) + y_f \\ z(t) &= \left(\frac{z_f - z_0}{x_f - x_0}\right)(x(t) - x_f) + z_f \end{aligned} \tag{7.17}$$

To plan a straight line trajectory in 3D space, we need to only obtain $x(t)$ and then we can use Eqn (7.17) to obtain $y(t)$ and $z(t)$. We plan a smooth trajectory for $x(t)$ using the same techniques as discussed for joint space schemes. Hence, we can write

$$x(t) = a_0 + a_1 t + a_2 t^2 + a_3 t^3 \tag{7.18}$$

where a_0, a_1, a_2, and a_3 are four constant coefficients.

Using the specified x_0, x_f, $\dot{x}_0$, and $\dot{x}_f$ at $t = 0$ and $t = t_f$, and following the approach used to obtain the coefficients for a joint trajectory, we can get

$$
\begin{aligned}
a_0 &= x_0 \\
a_1 &= \dot{x}_0 \\
a_2 &= \frac{3}{t_f^2}(x_f - x_0) - \frac{2}{t_f}\dot{x}_0 - \frac{1}{t_f}\dot{x}_f \\
a_3 &= -\frac{2}{t_f^3}(x_f - x_0) + \frac{1}{t_f^2}(\dot{x}_0 + \dot{x}_f)
\end{aligned}
\tag{7.19}
$$

Once $x(t)$ is planned as a smooth, C^2, function of time, substituting $x(t)$ in Eqn (7.17) gives us smooth and continuous $y(t)$ and $z(t)$. Since $x(t)$, $y(t)$, and $z(t)$ satisfy the equation of a straight line in 3D space, all trajectory points lie *exactly* on a straight line, up to computational accuracy and errors due to discretization at update rate during run time. As discussed in Section 7.3, we can also plan $x(t)$ to pass through via points and, if required, match accelerations at the via points. We can also use quintic polynomials to specify accelerations in addition to position and velocity.

7.4.2 Cartesian Circular Motion

Although a circle can be discretized by straight line segments, for smoothness it is desirable to provide for specifying circular arcs in any Cartesian trajectory planning scheme. A circle is specified by three points lying in a plane. To obtain the equation of an arc of a circle from three given points, ${}^0\mathbf{p}_1$, ${}^0\mathbf{p}_2$, ${}^0\mathbf{p}_3$, in $\Re^3$, we proceed as follows.

1. Compute the normal to the plane as

$$
{}^0\hat{\mathbf{n}} = \frac{({}^0\mathbf{p}_2 - {}^0\mathbf{p}_1) \times ({}^0\mathbf{p}_3 - {}^0\mathbf{p}_1)}{|({}^0\mathbf{p}_2 - {}^0\mathbf{p}_1) \times ({}^0\mathbf{p}_3 - {}^0\mathbf{p}_1)|}
\tag{7.20}
$$

2. Assign a new coordinate system with the ${}^0\widehat{\mathbf{X}}$ and ${}^0\widehat{\mathbf{Y}}$ axes lying in the plane and the ${}^0\widehat{\mathbf{Z}}$ axis along the normal to the plane. The unit vectors are given by

$$
\begin{aligned}
{}^0\widehat{\mathbf{X}} &= \frac{({}^0\mathbf{p}_2 - {}^0\mathbf{p}_1)}{|({}^0\mathbf{p}_2 - {}^0\mathbf{p}_1)|} \\
{}^0\widehat{\mathbf{Y}} &= {}^0\mathbf{n} \times {}^0\widehat{\mathbf{X}}
\end{aligned}
\tag{7.21}
$$

   Let this coordinate system be called $\{CIRC\}$.

3. Obtain the rotation matrix ${}^0_{CIRC}[R]$ with the first, second, and third columns as ${}^0\widehat{\mathbf{X}}$, ${}^0\widehat{\mathbf{Y}}$, and ${}^0\hat{\mathbf{n}}$, respectively.
4. Transform the given points, using ${}^{CIRC}_{0}[R]$, to the frame $\{CIRC\}$. Let these be (x_1, y_1), (x_2, y_2), and (x_3, y_3). It may be noted that the Z coordinates of the transformed points are the same, say c.

5. Compute centre, (a, b), and radius r of the circular arc. This can be done by substituting the known (x_i, y_i), where $i = 1, 2, 3$, in the equation of the circle

$$(x - a)^2 + (y - b)^2 = r^2 \tag{7.22}$$

and solve for a, b, and r from the three equations.

6. Transform ${}^0\widehat{\mathbf{X}}$, ${}^0\widehat{\mathbf{Y}}$, and ${}^0\widehat{\mathbf{n}}$ to $\{CIRC\}$. Compute the angle made by the transformed $\widehat{\mathbf{X}}$ axis with the lines joining the centre of the arc to the three points. Let this be ϕ_1, ϕ_2, and ϕ_3.
7. Plan a cubic trajectory for $\phi(t)$ such that ϕ_1, ϕ_2, and ϕ_3 are reached at the specified times and in the specified order. This can be done by using procedures discussed for joint space schemes.
8. The X and Y coordinates of the points on the circular arc in $\{CIRC\}$ are given as

$$\begin{aligned} x(t) &= a + r\cos[\phi(t)] \\ y(t) &= b + r\sin[\phi(t)] \\ z(t) &= c \end{aligned} \tag{7.23}$$

9. The path of the tool or end-effector in $\{0\}$ can be obtained by using ${}^0_{CIRC}[R]$.

Although the end-effector position *exactly* traces a straight line or a circle, to actuate the joints of a robot we have to use the manipulator *inverse kinematics* to compute the joint variables and their rates. This may lead to non-smooth or wildly fluctuating trajectories in joint space. This can be avoided if we are allowed to *approximately* trace a straight line or a circle with a smoother joint motion. We proceed as follows.

1. Discretize the straight line or the circle between the given initial and final positions into several smaller line segments, each with an initial and a final position and orientation (see Section 7.4.3 for planning the orientation). This can be done by simple linear interpolation.
2. Perform inverse kinematics and inverse Jacobian on the initial and final positions, orientation, and velocities at the start and end points of each segment, and obtain values of θ_i and $\dot{\theta}_i$ $(i = 1, \ldots, n)$ for each of the joint variables, at start and end of each segment.
3. Use joint space schemes between initial and final values of all the joint variables using θ_i and $\dot{\theta}_i$ calculated from the previous step.

In this scheme, the end-effector path inside a segment will deviate from a straight line or the circle, and we have to take a larger number of segments if the deviation from the straight line is significant.

7.4.3 Trajectory Planning for Orientation

To plan a trajectory for the orientation of the end-effector, we assume that the orientation of the end-effector at the initial and final times is given in terms of *Euler parameters* as $({}^0\epsilon_{Tool}(0), \epsilon_4(0))^T$, $({}^0\epsilon_{Tool}(t_f), \epsilon_4(t_f))^T$, respectively.[2] We assume that the space-fixed angular velocity vectors at initial and final times, ${}^0\omega_{Tool}(0)$, and ${}^0\omega_{Tool}(t_f)$, respectively, are specified. To plan the orientation trajectory, we need to obtain the relationship between ${}^0\omega_{Tool}(t)$ and $({}^0\epsilon_{Tool}(t), \epsilon_4(t))^T$ and the time derivative $({}^0\epsilon_{Tool}(t), \epsilon_4(t))^T$. This can be done by following the development in Section 2.2 in Chapter 2 and deriving the required expressions (see also Exercise 2.3). We present the relevant expressions for convenience [see pp. 58–60 in Kane et al. (1983) for the detailed derivation of the angular velocity vector in terms of Euler parameters and vice versa].

1. The space-fixed angular velocity ${}^0\omega_{Tool}(t)$ can be written in terms of the four Euler parameters and their derivatives as
$$ {}^0\omega_{Tool}(t) = 2[E(t)]({}^0\dot{\epsilon}_{Tool}(t), \dot{\epsilon}_4(t))^T \tag{7.24}$$
and the inverse relation for $({}^0\dot{\epsilon}_{Tool}(t), \dot{\epsilon}_4(t))$ in terms of ${}^0\omega_{Tool}(t)$ and the Euler parameter is given by
$$({}^0\dot{\epsilon}_{Tool}(t), \dot{\epsilon}_4(t))^T = \frac{1}{2}[E(t)]^T\, {}^0\omega_{Tool}(t) \tag{7.25}$$
where the matrix $[E(t)]$ is given in terms of the four Euler parameters as
$$[E(t)] = \begin{pmatrix} -\epsilon_1 & \epsilon_4 & -\epsilon_3 & \epsilon_2 \\ -\epsilon_2 & \epsilon_3 & \epsilon_4 & -\epsilon_1 \\ -\epsilon_3 & -\epsilon_2 & \epsilon_1 & \epsilon_4 \end{pmatrix} \tag{7.26}$$
In Eqn (7.26), ϵ_1, ϵ_2, and ϵ_3 are the components of ${}^0\epsilon_{Tool}$ and we have dropped the functional dependence of the four Euler parameters on time t for convenience.
From Eqns (7.25) and (7.26), we can obtain ${}^0\dot{\epsilon}_{Tool}$ and $\dot{\epsilon}_4$ at $t = 0$ and $t = t_f$ from the given ${}^0\omega_{Tool}(0)$ and ${}^0\omega_{Tool}(t_f)$.
2. Once $({}^0\epsilon_{Tool}, \epsilon_4)^T$ and $({}^0\dot{\epsilon}_{Tool}, \dot{\epsilon}_4)^T$ are known at $t = 0$ and $t = t_f$, we can plan smooth, C^2, cubic trajectories for ${}^0\epsilon_{Tool}(t)$ similar to the joint space trajectories.
3. The trajectory for $\epsilon_4(t)$ is computed from
$$\epsilon_4(t) = \pm\sqrt{1 - [{}^0\epsilon_{Tool}(t) \cdot {}^0\epsilon_{Tool}(t)]} \tag{7.27}$$

[2] If the orientation is given in terms of Euler angles or any other form, they can be converted to the angle-axis form and then to Euler parameters by the algorithms given in Section 2.2 in Chapter 2.

which ensures that the four Euler parameters always satisfy the constraint inherent in them. It may be noted that we must take the appropriate sign in the above equation so that the initial and final conditions are as specified.

4. Once (ϵ, ϵ_4) is known at each time instant, one can obtain the angle-axis form $(\phi, {}^0\widehat{\mathbf{k}}_{Tool})$ or any other required representation of the orientation of the end-effector at each instant of time.

It may be noted that, as in the case of joint space schemes, if required one can plan a smooth orientation trajectory with via points or as quintic polynomials to satisfy specified accelerations at initial and final time.

7.5 Some Additional Issues in Trajectory Planning

The joint space schemes discussed in Section 7.2 can be applied for arbitrary serial manipulators as long as any point in the planned trajectory does not violate joint limits. In parallel manipulators and closed-loop mechanisms, however, we must take into account the mobility (see Chapter 4, Section 4.6) of the manipulator or the mechanism. It must be ensured that at all points in the planned trajectory for the *actuated* joints we have real values of the *passive* joint variables, i.e., the direct kinematics problem of the parallel manipulator and the closed-loop mechanism is solvable. Hence, the joint space schemes in parallel manipulators and closed-loop mechanisms involve added computational complexities.

Trajectory planning in the Cartesian space has several difficulties which arise from the singularities and joint limits present in any manipulator. In a trajectory planning scheme these need to be addressed by careful programming. We list a few of them.

1. The initial and final points in a straight line (or circular arc) in the Cartesian space maybe within the workspace. However, one or more intermediate via points may be outside the workspace. This may happen if the workspace has holes or if there are joint limits. In such a situation, one or more via points may have to be introduced to ensure that the entire straight line or circular arc lies in the workspace. Similar care must be taken in planning orientation trajectories.
2. The initial and final joint rates for a planned Cartesian trajectory may be acceptable. However, if the planned trajectory goes near a singularity, joint rates may become unacceptably large. Simulation is required to check this sort of situation. To avoid such problems, we may have to resort to scaling down the velocities, but this may cause loss of time coordination.

3. Cartesian space schemes require performing inverse kinematics. As seen in Chapters 3 and 4, inverse kinematics often gives rise to multiple solutions. One needs to be careful to ensure that the 'same' solution branch is being followed during the entire trajectory. If the solutions are on different branches, then some warning should be given to the robot operator.

The joint and Cartesian space control schemes discussed in this chapter do not take into account any restrictions on available maximum (or minimum) actuator torques or forces. A planned trajectory may require torques which are not possible for the actuators in the robot. In such situations, the control system and actuators will supply the maximum (or minimum) torque available. This will result in loss of coordination or straying from the planned path. A user-friendly trajectory planning software should warn the user that required torques are greater than available actuator torques. This can be done by performing inverse dynamic computations using an assumed model of the robot. Several researchers have developed algorithms which can plan trajectories taking into account limitations on actuator torques and forces (Bobrow et al 1983). One can also plan trajectories which minimize the time or some other quantity of interest (see, for example, Shin & Mackay 1985). Finally, it is of immense importance to plan trajectories which avoid obstacles or other objects in the workspace of the robot. This is a vast topic and some of the original was work done by Lozano-Perez (1982, 1983) and Brooks (1983a, 1983b). Trajectory planning with obstacle avoidance is beyond the scope of this book and the interested reader is referred to literature listed in references.

Exercises

7.1 Obtain the coefficients of a cubic polynomial $\theta(t) = a_0 + a_1 t + a_2 t^2 + a_3 t^3$ if $\theta(0)$, $\dot{\theta}(0)$, $\ddot{\theta}(0)$, and $\theta(t_f)$ are specified.

7.2 Obtain expressions for the six coefficients of a quintic polynomial $\theta(t) = a_0 + a_1 t + a_2 t^2 + a_3 t^3 + a_4 t^4 + a_5 t^5$ when the position, velocity, and acceleration are specified at $t = 0$ and $t = t_f$.

†7.3 Using the numerical values of Example 7.3, assume that $\ddot{\theta}_i$ are specified at the initial, final and via points as $\ddot{\theta}_1(0) = -2.0834$, $\ddot{\theta}_1(2) = 11.6662$, $\ddot{\theta}_1(3) = -110.8336$. The velocity at the via point is also specified as $\dot{\theta}_1(2) = 19.5833$. Plan a trajectory using quintic polynomials between $t = 0$ and $t = 3$. What is the difference between this trajectory and the one obtained in Example 7.3?

†7.4 Assume that the accelerations at the initial and final times are $\ddot{\theta}_1(0) = -5.0$, and $\ddot{\theta}_1(3) = -100.0$ with all other specifications the same in as Exercise 7.3

and Example 7.3. Plan a trajectory using quintic polynomials between $t = 0$ and $t = 3$. What is the difference between this trajectory and the one obtained in Exercise 7.3?

†7.5 Write a MATLAB program to obtain coefficients of a cubic trajectory when more than one via point ($k > 1$) is specified and velocities and accelerations are to be matched at each via point. Try with various values of k and test how large a value of k can be handled in MATLAB. Note that several of the coefficients are already solved and should not be put in the matrix to be inverted.

†7.6 The tip of a planar 2R manipulator is to trace a straight line in its workspace. Write a MATLAB program to plan a smooth cubic trajectory for known link lengths and given initial and final (x, y) and $(\dot{x}, \dot{y})$ such that the tip of the planar 2R manipulator *exactly* traces a straight line. Plot x, y, θ_1, θ_2 and their derivatives as a function of time for the numerical values used in Example 6.3.

†7.7 For the θ_1, θ_2, and their derivatives obtained in Exercise 7.6, using the dynamic equations of motion obtained in Chapter 6, obtain the torques τ_1 and τ_2 as a function of time. Where is the torque largest? Of $[\mathrm{M}]\ddot{\Theta}$, the Coriolis/centripetal term and the gravity term, whose contribution is the largest?

†7.8 The tip of a planar 2R manipulator is to trace a full circle in its workspace with centre at (a, b) and radius r in 1.0 sec. Assume the full circle is inside the workspace and at the start and final times, the Cartesian velocities are zero. Write a MATLAB program to plan a smooth cubic trajectory for known link lengths. Plot x, y, θ_1, θ_2 and their derivatives as a function of time. Use numerical values of Example 6.3 and $(a, b) = (0.5, 0.5)$ and $r = 0.5$.

†7.9 As in Exercise 7.8, the tip of the 2R manipulator is to trace a circle in its workspace. Instead of a cubic trajectory, we can also assume the parameter $\theta(t)$ used to describe the circle in the X-Y plane as $A\sin(\omega t)$, where A and ω are chosen to satisfy initial and final conditions. Plot $\theta_1(t)$, $\theta_2(t)$, and their derivatives as a function of time for this case. Comment on the difference between the joint trajectories obtained this way and in Exercise 7.8.

†7.10 For the θ_1, θ_2 obtained from Exercise 7.9, using the dynamic equations of motion obtained in Chapter 6, obtain the torques τ_1 and τ_2 as a function of time. Where is the torque largest for the trajectory obtained using $A\sin(\omega t)$? Of $[\mathrm{M}]\ddot{\Theta}$, the Coriolis/centripetal term and the gravity term, whose contribution is the largest?

†7.11 As discussed in Section 7.4.2, divide the straight line in the workspace into 10 equal segments and obtain θ_1, θ_2, and their derivatives at only the end points of the segment by inverse kinematics and inverse Jacobian. Plan and generate cubic trajectories in the joint space in each of the segments. Plot θ_1, θ_2, and their derivatives over the full trajectory. Comment on the joint trajectories obtained as described above and as in Exercise 7.6.

†7.12 Write a MATLAB program to generate smooth cubic trajectories in the Cartesian space for the end-effector of the PUMA 560 robot discussed in Chapters 2 and 3. Use the dimensions given in Chapter 3 and test your program for simple linear and circular trajectories.

†7.13 Write a MATLAB program to discretise an arbitrary trajectory in the Cartesian space into linear and circular segments at desired points and obtain the coefficients of a smooth cubic trajectory for the arbitrary trajectory. Note that the consecutive linear and circular segments need not be in the same plane.

References and Suggested Additional Reading

Bobrow, J., S. Dubowsky, and J. Gibson 1983, 'On optimal control of robotic manipulators with actuator constraints', *Proceedings of American Control Conference*, San Francisco, California, pp. 782–7.

Brooks, R.A. 1983a, 'Solving find-path problem by good representation of free space', *IEEE Trans. Syst. Man. Cyb.*, vol. 13, pp. 190–97.

Brooks, R.A. 1983b, 'Planning collision-free motion for pick-and-place operations', *Int. J. Robotic. Res.*, vol. 2, pp. 19–44.

Craig, J.J., 1989, *Introduction to Robotics: Mechanics and Control*, 2nd edn, Addison-Wesley,

Kane, T.R., P.W. Likins, and D.A. Levinson 1983, *Spacecraft Dynamics*, McGraw-Hill Inc.

Lozano-Perez, T., 1982, 'Spatial planning: A configuration space approach', *IEEE Trans. Comput.*, vol. C-32, pp. 108–20.

Lozano-Perez, T. 1983, 'Task planning', in *Robot Motion: Planning and Control*, M. Brady et al. (eds), MIT Press, Cambridge, MA.

Mortenson, M. E., 1985, *Geometric Modeling*, Wiley,

Shin, K. and N. McKay 1985, 'Minimum-time control of robotic manipulators with geometric path constraints', *IEEE Trans. Autom. Control*, vol. AC-30, no. 6. pp. 531–41.

8

Position and Force Control of Manipulators

8.1 Introduction

In the preceding chapter on trajectory planning and generation, we presented algorithms to compute the desired position, velocity, and acceleration of one (or more) joint(s) or the end-effector of a manipulator. In this chapter, we will first discuss how a manipulator is actually made to *follow* these desired joint or end-effector trajectories in 3D space with minimum error. In many applications, a manipulator end-effector is in contact with the environment, and it is required that the end-effector *apply* a desired force or moment on the environment in addition to following a desired trajectory while maintaining contact. Following a desired trajectory in 3D (free) space is called *position control* and applying desired force or moment on the environment is called *force control*, and a combination of position and force control is called *hybrid* control. This chapter presents the basic mathematical tools and concepts required for the analysis and development of position, force, and hybrid controllers for manipulators.

The main technique used to ensure that the error between the desired and the actual performance of a device or a dynamical system is small is *feedback control*. Formally, the goal of feedback control is to ensure that the *output* of a *dynamical system* follows a *desired trajectory* accurately in spite of *external disturbances* or *internal parameter changes*. Feedback control is also used to *stabilize* an unstable system and to improve the *performance* of a system. In this chapter we first start with the notion of *linear control*, where the term 'linear' implies that the dynamical system to be controlled can be reasonably accurately modelled by linear differential equations. We study linear control for the following reasons: (a) it is very well understood, (b) several very good design procedures for linear controllers are available in literature, (c) a large number of industrial devices can be controlled reasonably well by linear control schemes, and (d) often a non-linear system is *linearized* about its operating point and techniques from linear control can be used to design controllers for them. It must be kept in mind that a serial or a parallel manipulator can by no means be modelled as a linear system—the dynamical equations of motion for a serial or a parallel manipulator, derived in Chapter 6, are highly non-linear. However, linear control is successfully used in a variety of industrial manipulators and forms the basis of several *non-linear control* schemes presented in this chapter.

We start this chapter by introducing some of the basic concepts of linear control by means of an example of the control of a single-link manipulator.

8.2 Feedback Control of a Single-link Manipulator

Figure 8.1 shows a sketch of a one-link robot being driven by a DC servo-motor through a gear[1] reducer. Although the large speed reduction required cannot be done in one stage, for the purpose of modelling and control, we assume that we have just two gears in the speed reducer.

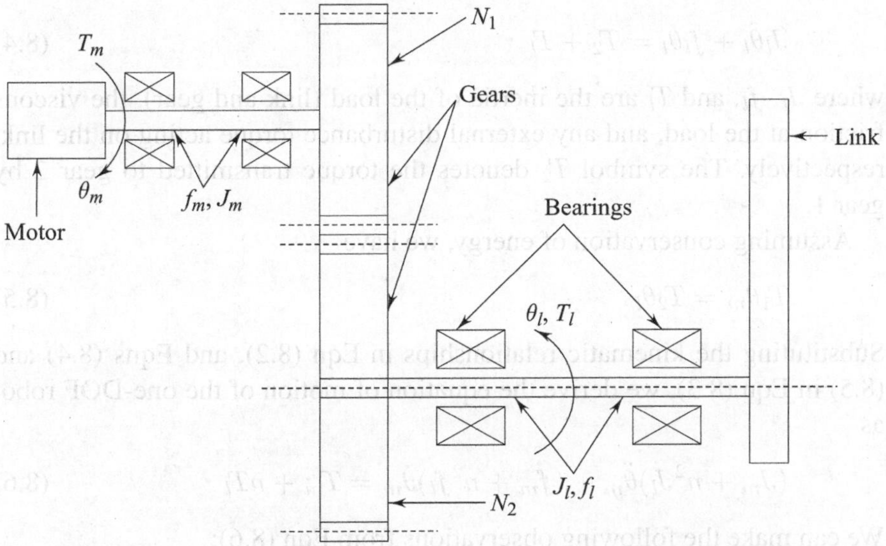

Fig. 8.1 Model of a single link

[1] Gears are needed for speed reduction since a typical DC servo-motor is efficient only at high rpm (2000 rpm or more) values, whereas the typical angular velocity required at the joint is around 60 rpm. Gear reduction also allows amplification of torques and reduction of the inertia of the load. Gears, however, introduce backlash, which makes the system non-linear and difficult to control. There are special motors which work efficiently at low speeds. These are used in *direct drive* robots.

Let us denote the rotation of the link by θ_l and that of the motor by θ_m. Thus, we have

$$\frac{\theta_l}{\theta_m} = n \tag{8.1}$$

where n is the gear ratio and is less than 1.0.

The kinematics of this one-DOF system is described by

$$\theta_l = n\theta_m, \qquad \dot{\theta}_l = n\dot{\theta}_m, \qquad \ddot{\theta}_l = n\ddot{\theta}_m \tag{8.2}$$

The equations of motion can be obtained by considering gear 1 and gear 2 and the link. For gear 1, we have

$$J_m\ddot{\theta}_m + f_m\dot{\theta}_m + T_1 = T_m \tag{8.3}$$

where J_m, f_m, and T_m are the inertia of the motor, the viscous friction at the motor shaft, and the torque output of the motor, respectively. The symbol T_1 denotes the torque acting on gear 1 from gear 2 and the link.

For gear 2, we have

$$J_l\ddot{\theta}_l + f_l\dot{\theta}_l = T_2 + T_l \tag{8.4}$$

where J_l, f_l, and T_l are the inertia of the load (link and gear), the viscous friction at the load, and any external disturbance torque acting on the link, respectively. The symbol T_2 denotes the torque transmitted to gear 2 by gear 1.

Assuming conservation of energy, we have

$$T_1\theta_m = T_2\theta_l \tag{8.5}$$

Substituting the kinematic relationships in Eqn (8.2), and Eqns (8.4) and (8.5) in Eqn (8.3), we derive the equation of motion of the one-DOF robot as

$$(J_m + n^2 J_l)\ddot{\theta}_m + (f_m + n^2 f_l)\dot{\theta}_m = T_m + nT_l \tag{8.6}$$

We can make the following observations from Eqn (8.6):

1. Since n is small (typically around 0.01), the effect of the load inertia and load friction, as seen from the motor, is significantly reduced by a factor of n^2.
2. The effect of T_l is also reduced by a factor of n. In a multi-link robot, the effect of the coupling torques due to the motion of other links, which contributes to T_l, is also reduced. This is one of the reasons why linear control schemes work for a non-linear robot particularly when it is operating at low speeds.

We next consider the model of the DC servo-motor. Figure 8.2 shows the circuit diagram of a typical permanent-magnet DC servo-motor. It consists of a stationary armature of resistance R_a and inductance L_a with the rotor being a permanent magnet made of rare earth materials. When current i_a flows through the circuit, the torque generated is given by

$$T_m = K_t i_a \tag{8.7}$$

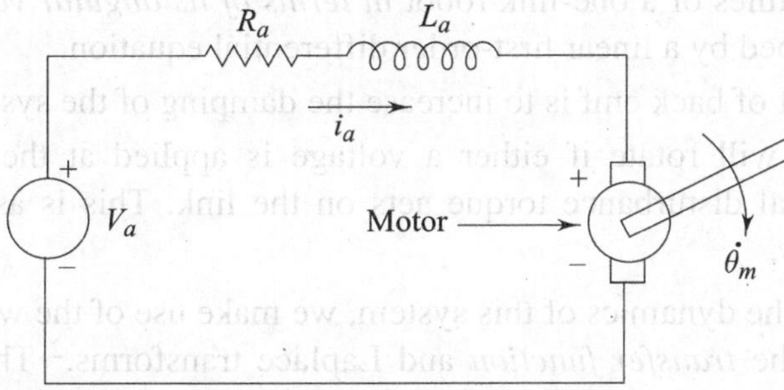

Fig. 8.2 Model of a permanent-magnet DC servo-motor

When the motor rotates at an angular velocity of $\dot{\theta}_m$, a *back emf* is generated, which is given by

$$V = K_g \dot{\theta}_m \tag{8.8}$$

Parameters K_t and K_g are called the *torque* and *back emf constant* of the DC servo-motor. These, together with R_a and L_a and other motor parameters, are available from the motor manufacturer and/or provided in motor catalogues specifications.

The dynamics of the circuit shown in Fig. 8.2 can be modelled by the first-order differential equation

$$L_a \dot{i}_a + R_a i_a + K_g \dot{\theta}_m = V_a \tag{8.9}$$

where V_a is the voltage applied at the terminals of the motor.

Typically, for small permanent-magnet DC servo-motors, the inductance L_a is small and can be ignored. Substituting Eqns (8.7) and (8.9) in Eqn (8.6) and setting L_a to zero, we get

$$(J_m + n^2 J_l)\ddot{\theta}_m + (f_m + n^2 f_l)\dot{\theta}_m = K_t \left(\frac{V_a - K_g \dot{\theta}_m}{R_a} \right) + nT_l$$

This equation can be written in a compact form as

$$J\dot{\Omega} + F\Omega = KV_a + T_d \tag{8.10}$$

where we have used the following symbols for convenience:

$$K = K_t/R_a, \qquad F = (f_m + n^2 f_l) + K_t K_g/R_a$$
$$J = J_m + n^2 J_l, \qquad T_d = nT_l, \qquad \Omega = \dot{\theta}_m \tag{8.11}$$

Equation (8.10) is a useful model of the single-link manipulator, including its mechanical and electrical components. We can make the following observations from this equation.

1. The dynamics of a one-link robot *in terms of its angular velocity* can be described by a linear first-order differential equation.
2. The effect of back emf is to increase the damping of the system.
3. The link will rotate if either a voltage is applied at the motor or an external disturbance torque acts on the link. This is as expected intuitively.

To analyse the dynamics of this system, we make use of the well-known technique of the *transfer function* and Laplace transforms.[2] The transfer function of a system is defined as the *ratio* of the *output to the input* in the Laplace or s domain. In our case, we can have two inputs, namely, voltage V_a and the external disturbance T_d, and one output Ω. To obtain the transfer functions, we first take the Laplace transform of Eqn (8.10) with the initial conditions zero. We get

$$Js\Omega(s) + F\Omega(s) = KV_a(s) + T_d(s) \tag{8.12}$$

The two transfer functions can be written as

$$\frac{\Omega(s)}{V_a(s)} = \frac{K}{Js+F}, \qquad \frac{\Omega(s)}{T_d(s)} = \frac{1}{Js+F} \tag{8.13}$$

Transfer functions are best represented by block diagrams,[3] and the two transfer functions given in Eqn (8.13) are shown in Fig. 8.3(a). Equation (8.12) can be represented in terms of these two blocks and a summing block as shown in Fig. 8.3(b). The transfer functions described in Eqn (8.13) are called *open-loop* transfer functions since there is no feedback. Figure 8.3(c) shows the block diagram of the motor with a controller and a sensor. This is called a *closed-loop* system since the output of the motor is measured and fed back as an input to the controller. In Fig. 8.3(c), the transfer function of the sensor is taken as 1 (also called *unity feedback*), but this is not a serious restriction. A non-unity feedback can be converted into an equivalent unity

[2] It is assumed that the reader is familiar with Laplace transforms, discussed in any undergraduate mathematics textbook.

[3] For more details on the use of Laplace transforms and block diagrams in control system analysis, the reader is referred to any textbook on linear control, such as Ogata (1987).

feedback by *block diagram* manipulation [see Ogata (1987)]. The quantity Ω_d is the desired angular velocity of the motor.

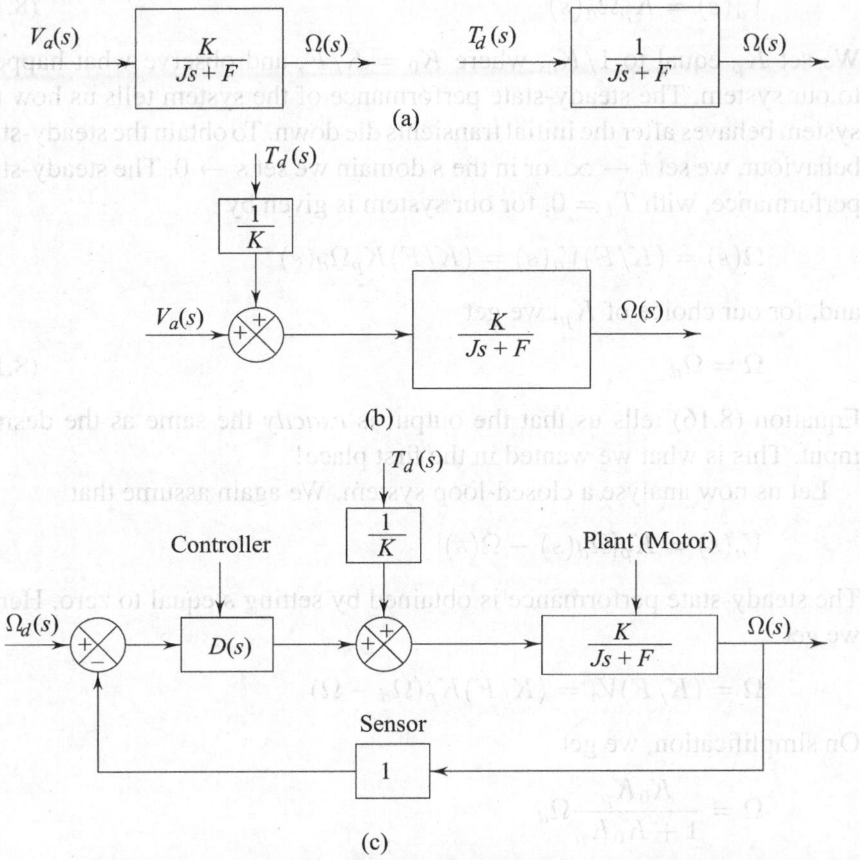

Fig. 8.3 Transfer functions of a single-link manipulator

We will now discuss the usefulness of feedback by comparing an open-loop and a closed-loop system. We will use the example of the single-link manipulator.

8.2.1 Usefulness of Feedback

We assume that the voltage $V_a(s)$ is proportional to the difference between Ω_d and $\Omega(s)$. Now we can write

$$V_a(s) = K_p[\Omega_d(s) - \Omega(s)] \tag{8.14}$$

where K_p is the controller transfer function (called the *controller gain*) and can be set. However, we assume that once the controller gain is set (as in a factory), it cannot be changed. Without any feedback, the controller output is given by

$$V_a(s) = K_p\Omega_d(s) \tag{8.15}$$

We set K_p equal to $1/K_0$, where $K_0 = K/F$, and observe what happens to our system. The steady-state performance of the system tells us how the system behaves after the initial transients die down. To obtain the steady-state behaviour, we set $t \to \infty$, or in the s domain we set $s \to 0$. The steady-state performance, with $T_d = 0$, for our system is given by

$$\Omega(s) = (K/F)V_a(s) = (K/F)K_p\Omega_d(s)$$

and, for our choice of K_p, we get

$$\Omega = \Omega_d \tag{8.16}$$

Equation (8.16) tells us that the output is *exactly* the same as the desired input. This is what we wanted in the first place!

Let us now analyse a closed-loop system. We again assume that

$$V_a(s) = K_p[\Omega_d(s) - \Omega(s)]$$

The steady-state performance is obtained by setting s equal to zero. Hence we get

$$\Omega = (K/F)V_a = (K/F)K_p(\Omega_d - \Omega)$$

On simplification, we get

$$\Omega = \frac{K_0K_p}{1 + K_0K_p}\Omega_d$$

where $K_0 = K/F$.

We choose the controller gain K_p such that $K_0K_p >> 1$. For this choice, $\Omega \approx \Omega_d$. Apparently, with feedback, the situation is worse since the output is only approximately equal to the input! However, let us look at the case where the model or K_0 is not known exactly. This can arise when R_a or friction or K_t or K_g is not known exactly. In the case of an open loop, we have

$$\Omega + \delta\Omega = (K_0 + \delta K_0)K_p\Omega_d$$

We have already set $K_p = 1/K_0$. Hence we have

$$\delta\Omega = (\delta K_0/K_0)\Omega_d$$

This equation implies that a 10% error in K_0 will result in a 10% error in Ω, or all the errors are passed through.

In the case of a closed loop, we first denote the unperturbed output by Ω', given by $[K_0K_p/(1+K_0K_p)]\Omega_d$. It may be noted that $\Omega' \approx \Omega$ since $K_0K_p >> 1$. We can write

$$\delta\Omega'/\Omega' = \frac{1}{1+K_0K_p}(\delta K_0/K_0)$$

Hence a 10% change in K_0 will result in a $[1/(1+K_0K_p)] \times 10\%$ change in Ω'. Since $1+K_0K_p$ is much greater than 1, the change in the output is greatly reduced. This is one of the main reasons why feedback is used, namely, *feedback reduces the output error due to internal parameter variation*. The usefulness of feedback in *reducing the effect of external disturbances* can also be shown as follows.

For the open-loop system, with $T_d \neq 0$, the steady-state output is given as

$$\Omega = K_0K_c\Omega_d + K_0(T_d/K)$$

Hence, if we choose a controller gain K_c such that $K_0K_c = 1$, we get

$$\Omega = \Omega_d + K_0(T_d/K)$$

This equation implies that the change in the output Ω is proportional to the disturbance T_d. In the case of the closed-loop system, the steady-state output is given by

$$\Omega = \frac{K_0K_c}{1+K_0K_c}\Omega_d + \frac{K_0}{1+K_0K_c}(T_d/K)$$

It is possible to choose $K_0K_c >> 1$ and $K_0K_c >> (K_0/K)$ (or $K_c >> 1/K$) and hence we can reduce the effect of T_d on the output.

We go back to our example of the control of a one-link robot to discuss the methods for obtaining the response of a feedback controlled system. There are several methods; however, we will restrict ourselves to the s-plane analysis.

8.2.2 First-order System

Figure 8.4 shows a block diagram representation of the feedback control of a one-link robot with $T_d = 0$. This is called a *first-order* system because the governing differential equation, Eqn (8.10), is first order. Here, we are interested in studying the control of the *angular velocity* of the link. Note that the equation in the Laplace domain is also first degree in s.

The closed-loop transfer function between the output $\Omega(s)$ and the input $\Omega_d(s)$ is given by

$$\frac{\Omega(s)}{\Omega_d(s)} = (KK_p/J)\left(\frac{1}{s+[(F+KK_p)/J]}\right) \tag{8.17}$$

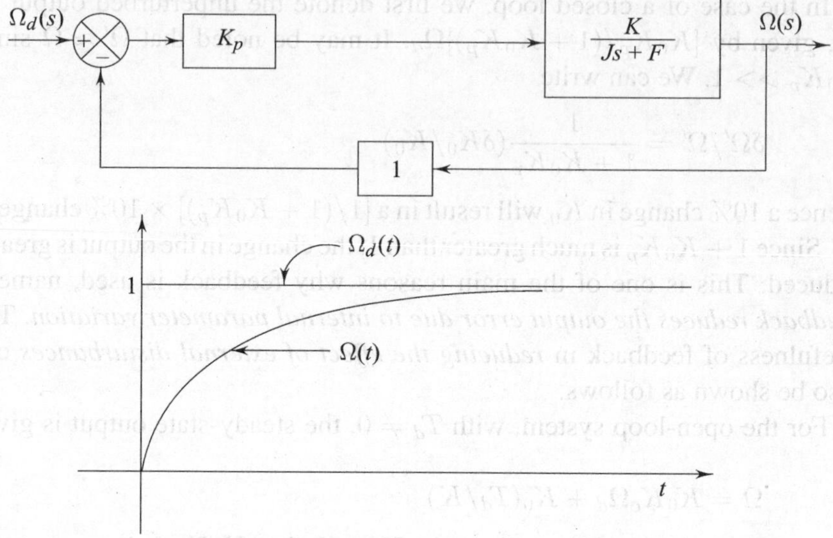

Fig. 8.4 First-order system and its step response

One way to analyse this system is to look at the step response. For the step response, we use $\Omega_d(s) = 1/s$. The output $\Omega(t)$ is obtained by taking the inverse Laplace transform with $\Omega_d(s) = 1/s$. Figure 8.4 shows the plot of $\Omega(t)$ and $\Omega_d(t)$ as functions of time. $\Omega(t)$ is of the form $1 - \exp\{-([(F + KK_p)/J])t\}$. Since F, K, K_p, and J are all positive, $\Omega(t)$ is always *bounded* and approaches $\Omega_d(t)$ as $t \to \infty$. In the control theory, a system having a bounded output for a bounded input is called *stable*; in this sense, the single-link manipulator is always stable. In the s domain, the stability of a linear system is determined by the roots (also called the *poles*) of the denominator polynomial of the closed-loop transfer function. If the poles lie in the left half of the s plane, the system is said to be stable. For the single-link manipulator, the pole is at $[-([(F + KK_p)/J]), 0]$ and since F, K, K_p, and J are all positive, the first-order system modelling the single-link manipulator is always stable.

8.2.3 Second-order System

We consider the same single-link manipulator except that now we are interested in studying the control of the *angular rotation* of the link. The open-loop transfer function with $T_d = 0$ is given as

$$\frac{\theta(s)}{V_a(s)} = \frac{K}{s(Js + F)}$$

The transfer function is called *second order* since the polynomial in the denominator of the transfer function is of degree 2. The closed-loop block diagram of the system is given in Fig. 8.5(a).

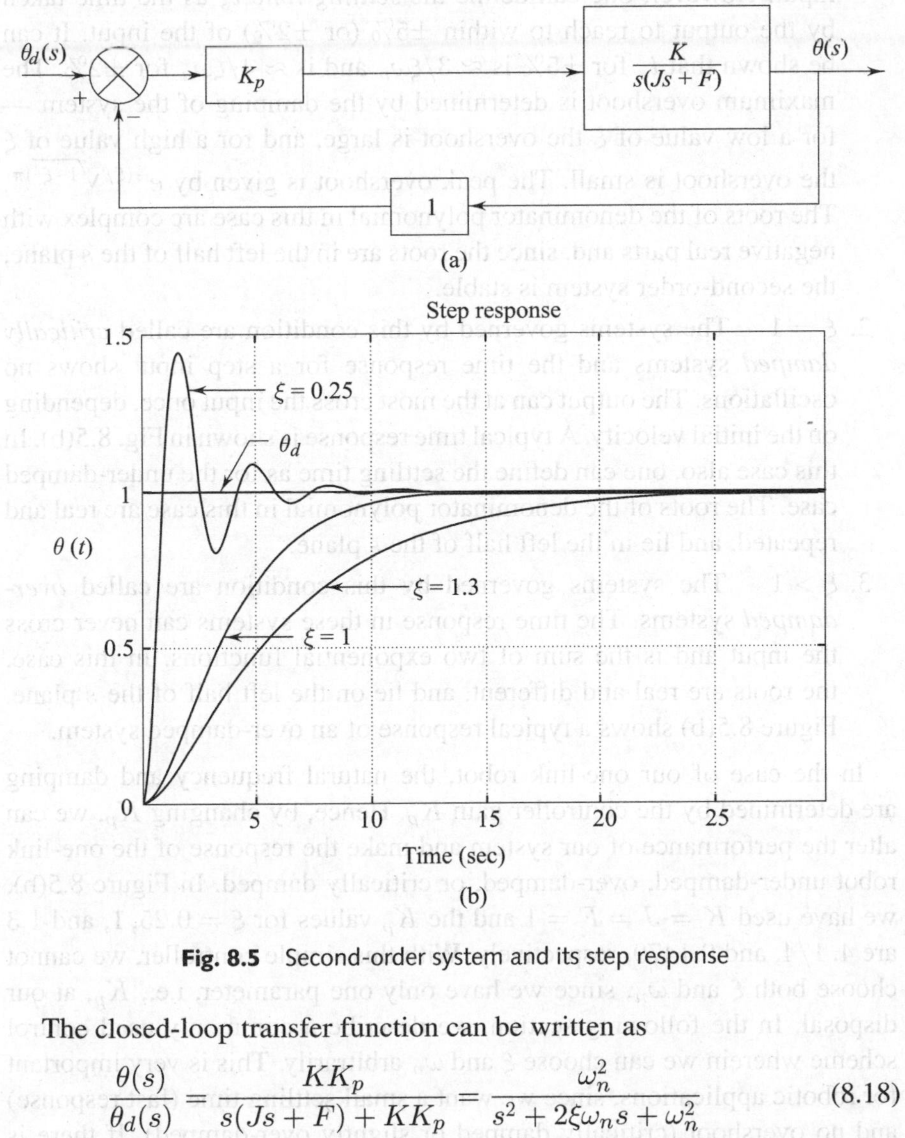

Fig. 8.5 Second-order system and its step response

The closed-loop transfer function can be written as

$$\frac{\theta(s)}{\theta_d(s)} = \frac{KK_p}{s(Js+F)+KK_p} = \frac{\omega_n^2}{s^2+2\xi\omega_n s+\omega_n^2} \tag{8.18}$$

where $\omega_n^2 = (KK_p/J)$, $F/J = 2\xi\omega_n$, and $\xi = F/(2\sqrt{JKK_p})$. For second-order systems, ω_n is called the *natural frequency* of the system and ξ is called the *damping*. There are three possibilities.

1. $0 < \xi < 1$ The systems governed by this condition are called *under-damped* systems. The time response of a second-order, under-damped

system for a step input is shown in Fig. 8.5(b). As can be seen from the figure, the output oscillates about the desired input before settling down. It may be mentioned that it takes an infinite time to reach the input. However, one can define the *settling time* t_s as the time taken by the output to reach to within $\pm 5\%$ (or $\pm 2\%$) of the input. It can be shown that t_s for $\pm 5\%$ is $\approx 3/\xi\omega_n$ and is $\approx 4/\xi\omega_n$ for $\pm 2\%$. The maximum overshoot is determined by the damping of the system — for a low value of ξ the overshoot is large, and for a high value of ξ the overshoot is small. The peak overshoot is given by $e^{-(\xi/\sqrt{1-\xi^2})\pi}$. The roots of the denominator polynomial in this case are complex with negative real parts and, since the roots are in the left half of the s plane, the second-order system is stable.

2. $\xi = 1$ The systems governed by this condition are called *critically damped* systems and the time response for a step input shows no oscillations. The output can at the most cross the input once, depending on the initial velocity. A typical time response is shown in Fig. 8.5(b). In this case also, one can define the settling time as for the under-damped case. The roots of the denominator polynomial in this case are real and repeated, and lie in the left half of the s plane.
3. $\xi > 1$ The systems governed by this condition are called *over-damped* systems. The time response in these systems can never cross the input and is the sum of two exponential functions. In this case, the roots are real and different, and lie on the left half of the s plane. Figure 8.5(b) shows a typical response of an over-damped system.

In the case of our one-link robot, the natural frequency and damping are determined by the controller gain K_p. Hence, by changing K_p, we can alter the performance of our system and make the response of the one-link robot under-damped, over-damped, or critically damped. In Figure 8.5(b), we have used $K = J = F = 1$ and the K_p values for $\xi = 0.25, 1$, and 1.3 are $4, 1/4$, and 0.1479, respectively. With this simple controller, we cannot choose both ξ and ω_n, since we have only one parameter, i.e., K_p, at our disposal. In the following section, we describe a commonly used control scheme wherein we can choose ξ and ω_n arbitrarily. This is very important for robotic applications, since we want a small settling time (fast response) and no overshoot (critically damped or slightly over-damped). If there is overshoot, the robot may hit the object or fixture that it is trying to grasp.

8.2.4 PID Control of a Single-link Manipulator

Figure 8.6 shows the closed-loop feedback control system for a one-link robot where $D(s)$ is the controller and we have assumed unity feedback. In Section 8.2.2, we considered $D(s) = K_p$. This is called the *proportional controller* and the symbol K_p is called the *proportional gain*. We can also choose $D(s)$ as

$$D(s) = K_p + K_v s \tag{8.19}$$

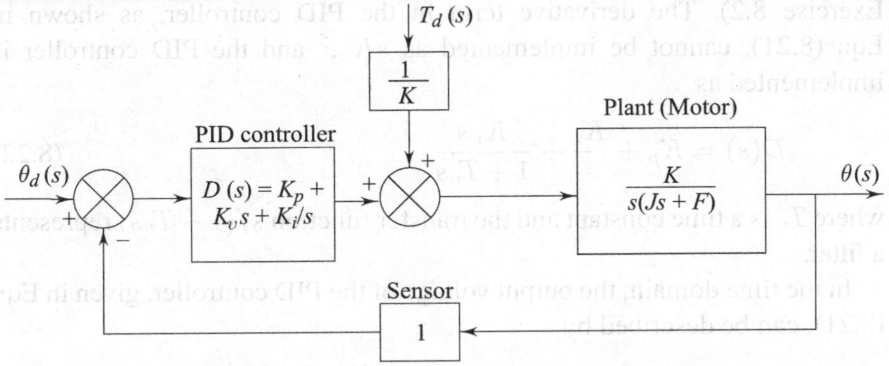

Fig. 8.6 PID control of a single-link manipulator

In this controller the symbol K_v is called the *derivative* or *velocity gain*. For this controller, the closed-loop transfer function is given by

$$\frac{\theta(s)}{\theta_d(s)} = \frac{KK_p + sKK_v}{Js^2 + s(F + KK_v) + KK_p} \tag{8.20}$$

The natural frequency and damping are now determined by K_p and K_v. Roughly speaking, if K_p is increased, the settling time reduces (the system becomes faster); if K_v is increased, the overshoot decreases and the system becomes slower (see Exercise 8.1). Hence by proper choice[4] of values for K_p and K_v we can choose any ξ and ω_n to satisfy our requirements of the settling time and no overshoot. For critical damping, we get $K_v = 2\sqrt{K_p}$.

The controller described above is called the *proportional plus derivative* (PD) controller. It is quite versatile and is used extensively for robot control. However, to enhance the performance of the controller and to remove steady-state errors arising out of backlash or unmodelled friction/stiction in the mechanical system, often an additional integral term is added. Thus we can write

$$D(s) = K_p + \frac{K_i}{s} + sK_v \tag{8.21}$$

[4] There exist analytical techniques by which one can arrive at proper choices of controller gains to satisfy ones requirements. These are available in many textbooks on control theory and the reader is referred to Ogata (1987) and Franklin et al. (1991) for details.

This is the well-known *proportional plus integral plus derivative* (PID) controller. It can be shown that the addition of an integral term makes the system *third order* and results in zero steady error for any constant disturbance. The integral term, however, needs to be used with care, since a large integral gain (value of K_i) can make the system unstable (see Exercise 8.2). The derivative term in the PID controller, as shown in Eqn (8.21), cannot be implemented as sK_v,[5] and the PID controller is implemented as

$$D(s) = K_p + \frac{K_i}{s} + \frac{K_v s}{1 + T_v s} \tag{8.22}$$

where T_v is a time constant and the transfer function $s/(1 + T_v s)$ represents a filter.

In the time domain, the output voltage of the PID controller, given in Eqn (8.21), can be described by

$$V_a(t) = K_p e(t) + K_v \dot{e}(t) + K_i \int_0^t e(t)dt \tag{8.23}$$

where the error $e(t)$ is given by $[\theta_d(t) - \theta(t)]$. This controller works well for step input. However, for following a trajectory an additional feed-forward term is added. The output of a modified PD controller with the feed-forward term is given by

$$V_a(t) = \ddot{\theta}_d(t) + K_p e(t) + K_v \dot{e}(t) \tag{8.24}$$

The control scheme given in Eqn (8.24) forms the basis of the *linear control* of robots, and is also used in the non-linear control of robots. The integral term is left out since it makes the analysis difficult; but in most actual robotic systems, the integral term is also present.

8.2.5 Digital Control of a Single-link Manipulator

The PID control scheme, given in Eqn (8.24), assumes that time is *continuous*. A majority of the modern controllers are, however, implemented digitally using microprocessors. This is a vast topic on its own. In this section, we give only a very brief introduction to the digital control of a DC servo-motor. For details on digital or discrete-time control, the reader is referred to Franklin et al. (1990).

One of the key differences between continuous-time control and discrete-time control is the notion of sampling time T_s. In the discrete-time

[5] According to the classical control theory, in any transfer function, the order of the numerator polynomial must be *less than* or *equal to* the denominator transfer function (Ogata 1987).

control, the input $\theta_d(t)$ and the output $\theta(t)$ are not continuous functions of time but discretized or sampled every T_s units of time. As a result, a continuous trajectory for $\theta_d(t)$, shown as a solid line in Fig. 8.7, will be actually available as shown schematically by the dashed lines in Fig. 8.7, and a similar discretization will occur for the measured output $\theta(t)$.[6] The analog to digital (A/D) conversion is done electronically and the difference between the two discretized values of the analog signal is determined by the number of bits used for the A/D conversion. The typical values of T_s used for motor control are between 1 and 10 msec, and typically 8 – 12 bits[7] are used in the A/D conversion. Clearly, as $T_s \to 0$, the discretized θ_d will approach $\theta_d(t)$. The discretization or the sampling of the analog signal is

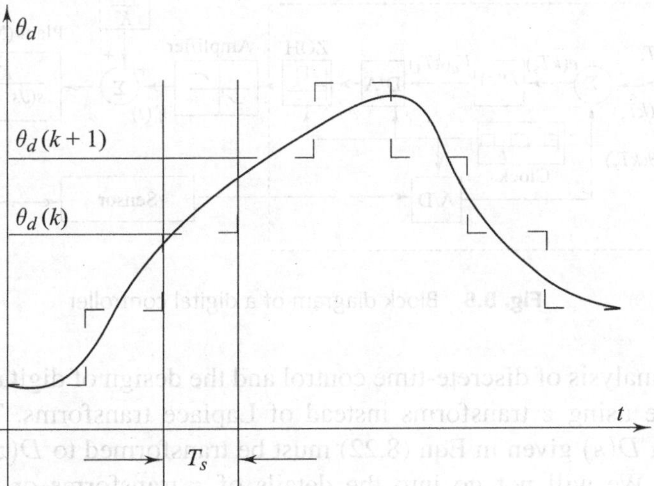

Fig. 8.7 Discretization of reference input

done by an independent clock which *interrupts* the microprocessor or the computer. The error $e(kT_s)$ is the difference between the kth value of the desired θ_d, denoted by $\theta_d(kT_s)$, and the kth value of the measured rotation θ, denoted by $\theta(kT_s)$. This is done in the digital computer and is available as a digital value. The error is multiplied by the controller transfer function $D(z)$ also implemented digitally in the digital computer. Since the motor

[6] In a robot the desired θ_d is generated by discretizing the generated cubic trajectory (see Chapter 7) and often an optical encoder is used to measure θ. An optical encoder directly generates a digital output and the A/D conversion may not be required.

[7] This implies that the difference between the two discretized values is $1/2^N$, where N is the number of bits in the A/D conversion.

needs analog voltage, the output of the controller must be converted back to analog or continuous values. This is done in the D/A converter and in the block called *zero order hold* (ZOH) implemented in the digital computer. The purpose of the ZOH is to hold constant the value of $V_a(t)$ over the time interval T_s, and this introduces a delay. This delay is the second key difference between continuous- and discrete-time controls. This delay often leads to complications in the analysis and design of discrete-time controllers. The output of the digital controller is typically in milliamperes and cannot drive the motor. As shown in Fig. 8.8 the output of the digital controller is amplified and then fed to the motors.

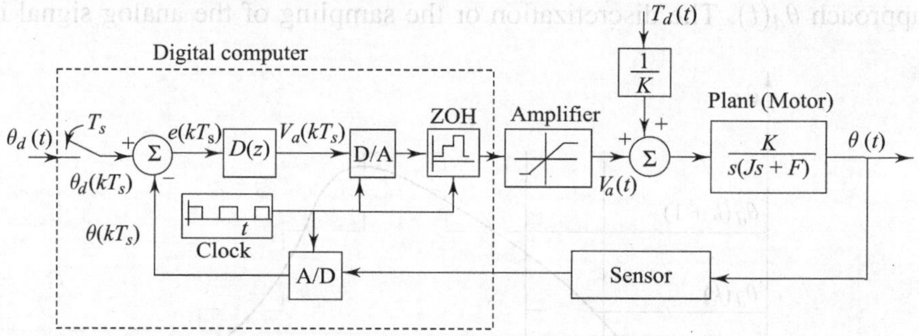

Fig. 8.8 Block diagram of a digital controller

The analysis of discrete-time control and the design of digital controllers are done using z transforms instead of Laplace transforms. The transfer function $D(s)$ given in Eqn (8.22) must be transformed to $D(z)$ for digital control. We will not go into the details of z transforms or the analysis and design of PID controllers in the discrete-time domain and the reader is referred to textbooks on digital control (such as Franklin et al. 1990). An equivalent of the PID controller, $D(s)$, given in Eqn (8.22), in terms of z transforms is given by

$$D(z) = K_p + \frac{K_i T_s}{1 - z^{-1}} + \frac{K_v(1 - z^{-1})}{T_s + T_v(1 - z^{-1})} \tag{8.25}$$

In the rest of the chapter, for the sake of simplicity, we will use continuous-time approximation for the analysis and development of control algorithms for multi-link robots.

8.3 PID Control of a Multi-link Manipulator

In the preceding section, we described the control algorithms for a single-link manipulator. In this section, we extend the continuous-time PD control to multi-link manipulators. It must be noted that a multi-link manipulator, unlike a single-link manipulator, is a non-linear system, and the linear control schemes do not perform as well as do non-linear control schemes described in the following section. In particular, we cannot guarantee critical damping or the same settling time uniformly at every point in the workspace of the manipulator. It is also not possible to analyse and predict the exact performance[8] of a PD (or PID) scheme for a non-linear plant such as a robot. However, because of its simplicity and due to the large gear reduction present in most industrial robots, a PD (or PID) control scheme works reasonably well and is thus used in most industrial robots.

We extend the PD control scheme described in the preceding section to the control of multi-link manipulators. Each joint or motor of the multi-link manipulator is assumed to be controlled independently. The output of the controller is now an $n \times 1$ vector, where n is the number of actuated joints. We assume that all the joints are actuated by DC permanent-magnet servomotors similar to that in the single-link manipulator.[9] The $n \times 1$ output of the controller is calculated as

$$\mathbf{V}_a(t) = \ddot{\mathbf{q}}_d(t) + [K_p]\mathbf{e}(t) + [K_v]\dot{\mathbf{e}}(\mathbf{t}) \tag{8.26}$$

where $[K_p]$ and $[K_v]$ are $n \times n$ positive-definite proportional and derivative gain matrices, respectively. The $n \times 1$ error vector between the desired joint rotation $\mathbf{q}_d(t)$ and the measured actual joint rotation $\mathbf{q}(t)$ is given as

$$\mathbf{e}(t) = \mathbf{q}_d(t) - \mathbf{q}(t) \tag{8.27}$$

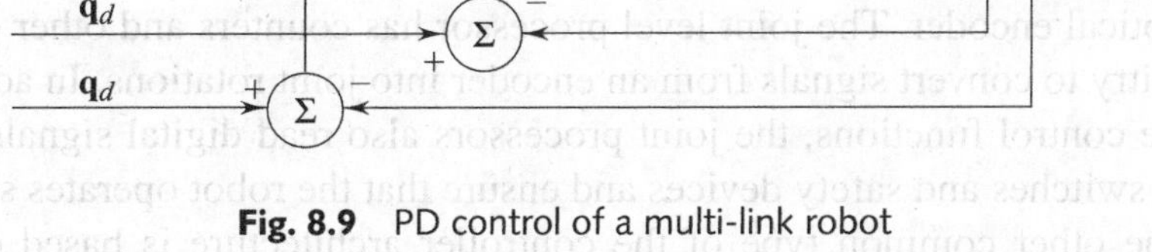

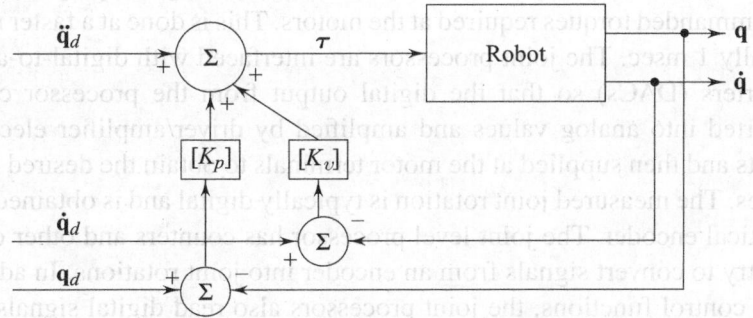

Fig. 8.9 PD control of a multi-link robot

[8] One can always simulate numerically for a particular trajectory and obtain the performance, but no general analytical results can be obtained.

[9] Even if a joint in a multi-link manipulator is prismatic and the joint variable is d_i, the actuation is usually a DC motor and a transmission mechanism is used to convert rotary motion into translatory motion.

Figure 8.9 shows the block diagram of the PD control scheme for a multi-link manipulator. In the figure, for simplicity, the output of the controller is shown as $\boldsymbol{\tau}(t)$, which is an $n \times 1$ vector of joint torques.[10] Another difference between Fig. 8.9 and Fig. 8.6 is the desired input. In traditional control of single-input/single-output (or SISO) systems, the desired trajectory is a single input θ_d and $\dot{\theta}_d$ is either not available or not used, and the PID control is implemented by the controller transfer function $D(s)$ given in Eqn (8.22). For a sophisticated robot, the trajectory planner generates $\theta_d(t)$, $\dot{\theta}_d(t)$, and even $\ddot{\theta}_d(t)$ for all the joints and, hence, we can readily use them as shown in Fig. 8.9. Note that in this figure we have used $\mathbf{q}_d$, $\dot{\mathbf{q}}_d$, and $\ddot{\mathbf{q}}_d$ to denote the vectors of the desired joint motion, the desired joint rate, and the desired joint acceleration, respectively.

The PID control scheme for a robot is typically implemented using several microprocessors. There are two common controller architectures or arrangements for microprocessors used in robots. In the *joint parallel control architecture*, each joint or a group of joints is controlled by means of a single processor. The overall control of the individual microprocessors is done by a 'master' or a coordinating processor. The coordinating processor also has user interfaces such as a keyboard, a display unit, data storage, and a teach pendant. The coordinating processor computes inverse kinematics, performs other computationally intensive and high-level tasks, and communicates the desired set points $[\mathbf{q}_d(t)]$ to the joint processors at an update rate of typically 10 msec. The joint processors acquire the desired and measured joint rotation, perform interpolation, compute the servo error, execute the PID control algorithms using the servo error, and output signals proportional to the commanded torques required at the motors. This is done at a faster rate of typically 1 msec. The joint processors are interfaced with digital-to-analog converters (DACs) so that the digital output from the processor can be converted into analog values and amplified by driver/amplifier electronic circuits and then supplied at the motor terminals to obtain the desired motor torques. The measured joint rotation is typically digital and is obtained from an optical encoder. The joint level processor has counters and other digital circuitry to convert signals from an encoder into joint rotations. In addition to the control functions, the joint processors also read digital signals from limit switches and safety devices and ensure that the robot operates safely.

The other common type of the controller architecture is based on the various *functions* of a robot controller and is called *functional parallel architecture*. In this architecture each function is handled by a processor.

[10] It may be noted that the joint torque is related to the applied voltage at the motor terminals since $T_m = K_t i_a = (K_t/R_a)(V_a - kg\dot{\theta}_m)$ and $\tau = T_m/n$.

Each processor needs to communicate with the other processors so that the robot operates in a desired manner.

In the original PUMA robot controller, the individual joint microprocessors were 6503 microprocessors and the coordinating processor was a DEC LSI-11. Every 28 msec, the desired set points were sent by the LSI-11 processor. The joint processor operated at a cycle time of 0.875 msec. The PUMA controller also offered a high-level robot programming language VAL for easy operation of the robot. The language VAL interpreted 'high-level' commands from a robot operator and performed the needed inverse kinematics, interpolation, and other tasks. Present-day robot controllers are equipped with vastly superior and advanced processors available in the market. With the advent of robust and industrial PC's, the robot controllers are often PC-based with add-on cards containing one or more microcontrollers implementing the PID control scheme for the joints and the PC serving as a master controller to compute higher level functions such as inverse kinematics, data logging, diagnostics, and other tasks. The PC also interfaces with the robot operator. One such card and associated software is available from dSPACE, and this has been used in many universities and R & D laboratories for developing various kinds of motion and other controllers.

8.4 Non-linear Control of Manipulators

In the previous sections, we discussed the linear control of single- and multi-link manipulators. The analysis was based on several assumptions — the most important being that the system was linear and the inertia of the links were constant. In Chapter 6, we observed that the reality is contrary to our assumptions and the manipulator mass matrix is dependent on the configuration of the manipulator. In addition, there are non-linear Coriolis and centripetal terms. A linear control scheme when used on a non-linear robot does not give uniform performance at every point in the workspace of the robot. In this section, we present advanced control schemes which give much better performances as compared to linear control schemes. The field of non-linear control of robots is quite vast (see, for example, Canudas de Wit 1996). In this text we restrict ourselves to a class of non-linear controllers which use *estimates of the model* of a robot to 'cancel' out the non-linearities, and in an 'ideal' case reduce the non-linear equations to a set of decoupled linear equations. These model-based control schemes are broadly called *computed torque* control schemes and were first developed by Freund (1982), whose work in turn was based on the work by Singh and Rugh (1972).

We start by rewriting the dynamic equations of motion, Eqn (6.30), for a serial manipulator from Chapter 6. The dynamic equations of motion of an n-DOF robot can be written as

$$\boldsymbol{\tau} = [\mathbf{M}(\mathbf{q})]\ddot{\mathbf{q}} + \mathbf{C}(\mathbf{q}, \dot{\mathbf{q}}) + \mathbf{G}(\mathbf{q}) + \mathbf{F}(\mathbf{q}, \dot{\mathbf{q}}) \tag{8.28}$$

where $[\mathbf{M}(\mathbf{q})]$ is an $n \times n$ mass matrix and $\mathbf{C}(\mathbf{q}, \dot{\mathbf{q}})$, $\mathbf{G}(\mathbf{q})$, and $\mathbf{F}(\mathbf{q}, \dot{\mathbf{q}})$ are $n \times 1$ vectors representing Coriolis/centripetal terms, gravity terms, and friction terms, respectively. The $n \times 1$ vector $\boldsymbol{\tau}$ represents joint torques, and $\mathbf{q}$, $\dot{\mathbf{q}}$, and $\ddot{\mathbf{q}}$ are $n \times 1$ vectors representing joint variables, joint velocities, and joint accelerations, respectively. The key step in the non-linear control of such a complicated system is the use of the concept of *control law partitioning*. We assume that the output of the controller can be written in the form

$$\boldsymbol{\tau} = [\boldsymbol{\alpha}]\boldsymbol{\tau}' + \boldsymbol{\beta} \tag{8.29}$$

We choose

$$\begin{aligned} [\boldsymbol{\alpha}] &= [\mathbf{M}(\mathbf{q})] \\ \boldsymbol{\beta} &= \mathbf{C}(\mathbf{q}, \dot{\mathbf{q}}) + \mathbf{G}(\mathbf{q}) + \mathbf{F}(\mathbf{q}, \dot{\mathbf{q}}) \end{aligned} \tag{8.30}$$

Substituting Eqns (8.29) and (8.30) in Eqn (8.28), we get

$$\boldsymbol{\tau}' = \ddot{\mathbf{q}} \tag{8.31}$$

This equation can be viewed as a unit inertia system with a new input $\boldsymbol{\tau}'$. In a sense, the non-linearities in the original non-linear system are 'cancelled' by the computed $[\boldsymbol{\alpha}]$ and $\boldsymbol{\beta}$, and the non-linear system is *transformed* into a linear system. We can now apply our previously developed PD controller to this unit inertia system. If $\boldsymbol{\tau}'$ is chosen as

$$\boldsymbol{\tau}' = \ddot{\mathbf{q}}_d(t) + [K_p]\mathbf{e}(t) + [K_v]\dot{\mathbf{e}}(\mathbf{t}) \tag{8.32}$$

with $\mathbf{e}(t) = \mathbf{q}_d - \mathbf{q}$, then we can write a linear error equation for the unit inertia system as

$$\ddot{\mathbf{e}}(t) + [K_p]\mathbf{e}(t) + [K_v]\dot{\mathbf{e}}(\mathbf{t}) = \mathbf{0} \tag{8.33}$$

For positive-definite diagonal matrices $[K_p]$ and $[K_v]$, we will get a *linear decoupled* error equation, and following the well-known linear control results, we can set $[K_p]$ and $[K_v]$ to get critically (or over) damped performance at *every* point in the workspace of the robot.

The control scheme described above uses the notion of *feedback linearization* and this is very different from *linearization* of a non-linear system. The key difference is that linearization is at *every* point as opposed to at only *an* operating point of the workspace. A block diagram of this control scheme is shown in Fig. 8.10. It can be seen that the controller consists of two parts: (a) an *error-driven* portion whose output is τ' as computed by the PD (or PID) control law given in Eqn (8.32) and (b) a *model-based* portion where $[\alpha]$ and β are computed.

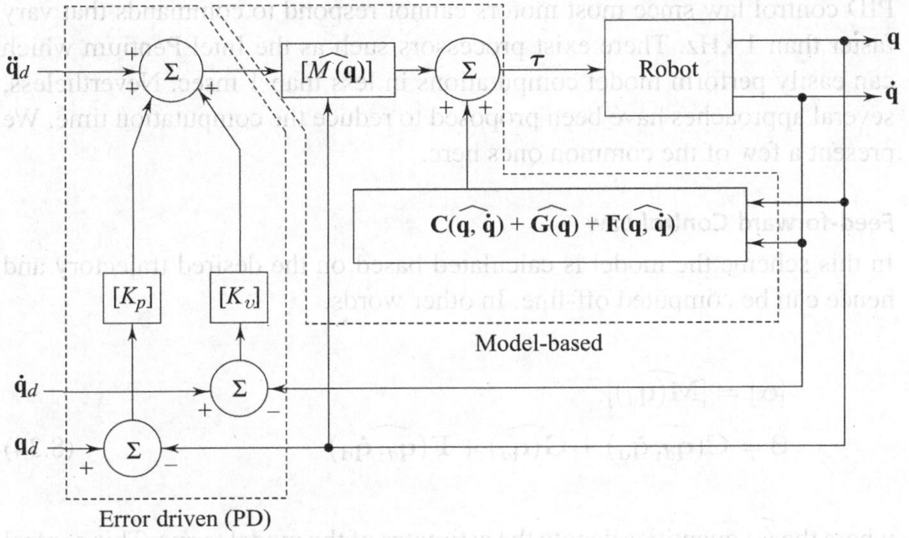

Fig. 8.10 Computed torque method for robot control

It may be noted that in Fig. 8.10 we have used the symbols $[\widehat{M(q)}]$, etc., since the 'ideal' performance represented by Eqn (8.33) is unachievable in practice due to two main reasons.

1. The time required to compute the model is non-zero and, hence, the model and the commanded torque to the controller are computed after a finite elapsed time, by which time the manipulator configuration may have changed (perhaps by a small amount).
2. Lack of the knowledge of manipulator parameters and hence the model calculated may not 'exactly cancel' the non-linear terms. This is particularly true in the case of hard-to-model things such as friction, backlash, and flexibilities at the links and joints. In addition, since the controller is implemented digitally and discretization is involved, the feedback will never exactly cancel the non-linearities.

We address these two issues in the following sections.

8.4.1 Time Required to Compute the Model

Implicit in the computed torque scheme is the notion that the models $[\alpha]$ and β are available instantaneously. With the advent of high-speed processors and algorithms capable of performing model computations efficiently, this is a reasonable assumption except perhaps for fast-moving direct drive robots. In a sense, we do not need 'instantaneous' computations of the model or the PID control law since most motors cannot respond to commands that vary faster than 1 kHz. There exist processors such as the Intel Pentium which can easily perform model computations in less than 1 msec. Nevertheless, several approaches have been proposed to reduce the computation time. We present a few of the common ones here.

Feed-forward Control Law

In this scheme the model is calculated based on the desired trajectory and hence can be computed off-line. In other words,

$$
\begin{aligned}
[\alpha] &= [\widehat{\mathbf{M}(\mathbf{q}_d)}] \\
\beta &= \widehat{\mathbf{C}(\mathbf{q}_d, \dot{\mathbf{q}}_d)} + \widehat{\mathbf{G}(\mathbf{q}_d)} + \widehat{\mathbf{F}(\mathbf{q}_d, \dot{\mathbf{q}}_d)}
\end{aligned}
\tag{8.34}
$$

where the $\widehat{(.)}$ quantities denote the estimates of the model terms. This control scheme works fairly well in simulation since the error-driven portion of the controller ensures that $\mathbf{q}_d(t)$ is fairly close to $\mathbf{q}(t)$. This control scheme was implemented on an actual robot and we show the results in Section 8.5.

Other Simplified Control Schemes

In the initial years of the robot controller development, many researchers suggested 'partial' cancellation schemes, wherein computation needs were not so large. We present two of these schemes.

1. $[\alpha] = [U]$ (identity matrix) and $\beta = \mathbf{G}(\mathbf{q})$. This is the same as the individual joint PD (or PID) control scheme with compensation only for gravity.
2. $[\alpha]$ contains only the diagonal terms of the mass matrix $[\mathbf{M}(\mathbf{q})]$ and $\beta = \mathbf{G}(\mathbf{q})$.

It may be noted that if $[\alpha] = [U]$ (identity) and $\beta = \mathbf{0}$, then we get our familiar individual joint control PD (or PID) scheme where no model is used.

8.4.2 Lack of Knowledge of Model Parameters

If we assume that only the estimates

$$[\alpha] = [\widehat{\mathbf{M}(\mathbf{q})}]$$
$$\beta = \widehat{\mathbf{C}(\mathbf{q}, \dot{\mathbf{q}})} + \widehat{\mathbf{G}(\mathbf{q})} + \widehat{\mathbf{F}(\mathbf{q}, \dot{\mathbf{q}})} \tag{8.35}$$

are available (and they are *not equal* to the actual model), then 'cancellation' does not take place, and the error equation is given by

$$\ddot{\mathbf{e}} + [K_p]\mathbf{e} + [K_v]\dot{\mathbf{e}}$$
$$= [\widehat{\mathbf{M}}]^{-1}[([\mathbf{M}] - [\widehat{\mathbf{M}}])\ddot{\mathbf{q}} + (\mathbf{C} - \widehat{\mathbf{C}}) + (\mathbf{G} - \widehat{\mathbf{G}}) + (\mathbf{F} - \widehat{\mathbf{F}})] \tag{8.36}$$

where we have dropped the arguments of the functions for brevity. Equation (8.36) is a complex non-linear equation, and it is not possible to predict the behaviour of the servo errors as easily as in exact cancellation. We, however, expect that if the errors between the model and estimates are small, then the right-hand side of Eqn (8.36) will be small and our control scheme, even with estimates, will perform better than a linear control scheme. This is indeed borne out by simulation and actual implementation results. We will present one such actual experimental result in Section 8.5.

8.5 Simulation and Experimental Results

In this section, we present simulation and experimental results comparing individual joint PD and model-based control schemes. We start with some numerical simulations.

8.5.1 Simulation Results

Figure 8.11 shows a planar 2R planar manipulator in two configurations. The model parameters of the robot are as follows:

Link	Link length	CG location	Mass	Inertia
1	1 m	0.773 m	12.456 kg	1.042 kg m^2
2	1 m	0.583 m	12.456 kg	1.042 kg m^2

We assume that the robot is carrying a payload of 2.5 kg. The tip of the manipulator moves along the $\hat{\mathbf{Y}}_0$ axis from 0.55 m, upward against gravity, to 1.45 m and then downwards, with the aid of gravity, back to 0.55 m. We perform simulation for two kinds of motion, namely, a 'fast' motion where the total duration is 2 sec and a 'slow' motion where the total duration is 2 min. The desired Cartesian trajectory for the fast motion of 2 sec is shown

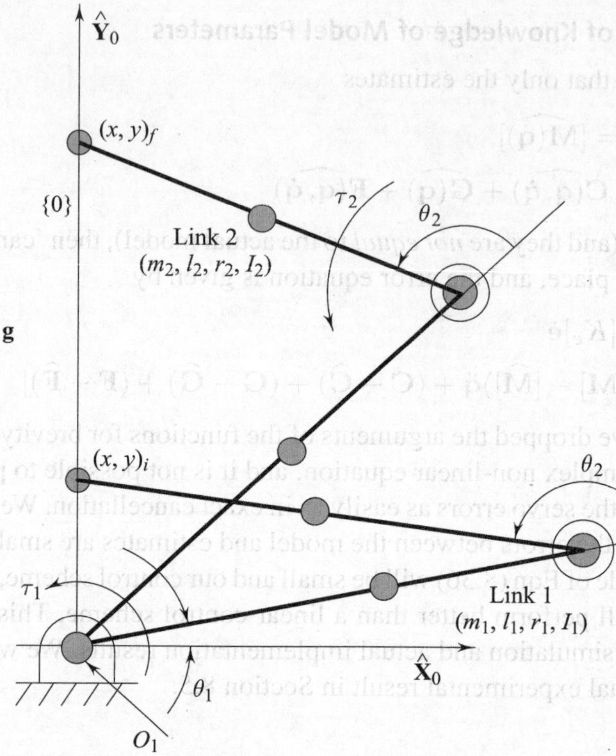

Fig. 8.11 A planar 2R robot

in Fig. 8.12 with the desired $x_d(t) = 0$ and the desired $y_d(t)$ is computed as a smooth cubic trajectory (see Chapter 7) between 0 and 2 sec. Using inverse kinematics (see Chapter 3), we can find out the desired θ_i, $\dot{\theta}_i$, and $\ddot{\theta}_i$ for $i = 1, 2$. The desired θ_i $(i = 1, 2)$ for $t \in [0, 2]$ are shown in Fig. 8.13.

We present simulation results for the following control schemes for both the slow and the fast motion: (1) the PD control scheme as described by Eqn (8.26) with $\mathbf{V}_a(t)$ assumed to be $\tau(t)$, (2) a feed-forward controller as described by Eqn (8.34) with an *exact* knowledge of the model parameters, (3) a model-based controller as described in Eqn (8.35) with 10% error in masses and 5% error in the location of the centre of gravity of each link, and (4) a Cartesian control scheme discussed in Section 8.7. For all these control schemes, we have used gain values K_{p_i}, K_{v_i} such that ω_i are 85.0 and 75.0, respectively, and ξ_i are 2.0 making the system over-damped. In all these cases, $i = 1, 2$ denote the two actuated joints.

Figure 8.14 shows the error in θ_1 and θ_2 and Fig. 8.15 shows the resulting error (obtained by the direct kinematics) in the Cartesian trajectory. For the

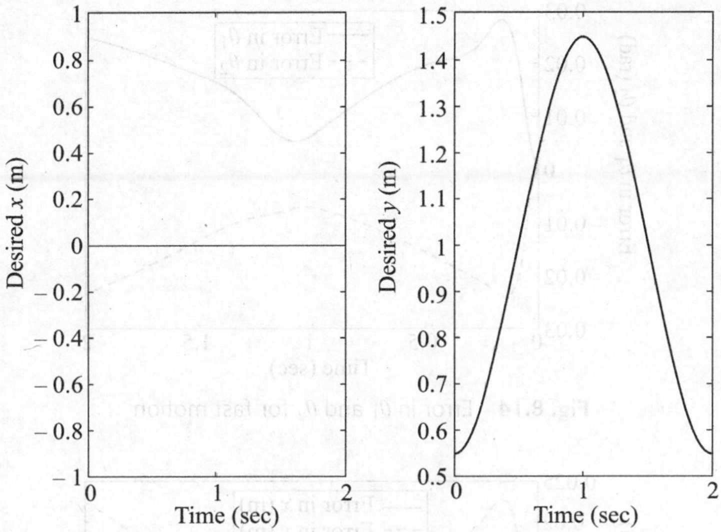

Fig. 8.12 Desired Cartesian trajectory

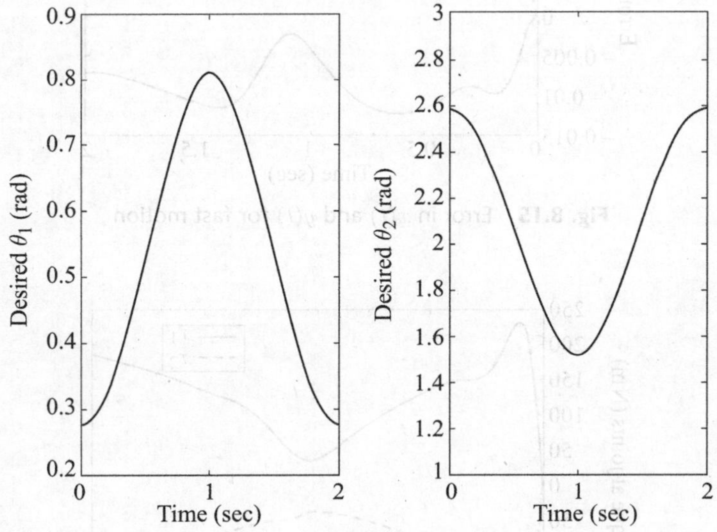

Fig. 8.13 Desired $\theta_1(t)$ and $\theta_2(t)$

PD control of the fast motion. Figure 8.16 shows the torque at the two joints required for the PD control of the planar 2R robot for the fast motion. For the

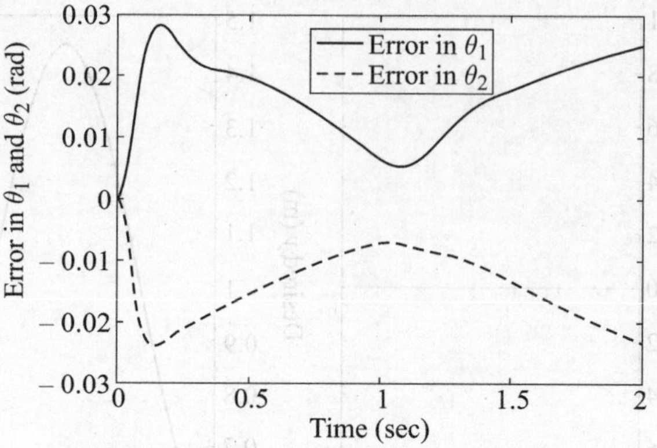

Fig. 8.14 Error in θ_1 and θ_2 for fast motion

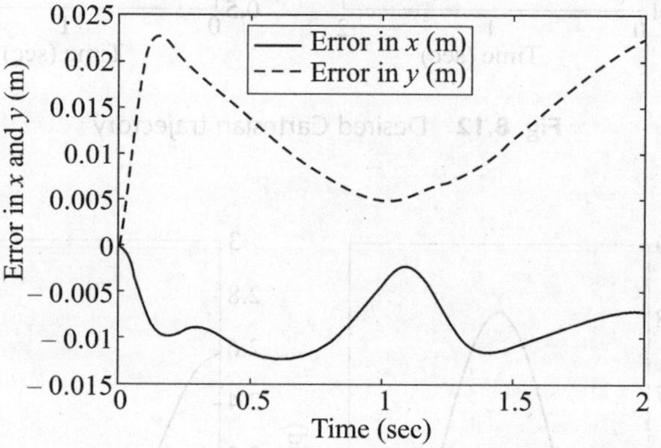

Fig. 8.15 Error in $x(t)$ and $y(t)$ for fast motion

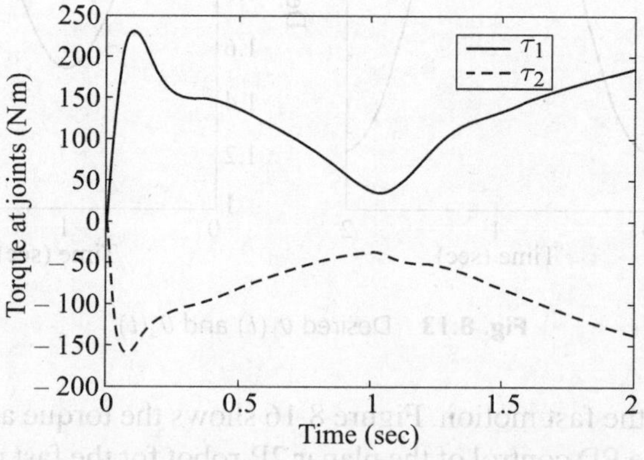

Fig. 8.16 Torque at the two joints for fast motion

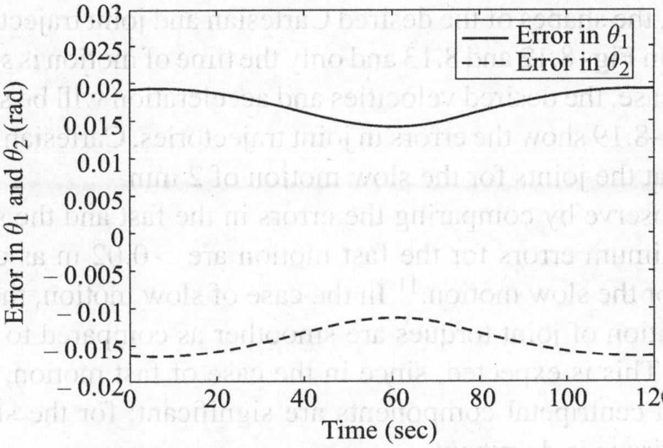

Fig. 8.17 Error in θ_1 and θ_2 for slow motion

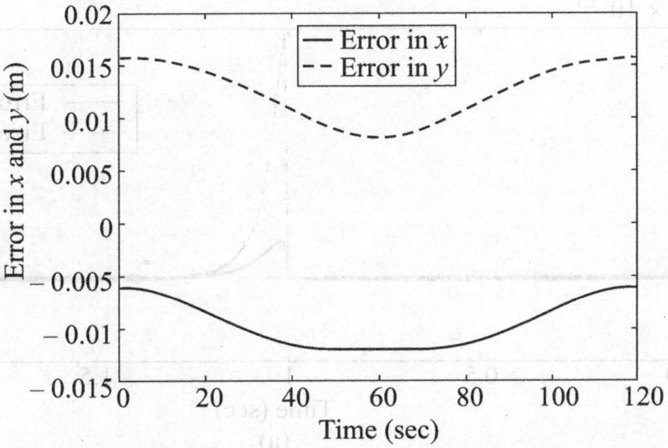

Fig. 8.18 Error in $x(t)$ and $y(t)$ for slow motion

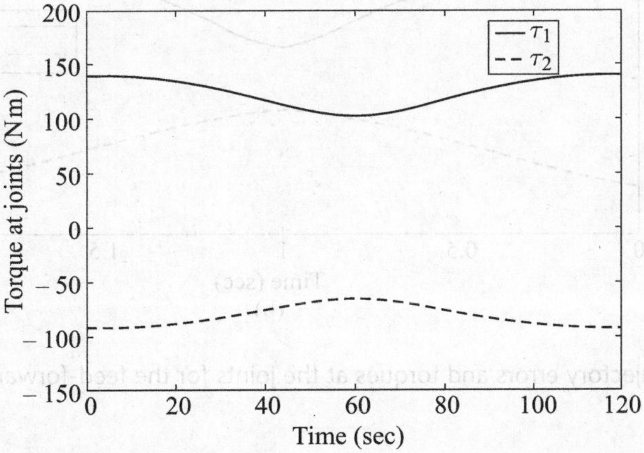

Fig. 8.19 Torque at the two joints for slow motion

slow motion, the shapes of the desired Cartesian and joint trajectories are the same as that in Figs 8.12 and 8.13 and only the time of motion is stretched to 2 min. In this case, the desired velocities and acceleration will be scaled down. Figures 8.17–8.19 show the errors in joint trajectories, Cartesian trajectories, and torques at the joints for the slow motion of 2 min.

We can observe by comparing the errors in the fast and the slow motion that the maximum errors for the fast motion are >0.02 m as compared to <0.015 m for the slow motion.[11] In the case of slow motion, the error plots and the variation of joint torques are smoother as compared to those in the fast motion. This is expected, since in the case of fast motion, the inertial, Coriolis, and centripetal components are significant; for the slow motion, the gravity torque is dominant.

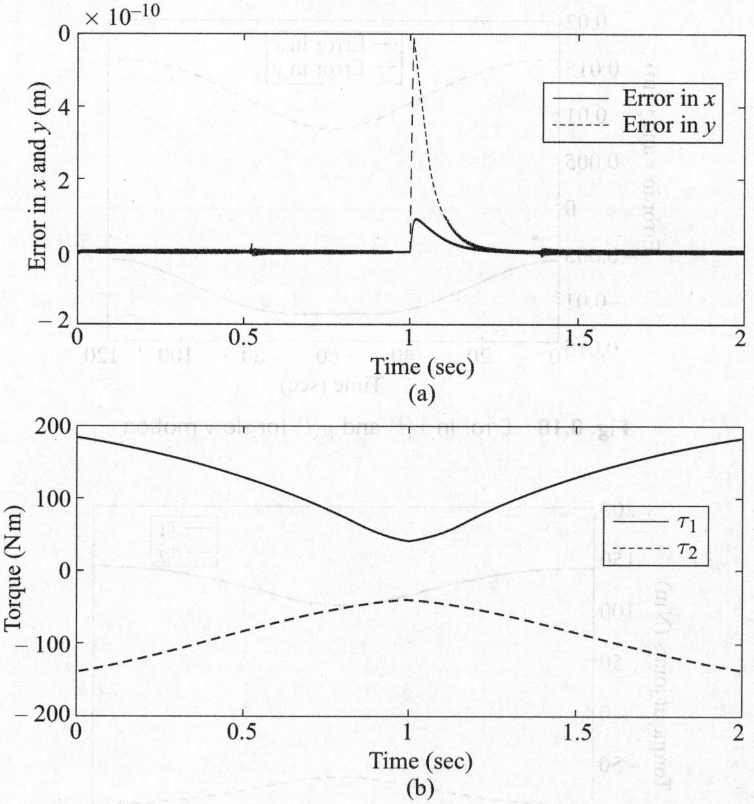

Fig. 8.20 Trajectory errors and torques at the joints for the feed-forward controller

[11] The errors in both cases are quite large and one can, by trial and error, obtain values of gains K_{p_i} and K_{v_i}, which can lower the errors.

To bring out the comparison and enhanced performance of the non-linear controllers, we perform simulation for a feed-forward controller with an exact knowledge of the model, a non-linear controller with uncertainty in the model and a Cartesian controller. The gains and the desired trajectory are chosen to be the same as those in the PD controller. We simulate for the fast motion of 2 sec. Figure 8.20(a) shows the errors in the trajectory of the tip, $x(t)$ and $y(t)$, for the feed-forward controller and Fig. 8.20(b) shows the variation in torques at the joints. It can be clearly seen that the trajectory errors are very small (almost zero). The torque values are also lower in comparison with the PD control and are smoother. Likewise, Figs 8.21 and 8.22 show the errors in the trajectory and the variation in torque at the joints for a model-based controller and a Cartesian controller wherein we have assumed a 10% error in masses and 5% error in the location of the centre of gravity. From Figs 8.21 and 8.22 we can observe that the trajectory errors are somewhat larger in comparison with the feed-forward controller with

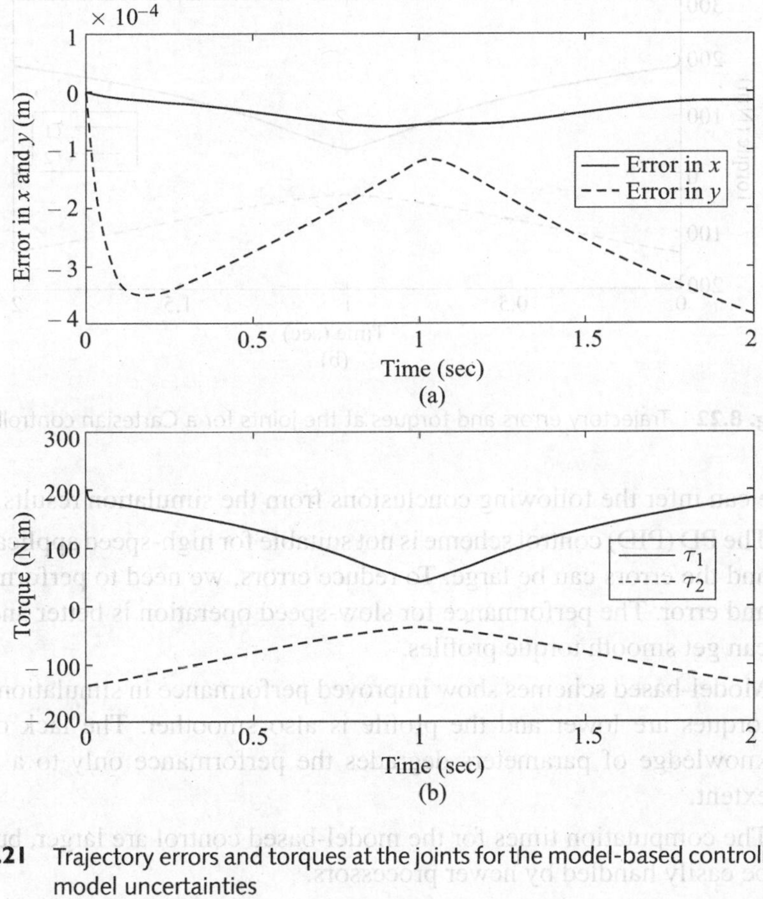

Fig. 8.21 Trajectory errors and torques at the joints for the model-based controller with model uncertainties

an exact knowledge of the model. However, in comparison with the PD controller, the errors are much smaller and are more acceptable. The torque profiles are also smoother.

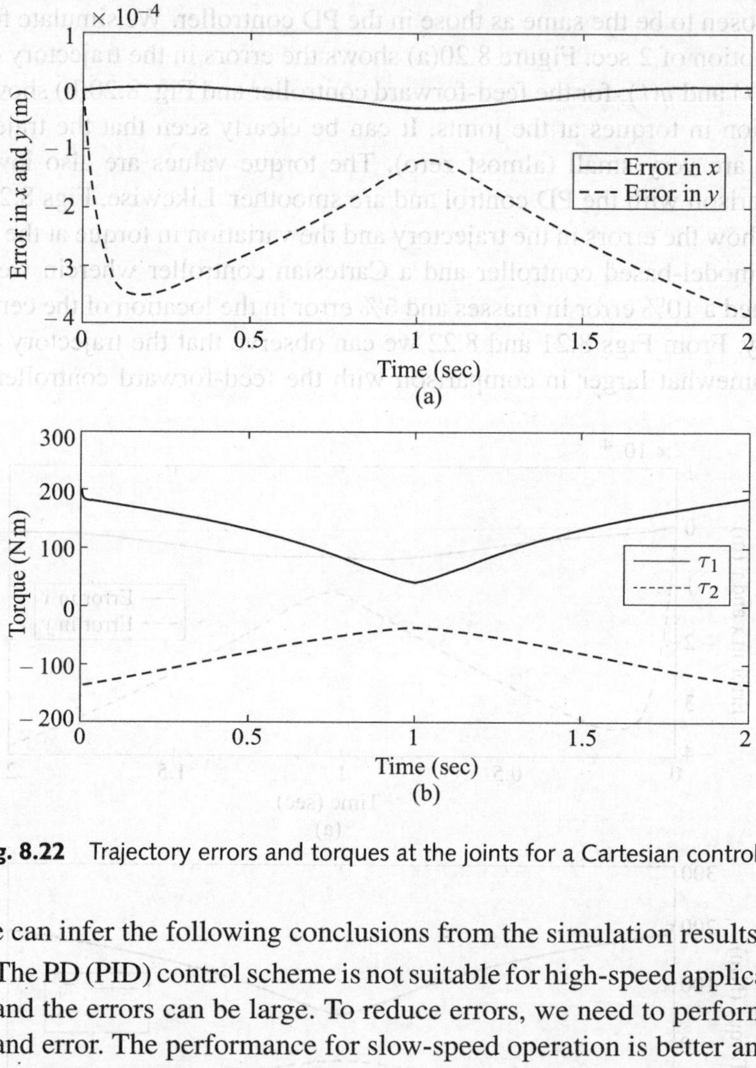

Fig. 8.22 Trajectory errors and torques at the joints for a Cartesian controller

We can infer the following conclusions from the simulation results.

1. The PD (PID) control scheme is not suitable for high-speed applications and the errors can be large. To reduce errors, we need to perform trial and error. The performance for slow-speed operation is better and one can get smooth torque profiles.
2. Model-based schemes show improved performance in simulation. The torques are lower and the profile is also smoother. The lack of the knowledge of parameters degrades the performance only to a small extent.
3. The computation times for the model-based control are larger, but can be easily handled by newer processors.

We next present experimental results which show that a model-based feed-forward controller shows improved performance in an actual experimental robot.

8.5.2 Experimental Results

The robotic manipulator used in our experiments is shown in Fig. 8.23 [more details about the robot and the experiments are available in Gopal (1995), Ravikiran (1995), and Ghosal (1996)]. It has five revolute joints with a four-bar linkage used to drive axis 3. The motor that actuates the third joint is located remotely from the joint and toward the base of the manipulator. The four-bar mechanism considerably complicates the model of the structure but reduces the overall inertia of the manipulator. Unlike most of the electronically driven robots, this manipulator is actuated by two-phase AC motors. The AC motors are relatively heavy and develop low driving torques at high speeds. Due to this, large speed reduction is required at each joint and gear trains are used to achieve the required speed reduction. The gears used for speed reduction are invariably accompanied by backlash and friction, which result in a poor system response. Encoders are connected to the joints to measure the rotation of the joint angle and tacho-generators are attached to the motors to read their angular velocity.

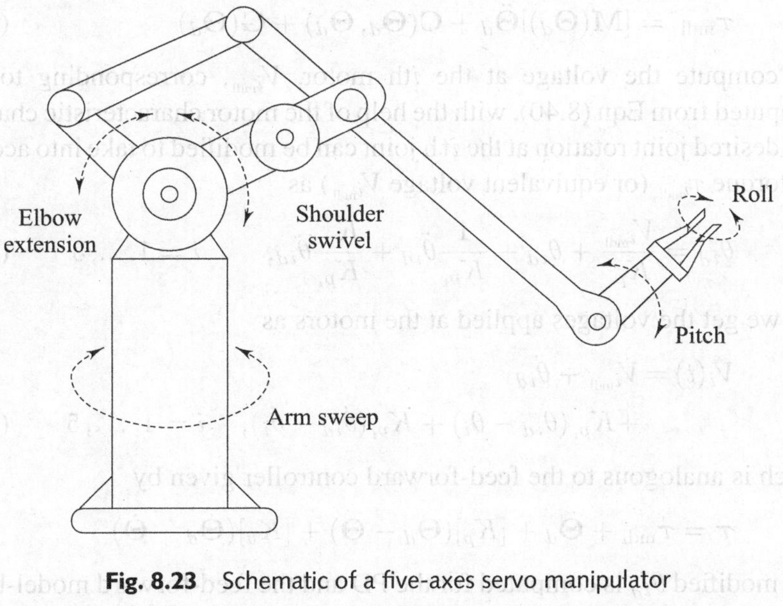

Fig. 8.23 Schematic of a five-axes servo manipulator

The existing controller for the robot implemented a control law of the form

$$V_i(t) = K_{p_i}(\theta_{id} - \theta_i) - K_{v_i}\dot{\theta}_i, \qquad i = 1, ..., 5 \tag{8.37}$$

where $V_i(t)$ is the voltage applied at motor i. This control scheme is a subset of the PD control law discussed in Section 8.3 and does not take into account the desired joint rates $\dot{\theta}_{id}$ or the desired joint acceleration $\ddot{\theta}_{id}$. To take into account the desired joint rate and the acceleration *without* altering the controller hardware, we modify the desired joint rotation θ_{id} as

$$\theta_{id}^* = \theta_{id} + \frac{1}{K_{p_i}}\ddot{\theta}_{id} + \frac{K_{v_i}}{K_{p_i}}\dot{\theta}_{id}, \qquad i = 1, ..., 5 \tag{8.38}$$

where $\dot{\theta}_{id}$ and $\ddot{\theta}_{id}$ are available from the trajectory planner. Since the error is now $\theta_{id}^* - \theta_i$, we get

$$\begin{aligned} V_i(t) &= K_{p_i}(\theta_{id}^* - \theta_i) - K_{v_i}\dot{\theta}_i \\ &= \ddot{\theta}_{id} + K_{p_i}(\theta_{id} - \theta_i) + K_{v_i}(\dot{\theta}_{id} - \dot{\theta}_i), \qquad i = 1, ..., 5 \end{aligned} \tag{8.39}$$

This is the PD control scheme discussed in Section 8.3, and by only *modifying* the desired joint rotation to θ_{id}^* given in Eqn (8.38), we implemented the PD controller on the existing controller.

In a similar manner, we also implemented a model-based feed-forward controller without altering the existing controller. Let $\boldsymbol{\tau}$ be the vector of torques for a given trajectory $\boldsymbol{\Theta}_d(t)$. This can be computed from a model of the robot and the chosen trajectory as

$$\boldsymbol{\tau}_{\text{mdl}} = [\mathbf{M}(\boldsymbol{\Theta}_d)]\ddot{\boldsymbol{\Theta}}_d + \mathbf{C}(\boldsymbol{\Theta}_d, \dot{\boldsymbol{\Theta}}_d) + \mathbf{G}(\boldsymbol{\Theta}_d) \tag{8.40}$$

We compute the voltage at the ith motor, $V_{i_{\text{mdl}}}$, corresponding to $\tau_{i_{\text{mdl}}}$ computed from Eqn (8.40), with the help of the motor characteristic charts.[12] The desired joint rotation at the ith joint can be modified to take into account the torque $\tau_{i_{\text{mdl}}}$ (or equivalent voltage $V_{i_{\text{mdl}}}$) as

$$\theta_{id}^* = \frac{V_{i_{\text{mdl}}}}{K_{p_i}} + \theta_{id} + \frac{1}{K_{p_i}}\ddot{\theta}_{id} + \frac{K_{v_i}}{K_{p_i}}\dot{\theta}_{id}, \qquad i = 1, ..., 5 \tag{8.41}$$

and we get the voltages applied at the motors as

$$\begin{aligned} V_i(t) = V_{i_{\text{mdl}}} &+ \ddot{\theta}_{id} \\ &+ K_{p_i}(\theta_{id} - \theta_i) + K_{v_i}(\dot{\theta}_{id} - \dot{\theta}_i), \quad i = 1, \ldots, 5 \end{aligned} \tag{8.42}$$

which is analogous to the feed-forward controller given by

$$\boldsymbol{\tau} = \boldsymbol{\tau}_{\text{mdl}} + \ddot{\boldsymbol{\Theta}}_d + [K_p](\boldsymbol{\Theta}_d - \boldsymbol{\Theta}) + [K_v](\dot{\boldsymbol{\Theta}}_d - \dot{\boldsymbol{\Theta}})$$

The modified θ_{id}^* is computed for the PD and the feed-forward model-based controllers and the modified θ_{id}^* are given as the reference input to the existing robot controller. The experiments are conducted with different

[12] Most motor manufacturers provide torque–speed curves at different applied voltages.

trajectories, sampling rates, time spans, and friction compensation. We present one representative result for the first three joints.

The manipulator is made to traverse from the home position (0°, 0°, –90°, 180°, 0°) to the goal position (30°, 40°, –60°, 180°, 0°) and back to the home position in a span of 4 sec. Smooth cubic trajectories are generated using the technique discussed in Chapter 7 and the joint values are computed at intervals of 5 ms (i.e., at a frequency of 200 Hz). Figure 8.24 shows the desired trajectory and the path followed by the first joint of the robot under the model-based control and PD control. It can be seen that the trajectory followed using the model-based controller is closer to the desired trajectory. To get a better idea, we plot the error for the model-based and PD controllers for the duration of the motion. Figure 8.25 shows the error between θ_{1d} and the actual measured θ_1 during the motion. It can be seen that the errors are between $+1°$ and $-2°$ for the model-based control as compared to a maximum of $-5°$ for the PD control. For joint 2, as shown in Fig. 8.26, the error in the model-based control is again less than that in the PD control, although the difference is not as significant as in joint 1. For joint 3 the performance is similar for both the controllers (see Fig. 8.27) since the effect of the dynamics, $\boldsymbol{\tau}_{\text{model}}$, is smaller as one goes away from the base. During experiments, it is observed that the manipulator while operating under the PD control often overshoots the goal position and results in the second joint impacting the mechanical stops. This is never observed when the manipulator is operating under the model-based control.

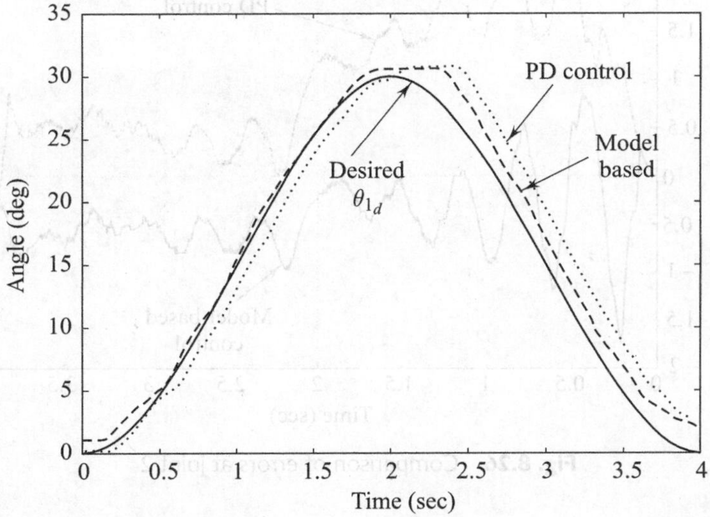

Fig. 8.24 Controller performance in following the desired trajectory of joint 1

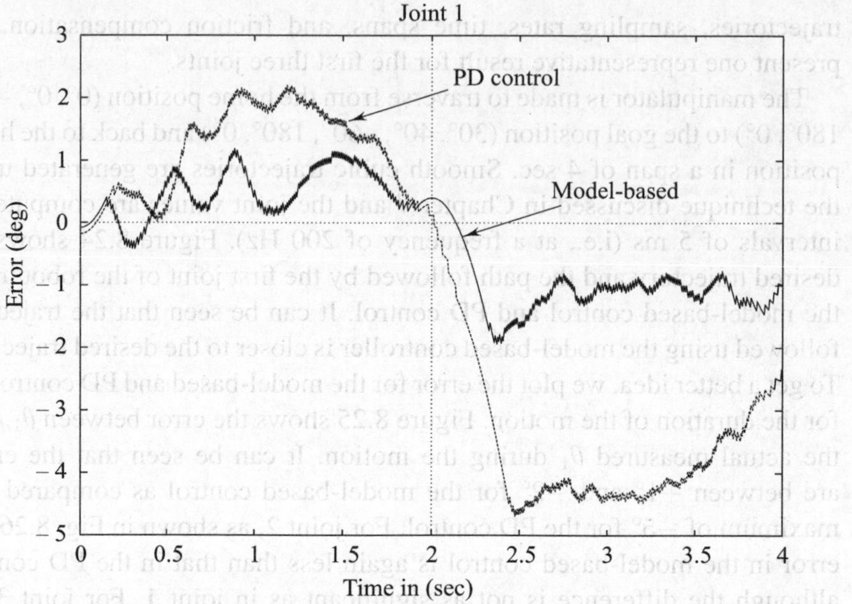

Fig. 8.25 Comparison of errors at joint 1

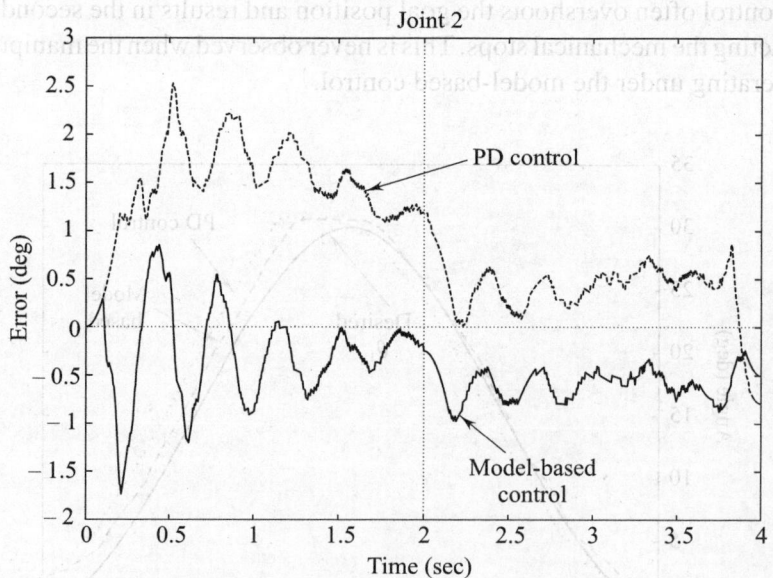

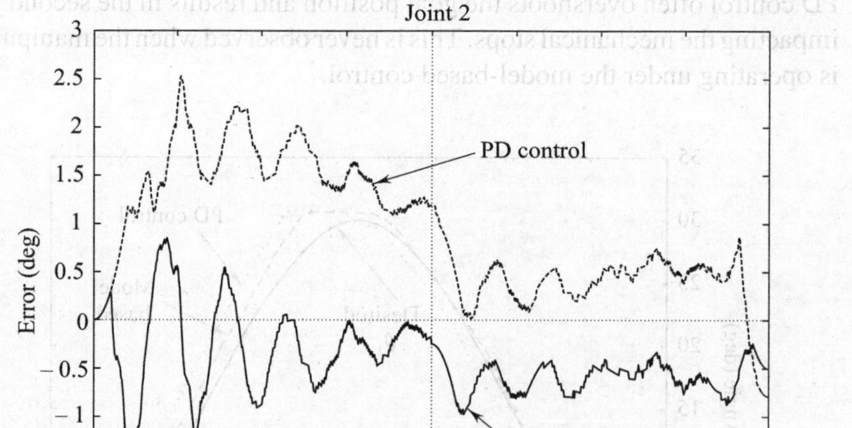

Fig. 8.26 Comparison of errors at joint 2

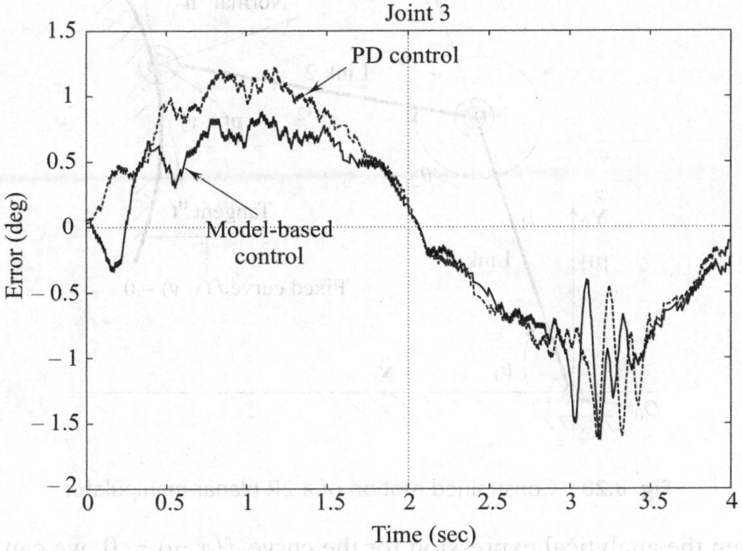

Fig. 8.27 Comparison of errors at joint 3

8.6 Non-linear Control of Constrained and Parallel Manipulators

In the preceding section, we discussed the control of a serial manipulator moving without any constraint on the joint variables $\mathbf{q}(t)$. In many applications, it is required that the end-effector of a serial manipulator traces a desired path while maintaining contact with a surface or applying a desired force normal to the surface. A two-dimensional example is that of a planar 2R manipulator whose tip must always be in contact with a curve $f(x, y) = 0$ as shown in Fig. 8.28. In this section, we will discuss the control of constrained manipulators with the help of the planar 2R example. In contrast to a serial manipulator moving under a constraint, in parallel manipulators, as discussed in Chapter 4, we have loop-closure constraints in terms of the actuated and passive joint variables. In this section, we will also discuss the control of parallel manipulators with loop-closure constraints.

For the tip of the planar 2R manipulator to keep in contact with the curve $f(x, y) = 0$, we must have

$$F(\theta_1, \theta_2) = f(l_1 c_1 + l_2 c_{12}, l_1 s_1 + l_2 s_{12}) = 0 \tag{8.43}$$

where $x = l_1 c_1 + l_2 c_{12}$ and $y = l_1 s_1 + l_2 s_{12}$ are the direct kinematic equations for the planar 2R manipulator.

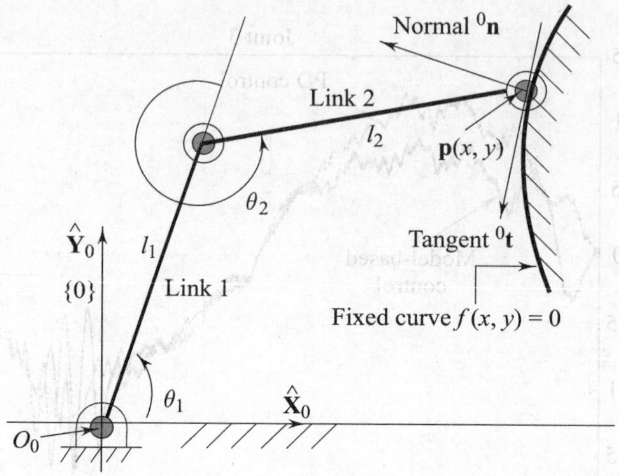

Fig. 8.28 Constrained motion of a 2R planar manipulator

Given the analytical expression for the curve $f(x, y) = 0$, we can obtain expressions for θ_1 and θ_2 in terms of a parameter ϕ, and we can write

$$\theta_1 = h_1(\phi), \qquad \theta_2 = h_2(\phi) \tag{8.44}$$

or in a vector form as $\boldsymbol{\Theta} = \mathbf{h}(\phi)$ with $\boldsymbol{\Theta}$ denoting $(\theta_1, \theta_2)^T$.

The equations of the form Eqn (8.44) can be obtained, at least conceptually, from the inverse kinematics equations of a planar 2R manipulator when the constraint $f(x, y) = 0$ is available in terms of a parametric ϕ in the form $x = f_1(\phi)$, $y = f_2(\phi)$. As shown in Chapter 3 (in Example 3.4), we can obtain θ_1 and θ_2 [see Eqns (3.9) and (3.8)] for a given X and Y (equivalent to x and y in this case) and then obtain θ_1 and θ_2 in terms of the parameter ϕ.

It is also possible to use the concepts from the solution of direct kinematics of parallel manipulators (see Chapter 4, Section 4.3) if the constraint $f(x, y) = 0$ is simple such as a circle centred at $(l_0, 0)$ of radius l_3. For this circle, we can consider the constrained motion of the planar 2R manipulator as an equivalent four-bar mechanism and write the loop-closure equations. As shown in Eqn (4.6), we can get

$$x = l_1 c_1 + l_2 c_{12} = l_0 + l_3 \cos\phi$$
$$y = l_1 s_1 + l_2 s_{12} = l_3 \sin\phi$$

where $x = l_0 + l_3 \cos\phi$ and $y = l_3 \sin\phi$ are the parametric equations of the circle. As discussed in Chapter 4 (in Example 4.1), we can obtain equations relating θ_1 and θ_2 (same as ϕ_2 in Example 4.1) with ϕ (same as ϕ_1 in Example 4.1) (see also Exercise 8.7 at the end of this chapter). Clearly, if $f(x, y) = 0$

is a complicated curve, it may not be easy or possible to obtain analytical expressions for θ_1 and θ_2 in terms of parameter ϕ. In Section 8.8 we present an alternate approach.

From Eqn (8.44) we can obtain

$$\dot{\theta}_i = \frac{\partial h_i}{\partial \phi}\dot{\phi}, \quad i = 1,2$$
$$\ddot{\theta}_i = \frac{\partial h_i}{\partial \phi}\ddot{\phi} + \left(\frac{\partial^2 h_i}{\partial \phi^2}\dot{\phi}\right)\dot{\phi}, \quad i = 1,2 \tag{8.45}$$

Substituting the expressions for θ_i, $\dot{\theta}_i$, and $\ddot{\theta}_i$ $(i = 1,2)$ in the equations of motion of a planar 2R manipulator, Eqn (6.34), we can get

$$[\mathbf{M}(\mathbf{\Theta})][J_{\mathbf{h}}]\ddot{\phi} + (\mathbf{C}(\mathbf{\Theta}, \dot{\mathbf{\Theta}}) + [\mathbf{M}(\mathbf{\Theta})][\dot{J}_{\mathbf{h}}]\dot{\phi}) + \mathbf{G}(\mathbf{\Theta}) = \boldsymbol{\tau} \tag{8.46}$$

where the matrix $[J_{\mathbf{h}}]$ denotes the Jacobian of the transformation $\mathbf{\Theta} = \mathbf{h}(\phi)$ (same as the column vector $[\partial h_1/\partial \phi, \partial h_2/\partial \phi]^T$), and $[\dot{J}_{\mathbf{h}}]$ are the time derivatives of the elements of $[J_{\mathbf{h}}]$ [yielding terms similar to the one inside parentheses in Eqn (8.45)]. Equation (8.46) can be reduced to a single differential equation by premultiplying the left- and the right-hand side by $[J_{\mathbf{h}}]^T$, and we can write

$$\bar{M}(\phi)\ddot{\phi} + \bar{C}(\phi, \dot{\phi}) + \bar{G}(\phi) = [J_{\mathbf{h}}]^T \boldsymbol{\tau} \tag{8.47}$$

where

$$\bar{M}(\phi) = [J_{\mathbf{h}}]^T [\mathbf{M}(\mathbf{h}(\phi))][J_{\mathbf{h}}]$$
$$\bar{C}(\phi, \dot{\phi}) = \mathbf{C}(\mathbf{h}(\phi), [J_{\mathbf{h}}]\dot{\phi}) + [\mathbf{M}(\mathbf{h}(\phi))][\dot{J}_{\mathbf{h}}]\dot{\phi}$$
$$\bar{G}(\phi) = \mathbf{G}(\mathbf{h}(\phi)) \tag{8.48}$$

A solution $\phi(t)$ of the differential equation (8.47) can be used to obtain $\theta_1(t)$ and $\theta_2(t)$ from Eqn (8.44) and for these $\theta_1(t)$, $\theta_2(t)$ the tip of the planar 2R manipulator is always in contact with the curve $f(x, y) = 0$. In this sense, the single second-order differential equation, Eqn (8.47), represents an unconstrained system equivalent to a planar 2R manipulator satisfying the constraint $f(x, y) = 0$.

The projection by $[J_{\mathbf{h}}]^T$ removes all information about the force acting along the normal ${}^0\mathbf{n}$ to the curve which is along the gradient $\nabla f(x, y) = [\partial f(x, y)/\partial x, \partial f(x, y)/\partial y]^T$. Due to this reason, the single differential equation is not suitable for developing control schemes for the planar 2R manipulator where the manipulator is required to trace a path satisfying the constraint *and apply* a normal force without violating the constraint.

To develop such a control scheme, we consider the joint torque vector $\boldsymbol{\tau}_{\mathbf{n}}$, corresponding to a force normal to $f(x, y) = 0$, of the form

$$\boldsymbol{\tau}_{\mathbf{n}} = \lambda \nabla F(\theta_1, \theta_2) \tag{8.49}$$

where $F(\theta_1, \theta_2) = 0$ is the equivalent expression for the constraint $f(x, y) = 0$ in terms of the joint variables θ_1 and θ_2 (obtained from the use of the direct kinematics equations) as given in Eqn (8.43), and $\nabla F(\theta_1, \theta_2)$ is the gradient of $F(\theta_1, \theta_2)$ given by $[\partial F(\theta_1, \theta_2)/\partial\theta_1, \partial F(\theta_1, \theta_2)/\partial\theta_2]^T$. Since

$$\begin{aligned}\boldsymbol{\tau}_{\mathbf{n}} \cdot \dot{\boldsymbol{\Theta}} &= \lambda \left(\frac{\partial F(\theta_1, \theta_2)}{\partial \theta_1} \dot{\theta}_1 + \frac{\partial F(\theta_1, \theta_2)}{\partial \theta_2} \dot{\theta}_2 \right) \\ &= \lambda \frac{d}{dt} [F(\theta_1, \theta_2)] = 0\end{aligned}$$

the torque $\boldsymbol{\tau}_{\mathbf{n}}$ does not do any work and hence does not cause any motion for the planar 2R manipulator in contact with the curve $f(x, y) = 0$. We assume that the desired $\lambda(t)$, representing the desired force along ${}^0\mathbf{n}$, is prescribed. We can now obtain a combined torque of the form

$$\boldsymbol{\tau} = \lambda(t) \nabla F(\theta_1, \theta_2) + \boldsymbol{\tau}_\phi \tag{8.50}$$

where $\lambda(t)\nabla F(\theta_1, \theta_2)$ yields the desired force along the normal ${}^0\mathbf{n}$ and the $\boldsymbol{\tau}_\phi$ term can be utilized to trace a desired path without violating the constraint $f(x, y) = 0$. Following the concept of the computed torque control of an unconstrained serial manipulator, we set

$$\boldsymbol{\tau}_\phi = [\boldsymbol{\alpha}]_\phi \boldsymbol{\tau}'_\phi + \boldsymbol{\beta}_\phi \tag{8.51}$$

with

$$\begin{aligned}[\boldsymbol{\alpha}]_\phi &= [\mathbf{M}(\boldsymbol{\Theta})][J_{\mathbf{h}}] \\ \boldsymbol{\beta}_\phi &= \{\mathbf{C}(\boldsymbol{\Theta}, \dot{\boldsymbol{\Theta}}) + [\mathbf{M}(\boldsymbol{\Theta})][\dot{J}_{\mathbf{h}}]\dot{\phi}\} + \mathbf{G}(\boldsymbol{\Theta}) \\ \tau'_\phi &= \ddot{\phi}_d + [K_v](\dot{\phi}_d - \dot{\phi}) + [K_p](\phi_d - \phi)\end{aligned} \tag{8.52}$$

As in the case of the control of a unconstrained serial manipulator, for a proper choice of the gain matrices $[K_p]$ and $[K_v]$, we can ensure that the tip of the manipulator always keeps in contact with $f(x, y) = 0$. It may be again noted that the terms $\lambda(t)\nabla F(\theta_1, \theta_2)$ and $\boldsymbol{\tau}_\phi$ do not affect each other—while the first term ensures that the desired force along the normal is achieved without affecting the motion, the second term achieves the desired motion without violating the constraint. It may be mentioned that the control scheme given by Eqn (8.51) assumes *exact* knowledge of the model and any uncertainties in the model parameters will affect the controller performance.

The above discussion on the control of the constrained planar 2R manipulator can be extended to a general six-DOF serial manipulator moving in $\Re^3$ under k $(k < 6)$ constraints. For k constraints of the form $f_i({}^0\mathbf{p}, {}^0_{Tool}[R]) = 0$, $i = 1, \ldots, k$, where $({}^0\mathbf{p}, {}^0_{Tool}[R])$ are the position and orientation of the end-effector, we can obtain (from the inverse kinematics of the six-DOF manipulator) expressions analogous to Eqn (8.44) relating the joint variables θ_i $(i = 1, \ldots, 6)$ in terms of parameters ϕ_i $(i = 1, \ldots, 6 - k)$. We can obtain the $6 \times (6 - k)$ Jacobian matrix $[J_{\rm h}]$ and proceed to obtain $\boldsymbol{\tau}_\phi$ and $\boldsymbol{\tau}_{\rm n}$ as shown above. However, this procedure can become quite complicated. In section 8.9 we present a better approach.

Similar to a serial manipulator with constraints, in the case of parallel manipulators, we have loop-closure constraints, $\eta_i(\mathbf{q}) = 0$ $(i = 1, ..., m)$. As discussed in Chapter 6, the equations of motion of a parallel manipulator can be derived, taking into account the m constraints, in terms of Lagrange multipliers. The general form of the equations of motion of a parallel manipulator, Eqn (6.21), reproduced here for convenience, is given by

$$[\mathbf{M}(\mathbf{q})]\ddot{\mathbf{q}} + [\mathbf{C}(\mathbf{q}, \dot{\mathbf{q}})]\dot{\mathbf{q}} + \mathbf{G}(\mathbf{q}) = \boldsymbol{\tau} + [\boldsymbol{\Psi}(\mathbf{q})]^T \boldsymbol{\lambda}$$

where $[\boldsymbol{\Psi}(\mathbf{q})]$ and $\boldsymbol{\lambda}$ are similar to the Jacobian matrix $[J_{\rm h}]$ and λ for 2R serial manipulators with constraints. A key difference is that we do not want to control the constraint forces which arise out of the loop-closure constraints and the torque vector $\boldsymbol{\tau}$ has non-zero elements *only* for the n actuated joints. Hence, we can directly use the equation of motion obtained by eliminating $\boldsymbol{\lambda}$, Eqn (6.25), to develop model-based control schemes for parallel manipulators. From Eqn (6.25), i.e.,

$$[\mathbf{M}]\ddot{\mathbf{q}} = \mathbf{f} - [\boldsymbol{\Psi}]^T([\boldsymbol{\Psi}][\mathbf{M}]^{-1}[\boldsymbol{\Psi}]^T)^{-1}\{[\boldsymbol{\Psi}][\mathbf{M}]^{-1}\mathbf{f} + [\dot{\boldsymbol{\Psi}}]\dot{\mathbf{q}}\}$$

with $\mathbf{f}$ denoting $(\boldsymbol{\tau} - [\mathbf{C}]\dot{\mathbf{q}} - \mathbf{G})$, we can write the n equations of motion in a compact form as

$$[\mathbf{M}]\ddot{\mathbf{q}} + \mathbf{B}(\mathbf{q}, \dot{\mathbf{q}}) = [\mathbf{A}(\mathbf{q})]\boldsymbol{\tau} \tag{8.53}$$

Following the concept of the control law partitioning discussed in Section 8.4, we can set

$$[\mathbf{A}(\mathbf{q})]\boldsymbol{\tau} = [\boldsymbol{\alpha}]\boldsymbol{\tau}' + \boldsymbol{\beta} \tag{8.54}$$

and choose $[\boldsymbol{\alpha}]$ and $\boldsymbol{\beta}$ as $[\mathbf{M}(\mathbf{q})]$ and $\mathbf{B}(\mathbf{q}, \dot{\mathbf{q}})$, respectively, to achieve feedback linearization. In addition, $\boldsymbol{\tau}'$ can be chosen as in Eqn (8.32) while ensuring that its non-zero elements are only for the actuated variables in the parallel manipulator. As in the case of a serial manipulator, by choosing appropriate gain matrices $[K_p]$ and $[K_v]$, the actuated joints in a parallel manipulator can be made to follow desired trajectories, and since the equations of motion incorporate the loop-closure constraints, the motion of the actuated joints will not violate the loop-closure constraints. For a parallel manipulator, it can be observed that model-based terms are significantly more complicated and are functions of *both* the passive and the active joint variables. Typically the passive variables are not measured and to form $\mathbf{q}$ and $\dot{\mathbf{q}}$ from only the measured actuated variables, we must use the direct kinematics of parallel manipulators (see Chapter 4) and Eqn (5.41) to compute the passive variables and their rates. Again in the above discussion, the effect of the lack of knowledge of model parameters is not taken into account and uncertainties in the model can make the task of control significantly more difficult for parallel manipulators.

8.7 Cartesian Control of Manipulators

In the control schemes discussed so far, the desired trajectory was available as time histories of the joint position, velocity, and acceleration. It is more natural to specify the motion of a manipulator in terms of the position and orientation of the end-effector and their velocities and accelerations. As shown in Section 7.4, we can plan and generate straight-line and circular trajectories, as well as the desired trajectories for the orientation, of the end-effector. Given the Cartesian position, orientation, and their derivatives, one possibility is to use the inverse kinematics and the inverse Jacobian of the serial manipulator (or direct kinematics in the case of parallel manipulators) to compute the desired joint trajectories and their rates, and use the desired joint space trajectories in model-based or PD control schemes discussed earlier. This intuitive concept is, however, very computation intensive as we have seen in Chapters 3 and 4. It is also not a good idea to attempt a reduction in the computational effort by only computing Θ_d by inverse or direct kinematics and obtain the rates by numerical differentiation since it is very 'noisy' and we would require complicated controllers. In this section, we discuss a method based on using the Cartesian equations of motion discussed in Section 6.4. This method was first developed by Khatib (1987) and has found widespread use in the Cartesian control of manipulators. The method is also referred to as *operational space* or *task space control* of manipulators.

We start with the equations of motion of a manipulator written in the Cartesian space, Eqn (6.35), reproduced below for convenience:

$$\mathcal{F} = [M_{\mathcal{X}}(\mathbf{q})]\ddot{\mathcal{X}} + \mathbf{C}_{\mathcal{X}}(\mathbf{q}, \dot{\mathbf{q}}) + \mathbf{G}_{\mathcal{X}}(\mathbf{q}) \tag{8.55}$$

where $\mathcal{F}$ is a 6×1 entity of forces and moments acting on the end-effector and $\mathcal{X}$ is an appropriate 6×1 entity representing the position and orientation of the end-effector. The Cartesian mass matrix is denoted by $[M_{\mathcal{X}}(\mathbf{q})]$, the Cartesian Coriolis and centripetal term is denoted by $\mathbf{C}_{\mathcal{X}}(\mathbf{q}, \dot{\mathbf{q}})$, and the term $\mathbf{G}_{\mathcal{X}}(\mathbf{q})$ is the Cartesian gravity term. It may be noted that the Cartesian mass matrix and other terms can be related to the joint space mass matrix and other terms and are given in Eqn (6.36). In principle, the joint variable $\mathbf{q}$ and its derivative $\dot{\mathbf{q}}$ can be removed by using the inverse kinematics and inverse Jacobian functions. However, this step is not required for control.

As in the model-based joint space schemes, we assume a control law of the form

$$\mathcal{F} = [\boldsymbol{\alpha}_{\mathcal{X}}]\mathcal{F}' + \boldsymbol{\beta}_{\mathcal{X}} \tag{8.56}$$

We choose

$$\begin{aligned}[\boldsymbol{\alpha}_{\mathcal{X}}] &= [M_{\mathcal{X}}(\mathbf{q})] \\ \boldsymbol{\beta}_{\mathcal{X}} &= \mathbf{C}_{\mathcal{X}}(\mathbf{q}, \dot{\mathbf{q}}) + \mathbf{G}_{\mathcal{X}}(\mathbf{q})\end{aligned} \tag{8.57}$$

Substituting Eqns (8.56) and (8.57) in Eqn (8.55), we get

$$\mathcal{F}' = \ddot{\mathcal{X}} \tag{8.58}$$

Equation (8.58) can be viewed as a unit mass plant with a new input $\mathcal{F}'$. If $\mathcal{F}'$ is chosen as

$$\mathcal{F}' = \ddot{\mathcal{X}}_d(t) + [K_v]_{\mathcal{X}}\dot{\mathbf{e}}(t) + [K_p]_{\mathcal{X}}\mathbf{e}(t) \tag{8.59}$$

with $\mathbf{e}(t) = \mathcal{X}_d(t) - \mathcal{X}(t)$, then we can write a linear error equation as

$$\ddot{\mathbf{e}}(t) + [K_v]_{\mathcal{X}}\dot{\mathbf{e}}(\mathbf{t}) + [K_p]_{\mathcal{X}}\mathbf{e}(t) = \mathbf{0} \tag{8.60}$$

If we choose diagonal $[K_p]_{\mathcal{X}}$ and $[K_v]_{\mathcal{X}}$, we will get a *linear decoupled* error equation, and following the well-known linear control results, we can set $[K_p]_{\mathcal{X}}$ and $[K_v]_{\mathcal{X}}$ to get critically (or over-damped) performance at *every* point in the workspace of the robot.

For this control scheme to work, the actuation must be the Cartesian force and moment $\mathcal{F}$ at the end-effector, whereas in a typical manipulator we have only actuators at the joints. We can, however, make use of the relationship between joint torques $\boldsymbol{\tau}$, the Cartesian force, moment $\mathcal{F}$ applied on the end-effector,

$$\boldsymbol{\tau} = [J(\mathbf{q})]^T\mathcal{F} \tag{8.61}$$

The Cartesian control scheme also requires the measurements $\mathcal{X}$ and $\dot{\mathcal{X}}$ to obtain the error and its derivative. We can either make use of the existing joint-based sensors (optical encoders for $\mathbf{q}$ and tachometers for $\dot{\mathbf{q}}$) or have vision-based sensors which can directly measure the Cartesian position or orientation and the velocities. In the first case we compute the direct kinematics and the Jacobian matrix and use them to obtain $\mathcal{X}$ and $\dot{\mathcal{X}}$. In either case, we do not use the computationally intensive inverse kinematics or the inverse Jacobian. It may be mentioned that the expressions for the Cartesian mass matrix, Coriolis/centripetal and gravity terms contain inverse Jacobian, but these do not have to be done in real time. We can, for once, compute the Cartesian model terms symbolically and use them in the control law implementation. However, in real time, we need the joint space variables $\mathbf{q}$ and $\dot{\mathbf{q}}$ to compute the model-based terms. Figure 8.29 shows a block diagram of the Cartesian-model-based control scheme.

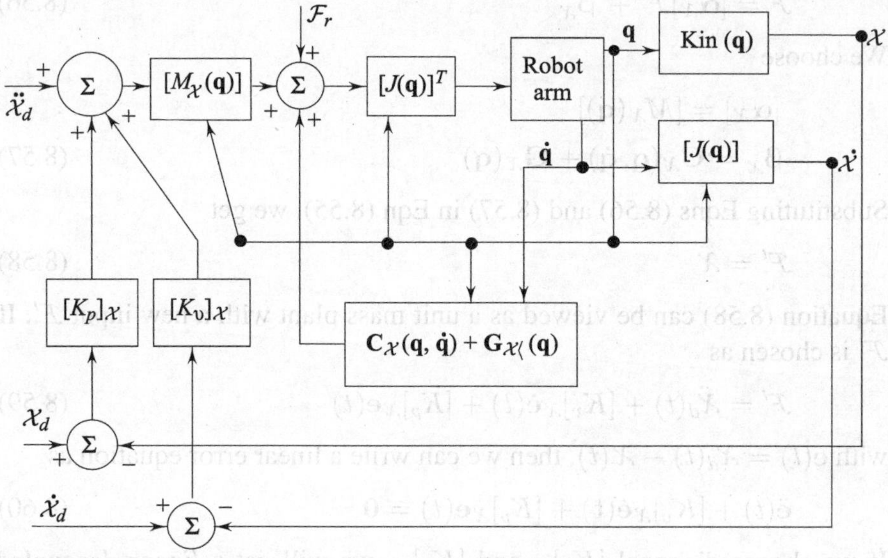

Fig. 8.29 Cartesian-model-based control scheme

The above Cartesian-model-based control scheme with a little modification can also be used for obstacle avoidance in real time at the level of control.[13] This was first proposed by Khatib (1986) and has been used with success in mobile robots and other applications. To understand real-time obstacle avoidance, we go back to Fig. 8.29. The output after the summing block (and before the $[J(\mathbf{q})]^T$ block) is a force acting on the end-effector, which causes the end-effector to follow the desired trajectory. Khatib (1986)

[13] There are numerous obstacle avoidance schemes which make use of search and AI techniques; however, they are beyond the scope of this text.

suggested adding another fictitious Cartesian force at this summing point. The force is computed as

$$\mathcal{F}_r = \sum_{i}^{N} \mathcal{F}_i \propto K_i / r_i^n \tag{8.62}$$

where N is the number of obstacles, r_i is the distance of the ith obstacle[14] from the robot and K_i, n are appropriately chosen constants such that the force $\mathcal{F}_i$ falls off rapidly as the distance increase and is large only when r_i is small. This *synthetic* force is *repulsive* in nature and its effect is to repulse the robot when it comes close to any obstacle. When there are no nearby obstacles, the nominal trajectory is followed by the action of the controller and this repulsive force is active only when the robot is close to an obstacle. One can also add a synthetic *attractive* force of the goal point, which helps in moving the robot towards the goal.

8.8 Force Control of Manipulators

When a manipulator is moving in free space, position control is used to follow a desired trajectory. However, when the manipulator is moving such that the end-effector is in *contact* with the environment, position control may not be sufficient. Consider a manipulator being used for grinding a surface with a grinding wheel. It is well known that for proper grinding an appropriate force should be applied *normal* to the surface. One way to achieve the required force would be to plan an appropriate trajectory in the Cartesian or joint space such that the trajectory is 'inside' the surface by a 'small' amount and expect that the passive compliance in the wrist will provide the required normal force. However, if the stiffness of the end-effector or the environment is high, it is very difficult to apply the required force. In addition, any uncertainty in the surface geometry or errors in the position control will make the task very difficult. Either the end-effector will be damaged by the application of excessive force or the grinding wheel may not touch the surface. In such tasks, it is appropriate to specify and control the force *normal* to the surface and, in addition, specify and control the position of the end-effector *tangent* to the surface. The requirement of force control is not limited to grinding or manufacturing. It is useful in any application where the manipulator is in contact with the environment and one very important application is the *robotic assembly*.

[14] The distance r_i can be from the centre of the obstacle to the robot and can be obtained from a CAD database or by a vision sensor.

As discussed in Section 8.6, it is possible to specify the force normal to a curve (or a surface) and at the same time follow a trajectory along the curve (or on the surface) by using τ_n and τ_ϕ [see Eqns (8.49) and (8.50)]. However, one main difficulty in such joint- space-based schemes is that the relationship between the Lagrange multiplier λ and the forces and moments acting on the end-effector is not obvious, and hence it is difficult to specify $\lambda(t)$ as done in Section 8.6. Another difficulty in the joint-space approach involves the use of the inverse kinematics to obtain the gradient [$\nabla F(\theta_1, \theta_2)$ in Section 8.6]. In this section we present a control scheme based on the Cartesian controller, developed in Section 8.7, which overcomes these difficulties. As in the grinding application, a mix of the position and the force control is required. In this section, we also present a *hybrid position/force* controller which allows switching between the position and the force control according to the requirement of the task. The force and hybrid position/force control of manipulators has been an extensive topic of research for its usefulness in the analysis of robotic tasks such as grasping, assembly, and manufacturing. An advanced treatment of the theory and application of force control in robot manipulators is available in Canudas de Wit (1996) and issues such as grasping are discussed in Murray et al. (1994). In this text we follow Craig (1989) for the developement of force control and hybrid position/force control in manipulators.

8.8.1 Force Control of a Single Mass

We begin by developing a scheme to control the force exerted by mass m along a direction on the environment. We model the environment as a spring of stiffness K_e. We assume that the mass is acted upon by a force $f(t)$ and a disturbance force $f_{\text{dist}}(t)$, and the displacement of the mass is $x(t)$ along the direction of $f(t)$. Figure 8.30 shows schematically the mass m attached to the spring K_e and the forces acting on it. The mass may be thought of as the end-effector and the disturbance force may be thought of as due to unknown frictional and other forces. The force exerted on the environment is given by

$$f_e(t) = K_e x(t) \tag{8.63}$$

We wish to control $f_e(t)$ such that $f_e(t)$ is a desired $f_{ed}(t)$ by applying a suitable $f(t)$. The equation of motion of the system is given by

$$f = m\ddot{x} + K_e x + f_{\text{dist}} \tag{8.64}$$

Written in terms of f_e, we have

$$f = mK_e^{-1}\ddot{f}_e + f_e + f_{\text{dist}} \tag{8.65}$$

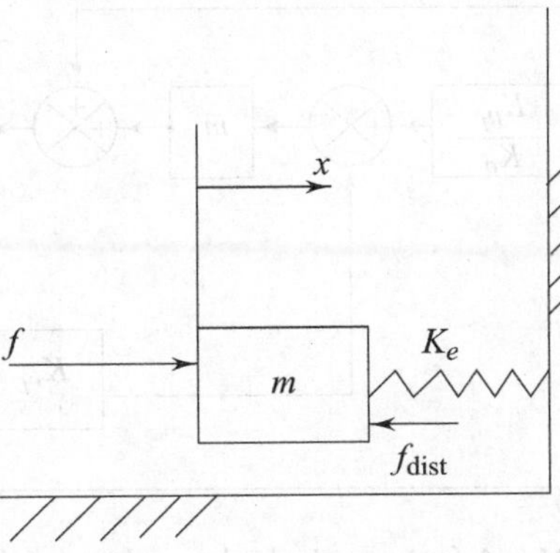

Fig. 8.30 Force control of mass along one direction

Equation (8.65) is a second-order linear differential equation and is very similar to the equation obtained for the single-link manipulator in Section 8.2. We can take Laplace transforms and obtain a PD (or PID) control scheme for this second-order system as discussed in Section 8.2.3, and by proper choice of the gains ensure that $f_e(t)$ approaches a desired $f_{ed}(t)$. But, we will use the control law partitioning technique used for the position control to obtain a force control scheme for the mass. We can write

$$\begin{aligned} f &= \alpha f' + \beta \\ \alpha &= m K_e^{-1} \\ \beta &= f_e + f_{\text{dist}} \\ f' &= \ddot{f}_{ed} + K_{v_f} \dot{e}_f + K_{p_f} e_f \end{aligned} \tag{8.66}$$

where $e_f = f_{ed} - f_e$, f_{ed} is the desired force, and f_e is the (measured) force acting on the environment. The quantities K_{v_f} and K_{p_f} are the derivative and proportional gains, respectively.

For the control law given in Eqn (8.66), the closed-loop error equation is given by

$$\ddot{e}_f + K_{v_f} \dot{e}_f + K_{p_f} e_f = 0 \tag{8.67}$$

and we can set the gains such that the error and its rate of change go to zero, i.e., the force acting on the environment, f_e, is the same as the desired force f_{ed}. However, we cannot use the knowledge of f_{dist} in the control scheme as this is unknown. To overcome this problem, we set $\beta = f_{ed}$. The steady-state error for this choice of β is given by

$$e_f = \frac{f_{\text{dist}}}{1 + m K_e^{-1}} \tag{8.68}$$

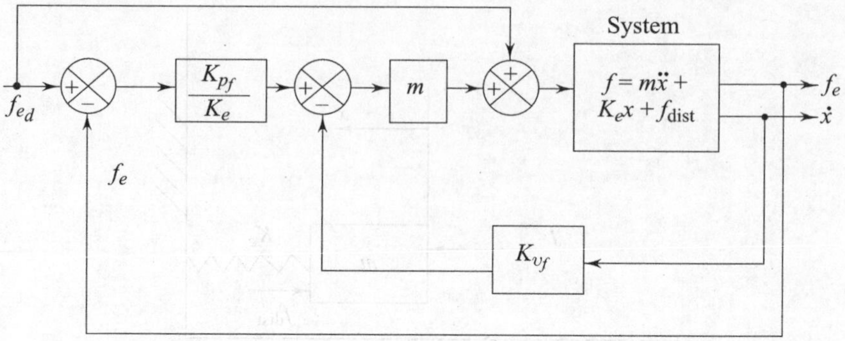

Fig. 8.31 A force control scheme for a spring–mass system

Since K_e is typically quite large, the denominator in Eqn (8.68) is close to 1 and the steady-state error is of the order of f_{dist}.

The control scheme given in Eqn (8.66) with $\beta = f_{ed}$ is impractical since $\dot{f}_{ed}$ and $\ddot{f}_{ed}$ are generally not specified—it also makes no physical sense to specify derivatives of the force. Hence, we drop the derivatives of f_{ed} from our control scheme. In addition, although f_e can be measured, the derivative of f_e is very difficult to measure or obtain by numerical differentiation. We can, however, obtain $\dot{f}_e$ as $K_e\dot{x}$ if the velocity of the mass can be measured. For these choices, the control law can be written as

$$f = m[K_{p_f} K_e^{-1} e_f - K_{v_f}\dot{x}] + f_{ed} \tag{8.69}$$

A possible force control scheme for a single mass applying a desired force along one direction is shown as a block diagram in Fig. 8.31. It may be noted that K_e appears in the control scheme and it is difficult to estimate the stiffness of the environment and, indeed, K_e could be changing with time. For most assembly tasks involving mating of nearly rigid parts, K_e should be chosen as a large number so that the system is reasonably robust for changes in K_e. It may be noted that due the dropping of terms involving derivatives of force and β and f_{ed}, the error equation will not be the same as given in Eqn (8.67) and the error in force will not go to zero as simply as in a second-order system.

For a six-DOF manipulator with force control in *all* directions, the force f and the displacement x become 6×1 entities—we use the symbol $\mathcal{F}$ to denote the Cartesian force and moment exerted by the end-effector on the environment, and $\mathcal{X}$ denotes the position and orientation of the end-effector. The gain matrices $[K_{p_f}]$ and $[K_{v_f}]$ are 6×6 positive-definite diagonal matrices, and the 6×6 stiffness matrix $[K_e]$ is also positive definite. In addition, for the orientation part, m should be viewed as an inertia and x should be viewed as some representation of orientation.

In the force control of multi-DOF manipulators, there is a central issue of *duality* of force and velocity. The principle of duality states that both force and velocity (or position) cannot be controlled in the same direction. The principle of duality basically arises from the fact that force and velocity are related through the power.[15] Hence, in tasks wherein force needs to be controlled in a certain direction, the position or velocity cannot be controlled in that direction. For example, in the grinding example we can control the force normal to the surface to be ground and not the position. Likewise, we would control the position of the grinding wheel tangent to the surface to be ground and not the forces and moments in the tangential direction. A similar notion of partitioning the control torque into two parts—a joint torque to apply a force normal to a curve and another to traverse a desired trajectory lying on the curve, with each part not interfering with the other—was discussed in the example of a planar 2R manipulator in Section 8.6. In the following section, we look at the notion of partitioning a manipulator task into position and force control in terms of Cartesian variables.

8.8.2 Partitioning a Task for Force and Position Control

Any manipulator task can be broken into subtasks, depending on whether the manipulator is in contact or moving in free space. For subtasks, when the manipulator is not in contact, the manipulator is under position control. We look in more detail at subtasks when the manipulator is in contact with the environment and some of the degrees of freedom, in the Cartesian space, are constrained.

When the manipulator is in contact with the environment, the contact places certain *natural* constraints on the manipulator. For example, if the manipulator is in contact with a surface, the manipulator cannot go through the surface, and hence a *natural position* constraint exists. If the surface is frictionless,[16] the manipulator cannot apply any force tangent to the surface, and hence a *natural force* constraint exists. In general, for any two surfaces in contact, natural position constraints exist normal to the surfaces and natural force constraints exist tangent to the surfaces. In contrast to the natural position and force constraints, all position and force variables that can be controlled during motion are called *artificial* constraints. Again, if the manipulator is in contact with the environment, the position variables in the

---

[15] A more accurate description can be given using advanced kinematic concepts of screws, wrenches, and the principle of reciprocity. However, this is beyond the scope of this text. The reader is referred to papers by Mason (1981), Raibert & Craig (1981) and others listed at the end of this chapter for a more detailed treatment.

[16] Even if the surface is not frictionless, the manipulator cannot apply arbitrary force tangent to the surface since after a while the manipulator will slip.

tangent plane between the two contacting surfaces can be controlled and the force normal to the two contacting surfaces can be controlled. The natural and artificial constraints partition the degrees of freedom of a manipulator into two complementary sets. The fact that they are complementary follows from the principle of duality mentioned earlier.

Figure 8.32(a) shows a manipulator holding a grinding wheel which is grinding a surface. We attach a constraint frame $\{C\}$ to the manipulator hand. For this example, we assume that the ${}^C\widehat{\mathbf{Z}}$ of $\{C\}$ is parallel to the normal $\mathbf{n}$ at the point of contact of the grinding wheel and the surface, and the axes ${}^C\widehat{\mathbf{X}}$ and ${}^C\widehat{\mathbf{Y}}$ determine the tangent plane at the point of contact on the surface. For grinding, the manipulator must apply a desired force along the normal and traverse the surface along a desired trajectory. We describe the natural constraints relative to $\{C\}$ using the symbols $V_x, V_y, V_z, \omega_x, \omega_y, \omega_z$, f_x, f_y, f_z, and n_x, n_y, n_z to denote the components of the end-effector linear velocity, angular velocity, force, and moment exerted by the end-effector, respectively.

Clearly, one of the natural position constraints is that there cannot be any motion along the ${}^C\widehat{\mathbf{Z}}$ direction, i.e., V_z is one of the variables subjected to the natural constraint. We assume that the grinding wheel has some finite width and, in that case, cannot have a non-zero ω_x and ω_y since again the wheel will interfere with the surface. For the force variables, the components f_x and f_y are determined by the friction between the wheel and the surface and hence they also come under natural constraints. Finally, the moment component n_z is also determined by the friction and this is also a natural constraint. We want to grind the surface by following some desired trajectory. Hence, clearly the V_x and V_y velocity components (in the tangent plane) are artificial constraints. We also want to apply a desired f_z normal to the surface and hence this is also an artificial constraint. The fact that ω_z, n_x, and n_y form the rest of the artificial constraints follows from the duality principle. The variables subjected to natural and artificial constraints for the example of robotic grinding are listed in Fig. 8.32(a).

Figure 8.32(b) shows an example of a crank being turned by a robot. In this example, the manipulator cannot move in the ${}^C\widehat{\mathbf{X}}$ or ${}^C\widehat{\mathbf{Z}}$ direction. It cannot also rotate about the ${}^C\widehat{\mathbf{X}}$ and ${}^C\widehat{\mathbf{Y}}$ axes. In addition, the manipulator cannot apply any force along the ${}^C\widehat{\mathbf{Y}}$ axis or apply moment about the ${}^C\widehat{\mathbf{Z}}$ axis. In the crank turning example, the artificial constraints or the controlled variables corresponding to the position are the linear velocity along ${}^C\widehat{\mathbf{Y}}$ and the angular velocity along ${}^C\widehat{\mathbf{Z}}$. The controlled force components are those along ${}^C\widehat{\mathbf{X}}$ and ${}^C\widehat{\mathbf{Z}}$ and the moments along ${}^C\widehat{\mathbf{X}}$ and ${}^C\widehat{\mathbf{Y}}$. The variables subjected to natural and artificial constraints for this example are given in Fig. 8.32(b).

Grinding wheel

Contact point

Surface

$c\hat{Y}$

$c\hat{X}$

$\{C\}$

$c\hat{Z}$

n

Variables subjected to natural constraints
$$V_z, \omega_x, \omega_y$$
$$f_x, f_y, n_z$$

Variables subjected to artificial constraints
$$V_x, V_y, \omega_z$$
$$f_z, n_x, n_y$$

(a)

$\{C\}$ $c\hat{Z}$

$c\hat{X}$

$c\hat{Y}$

Variables subjected to natural constraints
$$V_x, V_z, \omega_x, \omega_y$$
$$f_y, n_z$$

Variables subjected to artificial constraints
$$V_y, \omega_z$$
$$f_x, f_z, n_x, n_y$$

(b)

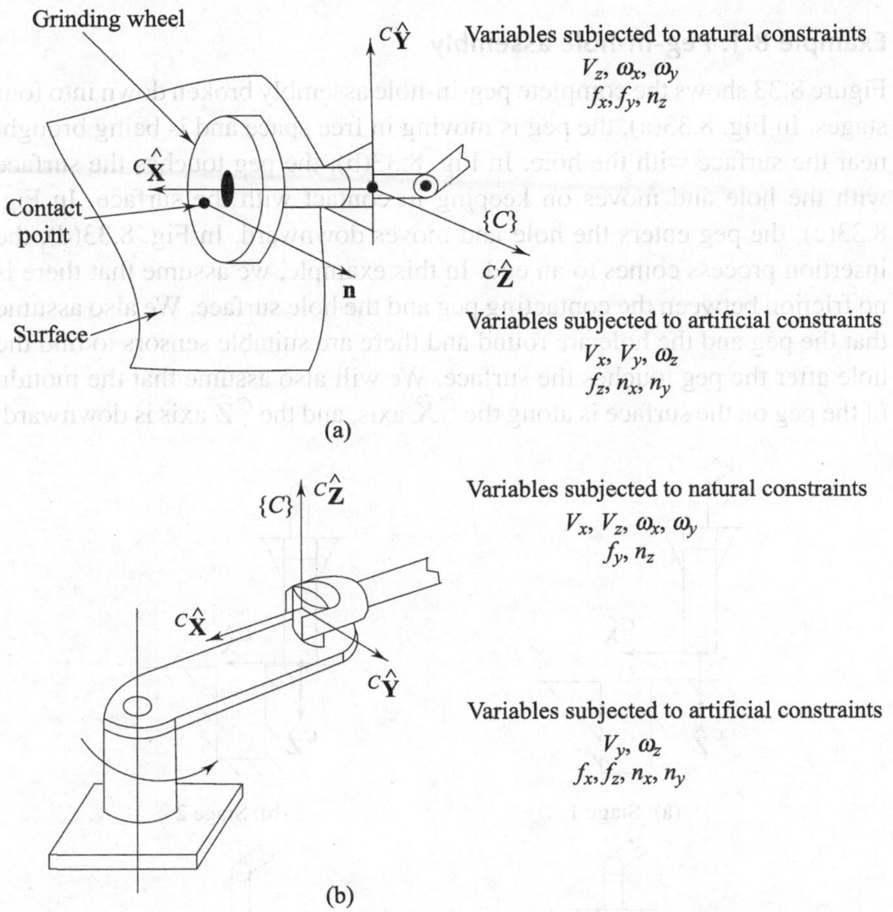

Fig. 8.32 Natural and artificial constraints for (a) grinding a surface and (b) turning a crank

We summarize the key concepts for partitioning of tasks as follows.

1. The natural and artificial constraints partition the degrees of freedom and constraints on the position and force are complementary.

2. The artificial constraints are complementary to the natural constraints. If a force is controlled along a direction, then the position or velocity in that direction is a natural constraint; likewise if the position or velocity is controlled in one direction, then the force is a natural constraint in that direction.

These concepts are a result of the *duality* of force and velocity mentioned earlier.

In the two examples given, the constraints remain the same throughout the operation. In an *assembly* operation, the natural and artificial constraints change during the complete task. This is illustrated next.

Example 8.1: Peg-in-hole assembly

Figure 8.33 shows the complete peg-in-hole assembly broken down into four stages. In Fig. 8.33(a), the peg is moving in free space and is being brought near the surface with the hole. In Fig. 8.33(b), the peg touches the surface with the hole and moves on keeping in contact with the surface. In Fig. 8.33(c), the peg enters the hole and moves downward. In Fig. 8.33(d), the insertion process comes to an end. In this example, we assume that there is no friction between the contacting peg and the hole surface. We also assume that the peg and the hole are round and there are suitable sensors to find the hole after the peg touches the surface. We will also assume that the motion of the peg on the surface is along the ${}^C\hat{\mathbf{X}}$ axis, and the ${}^C\hat{\mathbf{Z}}$ axis is downward.

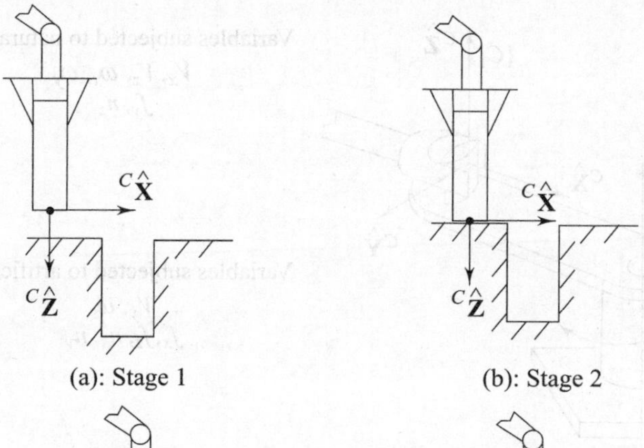

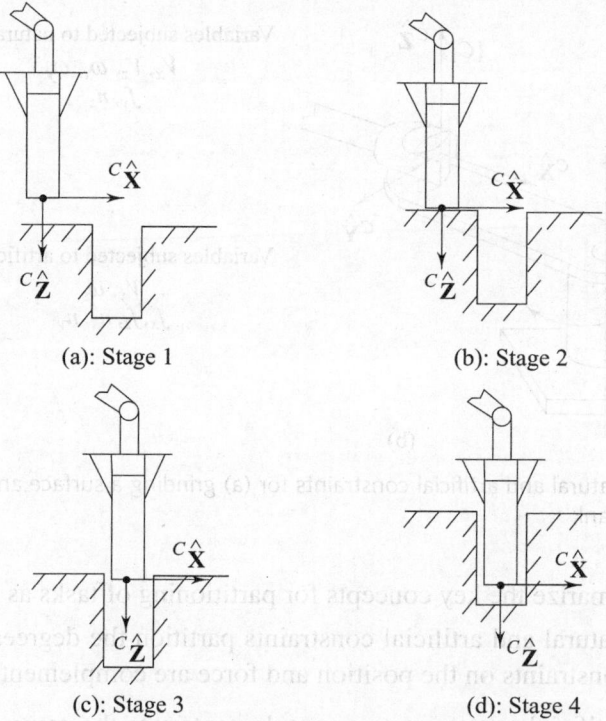

Fig. 8.33 Peg-in-hole assembly

Stage 1 The peg moves in free space. Hence the natural constraints are that all components of force/moment are zero.

$$ {}^C\mathcal{F} = [f_x,\ f_y,\ f_z;\ n_x,\ n_y,\ n_z]^T = 0 \tag{8.70} $$

Therefore, the artificial constraints are on the velocity (or position), and we have

$$^{C}\mathcal{V} = [0,\ 0,\ v_a;\ 0,\ 0,\ 0]^T \tag{8.71}$$

This implies that in the first stage, the manipulator is moving under position control and v_a is some constant approach velocity.

Stage 2 The peg touches the surface. To detect this, we monitor the force in the $^{C}\widehat{\mathbf{Z}}$ direction. When this crosses a threshold, we know that the peg is in contact with the surface. Once the peg is in contact with the surface, it cannot move in the $^{C}\widehat{\mathbf{Z}}$ direction or rotate about the $^{C}\widehat{\mathbf{X}}$ or $^{C}\widehat{\mathbf{Y}}$ axis. It cannot apply force along the direction of sliding. Hence, the natural constraints are

$$\begin{aligned} V_z &= 0, & \omega_x &= 0, & \omega_y &= 0 \\ f_x &= 0, & f_y &= 0, & n_z &= 0 \end{aligned}$$

The artificial constraints are the controlled variables. We will apply a small force along the $^{C}\widehat{\mathbf{Z}}$ axis to keep it in contact and control the velocity along the $^{C}\widehat{\mathbf{X}}$ axis. The artificial constraints are

$$\begin{aligned} V_x &= v_s, & V_y &= 0, & \omega_z &= 0 \\ f_z &= f_c, & n_x &= 0, & n_y &= 0 \end{aligned}$$

where v_s and f_c are the sliding velocity and the contact force, respectively.

Stage 3 After some motion along the $^{C}\widehat{\mathbf{X}}$ axis, the peg will fall into the hole. This can be detected by sensing the motion along the $^{C}\widehat{\mathbf{Z}}$ axis. Once the velocity along the $^{C}\widehat{\mathbf{Z}}$ axis crosses a threshold, we know that the natural constraints have changed. The new natural constraints are

$$\begin{aligned} V_x &= 0, & V_y &= 0, & \omega_x &= 0, & \omega_y &= 0 \\ f_z &= 0, & n_z &= 0 \end{aligned}$$

Since the peg has to be inserted into the hole, the artificial constraints are

$$\begin{aligned} V_z &= v_i, & \omega_z &= 0 \\ f_x &= 0, & f_y &= 0, & n_x &= 0, & n_y &= 0 \end{aligned}$$

where v_i is the insertion speed of the peg.

Stage 4 After a certain amount of motion, the peg has to come to a stop. At this stage the natural constraints are that no motion is possible and we get

$$^{C}\mathcal{V} = [V_x,\ V_y,\ V_z;\ \omega_x,\ \omega_y,\ \omega_z]^T = 0 \tag{8.72}$$

The artificial constraints are on the forces and moments and we get

$$^C\mathcal{F} = [f_x,\ f_y,\ f_z;\ n_x,\ n_y,\ n_z]^T = 0 \tag{8.73}$$

Again the change in the natural and artificial constraints is detected by monitoring the force along the $^C\widehat{\mathbf{Z}}$ direction. Once this crosses a threshold, we know that the peg has reached the end and we should stop.

It may be noted that the changes in the natural constraints are detected by monitoring the position or force variables which are *not* being controlled. For example, in stage 2, the force along the $^C\widehat{\mathbf{Z}}$ axis is being controlled (a minimum force is being applied to keep in contact) and the velocity along the $^C\widehat{\mathbf{Z}}$ axis is not being controlled. To detect the change from stage 2 to stage 3, we monitor the velocity (complement of the controlled force) along $^C\widehat{\mathbf{Z}}$ and when this changes we change to the control strategy of stage 3. To determine that the peg has reached the bottom, we monitor the force along the $^C\widehat{\mathbf{Z}}$ axis although we are controlling the velocity along this axis. Observations of this kind are used to switch between force and position control in the hybrid position/force control scheme discussed in the following section.

The peg-in-hole example is fairly simple. For any complex assembly task, the entire process will have to be subdivided into many more subtasks. At present, this is done manually by the robot programmer. The automatic generation of the assembly plan is still a topic of research.

8.9 Hybrid Position/Force Controller

In this section, we will describe a hybrid position/force controller similar to the concept in Craig (1989). We will assume that the artificial constraints are known for the tasks and that we have a six-axes force/torque sensor through which we can obtain feedback of the force and moments applied by the end-effector on the environment. We will use the Cartesian controller described in Section 8.7.

The top half of Fig. 8.34 shows a block diagram of a model-based Cartesian controller described in Section 8.7. For simplicity, the details of the error-driven and model-based portion are not shown. The output of the Cartesian position controller is $\mathcal{F}$ and it is multiplied by a 6×6 'switching' matrix $[S]$. The matrix $[S]$ is a diagonal matrix with 1 or 0 as the diagonal elements and zero off-diagonal terms. If the diagonal element is 1, then that particular component of position or orientation is to be controlled and the corresponding component of $\mathcal{F}$ is passed on to the summation block. If it is zero, then that component of position is filtered off and is not controlled.

The bottom half of the block diagram is the force control law described in Section 8.8.1 extended to a six-DOF motion. The output of the force control block is also $\mathcal{F}$ and is filtered with a matrix $[S']$. The matrix $[S']$ is also

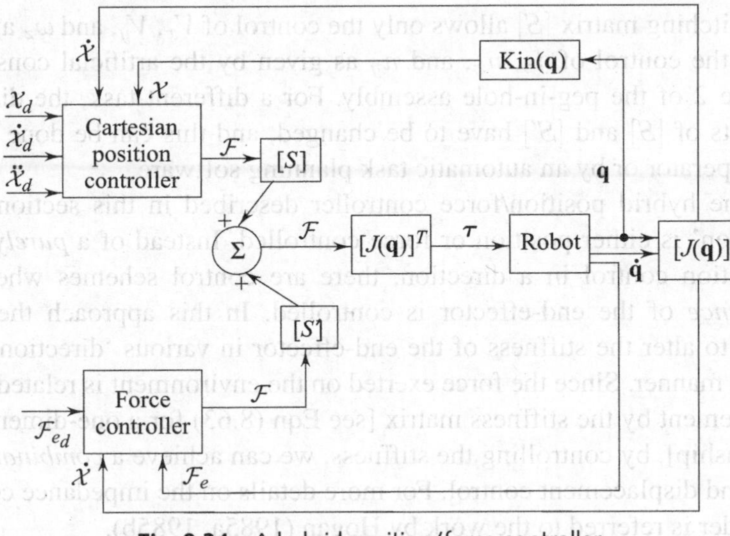

Fig. 8.34 A hybrid position/force controller

diagonal with the zero off-diagonal terms and it has a '1' where there is a '0' in $[S]$. This is due to the fact that the force- and position-controlled components are complementary to each other.

The outputs of both the Cartesian position controller and force controller are added up and multiplied by $[J(\mathbf{q})]^T$ to obtain the joint torques $\boldsymbol{\tau}$ which drive the robot. The feedback quantities $\mathcal{X}$ and $\dot{\mathcal{X}}$ are obtained from measured joint rotation $\mathbf{q}$ and the joint rate $\dot{\mathbf{q}}$ by using direct kinematics and the joint space Jacobian of the robot. The force and moment exerted on the environment $\mathcal{F}_e$ are assumed to be available from a six-axes force–torque sensor. The matrices $[S]$ and $[S']$ are directly obtained from the artificial constraints for the given task. To give an example, the $[S]$ and $[S']$ matrices for stage 2 of the peg-in-hole assembly task, described in the previous section, are given as

$$[S] = \begin{pmatrix} 1 & 0 & 0 & 0 & 0 & 0 \\ 0 & 1 & 0 & 0 & 0 & 0 \\ 0 & 0 & 0 & 0 & 0 & 0 \\ 0 & 0 & 0 & 0 & 0 & 0 \\ 0 & 0 & 0 & 0 & 0 & 0 \\ 0 & 0 & 0 & 0 & 0 & 1 \end{pmatrix}$$

$$[S'] = \begin{pmatrix} 0 & 0 & 0 & 0 & 0 & 0 \\ 0 & 0 & 0 & 0 & 0 & 0 \\ 0 & 0 & 1 & 0 & 0 & 0 \\ 0 & 0 & 0 & 1 & 0 & 0 \\ 0 & 0 & 0 & 0 & 1 & 0 \\ 0 & 0 & 0 & 0 & 0 & 0 \end{pmatrix}$$

The switching matrix $[S]$ allows only the control of V_x, V_y, and ω_z and $[S']$ allows the control of f_z, n_x, and n_y as given by the artificial constraints of stage 2 of the peg-in-hole assembly. For a different task, the diagonal elements of $[S]$ and $[S']$ have to be changed, and this can be done by the robot operator or by an automatic task planning software.

In the hybrid position/force controller described in this section, each 'direction' is either position or force controlled. Instead of a *purely* force or position control in a direction, there are control schemes where the *impedance* of the end-effector is controlled. In this approach the basic idea is to alter the stiffness of the end-effector in various 'directions' in a desired manner. Since the force exerted on the environment is related to the displacement by the stiffness matrix [see Eqn (8.63) for a one-dimensional relationship], by controlling the stiffness, we can achieve a *combination* of force and displacement control. For more details on the impedance control, the reader is referred to the work by Hogan (1985a, 1985b).

8.10 Stability Analysis of Non-linear Control Schemes

In Section 8.2, we mentioned about the stability of a single-link manipulator under a proportional control scheme. It was shown that $\Omega(t) \rightarrow \Omega_d(t)$ as $t \rightarrow \infty$ and hence the proportional controller was stable. The stability analysis of a linear system is a very mature and well-developed field [see, for example, textbooks by Ogata (1987) and Franklin et al. (1991)], and there exist several methods such as the root locus, Bode plots, and state-space approach to analyse and design stable controllers for linear systems. In contrast, the stability analysis of non-linear controllers is more complex and an open problem for research. In this section, we present a brief introduction to a method used for the stability analysis of non-linear systems and then apply the method to analyse the stability of non-linear control schemes discussed in Section 8.4.

8.10.1 Stability Analysis Using Lyapunov's Method

One of the important ways to analyse the stability of non-linear systems is by Lyapunov's second (or direct) method. This was first proposed by Lyapunov, a Russian mathematician, in the 19th century (Lyapunov 1892). This method is one of the most powerful and widely used methods to analyse the stability of non-linear systems [for a more recent detailed treatment of Lyapunov's method, the reader is referred to Khalil (1992) and Vidyasagar (1993)]. Lyapunov's method deals with the stability of non-linear systems described in the form

$$\dot{\mathbf{X}} = \mathbf{f}(\mathbf{X}, t) \tag{8.74}$$

where $\mathbf{X}$ is an n-dimensional state vector, $\mathbf{f}(\mathbf{X}, t)$ are n vector functions, and t denotes the time. It is also assumed that the set of ordinary differential equations, Eqn (8.74), has a unique solution starting at a given initial condition $\mathbf{X}_0$. In the context of robot manipulators, there is usually no explicit dependence on time t and the dynamic equations of motion, available as second-order ordinary differential equations [see Eqn (8.28)], can be easily converted to a state-space form as shown in Section 6.6. In the brief discussion here, we will drop the explicit dependence on time and consider non-linear systems of the form $\dot{\mathbf{X}} = \mathbf{f}(\mathbf{X})$.

The stability analysis of a non-linear system is performed at an equilibrium point or in an equilibrium state. The state $\mathbf{X}_e$ is called an *equilibrium state* when it satisfies

$$\mathbf{f}(\mathbf{X}) = \mathbf{0} \tag{8.75}$$

The determination of $\mathbf{X}_e$ does not involve the solution of differential equations; one needs to solve the n non-linear algebraic (or trigonometric in the case of robot manipulators) equations in Eqn (8.75). In a linear system of the form $\dot{\mathbf{X}} = [A]\mathbf{X}$, $[A]$ is a constant non-singular matrix and the only equilibrium point is $\mathbf{X} = 0$ (this follows from linear algebra). In contrast, in non-linear systems there can be several $\mathbf{X}_e$'s which satisfy Eqn (8.75). In that case, the stability has to be investigated at all the equilibrium states.

From the theory of vibration, we know that in a damped oscillatory system, such as single-DOF spring–mass–damper combination displaced from an equilibrium position, the amplitude of oscillation will slowly die down and it will end up in its equilibrium position. In such a system, the equilibrium point is known to be stable, and one can show that the total energy of the system is always *greater than* or *equal to zero* and *continuously decreasing* with time. Lyapunov's second method is a generalization of this concept. For a general non-linear system, as defined in Eqn (8.74), it may not be possible to define an 'energy function' and Lyapunov introduced the notion of a positive-definite function[17] which decreases continuously with time. A non-linear system $\dot{\mathbf{X}} = \mathbf{f}(\mathbf{X})$ is said to be asymptotically stable (in the sense of Lyapunov) if there exists a *positive-definite*, differentiable, scalar function of the state variables $V(\mathbf{X})$, with $\dot{V}(\mathbf{X})$ being *negative definite*.

The proof of the above theorem is available in Khalil (1992) and Vidyasagar (1993). We present a few key remarks related to this theorem.

[17] A function $f(\mathbf{x})$ is said to be positive- definite if $f(\mathbf{x}) > 0$ for all $\mathbf{x} \neq \mathbf{0}$ and is zero only when $\mathbf{x} = \mathbf{0}$. The function is said to be positive semidefinite if $f(\mathbf{x}) \geq 0$ (note the equality sign) and is said to be negative definite if $f(\mathbf{x}) < 0$. A simple example of positive-definite $f(\mathbf{x})$ is $x_1^2 + x_2^2$ while $(x_1 - x_2)^2$ is semidefinite and $-(x_1^2 + x_2^2)$ is negative definite. For more details on positive-definite functions, the reader is referred to any textbook on linear algebra.

1. The theorem presents a *sufficient condition* for stability, and it is one of the few of its kind (many theorems involve *necessary* conditions and a few are in the *necessary* and *sufficient* form). One implication of the sufficient condition is that if we find a *single* $V(\mathbf{X}) > 0$ such that $\dot{V}(\mathbf{X}) < 0$, it is enough to conclude the asymptotic stability for the non-linear system. Whereas for a given choice of $V(\mathbf{X}) > 0$, if $\dot{V}(\mathbf{X})$ is *not* less than zero, we *cannot* automatically conclude that the system is *not* stable (or unstable). All it means is that our choice of $V(\mathbf{X})$ was not proper.
2. If $V(\mathbf{X}) > 0$ and $\dot{V}(\mathbf{X})$ is negative semidefinite, i.e., $\dot{V}(\mathbf{X}) \leq 0$, then as shown by LaSalle and Lefschetz (1961), we can still conclude asymptotic stability under certain conditions.
3. An equilibrium point $\mathbf{X}_e$ is said to be asymptotically stable if any trajectory starting in a region around the point converges to $\mathbf{X}_e$ as $t \to \infty$ [see Khalil (1992) or Vidyasagar (1993) for a more formal definition]. Hence, it is a local concept and, as such, it is important to obtain the largest region of the asymptotic stability or the domain of attraction.
4. Lyapunov's second method is equally applicable for linear systems. It can be also used for the stability analysis of *non-autonomous* non-linear systems with t appearing explicitly as in Eqn (8.74) [see Khalil (1992) and Vidyasagar (1993)].

The main difficulty in applying Lyapunov's second method to non-linear systems is finding an appropriate Lyapunov function, $V(\mathbf{X})$, since there are no standard procedures. In many applications, it is easy to guess, but obtaining $V(\mathbf{X})$ may be difficult for complicated systems. To describe the method, we consider the example of a single-link manipulator (or a pendulum).

Example 8.2: Stability analysis of a single-link manipulator

Consider a single-link manipulator as shown in Fig. 8.35. We assume, without loss of generality, that all the mass is concentrated at the end of a mass-less link of length l_1, gravity is acting downward, and a motor can apply torque as shown. The equation of motion is given by

$$\ddot{\theta}_1 + (g/l_1)\sin\theta_1 = u(t) \tag{8.76}$$

where $u(t) = \tau_1(t)/(m_1 l_1^2)$ and θ_1 is as shown in Fig. 8.35. Using the state variables (X_1, X_2) defined as $(\theta_1, \dot{\theta}_1)$, we obtain

$$\dot{X}_1 = X_2 \tag{8.77}$$
$$\dot{X}_2 = u(t) - (g/l_1)\sin(X_1)$$

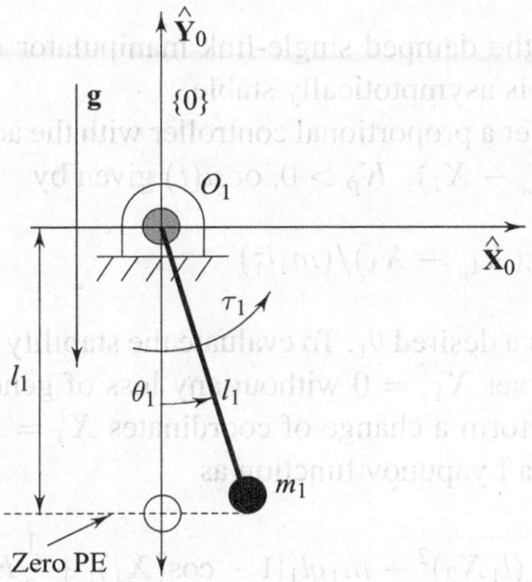

Fig. 8.35 A single-link manipulator

The equilibrium points for $u(t) = 0$ are $\theta_1 = 0$ and $\theta_1 = \pi$ (note θ_1 is defined with respect to negative $\widehat{\mathbf{Y}}_0$). To evaluate the stability of $\theta_1 = 0$, we consider a candidate Lyapunov function

$$V(X_1, X_2) = \frac{1}{2}m_1(l_1 X_2)^2 + m_1 g l_1[1 - \cos(X_1)] \tag{8.78}$$

The function $V(X_1, X_2)$ represents the total energy. Hence it is always greater than zero and is zero when $X_1 = X_2 = 0$—note that the potential energy is based on taking the zero potential energy at $y = -l_1$. The time derivative $\dot{V}(X_1, X_2)$, evaluated from Eqn (8.78), is given by

$$\dot{V}(X_1, X_2) = m_1 l_1^2 X_2 \dot{X}_2 + m_1 g l_1 \sin(X_1)\dot{X}_1 = 0 \tag{8.79}$$

where we have used Eqn (8.77) with $u(t) = 0$ to get the last equality.

Since $\dot{V}(X_1, X_2)$ is not *less* than zero, we cannot conclude asymptotic stability. Now we consider the manipulator with damping at the joint. We assume a viscous damping given by cX_2 ($c > 0$) at the joint. Then the state equations, with $u(t) = 0$, can be written as

$$\dot{X}_1 = X_2 \tag{8.80}$$
$$\dot{X}_2 = -(g/l_1)\sin(X_1) - cX_2, \quad c > 0$$

and we get

$$\dot{V}(X_1, X_2) = -m_1 l_1^2 c X_2^2 < 0 \tag{8.81}$$

which shows that the damped single-link manipulator with $u(t) = 0$ (or a simple pendulum) is asymptotically stable.

Next, we consider a proportional controller with the actuator output given by $\tau_1(t) = K_p(X_{1_d} - X_1)$, $K_p > 0$, or $u(t)$ given by

$$u(t) = K_p(X_{1_d} - X_1)/(m_1 l_1^2) \tag{8.82}$$

where X_{1_d} denotes a desired θ_1. To evaluate the stability of this proportional controller, we can set $X_{1_d} = 0$ without any loss of generality (if $X_{1_d} \neq 0$, we can always perform a change of coordinates $X_1' = X_{1_d} - X_1$). In such a case, we choose a Lyapunov function as

$$V(X_1, X_2) = \frac{1}{2} m_1 (l_1 X_2)^2 + m_1 g l_1 [1 - \cos(X_1)] + \frac{1}{2} K_p X_1^2 \tag{8.83}$$

and clearly $V(X_1, X_2)$ is positive definite. For the undamped state equations, Eqn (8.77), we obtain

$$\dot{V}(X_1, X_2) = m_1 l_1^2 X_2 u(t) + K_p X_1 X_2 \tag{8.84}$$

Substituting $u(t) = -K_p X_1/(m_1 l_1^2)$, we get $\dot{V}(X_1, X_2) = 0$ and we cannot conclude asymptotic stability. Again, for the damped state equations, Eqn (8.80), we obtain $\dot{V}(X_1, X_2) < 0$ and we can conclude asymptotic stability.

Finally, we consider a PD control scheme where the actuator output is $\tau_1(t) = -K_p X_1 - K_v \dot{X}_1$, $K_p, K_v > 0$. For the Lyapunov function given in Eqn (8.83) and the undamped state equations, Eqn (8.77), we get

$$\dot{V}(X_1, X_2) = -K_v X_2^2 \tag{8.85}$$

and since $K_v > 0$, we get asymptotic stability.

It may be noted that the control scheme is different from the PD control described in Section 8.2 for a single-link manipulator and in Section 8.3 for a multi-link manipulator. The main difference is that we have assumed that the desired joint velocity, $\dot{X}_{1_d}$, is zero (and the desired joint acceleration is not taken into account) and hence the stability for *trajectory following* is not proven. In the following section, we discuss the stability of PD- and model-based control schemes in multi-link manipulators in detail.

8.10.2 Stability Analysis of PD- and Model-based Control

The equation of motion of an n-DOF, multi-link manipulator, *without* the gravity term, can be written as [see Eqn (6.14)]

$$\boldsymbol{\tau} = [\mathbf{M}(\mathbf{q})]\ddot{\mathbf{q}} + [\mathbf{C}(\mathbf{q}, \dot{\mathbf{q}})]\dot{\mathbf{q}} \tag{8.86}$$

where $[\mathbf{C}(\mathbf{q}, \dot{\mathbf{q}})]$ is an $n \times n$ matrix and $[\mathbf{C}(\mathbf{q}, \dot{\mathbf{q}})]\dot{\mathbf{q}}$ represents the Coriolis/centripetal terms. To investigate the stability of the multi-link manipulator under feedback control, we start with a modified PD control scheme given by

$$\boldsymbol{\tau} = [K_p]\mathbf{e}(t) - [K_v]\dot{\mathbf{q}}(t) \tag{8.87}$$

where $[K_p]$ and $[K_v]$ are positive-definite matrices and $\mathbf{e} = \mathbf{q}_d - \mathbf{q}$. Note that we have set $\dot{\mathbf{q}}_d(t) = \mathbf{0}$. Without loss of generality, we assume $\mathbf{q}_d = \mathbf{0}$ (we can always perform a coordinate transformation $\mathbf{q}' = \mathbf{q}_d - \mathbf{q}$ and investigate the stability at $\mathbf{q}' = 0$) and get

$$\boldsymbol{\tau} = -[K_p]\mathbf{q}(t) - [K_v]\dot{\mathbf{q}}(\mathbf{t}) \tag{8.88}$$

Consider the Lyapunov function

$$V(\mathbf{q}, \dot{\mathbf{q}}) = \frac{\mathbf{1}}{\mathbf{2}}\dot{\mathbf{q}}^T[\mathbf{M}(\mathbf{q})]\dot{\mathbf{q}} + \frac{\mathbf{1}}{\mathbf{2}}\mathbf{q}^T[K_p]\mathbf{q} \tag{8.89}$$

It is clear that $V(\mathbf{q}, \dot{\mathbf{q}})$ is positive-definite. Evaluating $\dot{V}(\mathbf{q}, \dot{\mathbf{q}})$, we get

$$\begin{aligned} \dot{V}(\mathbf{q}, \dot{\mathbf{q}}) &= \dot{\mathbf{q}}^T[\mathbf{M}(\mathbf{q})]\ddot{\mathbf{q}} + \frac{\mathbf{1}}{\mathbf{2}}\dot{\mathbf{q}}^T[\dot{\mathbf{M}}(\mathbf{q})]\dot{\mathbf{q}} + \dot{\mathbf{q}}^T[K_p]\mathbf{q} \\ &= -\dot{\mathbf{q}}^T[K_v]\dot{\mathbf{q}} + \frac{1}{2}\dot{\mathbf{q}}^T\{[\dot{\mathbf{M}}(\mathbf{q})] - \mathbf{2}[\mathbf{C}(\mathbf{q}, \dot{\mathbf{q}})]\}\dot{\mathbf{q}} \end{aligned} \tag{8.90}$$

where $[\dot{\mathbf{M}}]$ denotes the derivative of the mass matrix $[\mathbf{M}]$ with respect to time. Since $([\dot{\mathbf{M}}(\mathbf{q})] - \mathbf{2}[\mathbf{C}(\mathbf{q}, \dot{\mathbf{q}})])$ is skew-symmetric (see Exercise 8.12), the second quadratic form is zero and we get

$$\dot{V}(\mathbf{q}, \dot{\mathbf{q}}) = -\dot{\mathbf{q}}^T[K_v]\dot{\mathbf{q}} \tag{8.91}$$

Since $\dot{V}(\mathbf{q}, \dot{\mathbf{q}})$ can be zero even for non-zero $\mathbf{q}$, $\dot{V}(\mathbf{q}, \dot{\mathbf{q}})$ is negative semidefinite and we cannot conclude asymptotic stability. Asymptotic stability can be shown by LaSalle's invariance principle (LaSalle & Lefschetz 1961), by which we can show that the largest invariant set is the single point $\mathbf{q} = \mathbf{0}$ and hence the equilibrium point $(\mathbf{q}, \dot{\mathbf{q}}) = \mathbf{0}$ is asymptotically stable. It may be noted that we have assumed $\ddot{\mathbf{q}}_d = \dot{\mathbf{q}}_d = \mathbf{0}$ and hence the PD control scheme is *not* proven asymptotically stable for trajectory following.

If the gravity term is included in the equations of motion, as in Eqn (6.14), we can use a model-based control scheme with gravity compensation (see Section 8.4.1). We use

$$\boldsymbol{\tau} = -[K_p]\mathbf{q}(t) - [K_v]\dot{\mathbf{q}}(\mathbf{t}) + \mathbf{G}(\mathbf{q}) \tag{8.92}$$

and using the Lyapunov function given in Eqn (8.89), we can again show that the equilibrium point $(\mathbf{q}, \dot{\mathbf{q}}) = \mathbf{0}$ is asymptotically stable. It may be noted that the demonstration of asymptotic stability depends on 'cancellation' of the gravity term.

In the *most ideal* situation of exact knowledge of the model and for the computed torque scheme for an n-DOF manipulator, as shown in Eqn (8.32), we can get n decoupled linear error equations of the form

$$\ddot{e}_i + K_{v_i}\dot{e}_i + K_{p_i}e = 0, \qquad i = 1, \ldots, n \tag{8.93}$$

where $e_i(t) = q_{d_i}(t) - q_i(t)$ and $i = 1, \ldots, n$. One can easily show, by an appropriate choice of the Lyapunov function (see Exercise 8.13) that for positive K_{p_i} and K_{v_i}, the state $(e_i, \dot{e}_i)$, (where $i = 1, \ldots, n$) is asymptotically stable. Since $(\mathbf{e}, \dot{\mathbf{e}}) \to \mathbf{0}$ as $t \to \infty$, $[\mathbf{q}(t), \dot{\mathbf{q}}(t)]$ approaches $[\mathbf{q}_d(t), \dot{\mathbf{q}}_d(t)]$ asymptotically and we can achieve *trajectory following* control with a computed torque controller with *exact* cancellation. As mentioned earlier, this is an 'ideal' situation and exact cancellation is not achievable in practice. When only the estimates of the model are available and exact cancellation does not happen, the closed-loop error equation is much more complicated. The stability analysis of model-based controllers with model estimates is an open research problem. This and a few other advanced topics in manipulator control are discussed, in brief, in the following section.

8.11* Advanced Topics in Non-linear Control of Manipulators

We end the chapter with a brief discussion on some of the advanced and active research topics in feedback control of serial and parallel manipulators. The list is by no means comprehensive, and the interested reader is referred to the current literature on robotics and control.

One important issue in manipulator control is the lack of the knowledge of model parameters. In case the model is not known exactly, there is no cancellation and the right-hand side of the error equation is non-zero (see Section 8.4.2). As shown in Eqn (8.36), the right-hand side is fairly complicated and it is not possible to show the asymptotic stability for this ordinary differential equation. The right-hand side is small if the error between the actual model and the estimates is small, and several simulation and experimental results show asymptotic stability. However, recent research work [see Shrinivas and Ghosal (1997), Ravishankar and Ghosal (1999)] shows that one can get chaotic behaviour[18] in certain robots performing repetitive motions under PD- and model-based control schemes, and asymptotic stability is seen (in numerical simulations) only when K_p and K_v are 'large'. They present numerical simulation results for planar two-DOF manipulators with rotary (R) and prismatic (P) joints, and present regions in the K_p and K_v space where the feedback-controlled (both PD and model based with estimates) manipulators exhibit chaotic motion with the desired manipulator trajectory *not* being tracked although remaining bounded. These results are not *inconsistent* with the asymptotic stability results shown in the preceding section since asymptotic stability was not shown for trajectory following (recollect that we set $\ddot{\mathbf{q}}_d = \dot{\mathbf{q}}_d = \mathbf{0}$) and indicate that there may be a lower bound on the controller gains for asymptotic stability in trajectory following. A reader interested in more details on possible chaotic behaviour in feedback-controlled robots is referred to papers by Shrinivas and Ghosal (1997), Ravishankar and Ghosal (1999), and the references contained therein.

One way of getting better estimates of model parameters (and better performance from a model-based controller) is by using *adaptive* control. The basic idea in adaptive control of robots is to use $\mathbf{e}(t)$ and $\dot{\mathbf{e}}(t)$, available in the error-driven portion of the model-based controller (see Fig. 8.10), to change the model parameters in a way so as to drive $\mathbf{e}(t)$ and $\dot{\mathbf{e}}(t)$ to zero. The adaptive schemes are usually very computation intensive and the model parameters need to be changed till the errors are acceptable. Once the updated model parameters are available, the adaptation scheme is switched off and a model-based controller, now with better estimates, takes over. For the adaptation scheme to work efficiently, appropriate trial trajectories are chosen. An analysis of adaptive controllers for manipulators is outside the scope of this book—the interested reader is referred to Craig (1988) and the review article by Ortega and Spong (1989).

[18] Chaotic behaviour is seen in non-linear mathematical and deterministic physical systems whose time history exhibits a sensitive dependence on the initial conditions. It is exhibited by non-linear differential equations for certain values of parameters and initial conditions, and time evolution of such non-linear differential equations, obtained by the numerical integration on finite precision computers, cannot always be predicted far into the future. A classic example of a chaotic system is Duffing's equation, which is a non-linear oscillator with a cubic stiffness term. A passive double pendulum or a planar 2R manipulator moving under gravity is another example of the physical system that can exhibit chaos. A review paper (Sekar & Narayanan 1992) lists many mechanical systems which can exhibit chaos. A good starting textbook on chaos is by Thompson and Stewart (1986) and a more advanced textbook is by Guckenheimer and Holmes (1983).

An important aspect in the control of a system is the mathematical notion of controllability. A system is said to be *controllable* if it is possible to transfer the system from any initial state $\mathbf{X}(0)$ to any desired state $\mathbf{X}(t_f)$ in finite time t_f by the application of the control input $\mathbf{u}(t)$. In a linear system with n state variables and m inputs and described by the n linear ordinary differential equations

$$\dot{\mathbf{X}} = [F]\mathbf{X} + [G]\mathbf{u} \tag{8.94}$$

controllability can be obtained from the rank of a controllability matrix

$$[Q_c] = [\,[G]|[F][G]|[F]^2G|\cdots|[F]^{n-1}[G]\,] \tag{8.95}$$

The linear system given by Eqn (8.94) is controllable if the $n \times (nm)$ matrix $[Q_c]$ has rank n (Franklin et al. 1991) and this concept (together with the concept of observability) was developed by Kalman and others in 1960s. In a non-linear system, the analog is the notion of small-time local controllability (STLC) and small-time local accessibility (STLA), and, instead of simple matrix algebra, the determination of controllability involves the use of the advanced techniques from Lie algebra. These concepts are outside the scope of this textbook and the reader is referred to the textbooks by Isidori (1995), Nijmeijer and van der Schaft (1990), or the papers by Sussman and Jurdjevic (1972) and Hermann and Krenner (1977), and the references contained in them. In the context of non-linear serial and parallel manipulators, the controllability issue must be limited to points inside the workspace—clearly a state that is outside the workspace cannot be reached by applying actuator torques. It is also clear that if the Jacobian matrix is singular at a point, a serial manipulator losing one or more degrees of freedom cannot be controlled. Parallel manipulators, in addition to points where they lose one or more degrees of freedom, can also gain one or more degrees of freedom (see Chapter 5, Section 5.6). In such a situation, the Lagrange multipliers in the equations of motion [see Section 6.3, Eqn (6.24)] of a parallel manipulator can go to infinity, and the parallel manipulator cannot be controlled. The relationship between kinematic singularities and STLC (and STLA) for parallel manipulators and closed-loop mechanisms are discussed in more detail in Chowdhury and Ghosal (2000).

Exercises

†8.1 For the single-link manipulator discussed in Section 8.2, choose $J = K = F = 1$, $K_p = 1$, and $\theta_d(t)$ as a step input. Vary K_v between 1 and 3 and numerically obtain the plots of $\theta(t)$. What happens at $K_v = 2$?

Note: This problem can be done by numerically solving the ordinary differential equation or by using tool boxes in MATLAB (see Control Tool Box or Simulink).

†8.2 For the single-link manipulator in Section 8.2, choose $J = K = F = 1, K_p = 1$, and $K_v = 2$. Assume a step input for θ_d and add a constant disturbance $T_d = 0.1$. What is the plot of $\theta(t)$? Next consider a PID controller with everything else remaining the same. Vary the integral gain K_i and plot $\theta(t)$ for different K_i's. For what value of K_i is the system unstable?
Note: As in Exercise 8.1, one can use MATLAB and its tool boxes for numerical simulation.

8.3 Consider the non-linear dynamical system

$$\ddot{x} + 7\dot{x}^2 + x\dot{x} + x^3 = u(t)$$

where $u(t)$ is the control input. Design a control system using the concepts given in Section 8.4 such that the error response is critically damped and the natural frequency ω_n is 1 rad/sec. Draw a block diagram of the system.

8.4 A researcher has proposed the following model-based control scheme for a serial manipulator:

$$\boldsymbol{\tau} = [\mathbf{M}(\mathbf{q})]\ddot{\mathbf{q}}_d + \mathbf{C}(\mathbf{q}, \dot{\mathbf{q}}) + \mathbf{G}(\mathbf{q})$$
$$+[K_p](\mathbf{q}_d - \mathbf{q}) + [K_v](\dot{\mathbf{q}}_d - \dot{\mathbf{q}})$$

where $[K_p]$ and $[K_v]$ are positive-definite gain matrices and the other symbols have the same meaning as in Section 8.4. Draw a block diagram of the proposed controller along the lines of Fig. 8.10. (a) What is the error equation? (b) What is the possible advantage of this scheme? (c) What are the possible disadvantages?

†8.5 Choose a circular trajectory for the planar 2R manipulator used in Section 8.5 as discussed in Exercise 7.9. Using the numerical data given in Section 8.5 for the manipulator, simulate its motion for a PD- and model-based (with estimates) controller.
Hint: Use the symbolic equations of motion derived for a planar 2R manipulator in Chapter 6.

8.6 The tip of a planar 2R manipulator is to move along a slot as shown in Fig. 8.36. Following the developments in Section 8.6 determine: (a) the symbolic expressions for the Jacobian $[J_{\mathrm{h}}]$ and (b) the symbolic expressions for τ_{n} and τ_ϕ.

†8.7 A planar 2R manipulator is to trace an arc of a circle whose parametric equation is given by $x = l_0 + l_3 \cos\phi$ and $y = l_3 \sin\phi$, where l_0 and l_3 are constants, and $\pi/2 \leq \phi \leq \pi$. Following the developments in Section 8.6, determine the following: (a) The symbolic expressions of the terms in the equation of motion as a function of ϕ [as in Eqns (8.47) and (8.48)]. (b) For an arbitrarily chosen $\boldsymbol{\tau}$, verify numerically that the tip actually traces a circle. The numerical data pertaining to the planar 2R manipulator given in Section 8.5 can be used for

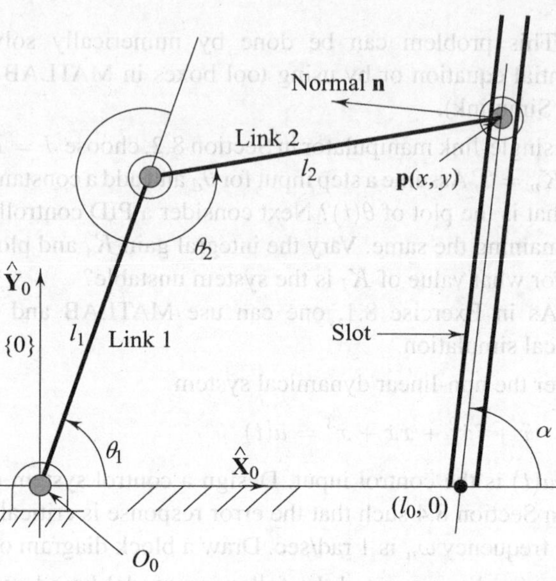

Fig. 8.36 A planar 2R manipulator

simulations. Choose an appropriate l_0 and l_3 such that the arc of the circle can be traced by the tip of the 2R manipulator

Hint: See Example 4.4 (in Chapter 4) for conditions on link lengths of a planar four-bar mechanism.

8.8 For the four-bar mechanism simulated in Chapter 6, (Example 6.5), we wish to rotate θ_1 from 0° to 150° in 15 sec with $\dot{\theta}_1$ being zero at $t = 0$ sec and $t =$ 15 sec. Instead of the spring, assume that the actuation is by a DC permanent-magnet motor and all other mass and geometrical data remain the same as in Example 6.5. Design a model-based controller such that the motion is critically damped along the trajectory. From the simulation, what is the maximum torque during the motion? If a simple PD controller is used with the same gains as in the model-based controller, what is the maximum torque during the simulation?

†8.9 For the planar 2R manipulator discussed in Section 8.5, obtain the symbolic expressions for the Cartesian mass matrix, the Cartesian Coriolis/centripetal term, and the Cartesian gravity term.

8.10 Figure 8.37 shows a manipulator tightening a screw. What are the natural and artificial constraints?

8.11 Show that $[\dot{\mathbf{M}}(\mathbf{q})] - \mathbf{2}[\mathbf{C}(\mathbf{q}, \dot{\mathbf{q}})]$ is skew-symmetric by considering Eqn (6.16) in Chapter 6.

8.12 Consider the linear ordinary differential equation $\ddot{e} + K_v\dot{e} + K_p e = 0$. Define state variables as $X_1 = e$ and $X_2 = \dot{e}$. By choosing an appropriate Lyapunov

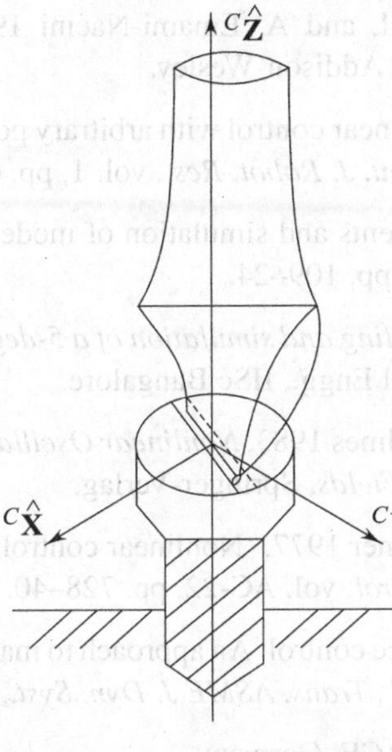

Fig. 8.37 A manipulator tightening a screw

function, show that the system is asymptotically stable for positive K_v and K_p.

8.13 Show that a multi-link manipulator modelled by Eqn (6.14) is asymptotically stable at **0** under the gravity compensated control scheme, Eqn (8.92).

8.14 Show that the control scheme $\boldsymbol{\tau} = -[\widehat{\mathbf{M}(\mathbf{q})}][\,[K_p]\mathbf{q} + [K_v]\dot{\mathbf{q}}] + \widehat{\mathbf{G}(\mathbf{q})}$ for a multi-DOF manipulator gives asymptotic stability. The matrix $[\widehat{\mathbf{M}(\mathbf{q})}]$ is positive-definite and is an estimate of the mass matrix for the multi-DOF manipulator.

References and Suggested Additional Reading

Canudas de Wit, C., B. Siciliano, and G. Bastin (eds) 1996, *Theory of Robot Control*, Springer Verlag.

Chowdhury, P. and A. Ghosal 2000, 'Singularity and controllability of closed loop mechanisms and parallel manipulators', *Mech. Mach. Theory*, vol. 35, pp. 1455–79.

Craig, J.J. 1988, *Adaptive Control of Mechanical Manipulators*, Addison-Wesley.

Craig, J.J. 1989, *Introduction to Robotics: Mechanics and Control*, 2nd edn, Addison-Wesley.

Franklin, G.F., J.D. Powell, and M.L. Workman 1990, *Digital Control of Dynamic Systems*, 2nd edn, Addison-Wesley.

Franklin, G.F., J.D. Powell, and A. Emami-Naeini 1991, *Feedback Control of Dynamic Systems*, 2nd edn, Addison-Wesley.

Freund, E. 1982, 'Fast nonlinear control with arbitrary pole placement for industrial robots and manipulators', *Int. J. Robot. Res.*, vol. 1, pp. 65–78.

Ghosal, A. 1996, 'Experiments and simulation of model based controlled robots', *J. Indian Inst. Sci.*, vol. 76, pp. 109–24.

Gopal, T.K.V., 1995, *Modelling and simulation of a 5-degree-of-freedom robot*, ME Thesis, Dept. of Mechanical Engg., IISc Bangalore.

Guckenheimer, J. and P. Holmes 1983, *Nonlinear Oscillations, Dynamical Systems, and Bifurcations of Vector Fields*, Springer-Verlag.

Hermann, R. and A.J. Krenner 1977, 'Nonlinear controllability and observability', *IEEE Trans. Automat. Control*, vol. AC-22, pp. 728–40.

Hogan, N., 1985, 'Impedance control: An approach to manipulation: Part I—Theory & Part II—Implementation', *Trans. ASME J. Dyn. Syst.*, vol. 107, pp. 1–16.

http://www.dspace.de, dSPACE, Germany.

Isidori, A. 1995, *Nonlinear Control Systems*, Springer-Verlag.

Khalil, H.K., 1992, *Nonlinear Systems*, Macmillan.

Khatib, O. 1986, 'Real-time obstacle avoidance for manipulators and mobile robots', *Int. J. Robot. Res.*, vol. 5, pp. 90–98.

Khatib, O. 1987, 'A unified approach for motion and force control of robot manipulators: The operational space formulation', *IEEE Trans. J. Robotic Autom.*, vol. 3, pp. 43–53.

LaSalle, J. and S. Lefschetz 1961, *Stability by Liapunov's Direct Method with Applications*, Academic Press.

Lyapunov, A.M., 1892, *On the general problem of motion stability* (in Russian), Kharkov Mathematical Society.

Mason, M.T., 1981, 'Compliance and force control for computer controlled manipulators', *IEEE Trans. J. Syst. Man Cyb.*, vol. 6, pp. 418–32.

Murray, R.M., Z. Li, and S.S. Sastry 1994, *A Mathematical Introduction to Robotic Manipulation*, CRC Press.

Nijmeijer, H. and J.A. van der Schaft 1990, *Nonlinear Dynamical Control System*, Springer-Verlag.

Ogata, K., 1987, *Modern Control Engineering*, Prentice-Hall of India.

Ortega, R. and M.W. Spong 1989, 'Adaptive motion control of rigid robots: A tutorial', *Automatica*, vol. 25, pp. 877–88.

Raibert, M. and J.J. Craig 1981, 'Hybrid position/force control of manipulators', *Trans. ASME J. Dyn. Syst.*, vol. 103, pp. 126–33.

Ravikiran, J. 1995, *Implementation of model based control on a 5-degree-of-freedom robot*, ME Thesis, Dept. of Mechanical Engg., IISc Bangalore.

Ravishankar, A.S. and A. Ghosal 1999, 'Nonlinear dynamics and chaotic motions in feedback controlled robots', *Int. J. Robot. Res.*, vol. 18, pp. 93–108.

Salisbury, J.K., 1980, 'Active stiffness control of a manipulator in Cartesian coordinates', *19th IEEE Conf. on Decision and Control*, pp. 95–100.

Sekar, P. and S. Narayanan 1992, 'Chaos in mechanical systems—A review', *Sadhana*, vol. 20, pp. 529–82.

Shrinivas, L. and A. Ghosal 1997, 'Chaos in robot control equations', *Int. J. Bifurcat. Chaos*, vol. 7, pp. 707–20.

Singh, S.N., and W.J. Rugh 1972, 'Decoupling in a class of nonlinear systems by state variable feedback', *Trans. ASME J. Dyn. Syst.*, vol. 94, pp. 323–29.

Sussman, H.J., and V. Jurdjevic 1972, 'Controllability of nonlinear systems', *J. Differ. Equations*, vol. 12, pp. 95–116.

Thompson, J.M.T., and H.B. Stewart 1986, *Nonlinear Dynamics and Chaos: Geometrical Methods for Engineers and Scientists*, John Wiley.

Vidyasagar, M. 1993, *Nonlinear Systems Analysis*, 2nd edn, Prentice-Hall.

Whitney, D.E., 1987, 'Historical perspective and state of the art in robot force control', *Int. J. Robot. Res.*, vol. 6, pp. 3–14.

9

Modelling and Control of Flexible Manipulators*

9.1 Introduction

The evolution of the modern industrial robot from computer-controlled machine tools and its intended use in manufacturing automation led to a rugged structural design. The demands of high accuracy and repeatability resulted in designs with large masses and high stiffness, which in turn made industrial robots slow, heavy, and with fairly low payload-to-robot weight ratios. As a typical example, a PUMA 761/762 series industrial robot weighed about 600 kg; it could carry a payload of only 10 kg at its rated speed, repeatability, and accuracy. Due to this fact most industrial robots can be accurately modelled and analysed as consisting of *rigid* links and joints, and, till now in this textbook, we have discussed modelling and analysis of rigid robots. In recent years, applications of robots in space have led to the design of lightweight manipulator arms—a classic example is the space shuttle remote manipulator system, which is so flexible that it can be operated safely only in a gravity-free environment. In addition, demands of increased operational speeds and higher payload capacities also called for relatively lightweight structures and trimmer designs. Lightweight robots, being inherently *flexible*, can no longer be modelled with rigid body assumptions. During the gross motion of such robots or due to external disturbances, the links vibrate. These vibrations must be damped out or controlled to achieve the required positioning accuracy and repeatability in manipulation tasks. In this chapter, we discuss modelling and control of flexible manipulators.

The overall flexibility in a robot arm is due to the flexibility at the joint and the flexibility of the link. The flexibility at the joint is due to the lack of rigidity in the drive (as in a harmonic drive), deformation of the gear teeth and shaft, backlash in mating gears, and due to the control action.[1] We will

[1] We may recall from Section 8.2 that a feedback-controlled single-link manipulator can be modelled as a second-order system with its associated natural frequency and damping. The natural frequency, an indicator of the 'stiffness', is dependent on the choice of the proportional gain in the controller.

discuss modelling of a flexible-jointed rigid-link manipulator in Section 9.2.

The flexibility in the link of a robot makes it a continuous dynamical system with infinite degrees of freedom. The motion of a flexible link can no longer be described by an ordinary differential equation (as in the case of a rigid link), and a partial differential equation is required. For completeness, the partial differential equation describing the transverse free bending vibration of a beam and the boundary conditions are presented in Section 9.3. It is much more difficult to analyse mathematical models or design controllers using the exact partial differential equations, and it is necessary that the continuous system be *discretized* to a finite-dimensional system. In Section 9.5, we present two commonly used approaches, namely, the *assumed modes method* (AMM) and the *finite element method* (FEM), to discretize a continuous system. After discretization, we get a set of ordinary differential equations, whose number is determined by the number of generalized flexible variables used for the discretization.

The kinematic modelling of rigid manipulators has been done using the Denavit–Hartenberg parameters. In Section 9.4, we discuss the kinematic modelling of a flexible manipulator with the assumption that flexible deformations are *small* and *linear elasticity* equations can be used. From the kinematic modelling, we can derive the position and velocity of points on the links. In Section 9.6, we use the Lagrangian formulation to derive the equations of motion of a flexible-link manipulator. One of the key differences between rigid and flexible manipulators is the strain energy in the vibrating link, which must be taken into account in the potential energy of the system. Due to the flexibility in the links, in addition to the equations of motion describing the rigid or gross motion of the manipulator, we also get ordinary differential equations involving the generalized flexible variables and a non-linear *coupling* between the rigid and flexible variables. In Section 9.7, we present control schemes for joint trajectory following and end-effector positioning in flexible manipulators. Finally, in Section 9.8, we list some of the other important topics in the area of modelling and control of flexible manipulators. The treatment on kinematic modelling and development of controllers for flexible-link manipulators draws heavily from the work of Theodore (1995) and research papers written by Theodore and Ghosal (1995; 1997; 2003).

9.2 Modelling of a Flexible Joint

A common way to model the flexibility at a joint is to assume that the motor or the actuator at the joint is connected to the link by means of a torsional spring with a spring constant K_s. This is shown schematically in Fig. 9.1. The rotation of the motor is denoted by θ_m and the rotation of the *rigid link* is denoted by θ_l. The link is assumed to move in a plane parallel to the plane determined by the $\widehat{\mathbf{X}}_0$-$\widehat{\mathbf{Y}}_0$ axes, i.e., the torsional spring does not cause any motion out of the plane.

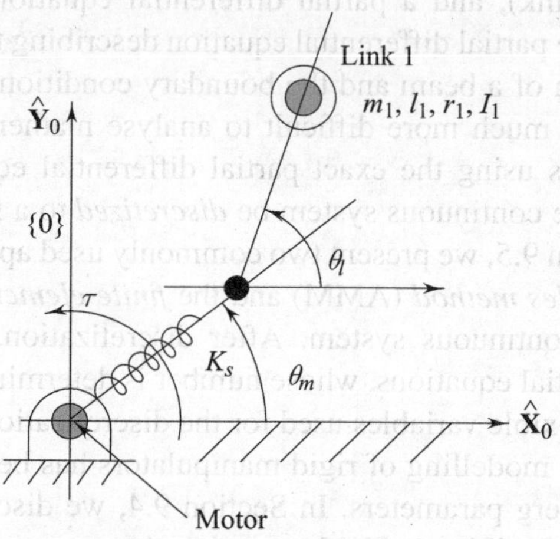

Fig. 9.1 A flexible-joint link

The equation of motion of the flexible-joint rigid-link system, schematically shown in Fig. 9.1, can be derived using the Lagrangian formulation presented in Section 6.3. In the absence of gravity and friction at the joint, these are given by

$$J_m\theta_m + K_s(\theta_m - \theta_l) = \tau$$
$$J_l\ddot{\theta}_l + K_s(\theta_l - \theta_m) = 0 \tag{9.1}$$

where $J_l = I_1 + m_1 r_1^2$ is the load inertia, τ is the torque from the motor, and J_m is the inertia of the motor.

A block diagram of the linear system (open loop) described in Eqn (9.1) is given in Fig. 9.2. We can observe that the actuator torque, τ, controls two outputs, θ_m and θ_l, and hence the model is more complicated than the model

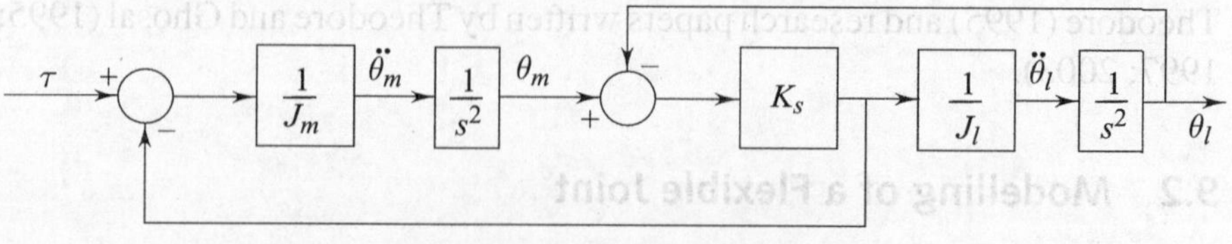

Fig. 9.2 A block diagram of the flexible-joint link

of a rigid link connected to a motor discussed in Section 8.2.3. To test if control of θ_l or θ_m is at all possible with τ, we can obtain the controllability matrix for the linear system [see Eqns (8.94) and (8.95) in Section 8.11)]. For the linear system in Eqn (9.1), a convenient choice of state variables is $(\theta_m, \theta_l, \dot{\theta_m}, \dot{\theta}_l)^T$ and the resulting matrices $[F]$ and $[G]$ are given by

$$[F] = \begin{pmatrix} 0 & 0 & 1 & 0 \\ 0 & 0 & 0 & 1 \\ -K_s/J_m & K_s/J_m & 0 & 0 \\ K_s/J_l & -K_s/J_l & 0 & 0 \end{pmatrix}, \quad [G] = \begin{pmatrix} 0 \\ 0 \\ \frac{1}{J_m} \\ 0 \end{pmatrix} \tag{9.2}$$

The controllability matrix $[Q_c]$ obtained from $[F]$ and $[G]$ and as given in Eqn (8.95) has determinant $-K_s^2/(J_m^4 J_l^2)$ which can never be zero. Hence, the linear system is controllable with $\tau(t)$ and one can develop proportional or PID controllers to achieve a desired $\theta_l(t)$.

In the presence of gravity, the equations of motion become non-linear, and we get

$$J_m\ddot{\theta}_m + K_s(\theta_m - \theta_l) = \tau$$

$$J_l\ddot{\theta}_l + K_s(\theta_l - \theta_m) + m_1 g r_1 \sin\theta_l = 0 \tag{9.3}$$

To control such a non-linear system, we could attempt to use the technique of feedback linearization discussed in Section 8.4. However, in contrast to rigid manipulators, the non-linearity is in the second equation in Eqn (9.2), where as the actuator torque is in the first equation. Hence, it is not obvious how to use control-law partitioning [see Eqn (8.29) in Section 8.4] to 'cancel' out the non-linear terms in the case of a flexible-joint single-link manipulator. Using advanced techniques from Lie algebra, Marino and Spong (1986) were able to derive a $[\alpha]$ and a β which could be used for feedback linearization. Later Spong (1987) presented a general theory of feedback linearization of multi-link flexible-jointed robots. Readers interested in more details are referred to these two cited papers. In the following section, we look at modelling of a flexible link.

9.3 Euler–Bernoulli Beam Model

We consider the free bending vibration equation of a single flexible link and begin by stating the assumptions.

1. The first main assumption is that linear elasticity theory adequately describes the deformation of a link. This implies that link deflections, measured with respect to the undeformed link coordinate system, are small. This assumption is fairly reasonable since it has been reported in

literature that the maximum deflection in a flexible link is usually less than one tenth of the total link length (Cannon & Schmitz 1984).

2. Each flexible link is made up of homogeneous, isotropic, and elastic material, and linear stress–strain relationships can be used.
3. We assume that planar cross sections remain plane and the cross section and the longitudinal axis remain mutually perpendicular after deformation. This is essentially the Euler–Bernoulli hypothesis for flexible slender beams, and it does not account for the shear deformation and rotary inertia effects of the beam (Meirovitch 1967).
4. We also assume that the longitudinal deformation is negligible. In addition, the cross section of the flexible link is geometrically symmetric and the shear-centre axis coincides with the centroidal axis of the link so that the transverse load does not give rise to torsion.

Figure 9.3 shows a slender beam of length l satisfying the above-mentioned assumptions. Let an arbitrary material point along the neutral axis of the beam be located by s. The transverse free bending vibration of the beam is given by the partial differential equation (PDE)

$$\frac{\partial^2}{\partial s^2}\left(EI(s)\frac{\partial^2 u(s,t)}{\partial s^2}\right)+\rho A(s)\frac{\partial^2 u(s,t)}{\partial t^2}=0 \tag{9.4}$$

where $u(s,t)$ denotes the transverse displacement, with reference to the neutral axis, of the material point s at time t. The quantity $EI(s)$ represents *flexural rigidity*, with E being the Young's modulus and $I(s)$ is the cross-sectional area moment of inertia. The quantity $\rho A(s)$ represents mass per unit length, with ρ denoting the density of the material and $A(s)$ denoting the cross-sectional area of the beam.

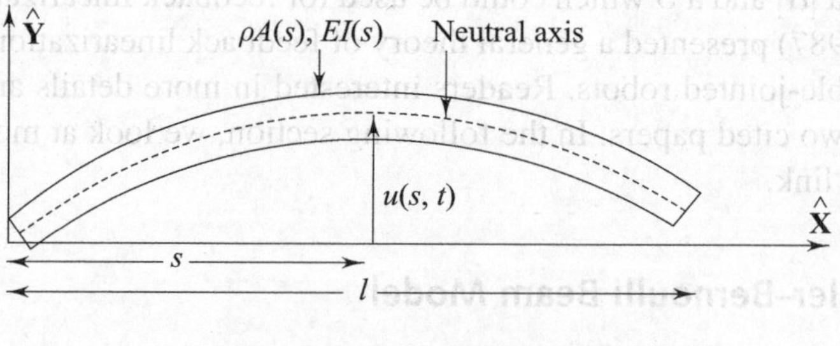

Fig. 9.3 A beam in flexure

To uniquely solve Eqn (9.4) we need to specify boundary conditions at $s = 0$ and $s = l$ at the initial time $t = 0$. The governing PDE, Eqn (9.4), is second order in t and we will need two initial conditions, $u(s,t)|_{t=0}$ and $[\partial u(s,t)/\partial t]|_{t=0}$. Since the PDE is fourth order in s, we need to specify four boundary conditions at $s = 0$ and $s = l$. These boundary conditions are classified into two distinct classes. The boundary conditions resulting from pure geometry are called *geometric* boundary conditions and are associated with deflection $[u(s,t)]$ or slope $[\partial u(s,t)/\partial s]$ at the boundaries. The boundary conditions resulting from moment $[(EI(s)[\partial^2 u(s,t)/\partial s^2])]$ or shear force $[\partial/\partial s\,(EI(s)[\partial^2 u(s,t)/\partial s^2])]$ are called *natural* boundary conditions. To choose the appropriate boundary conditions, we need to consider the type of joint in the flexible-link manipulator. We consider two of the most common joints, namely, a rotary (R) joint and a sliding or prismatic (P) joint.

9.3.1 Rotating Flexible Link

Consider a flexible link rotated by a motor in a horizontal plane. The flexible link, modelled as a slender beam, is attached to the motor shaft at one end and carries a general payload, characterized by both mass M_p and inertia J_p, at the other end, as shown in Fig. 9.4. Let $\{0\}$ denote the inertial frame with $\widehat{\mathbf{X}}_0$, $\widehat{\mathbf{Y}}_0$ axes lying in the horizontal plane and $\widehat{\mathbf{Z}}_0$ along the motor shaft. In most flexible manipulator models, the connection at the motor end is either considered *clamped* or *pinned*. In the clamped case, the non-inertial frame $\{1\}$, rotating with the link, is chosen such that $\widehat{\mathbf{X}}_1$ is tangent to the link at the origin, as shown in Fig. 9.4. This requires that the deflection and the slope at

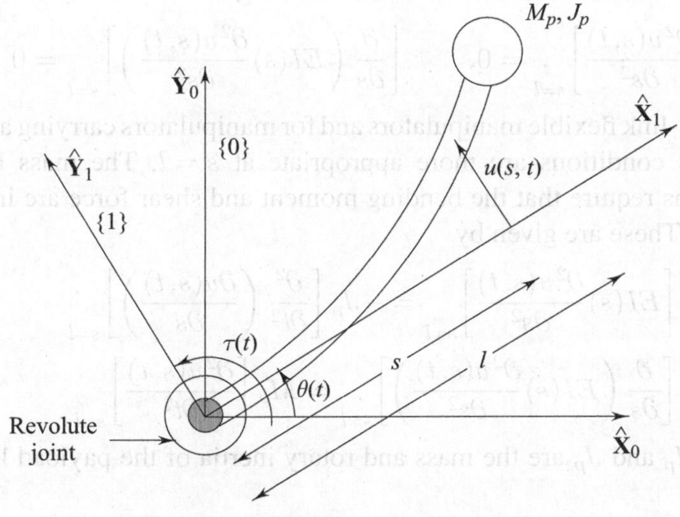

Fig. 9.4 A flexible link with a rotary joint

the connection point of the link to the motor shaft are zero, so we can write

$$[u(s,t)]_{s=0} = 0, \qquad \left[\frac{\partial u(s,t)}{\partial s}\right]_{s=0} = 0 \tag{9.5}$$

In the pinned case, the non-inertial frame $\{1\}$ is chosen such that $\widehat{\mathbf{X}}_1$ passes through the centre of mass of the flexible link at all times, and hence the slope at $s = 0$ need not be zero. In the case of the pinned condition, we must have

$$\begin{aligned}
&[u(s,t)]_{s=0} = 0, \\
&\left[EI(s)\frac{\partial^2 u(s,t)}{\partial s^2}\right]_{s=0} = J_a\left[\frac{\partial^2}{\partial t^2}\left(\frac{\partial u(s,t)}{\partial s}\right)\right]_{s=0}
\end{aligned} \tag{9.6}$$

where J_a is the total inertia of the rotor and the hub of the joint actuator. There has been extensive research to figure out which of the two conditions, clamped or pinned, is more appropriate for modelling flexible links at $s = 0$. It is clear that the condition at $s = 0$ cannot be purely clamped since the actuator end is *not* as in a traditional built-in vibrating cantilever beam. It is also not purely pinned since the effect of the control torque from the motor is to provide a non-zero stiffness. It has been shown that if J_a is much larger than the flexible beam inertia, with the ratio of beam inertia to J_a of the order of 0.1 or less, clamped conditions must be enforced for the controlled end (Cetinkunt & Yu 1991). It is now a generally accepted practice to use clamped conditions, as given in Eqn (9.5), for the end attached to the motor.

The other end of the flexible link ($s = l$) is usually treated as *free* end or with *mass* end-conditions. For a free end we get

$$\left[EI(s)\frac{\partial^2 u(s,t)}{\partial s^2}\right]_{s=l} = 0, \qquad \left[\frac{\partial}{\partial s}\left(EI(s)\frac{\partial^2 u(s,t)}{\partial s^2}\right)\right]_{s=l} = 0 \tag{9.7}$$

For multi-link flexible manipulators and for manipulators carrying a payload, the mass conditions are more appropriate at $s = l$. The mass boundary conditions require that the bending moment and shear force are in balance at $s = l$. These are given by

$$\begin{aligned}
&\left[EI(s)\frac{\partial^2 u(s,t)}{\partial s^2}\right]_{s=l} = -J_p\left[\frac{\partial^2}{\partial t^2}\left(\frac{\partial u(s,t)}{\partial s}\right)\right]_{s=l} \\
&\left[\frac{\partial}{\partial s}\left(EI(s)\frac{\partial^2 u(s,t)}{\partial s^2}\right)\right]_{s=l} = M_p\left[\frac{\partial^2 u(s,t)}{\partial t^2}\right]_{s=l}
\end{aligned} \tag{9.8}$$

where M_p and J_p are the mass and rotary inertia of the payload located at $s = l$.

To further analyse the PDE in Eqn (9.4), we assume that I and A are constant. We also introduce the non-dimensional variables:

$$\widetilde{u}(s,t) = u(s,t)/l, \quad \eta = s/l, \quad \tau = t/(l/U_g) \quad \left(\text{with } U_g \triangleq \frac{1}{l}\sqrt{\frac{EI}{\rho A}}\right)$$

It can be verified that the quantity U_g has units of speed (m/sec) and is a characteristic velocity associated with transverse bending vibrations of a beam of length l. The quantity l/U_g has units of time and can be thought of as the time taken for a disturbance induced at one end of the link to travel the length of the link. We observe that for a rigid link ($EI \to \infty$), the time (l/U_g) taken for the disturbance to travel l approaches zero. On the other hand, for a highly flexible beam (EI is small), or for a very long beam (l is large), the time for the disturbance to travel the length of the beam will be large. The quantity EI is the flexural rigidity of a beam.

Rewriting Eqn (9.4) and the boundary conditions in Eqn (9.8) in terms of the non-dimensional variables $\widetilde{u}$, η, and τ, we get,

$$\frac{\partial^4 \widetilde{u}(\eta,\tau)}{\partial \eta^4} + \frac{\partial^2 \widetilde{u}(\eta,\tau)}{\partial \tau^2} = 0, \qquad 0 < \eta < 1 \tag{9.9}$$

and

$$\begin{aligned}
&[\widetilde{u}(\eta,\tau)]_{\eta=0} = 0, \\
&\left[\frac{\partial^2 \widetilde{u}(\eta,\tau)}{\partial \eta^2}\right]_{\eta=1} = -\frac{J_p}{\rho A l^3}\left[\frac{\partial^2}{\partial \tau^2}\left(\frac{\partial \widetilde{u}(\eta,\tau)}{\partial \eta}\right)\right]_{\eta=1} \\
&\left[\frac{\partial \widetilde{u}(\eta,\tau)}{\partial \eta}\right]_{\eta=0} = 0, \\
&\left[\frac{\partial^3 \widetilde{u}(\eta,\tau)}{\partial \eta^3}\right]_{\eta=1} = \frac{M_p}{\rho A l}\left[\frac{\partial^2 \widetilde{u}(\eta,\tau)}{\partial \tau^2}\right]_{\eta=1}
\end{aligned} \tag{9.10}$$

In the non-dimensionalized form, one can more easily decide whether to use mass or free boundary conditions at $s = l$. We can observe that it is reasonable to use free end-conditions if the rotary inertia J_p and mass M_p of the payload are much smaller compared to the rotary inertia $\rho A l^3$ and mass $\rho A l$ of the flexible link. On the other hand, if the rotary inertia and mass of the payload are comparable to that of the link, it is more appropriate to use the mass end-conditions for the flexible link. In multi-link flexible manipulator links, the links *after* the flexible link j can be modelled as an effective M_{p_j} and J_{p_j} acting on the $s = l$ end of the flexible link j and hence the use of the mass end-condition is even more appropriate.

The PDE in Eqn (9.9) together with the boundary conditions given in Eqn (9.10) can be solved by the method of *separation* of variables. We assume that $\widetilde{u}(\eta, \tau)$ is separable in space (η) and time (τ) and we can write

$$\widetilde{u}(\eta, \tau) = \boldsymbol{\psi}(\eta)\mathbf{q}_f(\tau) \tag{9.11}$$

where $\boldsymbol{\psi}(\eta)$ are called *mode shape functions* and $\mathbf{q}_f(t)$ are the flexible generalized coordinates. Using Eqn (9.11) in Eqn (9.9), we get after rearranging,

$$\frac{1}{\mathbf{q}_f(\tau)}\frac{d^2\mathbf{q}_f(\tau)}{d\tau^2} = \frac{1}{\boldsymbol{\psi}(\eta)}\frac{d^4\boldsymbol{\psi}(\eta)}{d\eta^4} \tag{9.12}$$

From the separation of variables principle, Eqn (9.12) can have a solution if and only if both sides of the equation are equal to a constant. On the basis of physical considerations and to have only bounded solutions for the problem, this constant must be a real value and is usually chosen to be $-\omega^2$, so that Eqn (9.12) leads to

$$\frac{d^2\mathbf{q}_f(\tau)}{d\tau^2} + \omega^2\mathbf{q}_f(\tau) = 0 \tag{9.13}$$

$$\frac{d^4\boldsymbol{\psi}(\eta)}{d\eta^4} - \omega^2\boldsymbol{\psi}(\eta) = 0, \qquad 0 < \eta < 1 \tag{9.14}$$

and the boundary conditions (9.10) reduce to

$$\begin{aligned} &[\boldsymbol{\psi}(\eta)]_{\eta=0} = 0, && \left[\frac{d^2\boldsymbol{\psi}(\eta)}{d\eta^2}\right]_{\eta=1} = \frac{J_p\omega^2}{\rho A l^3}\left[\frac{d\boldsymbol{\psi}(\eta)}{d\eta}\right]_{\eta=1} \\ &\left[\frac{d\boldsymbol{\psi}(\eta)}{d\eta}\right]_{\eta=0} = 0, && \left[\frac{d^3\boldsymbol{\psi}(\eta)}{d\eta^3}\right]_{\eta=1} = -\frac{M_p\omega^2}{\rho A l}[\boldsymbol{\psi}(\eta)]_{\eta=1} \end{aligned} \tag{9.15}$$

The problem of determining the values of the parameter ω^2 for which non-trivial solutions $\boldsymbol{\psi}(\eta)$ of Eqn (9.14) exist, where the solutions are subject to boundary conditions (9.15), is called the *eigenvalue problem*. The corresponding values of ω^2 are known as *eigenvalues* and the associated functions $\boldsymbol{\psi}(\eta)$ as *eigenfunctions* (Meirovitch 1967). The eigenvalue problem gives rise to a *characteristic equation*, also called the *frequency equation*. The solution of the characteristic equation consists of an infinite set of discrete characteristic values, the square roots of which are the system *natural frequencies* ω_i $(i = 1, 2, \dots)$. To each of the natural frequencies corresponds an eigenfunction, or *natural mode* $\psi_i(\eta)$. We will discuss these in detail in Section 9.5.

9.3.2 Translating Flexible Link

Instead of a rotary joint at the actuator end of the flexible link, we can also have a prismatic joint as shown in Fig. 9.5.

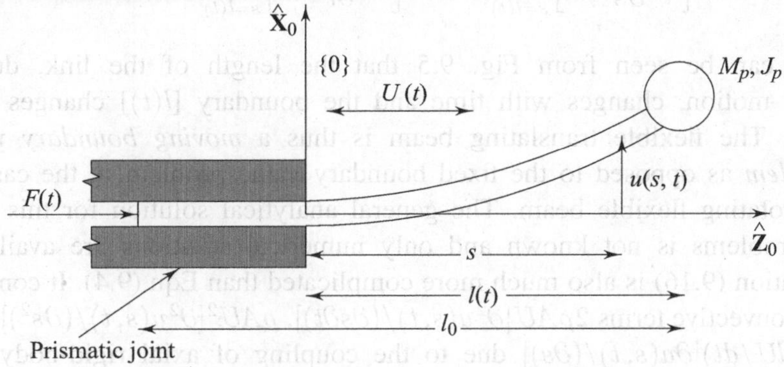

Fig. 9.5 A flexible link with a prismatic joint

In this figure, $\{0\}$ is a fixed coordinate system with $\widehat{\mathbf{X}}_0$ and $\widehat{\mathbf{Z}}_0$ spanning the horizontal plane. The $\widehat{\mathbf{Z}}_0$ axis is along the translation of the link (following the Denavit–Hartenberg convention for rigid manipulators) and is tangent to the neutral axis of the flexible link at the joint. The link is assumed to have a total length of l_0, with $l(t)$ denoting the length outside the rigid joint hub at time t, and is assumed to be vibrating in the horizontal plane. The length of the beam inside the hub, $[l_0 - l(t)]$, is assumed not to be vibrating. Any arbitrary material point can be located by s from the fixed end, and the axial velocity $U(t)$ is assumed to be independent of s. The free bending vibration of the translating flexible link with the Euler–Bernoulli assumptions can be written as

$$\frac{\partial^2}{\partial s^2}\left(EI\frac{\partial^2 u(s,t)}{\partial s^2}\right)+\rho A\left(\frac{\partial^2 u(s,t)}{\partial t^2}+2U\frac{\partial^2 u(s,t)}{\partial s \partial t}\right.$$
$$\left.+U^2\frac{\partial^2 u(s,t)}{\partial s^2}+\frac{dU}{dt}\frac{\partial u(s,t)}{\partial s}\right)=0 \tag{9.16}$$

where $0 < s < l(t)$, and the clamped-mass boundary conditions are given by

$$[u(s,t)]_{s=0} = 0,$$
$$EI\left[\frac{\partial^2 u(s,t)}{\partial s^2}\right]_{s=l(t)} = -J_p\left[\frac{\partial^2}{\partial t^2}\left(\frac{\partial u(s,t)}{\partial s}\right)\right]_{s=l(t)} \tag{9.17}$$

$$\left[\frac{\partial u(s,t)}{\partial s}\right]_{s=0} = 0,$$

$$EI\left[\frac{\partial^3 u(s,t)}{\partial s^3}\right]_{s=l(t)} = M_p\left[\frac{\partial^2 u(s,t)}{\partial t^2}\right]_{s=l(t)}$$

It can be seen from Fig. 9.5 that the length of the link, during axial motion, changes with time and the boundary $[l(t)]$ changes with time. The flexible translating beam is thus a *moving boundary value problem* as opposed to the fixed boundary value problem in the case of the rotating flexible beam. The general analytical solution for this kind of problems is not known and only numerical solutions are available. Equation (9.16) is also much more complicated than Eqn (9.4). It contains the convective terms $2\rho AU[\partial^2 u(s,t)/(\partial s \partial t)]$, $\rho AU^2[\partial^2 u(s,t)/(\partial s^2)]$, and $\rho A(dU/dt)[\partial u(s,t)/(\partial s)]$ due to the coupling of axial rigid-body and transverse vibratory motions. The centripetal term $\rho AU^2[\partial^2 u(s,t)/(\partial s^2)]$ represents an axially applied load, and hence it will alter the the 'stiffness' of the system. Since the axial velocity U appears in the centripetal term, for sufficiently large U the centripetal force may overcome the flexural restoring force. Likewise, it is also expected that the system's oscillatory frequencies would decrease with increasing axial velocity (Stylianou & Tabarrok 1994).

To analyse Eqn (9.16) further, we use the non-dimensional variables: $\widetilde{u}(s,t) = u(s,t)/l_0, \eta = s/l_0$, and $\tau = t/(l_0/U_g)$ with $U_g \triangleq 1/l_0\sqrt{EI/\rho A}$. It may be noted that U_g is based on the total length l_0 of the beam, and thus the smallest U_g value during the motion (which also represent the 'worst' case).

Rewriting the PDE in Eqn (9.16) and the corresponding clamped-mass boundary conditions in Eqn (9.17) in terms of the non-dimensionalized variables, we get

$$\frac{\partial^4 \widetilde{u}(\eta,\tau)}{\partial \eta^4} + \frac{\partial^2 \widetilde{u}(\eta,\tau)}{\partial \tau^2} + 2\left(\frac{U}{U_g}\right)\frac{\partial^2 \widetilde{u}(\eta,\tau)}{\partial \eta \partial \tau} + \left(\frac{U}{U_g}\right)^2 \frac{\partial^2 \widetilde{u}(\eta,\tau)}{\partial \eta^2} + \left[\frac{d}{d\tau}\left(\frac{U}{U_g}\right)\right]\frac{\partial \widetilde{u}(\eta,\tau)}{\partial \eta} = 0 \qquad (9.18)$$

and

$$[\widetilde{u}(\eta,\tau)]_{\eta=0} = 0,$$

$$\left[\frac{\partial^2 \widetilde{u}(\eta,\tau)}{\partial \eta^2}\right]_{\eta=l(t)/l_0} = -\frac{J_p}{\rho A l^3}\left[\frac{\partial^2}{\partial \tau^2}\left(\frac{\partial \widetilde{u}(\eta,\tau)}{\partial \eta}\right)\right]_{\eta=l(t)/l_0}$$

$$\left[\frac{\partial \tilde{u}(\eta,\tau)}{\partial \eta}\right]_{\eta=0} = 0,$$
$$\left[\frac{\partial^3 \tilde{u}(\eta,\tau)}{\partial \eta^3}\right]_{\eta=l(t)/l_0} = \frac{M_p}{\rho A l}\left[\frac{\partial^2 \tilde{u}(\eta,\tau)}{\partial \tau^2}\right]_{\eta=l(t)/l_0} \tag{9.19}$$

Using Eqn (9.11), we can write Eqn (9.18) as

$$\psi(\eta)\frac{d^2 \mathbf{q}_f(\tau)}{d\tau^2} + 2\frac{U}{U_g}\frac{d\psi(\eta)}{d\eta}\frac{d\mathbf{q}_f(\tau)}{d\tau}$$
$$= -\left[\frac{d^4\psi(\eta)}{d\eta^4} + \left(\frac{U}{U_g}\right)^2\frac{d^2\psi(\eta)}{d\eta^2} + \frac{d}{d\tau}\left(\frac{U}{U_g}\right)\frac{d\psi(\eta)}{d\eta}\right]\mathbf{q}_f(\tau) \tag{9.20}$$

Equation (9.20) is not separable in η and τ in the classical sense of separation of variables. However, if $U \ll U_g$ and the axial velocity is constant $[d/d\tau\,(U/U_g) = 0]$, the convective terms can be dropped and Eqn (9.20) reduces to Eqn (9.12) and one can analyse Eqn (9.20) in a manner similar to Eqn (9.12). There are, however, major differences with the case of a rotating flexible link—the mode shape functions $\psi(\eta)$ and the natural frequencies are *time dependent*. In addition, the boundary conditions are also time dependent and the solutions to the resultant eigenvalue problems must be solved using advanced integral transform methods (Meirovitch 1967). In Section 9.5, we present a simpler method based on Theodore and Ghosal (1995), where a differential equation is derived to account for the time-varying boundary conditions.

As can be sensed from above, the modelling and analysis of prismatic jointed flexible manipulators is significantly more complicated than flexible-link manipulators with only rotary joints. In this text, we will mainly model and analyse flexible manipulators with rotary joints. The reader interested in more details on flexible manipulators with prismatic joints is referred to Theodore (1995), Theodore and Ghosal (1997), the references contained in them. In the following section, we discuss methods to extend the approach applied to rigid manipulators in Chapter 2, namely, the use of Denavit–Hartenberg parameters and 4×4 homogeneous transformation matrices, to model flexible-link manipulators.

9.4 Kinematic Modelling of Multi-link Flexible Manipulators

In Section 2.6, we had introduced the Denavit–Hartenberg parameters to model rigid links in a manipulator. In this section, we extend the Denavit–Hartenberg convention to flexible links. We consider the case of flexible links joined by revolute (R) and prismatic (P) joints undergoing only *transverse* bending vibrations—the axial and torsional deformation of a link are not considered in this treatment. In addition, the flexible links are assumed to satisfy the basic assumptions listed in Section 9.3.

Similar to the Denavit–Hartenberg convention for rigid links in Chapter 2, we assign coordinate system $\{j\}$ to link j with $\{0\}$ as the fixed link and $\{n\}$ as the last link. The coordinate axes $(\widehat{\mathbf{X}}_j, \widehat{\mathbf{Y}}_j, \widehat{\mathbf{Z}}_j)$ are assigned to link j and the origin O_j is on the joint axis j. Again, similar to the case of rigid manipulators, $\widehat{\mathbf{Z}}_j$ is along the axis of joint j. We also define a coordinate system $\{j_*\}$ in such a way that when the link $j-1$ is in its *undeformed configuration*, the $\{j\}$ and $\{j_*\}$ are *coincident* (see Fig. 9.6).

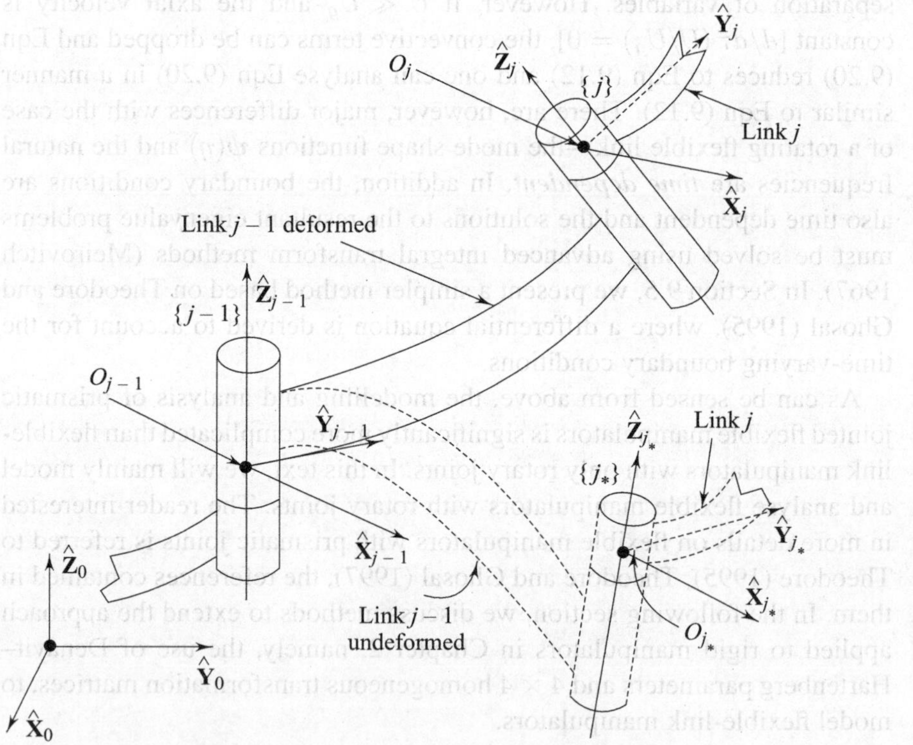

Fig. 9.6 Assignment of frames for the flexible links

The 4×4 homogeneous transformation matrix describing $\{j_*\}$ with respect to $\{j-1\}$ is exactly similar to that of a rigid manipulator discussed in Section 2.6. Replacing i by j_*, $i-1$ by $j-1$ in Eqn (2.51) and using $[T_r]$

to denote the rigid transformation, we can get

$$ {}^{j-1}_{j_*}[T_r] = \begin{pmatrix} c\theta_j & -s\theta_j & 0 & a_{j-1} \\ s\theta_j c\alpha_{j-1} & c\theta_j c\alpha_{j-1} & -s\alpha_{j-1} & -s\alpha_{j-1} d_j \\ s\theta_j s\alpha_{j-1} & c\theta_j s\alpha_{j-1} & c\alpha_{j-1} & c\alpha_{j-1} d_j \\ 0 & 0 & 0 & 1 \end{pmatrix} \tag{9.21} $$

where $\alpha_{j-1}, a_{j-1}, d_j$, and θ_j are the Denavit–Hartenberg parameters required to describe $\{j_*\}$ with respect to $\{j-1\}$. Similar to the rigid manipulator case, θ_j is the joint variable for a revolute (R) joint and d_j is the joint variable if the joint is prismatic (P). We will denote the joint variable, either θ_j or d_j, by q_{j_r} (the subscript r denoting the rigid motion at the joint) in this chapter, and for multi-link manipulator the rigid joint variable vector will be denoted by a $n \times 1$ vector $\mathbf{q}_r$. The variables representing the flexibility in the link j will be denoted by $\mathbf{q}_{f_j}$.

To obtain the 4×4 transformation matrix between $\{j_*\}$ and $\{j\}$, we recall that any general 3D spatial transformation can be represented by a sequence of three rotations and three translations (see Section 2.2). Hence, the $\{j_*\}$ can be taken to $\{j\}$ by the sequence

$$ \mathrm{rot}(\hat{Z}, \phi_{z_{j-1}})\mathrm{trans}(\hat{Z}, \delta_{z_{j-1}})\mathrm{rot}(\hat{Y}, \phi_{y_{j-1}})\mathrm{trans}(\hat{Y}, \delta_{y_{j-1}}) \\ \mathrm{rot}(\hat{X}, \phi_{x_{j-1}})\mathrm{trans}(\hat{X}, \delta_{x_{j-1}}) $$

where $\delta_{j-1} = (\delta_{x_{j-1}}, \delta_{y_{j-1}}, \delta_{z_{j-1}})^T$ is the translation vector from the origin O_{j_*} to O_j and $\phi_{j-1} = (\phi_{x_{j-1}}, \phi_{y_{j-1}}, \phi_{z_{j-1}})$ are the three Z-Y-X Euler angle rotations describing the orientation of $\{j\}$ with respect to $\{j_*\}$ (see Fig. 9.6).

Under the assumption of *small* elastic deformation, we can linearize the trigonometric functions (cosine and sine of ϕ_{j-1}'s) appearing in the rotation matrix obtained from the Z-Y-X Euler angles. After linearization, the 4×4 transformation matrix relating $\{j\}$ with respect to $\{j_*\}$, denoted by ${}^{j_*}_{j}[T_e]$, is given by (Book 1984)

$$ {}^{j_*}_{j}[T_e] = \begin{pmatrix} 1 & -\phi_{z_{j-1}} & \phi_{y_{j-1}} & \delta_{x_{j-1}} \\ \phi_{z_{j-1}} & 1 & -\phi_{x_{j-1}} & \delta_{y_{j-1}} \\ -\phi_{y_{j-1}} & \phi_{x_{j-1}} & 1 & \delta_{z_{j-1}} \\ 0 & 0 & 0 & 1 \end{pmatrix} \tag{9.22} $$

and if link $j-1$ is rigid, ${}^{j_*}_{j}[T]$ is a 4×4 identity matrix.

The 4×4 homogeneous transformation matrix relating $\{j\}$ with respect to $\{j-1\}$ is given by

$$ {}^{j-1}_{j}[T] = {}^{j-1}_{j_*}[T_r]\,{}^{j_*}_{j}[T_e] \tag{9.23} $$

and one can obtain ${}^{0}_{j}[T]$ by a sequence of matrix multiplication as

$$ {}^{0}_{j}[T] = {}^{0}_{1_*}[T_r]\; {}^{1_*}_{1}[T_e]\; {}^{1}_{2_*}[T_r]\; {}^{2_*}_{2}[T_e] \cdots {}^{j-1}_{j_*}[T_r]\; {}^{j_*}_{j}[T_e] \tag{9.24} $$

As in the case of the rigid-link transformation matrix, the matrix ${}^{0}_{j}[T]$ gives the position vector ${}^{0}\mathbf{O}_j$ of the origin O_j and the rotation matrix ${}^{0}_{j}[R]$ describing the orientation of $\{j\}$ with respect to the fixed frame $\{0\}$—the rigid-joint variables and flexible deformations only up to the *start* of link j are taken into account. To obtain the position vector of a material point along the neutral axis of link j, after the joint j, with respect to $\{0\}$, we can write

$$ {}^{0}\mathbf{p}_j = {}^{0}\mathbf{O}_j + {}^{0}_{j}[R]\mathbf{r}_j \tag{9.25} $$

To arrive at the expression for $\mathbf{r}_j$, we start by assuming that the link j can deflect in 3D space and denote the elastic deformation along the X, Y, and Z axes by $u_j(s,t)$, $v_j(s,t)$, and $w_j(s,t)$, respectively. It may be noted that $u_j(s,t)$, $v_j(s,t)$, and $w_j(s,t)$ are the displacements of a material point on the neutral axis of a flexible link j at a distance s and at time t; these are with respect to the undeformed link j. Since only the transverse deformation of the link is being considered, for a revolute joint $u_j(s,t) = s$ and $v_j(s,t)$, $w_j(s,t)$ represent the Y and Z transverse deformations (see Fig. 9.4 for the 2D case), respectively. For a prismatic joint, $u_j(s,t)$ and $v_j(s,t)$ represent the X and Y transverse deformations, respectively, and $w_j(s,t) = s$ (see Fig. 9.5 for the 2D case). Hence, the local position vector $\mathbf{r}_j$ is given by

$$ \mathbf{r}_j = \begin{cases} \begin{pmatrix} s \\ 0 \\ 0 \end{pmatrix} + \begin{pmatrix} 0 \\ v_j(s,t) \\ w_j(s,t) \end{pmatrix} & \text{if joint } j \text{ is revolute} \\ \begin{pmatrix} 0 \\ 0 \\ s \end{pmatrix} + \begin{pmatrix} u_j(s,t) \\ v_j(s,t) \\ 0 \end{pmatrix} & \text{if joint } j \text{ is prismatic} \end{cases} \tag{9.26} $$

The velocity of the material point ${}^{0}\mathbf{p}_j$ on link j can be obtained from the time derivative of the position vector in the inertial frame $\{0\}$, and is given by

$$ {}^{0}\mathbf{V}_p \triangleq \frac{d}{dt}({}^{0}\mathbf{p}_j) = \frac{d}{dt}({}^{0}\mathbf{O}_j) + \frac{d}{dt}({}^{0}_{j}[R])\mathbf{r}_j + {}^{0}_{j}[R]\frac{d}{dt}(\mathbf{r}_j) \tag{9.27} $$

where $(d/dt)(\mathbf{r}_j)$ is given by

$$\dot{\mathbf{r}}_j = \begin{cases} \begin{pmatrix} 0 \\ \dot{v}_j(s,t) \\ \dot{w}_j(s,t) \end{pmatrix} & \text{if joint } j \text{ is revolute} \\ \begin{pmatrix} 0 \\ 0 \\ U_j(t) \end{pmatrix} + \begin{pmatrix} \dot{u}_j(s,t) + \dfrac{\partial u_j(s,t)}{\partial s} U_j(t) \\ \dot{v}_j(s,t) + \dfrac{\partial v_j(s,t)}{\partial s} U_j(t) \\ 0 \end{pmatrix} & \text{if joint } j \text{ is prismatic} \end{cases} \tag{9.28}$$

where $U_j(t) \triangleq \dot{s}$ defines the translational velocity of the prismatic jointed link j.

The elastic displacements $u_j(s,t), v_j(s,t)$, and $w_j(s,t)$ are along the X, Y, and Z directions, respectively, and each of them are governed by an equation similar to the PDE given in Eqn (9.4) and the boundary conditions discussed in Section 9.3. As discussed in Section 9.1, the continuous dynamical system described by Eqn (9.4) needs to be discretized for analysis, simulation, and development of controllers. In the following section, we discuss methods to discretize the PDE given in Eqn (9.4) and present expressions for ${}^{j_*}_{j}[T_e]$ obtained after discretization.

9.5 Discretization Methods

There are two classes of discretization schemes based on series expansions, namely, Rayleigh–Ritz type methods and weighted residual methods (Meirovitch 1967). In this section, we discuss two most widely used models that belong to the class of the Rayleigh–Ritz method, namely, the assumed modes model and the finite element model, to discretize a flexible link of a multi-link manipulator.[2]

9.5.1 Assumed Modes Method

In the assumed modes model, the elastic displacements $u_j(s,t)$, $v_j(s,t)$, and $w_j(s,t)$ are usually described by a truncated modal series, in terms of spatial

[2] For modelling of flexible manipulators using the assumed modes method, see, for example, Book (1984), De Luca and Siciliano (1991), Li and Sankar (1993), Hu and Ulsoy (1994), and for the use of the finite element method see, for example, Sunada and Dubowsky (1981), Usoro, Nadira, and Mahil (1986), Bayo (1987), Jonker (1990), Chedmail, Aoustin, and Chevallereau (1991).

mode shape functions and time-dependent mode amplitudes, as

$$
\begin{aligned}
u_j(\eta,t) &= \sum_{i=1}^{N_j} \psi_i^{u_j}(\eta)\xi_i^{u_j}(t) \\
v_j(\eta,t) &= \sum_{i=1}^{N_j} \psi_i^{v_j}(\eta)\xi_i^{v_j}(t) \\
w_j(\eta,t) &= \sum_{i=1}^{N_j} \psi_i^{w_j}(\eta)\xi_i^{w_j}(t)
\end{aligned}
\tag{9.29}
$$

where $\eta = s/l_j$ (l_j is the length of flexible link j) and N_j is the number of modes used to describe the deflection of link j (see also Section 9.5.3 for a discussion on the choice of N_j). It should be mentioned that for the flexible link with a revolute joint, l_j is the link length and remains constant (see Section 9.3.1), while for a flexible link with prismatic joint, l_j is the joint variable of the link (i.e., d_j) and since d_j is a function of time, the mode shape functions are also time dependent for a prismatic-jointed link.

For the assumed mode shapes and time-dependent mode amplitudes, the 4×4 homogeneous transformation matrix ${}^{j_*}_{j}[T_e]$ is given by the following equations.

If joint $j-1$ is revolute,

$$
{}^{j_*}_{j}[T_e] = \sum_{i=1}^{N_{j-1}} \begin{pmatrix} 1 & -\dfrac{\partial\psi_i^v}{\partial\eta}(1)\xi_i^v(t) & \dfrac{\partial\psi_i^w}{\partial\eta}(1)\xi_i^w(t) & 0 \\ \dfrac{\partial\psi_i^v}{\partial\eta}(1)\xi_i^v(t) & 1 & 0 & \psi_i^v(1)\xi_i^v(t) \\ -\dfrac{\partial\psi_i^w}{\partial\eta}(1)\xi_i^w(t) & 0 & 1 & \psi_i^w(1)\xi_i^w(t) \\ 0 & 0 & 0 & 1 \end{pmatrix}
\tag{9.30}
$$

If joint $j-1$ is prismatic,

$$
{}^{j_*}_{j}[T_e] = \sum_{i=1}^{N_{j-1}} \begin{pmatrix} 1 & 0 & \dfrac{\partial\psi_i^u}{\partial\eta}(1)\xi_i^u(t) & \psi_i^u(1)\xi_i^u(t) \\ 0 & 1 & -\dfrac{\partial\psi_i^v}{\partial\eta}(1)\xi_i^v(t) & \psi_i^v(1)\xi_i^v(t) \\ -\dfrac{\partial\psi_i^u}{\partial\eta}(1)\xi_i^u(t) & \dfrac{\partial\psi_i^v}{\partial\eta}(1)\xi_i^v(t) & 1 & 0 \\ 0 & 0 & 0 & 1 \end{pmatrix}
\tag{9.31}
$$

where the subscripts $j-1$ on u, v, and w are dropped for clarity and the mode shape functions and its derivatives with respect to η are evaluated at $\eta = 1$.

Due to the convention and choice of $\{j-1\}$, $\{j_*\}$, and $\{j\}$, the generalized flexible variables contained in ${}^0_j[T]$ are the j rigid-joint variables $\mathbf{q}_{r_j}$ and the flexible variables $(\mathbf{q}_{f_1}, \mathbf{q}_{f_2}, \ldots, \mathbf{q}_{f_{j-1}})$ where in each $\mathbf{q}_{f_k}$ we have $2 \times N_k$ variables—$[\xi_1^{v_k}(t), \xi_1^{w_k}(t), \ldots, \xi_{N_k}^{v_k}(t), \xi_{N_k}^{w_k}(t)]$ for the revolute joint and $[\xi_1^{u_k}(t), \xi_1^{v_k}(t), \ldots, \xi_{N_k}^{u_k}(t), \xi_{N_k}^{v_k}(t)]$ for the prismatic joint. For a material point on link j, in addition, from $\mathbf{r}_j$, we will get another $2 \times N_j$ flexible variables.

From Eqn (9.29), the derivative of the local vector $\mathbf{r}_j$ is given by

$$\dot{\mathbf{r}}_j = \begin{cases} \begin{pmatrix} 0 \\ \displaystyle\sum_{i=1}^{N_j} \psi_i^v(\eta) \frac{d\xi_i^v(t)}{dt} \\ \displaystyle\sum_{i=1}^{N_j} \psi_i^w(\eta) \frac{d\xi_i^w(t)}{dt} \end{pmatrix} & \text{if joint } j \text{ is revolute} \\[2ex] \begin{pmatrix} \displaystyle\sum_{i=1}^{N_j} \left[\psi_i^u(\eta) \frac{d\xi_i^u(t)}{dt} - \frac{\partial \psi_i^u(\eta)}{\partial \eta} \xi_i^u(t) \frac{\eta U_j(t)}{l_j(t)} \right] \\ \displaystyle\sum_{i=1}^{N_j} \left[\psi_i^v(\eta) \frac{d\xi_i^v(t)}{dt} - \frac{\partial \psi_i^v(\eta)}{\partial \eta} \xi_i^v(t) \frac{\eta U_j(t)}{l_j(t)} \right] \\ U_j(t) \end{pmatrix} & \text{if joint } j \text{ is prismatic} \end{cases} \tag{9.32}$$

where the subscripts on u, v, and w are dropped for clarity.

To discretize Eqn (9.29) in the assumed modes approach, we choose the mode shape functions $\psi_i(\eta)$ as[3]

$$\begin{aligned} \psi_i(\eta) = {} & C_{1_i} \cos(\beta_i \eta) + C_{2_i} \sin(\beta_i \eta) + C_{3_i} \cosh(\beta_i \eta) \\ & + C_{4_i} \sinh(\beta_i \eta) \end{aligned} \tag{9.33}$$

where $\beta_i^4 \triangleq [(\rho_j A_j l_j^4)/(E_j I_j)]\omega_i^2$ and ω_i is the ith natural angular frequency of the eigenvalue problem for link j. Then using the Euler–Bernoulli bending vibration equation, we can write the associated boundary conditions for link j as follows.

[3] Note that the superscripts on ψ_i have been dropped for clarity.

Clamped conditions at $\eta = 0$ end:

$$[\psi_i(\eta)]_{\eta=0} = 0, \qquad \left[\frac{d\psi_i(\eta)}{d\eta}\right]_{\eta=0} = 0 \tag{9.34}$$

Mass conditions at $\eta = 1$ end:

$$\begin{aligned}\left[\frac{d^2\psi_i(\eta)}{d\eta^2}\right]_{\eta=1} &= \frac{J_{p_j}\beta_i^4}{\rho_j A_j l_j^3}\left[\frac{d\psi_i(\eta)}{d\eta}\right]_{\eta=1} + \frac{M_{Dp_j}\beta_i^4}{\rho_j A_j l_j^2}[\psi_i(\eta)]_{\eta=1} \\ \left[\frac{d^3\psi_i(\eta)}{d\eta^3}\right]_{\eta=1} &= -\frac{M_{p_j}\beta_i^4}{\rho_j A_j l_j}[\psi_i(\eta)]_{\eta=1} - \frac{M_{Dp_j}\beta_i^4}{\rho_j A_j l_j^2}\left[\frac{d\psi_i(\eta)}{d\eta}\right]_{\eta=1}\end{aligned} \tag{9.35}$$

where ρ_j, A_j, l_j are the density of the material, the area of the cross section, and the length of the flexible link j, respectively. The quantity M_{p_j} denotes the constant sum of all masses beyond the flexible link j, and J_{p_j} is the moment of inertia due to other links and payload 'seen' at the $\eta = 1$ end of the flexible link j. The quantity M_{Dp_j} accounts for the contributions of masses non-collocated at the end of link j, weighted by the relative distance from the shearing axis at the end of flexible link j.

The clamped conditions at the link base yield

$$C_{3_i} = -C_{1_i}, \qquad C_{4_i} = -C_{2_i} \tag{9.36}$$

while the mass conditions at the $\eta = 1$ end of the link lead to a homogeneous equations of the form

$$[\mathbf{F}(\beta_i)]\begin{pmatrix} C_{1_i} \\ C_{2_i} \end{pmatrix} = \mathbf{0} \tag{9.37}$$

From linear algebra, non-trivial solutions exist only if the determinant of the 2×2 matrix $[\mathbf{F}(\beta_i)]$ is equal to zero. Setting $\det[\mathbf{F}] = 0$ results in a transcendental equation, called the 'frequency equation' and this equation depends explicitly on the values of M_{p_j}, J_{p_j}, and M_{Dp_j} (Oakley & Cannon 1989). The equation is given by

$$\begin{aligned}&(1 + \cosh\beta_i\cos\beta_i) - M_j\beta_i(\cosh\beta_i\sin\beta_i - \sinh\beta_i\cos\beta_i) \\ &\quad - J_j{\beta_i}^3(\cosh\beta_i\sin\beta_i + \sinh\beta_i\cos\beta_i) \\ &\quad + M_j J_j{\beta_i}^4(1 - \cosh\beta_i\cos\beta_i) \\ &\quad - D_j^2{\beta_i}^4(1 - \cosh\beta_i\cos\beta_i) - 2D_j{\beta_i}^2\sinh\beta_i\sin\beta_i = 0\end{aligned} \tag{9.38}$$

where $M_j = M_{p_j}/(\rho_j A_j l_j)$, $J_j = J_{p_j}/(\rho_j A_j l_j^3)$, and $D_j = M_{Dp_j}/(\rho_j A_j l_j^2)$.

The transcendental Eqn (9.38) can have an infinite number of roots. A truncation is done by solving for the first N_j roots of this equation and the first N_j positive values of β_i (and thus ω_i) are used in Eqn (9.33). Since Eqn (9.37) is a homogeneous equation, both the coefficients C_{1_i} and C_{2_i} cannot be determined uniquely and hence the mode shape function $\psi_i(\eta)$ can only be obtained up to a scale factor and we have to use a suitable normalization. The implication is that the mode shape is unique but the amplitude is not unique.

In the case of clamped-mass boundary conditions, we get

$$\psi_i(\eta) = C_{2_i}\{\cos(\beta_i\eta) - \cosh(\beta_i\eta) + \nu_i\,[\sin(\beta_i\eta) - \sinh(\beta_i\eta)]\} \tag{9.39}$$

where

$$\nu_i = \frac{\sin\beta_i - \sinh\beta_i + M_j\beta_i(\cos\beta_i - \cosh\beta_i) - D_j\beta_i^2(\sin\beta_i + \sinh\beta_i)}{\cos\beta_i + \cosh\beta_i - M_j\beta_i(\sin\beta_i - \sinh\beta_i) - D_j\beta_i^2(\cos\beta_i - \cosh\beta_i)} \tag{9.40}$$

and C_{2_i} are the normalization constants.

Equation (9.38) can be readily solved for flexible manipulators with one link having a revolute joint—M_{p_1}, J_{p_1} are directly the mass and rotary inertia of the payload, respectively. In addition, M_{Dp_1} vanishes if the payload is balanced at the tip, and if there are mechanical offsets we would then have a constant M_{Dp_1} (Oakley & Cannon 1989). In the case of prismatic jointed flexible link with payload, the length of the vibrating portion of translating link l_j is a function of time.[4] In the case of flexible-link manipulators having more than one link with revolute joints, for a generic intermediate jth link, M_{Dp_j} and J_{p_j} depend on the position of successive links and thus are functions of time. For instance, in the case of a planar 2R flexible robot, M_{Dp_1} for link 1 is given by (De Luca & Siciliano 1991)

$$M_{Dp_1} = \left[m_2\frac{l_2}{2} + m_p l_2\right]\cos\theta_2 - \left[\sum_{i=1}^{N_2}\left(m_2\int_0^1 \psi_{2_i}(\eta)d\eta + m_p\psi_{2_i}(1)\right)\right.$$
$$\left.\xi_{2_i}(t)\right]\sin\theta_2$$

where m_2 and m_p are masses of link 2 and the payload, respectively, and l_2 is the length of the link 2. Note that M_{Dp_1} is a function of both the joint variable θ_2 and the elastic generalized variables ξ_{2_i} of link 2. For more than two links, J_{p_1} will also become a function of the generalized coordinates of link 3 and the following ones. For example, for a planar 3R flexible robot,

[4] For a prismatic joint, we must, however, ensure that the vibrating length l_j satisfies Euler–Bernoulli beam assumptions (see also assumption 3 of Section 9.2), viz., the length of the beam l_j must always be much greater than the beam cross-sectional dimensions.

the rotary inertia due to last two links 'seen' at the $\eta = 1$ end of the link 1 is given by

$$J_{p_1} = I_2 + I_3 + m_2 l_2^2 + m_3\left(l_2^2 + l_3^2 + 2l_2 l_3 \cos\theta_3\right)$$

where I_2 and I_3 are the inertias of links 2 and 3, respectively, m_3 and l_3 are the mass and the length of link 3, respectively.

Due to the above reasoning, the frequency equation, Eqn (9.38), and the mode shapes are time dependent, and for accurate mode shapes computation, M_{p_j}, J_{p_j}, and M_{Dp_j} should be updated as functions of the flexible manipulator configuration. In the following, we present an approach (Theodore & Ghosal 1995) to solve for the time-dependent β_i.

Time-dependent frequency equation The clamped-mass frequency equation, Eqn (9.38) can be written as

$$\begin{aligned} f(\beta_i, M_j, J_j, D_j) = {} & (1 + \cosh\beta_i \cos\beta_i) \\ & - M_j\beta_i(\cosh\beta_i \sin\beta_i - \sinh\beta_i \cos\beta_i) \\ & - J_j{\beta_i}^3(\cosh\beta_i \sin\beta_i + \sinh\beta_i \cos\beta_i) \\ & + M_j J_j {\beta_i}^4(1 - \cosh\beta_i \cos\beta_i) \\ & - {D_j}^2{\beta_i}^4(1 - \cosh\beta_i \cos\beta_i) \\ & - 2D_j{\beta_i}^2 \sinh\beta_i \sin\beta_i = 0 \end{aligned} \tag{9.41}$$

We note that when the terms M_j, J_j, and D_j change continuously in time, the solutions (β_i) for the frequency equation also change continuously. Since the roots of the frequency equation are all distinct, we can differentiate Eqn (9.41) with respect to time,

$$\begin{aligned} \frac{df(\beta_i, M_j, J_j, D_j)}{dt} = {} & \frac{\partial f}{\partial \beta_i}\frac{d\beta_i}{dt} + \frac{\partial f}{\partial M_j}\frac{dM_j}{dt} \\ & + \frac{\partial f}{\partial J_j}\frac{dJ_j}{dt} + \frac{\partial f}{\partial D_j}\frac{dD_j}{dt} = 0 \end{aligned} \tag{9.42}$$

and rearrange to obtain

$$\frac{d\beta_i}{dt} = \frac{-\left(\dfrac{\partial f}{\partial M_j}\dfrac{dM_j}{dt} + \dfrac{\partial f}{\partial J_j}\dfrac{dJ_j}{dt} + \dfrac{\partial f}{\partial D_j}\dfrac{dD_j}{dt}\right)}{\dfrac{\partial f}{\partial \beta_i}} \tag{9.43}$$

where

$$\frac{\partial f}{\partial M_j} = \beta_i[\sinh\beta_i\cos\beta_i - \cosh\beta_i\sin\beta_i + \beta_i^3 J_j(1-\cosh\beta_i\cos\beta_i)]$$

$$\frac{\partial f}{\partial J_j} = -\beta_i^3[\sinh\beta_i\cos\beta_i + \cosh\beta_i\sin\beta_i - \beta_i M_j(1-\cosh\beta_i\cos\beta_i)]$$

$$\frac{\partial f}{\partial D_j} = -2\beta_i^2\left[\sinh\beta_i\sin\beta_i + \beta_i^2 D_j(1-\cosh\beta_i\cos\beta_i)\right] \quad (9.44)$$

and

$$\begin{aligned}\frac{\partial f}{\partial \beta_i} &= \sinh\beta_i\cos\beta_i\left[1+M_j-\beta_i^2(2D_j+3J_j)+\beta_i^4(D_j^2-M_jJ_j)\right]\\ &\quad - \cosh\beta_i\sin\beta_i\left[1+M_j+\beta_i^2(2D_j+3J_j)+\beta_i^4(D_j^2-M_jJ_j)\right]\\ &\quad - 2\beta_i^3\cosh\beta_i\cos\beta_i\left[J_j-2(D_j^2-M_jJ_j)\right]\\ &\quad - 2\beta_i\sinh\beta_i\sin\beta_i(M_j+2D_j) - 4\beta_i^3\left[D_j^2-M_jJ_j\right] \quad (9.45)\end{aligned}$$

It should be noted that Eqn (9.43) is also a function of the generalized position and velocity variables of the flexible manipulator system.

The initial value $[\beta_i(t=0)]$ can be computed once from the frequency equation, Eqn (9.41), by using numerical root finding techniques, and β_i as a function of time can be solved by the numerical integration of the ODE in Eqn (9.43). The accuracy of β_i at time t will be determined by the initial value and the accuracy of the integration scheme used. The main advantage of this approach over other approaches, such as table look-up or using root finding at each configuration of the flexible manipulator, is that $\beta_i(t)$ can be solved together with the equations of motion and the computational effort will not be significantly large.

9.5.2 Finite Element Method

Another very popular method used for modelling and analysing flexible links is the *finite element method.*[5] In this method each flexible link is considered as an assemblage of a finite number of elements. By requiring that the displacements are continuous inside the element and compatible

[5] The finite element method is widely used in structural dynamics, fluid mechanics, and a variety of other fields. The development here is different from the traditional finite-element-based approaches in structural dynamics in the sense that the links of a flexible manipulators can have large rigid-body motion in addition to flexible deformation. This gives rise to coupling between the rigid-body motion and flexible deformation, which is typically not considered in structural dynamics.

across elements and the internal forces in balance at certain points (called 'nodes' of the elements), the entire link is made to act as one entity. The displacement at any point of the element is then expressed in terms of a finite number of displacements at the nodal points multiplied by simple low-degree polynomials called the 'interpolation functions'. Let PQ be one such element i on link j, with nodes i and $i+1$ as shown in Fig. 9.7. The

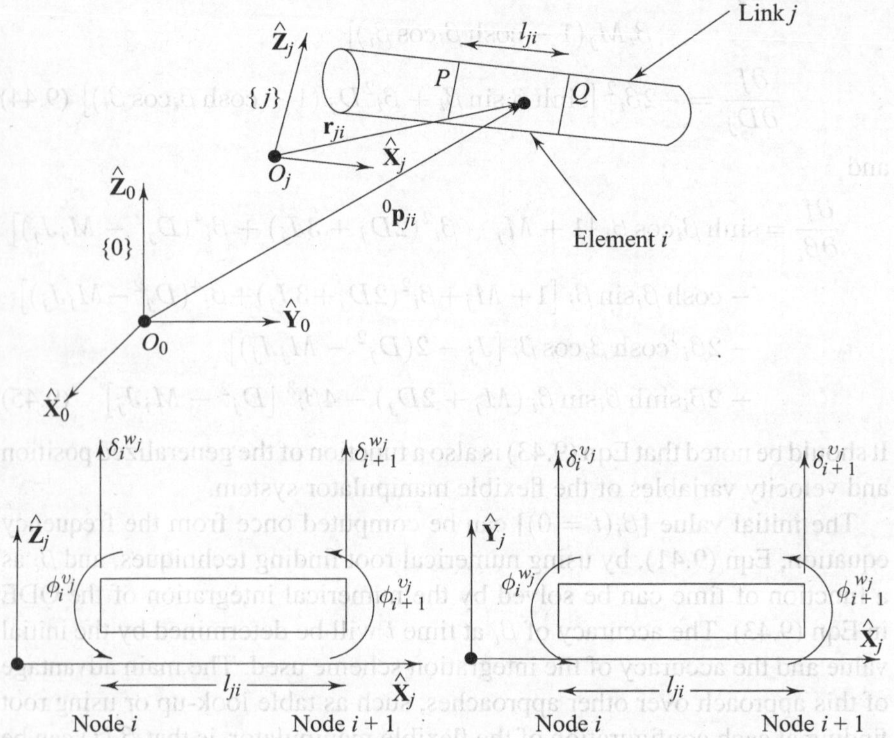

Fig. 9.7 A finite element discretization of a link j with beam element i and its nodal displacement variables

local position vector $\mathbf{r}_{ji}$ of any point along the neutral axis of the ith element, expressed in the undeformed link coordinate system, is given by

$$\mathbf{r}_{ji} = \begin{cases} \begin{pmatrix} (i-1)l_{ji} + s \\ 0 \\ 0 \end{pmatrix} + \begin{pmatrix} 0 \\ v_{ji}(s,t) \\ w_{ji}(s,t) \end{pmatrix} & \text{if joint } j \text{ is revolute} \\ \begin{pmatrix} 0 \\ 0 \\ (i-1)l_{ji} + s \end{pmatrix} + \begin{pmatrix} u_{ji}(s,t) \\ v_{ji}(s,t) \\ 0 \end{pmatrix} & \text{if joint } j \text{ is prismatic} \end{cases} \tag{9.46}$$

where l_{ji} is the length of element i, and the elastic displacements are expressed as follows.

If joint j is revolute,

$$v_{ji}(s,t) = \boldsymbol{\varphi}_i^{v_j}(s)^T \mathbf{q}_{f_{ji}}^{v_j}(t), \quad w_{ji}(s,t) = \boldsymbol{\varphi}_i^{w_j}(s)^T \mathbf{q}_{f_{ji}}^{w_j}(t) \tag{9.47}$$

with $\mathbf{q}_{f_{ji}}^{v_j}(t)$ denoting the vector $\left[\delta_i^{v_j}(t), \phi_i^{w_j}(t), \delta_{i+1}^{v_j}(t), \phi_{i+1}^{w_j}(t)\right]^T$ and $\mathbf{q}_{f_{ji}}^{w_j}(t)$ denoting the vector $\left[\delta_i^{w_j}(t), \phi_i^{v_j}(t), \delta_{i+1}^{w_j}(t), \phi_{i+1}^{v_j}(t)\right]^T$.

If joint j is prismatic,

$$u_{ji}(s,t) = \boldsymbol{\varphi}_i^{u_j}(s)^T \mathbf{q}_{f_{ji}}^{u_j}(t), \quad v_{ji}(s,t) = \boldsymbol{\varphi}_i^{v_j}(s)^T \mathbf{q}_{f_{ji}}^{v_j}(t) \tag{9.48}$$

with $\mathbf{q}_{f_{ji}}^{u_j}(t)$ denoting the vector $\left[\delta_i^{u_j}(t), \phi_i^{v_j}(t), \delta_{i+1}^{u_j}(t), \phi_{i+1}^{v_j}(t)\right]^T$ and $\mathbf{q}_{f_{ji}}^{v_j}(t)$ denoting the vector $\left[\delta_i^{v_j}(t), \phi_i^{u_j}(t), \delta_{i+1}^{v_j}(t), \phi_{i+1}^{u_j}(t)\right]^T$.
The quantities $[\delta_i^{u_j}(t), \delta_i^{v_j}(t), \delta_i^{w_j}(t)]$ and $[\phi_i^{u_j}(t), \phi_i^{v_j}(t), \phi_i^{w_j}(t)]$ denote the transverse flexural displacements and the flexural rotations of node i, along or about the X, Y, and Z axes, respectively. The interpolation functions are assumed to be the *same* in all three directions and there are various possible choices for them. We will use the simple cubic polynomials given by

$$\boldsymbol{\varphi}_i^{u_j}(s) = \boldsymbol{\varphi}_i^{v_j}(s) = \boldsymbol{\varphi}_i^{w_j}(s) = \begin{pmatrix} 1 - 3\left(\dfrac{s}{l_{ji}}\right)^2 + 2\left(\dfrac{s}{l_{ji}}\right)^3 \\ s\left(\dfrac{s}{l_{ji}} - 1\right)^2 \\ \left(\dfrac{s}{l_{ji}}\right)^2\left(3 - 2\dfrac{s}{l_{ji}}\right) \\ \dfrac{s^2}{l_{ji}}\left(\dfrac{s}{l_{ji}} - 1\right) \end{pmatrix} \tag{9.49}$$

It should be noted that for the flexible link j with a revolute joint the element length l_{ji} is a constant given by l_j/N_j, where N_j is the number of elements in link j. However, for a flexible link with a prismatic joint the element length (or the size of the elements) changes with time if the number of elements is kept constant. Clearly the prismatic-jointed flexible link is more difficult to model.

The 4×4 homogeneous transformation matrix ${}^{j_*}_{j}[T_e]$ in the finite element model reduces to

$$
{}^{j_*}_{j}[T_e] = \begin{pmatrix} 1 & -\phi^w_{N+1} & \phi^v_{N+1} & 0 \\ \phi^w_{N+1} & 1 & 0 & \delta^v_{N+1} \\ -\phi^v_{N+1} & 0 & 1 & \delta^w_{N+1} \\ 0 & 0 & 0 & 1 \end{pmatrix} \quad \text{(if joint } j - 1 \text{ is revolute)} \tag{9.50}
$$

$$
{}^{j_*}_{j}[T_e] = \begin{pmatrix} 1 & 0 & \phi^v_{N+1} & \delta^u_{N+1} \\ 0 & 1 & -\phi^u_{N+1} & \delta^v_{N+1} \\ -\phi^v_{N+1} & \phi^u_{N+1} & 1 & 0 \\ 0 & 0 & 0 & 1 \end{pmatrix} \quad \text{(if joint } j - 1 \text{ is prismatic)} \tag{9.51}
$$

where the subscript $j-1$ on u, v, w, and N has been dropped for clarity and compactness.

To enforce clamped (geometric) boundary conditions for the controlled or joint end of link j, the nodal displacement variables of the element 1, δ_{j1}, and ϕ_{j1} are set to zero. The natural boundary conditions for the flexible link on the other hand are incorporated in the model by considering the proper energy expressions due to the additional masses and the moments of inertia at the far end of the link (see Section 9.6).

The velocity of any point on the neutral axis of the ith element in the jth link in the local undeformed coordinate system is given by

$$
\dot{\mathbf{r}}_{ji} = \begin{cases} \begin{pmatrix} 0 \\ \displaystyle\sum_{k=1}^{4} \varphi^v_{ik}(s, l_{ji}) \frac{dq^v_{f_{jik}}(t)}{dt} \\ \displaystyle\sum_{k=1}^{4} \varphi^w_{ik}(s, l_{ji}) \frac{dq^w_{f_{jik}}(t)}{dt} \end{pmatrix} & \text{if joint } j \text{ is revolute} \\[2ex] \begin{pmatrix} \displaystyle\sum_{k=1}^{4} \left[\varphi^u_{ik}(s, l_{ji}) \frac{dq^u_{f_{jik}}(t)}{dt} + \frac{\partial \varphi^u_{ik}(s, l_{ji})}{\partial l_{ji}(t)} q^u_{f_{jik}}(t) \frac{U_j(t)}{N_j} \right] \\ \displaystyle\sum_{k=1}^{4} \left[\varphi^v_{ik}(s, l_{ji}) \frac{dq^v_{f_{jik}}(t)}{dt} + \frac{\partial \varphi^v_{ik}(s, l_{ji})}{\partial l_{ji}(t)} q^v_{f_{jik}}(t) \frac{U_j(t)}{N_j} \right] \\ \dfrac{iU_j(t)}{N_j} \end{pmatrix} & \text{if joint } j \text{ is prismatic} \end{cases} \tag{9.52}
$$

where the subscripts on u, v, and w have been dropped for clarity.

It may be recalled that in the case of the prismatic joint, the interpolation functions are time dependent because the element length l_{ji} changes in time. This implies that for the calculation of the time derivative of the interpolation functions, s is held fixed and $l_{ji}(t)$ is differentiated (Stylianou & Tabarrok 1994; Theodore & Ghosal 1995).

9.5.3 Comparison of Discretization Methods

In this section, we have presented two discretization methods, namely, the assumed modes method (AMM) and the finite element method (FEM). We end this section by comparing these two methods.

Before we can compare AMM and FEM, we must decide on the number of modes or elements (N_j) that must be considered to obtain an acceptable discretized model of a flexible link j. Most discretized models retain only the first few modes in the model and this is justified on the basis of the frequency—recall that each mode has an associated natural frequency and higher modes have higher natural frequency. The high frequency modes are also associated with low amplitudes and are usually dropped. From experiments, it has also been seen that actuators and sensors cannot operate beyond a certain frequency (Cannon & Schmitz 1984), and hence they can be dropped from the model; for robot manipulators, two or three modes suffices to represent the flexible dynamics reasonably accurately.[6] In the case of FEM, the first m calculated vibratory modes can be considered within acceptable accuracy when m-element model is used with minimal order polynomial interpolation functions (Przemieniecki 1968). For example, consider the natural frequencies (see Table 9.1) of a *clamped-free* beam of mass 0.33 kg, length 1.0 m, rotary inertia of joint (I_{joint_2}) 3.2 kg/m^2, and EI as 1165.5 N/m^2.

Table 9.1 Natural frequencies (Hz) of a clamped-free beam

Mode no.	No. of elements 1	2	3	Exact values
1	$2.0963e+2$	$2.0873e+2$	$2.0864e+2$	$2.0864e+2$
2	$2.0654e+3$	$1.3186e+3$	$1.3118e+3$	$1.3075e+3$
3		$4.4597e+3$	$3.7067e+3$	$3.6611e+3$
4		$1.2944e+4$	$8.3473e+3$	$7.1742e+3$
5			$1.5709e+4$	$1.1860e+4$
6			$3.1318e+4$	$1.7716e+4$

[6] For space structures and situations where there are many vibration modes with nearby frequencies, it is more difficult to decide how many modes should be included in the model. In addition, one may have to keep higher frequency modes if these modes excite servo-loop frequencies.

We observe that the first m-exact values of the natural frequencies of the beam (Meirovitch 1986) are within the acceptable accuracy when an m-element model with cubic interpolation functions [see Eqn (9.49)] is used for the clamped-free beam.

In the assumed modes and finite element models, the flexible deformation is represented by a linear combination of spatial functions multiplying time-dependent mode amplitudes or nodal coordinates, respectively. In the assumed modes model, these spatial functions are *global*, in the sense that they are defined over the entire length of the beam, and they belong to the class of *trigonometric functions*. As we will see in Section 9.6, this leads to complicated integrations to obtain the mass and stiffness matrices, especially if the link geometries are complicated. The orthogonal assumption of these functions however simplifies the number of terms in the mass and stiffness matrices and results in a *diagonal* matrix structure.

In the finite element model, on the other hand, the spatial functions are *local* because they are defined over small sub-domains of the link and they belong to the class of *polynomial functions*. The polynomial functions (cubics in our case) are simple and easy to work with and easier to integrate even for manipulator links with a complex geometry. The functions are the same for every element and they are 'nearly' orthogonal—the shape functions of an element i only contribute to the neighboring elements $i-1$ and $i+1$ and this, in turn, leads to *banded* mass and stiffness matrices.

Any constraint imposed on an infinite-dimensional dynamical system reduces the number of degrees of freedom and tends to make the system stiffer. A flexible beam in a lightweight flexible manipulator is an infinite-dimensional system governed by the Euler–Bernoulli beam equation (9.4). Any discretization procedure, whether AMM or FEM, imposes constraints in the sense that the neglected modes are zero and hence the natural frequencies predicted by a discretized model will always be *greater* than the true natural frequencies.[7] One theoretical advantage of AMM (over FEM) is that since the global mode shapes used in AMM are from a complete set (Meirovitch 1986), increasing the number of modes in the discretized model will result in monotonic convergence (from above) to the true natural frequency. Whereas, in FEM, since the local interpolation functions do not fall within the definition of a complete set, the monotonic convergence cannot always be guaranteed (Meirovitch 1986), and the natural frequencies of the FEM model tend to always overestimate the true natural frequencies of a flexible-link manipulator system—in structural dynamics applications, increasing the number of finite elements sometimes lead to a phenomenon

[7] This is also seen in the numerical results for the clamped-free beam given in Table 9.1.

of 'locking' where one can get very large or infinite natural frequencies. In the context of flexible manipulators, overestimation of natural frequencies may lead to unstable behaviour when model-based control techniques are used for flexible manipulator control (see also Section 9.7).

The local nature of the interpolation functions and the use of simple polynomials in FEM make it easier to obtain discretized models for flexible links with complex geometries. We can also use 3D beam elements for the efficient link design of the flexible manipulator arms. Moreover, the symbolic computation of the equations of motion (see Section 9.6) can be readily adapted to automation, as it merely consists of suitably assembling the element matrices for the manipulator system. These advantages outweigh the theoretical difficulties associated with the finite element method, and FEM is most commonly used for modelling flexible manipulator systems with complex link geometries.

9.6 Equations of Motion of Multi-link Flexible Manipulators

The dynamic equations of motion of a flexible manipulator are best obtained using Lagrange's formulation of dynamics presented in Section 6.3. In this section, we present the Lagrangian formulation for an n-link flexible manipulator and present an algorithm to obtain the equations of motion using a symbolic software tool such as MATHEMATICA or MAPLE.

The general form of Lagrange's equations for the clamped condition are given by

$$\frac{d}{dt}\left(\frac{\partial \mathrm{KE}}{\partial \dot{q}_{r_j}}\right)-\frac{\partial \mathrm{KE}}{\partial q_{r_j}}+\frac{\partial \mathrm{PE}}{\partial q_{r_j}}=\tau_j \quad \text{(for joint variable } q_{r_j}) \tag{9.53}$$

$$\frac{d}{dt}\left(\frac{\partial \mathrm{KE}}{\partial \dot{q}_{f_{ji}}}\right)-\frac{\partial \mathrm{KE}}{\partial q_{f_{ji}}}+\frac{\partial \mathrm{PE}}{\partial q_{f_{ji}}}=0 \quad \text{(for flexible variable } q_{f_{ji}}) \tag{9.54}$$

where KE is the total kinetic energy of the flexible manipulator system and *PE* is the total potential energy due to *elastic deformations and gravity*.[8] The generalized force corresponding to joint variable q_{r_j} is the joint input τ_j (torque for a revolute joint, or force for a prismatic joint). For the flexible variables ($\mathbf{q}_f$) the corresponding generalized force will be zero for the clamped condition at the controlled end (Book 1984). It should be noted

[8] The Lagrangian $\mathcal{L}(\mathrm{q}, \dot{\mathrm{q}})$ is defined as the difference between the total kinetic energy and the total potential energy of the system. Since the total potential energy PE is not a function of $\dot{q}_i$'s, the form in Eqns (9.53) and (9.54) is identical to the form in Eqn (6.12).

that other conditions for the controlled end of the link, such as 'pinned' or 'free', will yield non-zero generalized force for the q_{f_j} variable.

9.6.1 Kinetic Energy

The total kinetic energy of the flexible-link manipulator system is due to the motions of joints,[9] links, and the kinetic energy due to the payload. The kinetic energy for a revolute joint j, if considered as mass with rotary inertia about the axis of revolution, is given by

$$\mathrm{KE}_{\mathrm{joint}_j} = \frac{1}{2}\,{}^0\Omega_j^T\,{}^0[I_{\mathrm{joint}}]_j\,{}^0\Omega_j + \frac{1}{2}m_{\mathrm{joint}_j}\left(\frac{d\,{}^0\mathbf{O}_j}{dt}\right)^T\left(\frac{d\,{}^0\mathbf{O}_j}{dt}\right) \tag{9.55}$$

where m_{joint_j} is the mass (including actuation, transmission, and other mechanical elements), ${}^0\mathbf{O}_j$ is the position vector, and ${}^0[I_{\mathrm{joint}}]_j$ and ${}^0\Omega_j$ are the joint inertia matrix and the angular velocity vector of joint j, respectively.

The kinetic energy of a differential mass dm on flexible link j of the manipulator can be expressed as

$$d(\mathrm{KE}_{\mathrm{link}_j}) = \frac{1}{2}dm\left(\frac{d\,{}^0\mathbf{p}_j}{dt}\right)^T\left(\frac{d\,{}^0\mathbf{p}_j}{dt}\right) \tag{9.56}$$

where ${}^0\mathbf{p}_j$ is the position vector locating the differential mass dm on the link and is given by Eqn (9.25).

Under the assumption that the links are slender beams, the kinetic energy of the flexible link j can be obtained by integrating Eqn (9.56) over the length of the link as

$$\mathrm{KE}_{\mathrm{link}_j} = \begin{cases} \dfrac{1}{2}\displaystyle\int_0^{l_j} \rho_j A_j \left(\frac{d\,{}^0\mathbf{p}_j}{dt}\right)^T\left(\frac{d\,{}^0\mathbf{p}_j}{dt}\right) ds & \text{(for the assumed modes model)} \\ \dfrac{1}{2}\displaystyle\sum_{i=1}^{N_j}\int_0^{l_{ji}} \rho_j A_j \left(\frac{d\,{}^0\mathbf{p}_{ji}}{dt}\right)^T\left(\frac{d\,{}^0\mathbf{p}_{ji}}{dt}\right) ds & \text{(for the finite element model)} \end{cases} \tag{9.57}$$

where ρ_j is the density of the material, A_j is the cross-sectional area of the link j, N_j is the number of elements chosen for link j and ${}^0\mathbf{p}_{ji}$ is as shown in Fig. 9.7.

---

[9] By a joint we mean the mechanical structure, containing the actuation, transmission, and other mechanical elements, which is used to move the flexible link. Often this forms a significant percentage of the mass and inertia of the flexible manipulator system and is often considered as a lumped mass or inertia. By the 'motion of a joint' we mean the motion of this mechanical structure and *not* the rotation or translation of a flexible or rigid link about or along the joint axis.

If link j is considered rigid, the kinetic energy, Eqn (9.56), reduces to

$$\mathrm{KE}_{\mathrm{link}_j} = \frac{1}{2} m_j \left(\frac{d\,{}^0\mathbf{p}_{c_j}}{dt} \right)^T \left(\frac{d\,{}^0\mathbf{p}_{c_j}}{dt} \right) \tag{9.58}$$

where

$${}^0\mathbf{p}_{c_j} = {}^0\mathbf{O}_j + {}^0_j[R]\mathbf{r}_{c_j} \tag{9.59}$$

and m_j and $\mathbf{r}_{c_j}$ are the mass and the local position vector of the centre of mass of the rigid link j, respectively.

The kinetic energy due to the payload is given by

$$\mathrm{KE}_{\mathrm{payload}} = \frac{1}{2} m_p \left(\frac{d\,{}^0\mathbf{p}_{Tool}}{dt} \right)^T \left(\frac{d\,{}^0\mathbf{p}_{Tool}}{dt} \right) + \frac{1}{2} {}^0\Omega_{Tool}{}^T {}^0[J_p] {}^0\Omega_{Tool} \tag{9.60}$$

where ${}^0\mathbf{p}_{Tool}$ is the position vector of the centre of mass of the payload, m_p is mass of the payload, ${}^0[J_p]$ and ${}^0\Omega_{Tool}$ are the moment of inertia matrix of the payload and the angular velocity vector of the payload, respectively.

The total kinetic energy of the flexible manipulator system is

$$\mathrm{KE} = \sum_{j=1}^{n} (\mathrm{KE}_{\mathrm{joint}_j} + \mathrm{KE}_{\mathrm{link}_j}) + \mathrm{KE}_{\mathrm{payload}} \tag{9.61}$$

where n is the number of links.

9.6.2 Potential Energy

In rigid manipulators the potential energy is mainly due to gravity. Whereas in flexible manipulators, the potential energy due to the elastic deformation of the links must also be taken into account— due to the elastic deformation, *strain* energy is stored in the links and this must be added to the potential energy due to gravity. Assuming that linear elasticity relations hold good,[10] the links are slender beams, and neglecting the axial and torsional vibration components, the potential energy due to bending deformations of link j, with joint j as revolute, is given by

$$\mathrm{PE}_{f_j} = \int_0^1 \left(\frac{E_j I_{jy}}{2l_j^3} \left[\sum_{i=1}^{N_j} \frac{\partial^2 \psi_i^{v_j}(\eta)}{\partial \eta^2} \xi_i^{v_j}(t) \right]^2 + \frac{E_j I_{jz}}{2l_j^3} \left[\sum_{i=1}^{N_j} \frac{\partial^2 \psi_i^{w_j}(\eta)}{\partial \eta^2} \xi_i^{w_j}(t) \right]^2 \right) d\eta \quad \text{(for the assumed modes model)} \tag{9.62}$$

[10] This is true if elastic deformations are small, see assumption 1 in Section 9.3.

$$\text{PE}_{f_j} = \sum_{i=1}^{N_j} \int_0^{l_{ji}} \left(\frac{E_j I_{jy}}{2} \left[\sum_{k=1}^{4} \frac{\partial^2 \varphi_{ik}^{v_j}(s)}{\partial s^2} q_{f_{jik}}^{v_j}(t) \right]^2 + \frac{E_j I_{jz}}{2} \left[\sum_{k=1}^{4} \frac{\partial^2 \varphi_{ik}^{w_j}(s)}{\partial s^2} q_{f_{jik}}^{w_j}(t) \right]^2 \right) ds \quad \text{(for the finite element model)} \tag{9.63}$$

where E_j is Young's modulus, I_{jx}, I_{jy}, I_{jz} are the area moments of inertia about the respective axes of link j. It may be noted that in the expression for strain energy, the integrations can be performed once the number of mode shapes or elements (N_j) is chosen and the second partial derivatives of the associated mode shape functions (for AMM) or shape functions (for FEM) are used.

The gravitational potential energy due to the mass of joint and due to the elastic link j is of the form

$$\text{PE}_{g_j} = m_{\text{joint}_j} \mathbf{g}^T\, {}^0\mathbf{O}_j + \int_0^{l_j} \rho_j A_j \mathbf{g}^T\, {}^0\mathbf{p}_j ds \tag{9.64}$$

where g is the gravity vector in the inertial coordinate system $\{0\}$. It should be noted that if link j is rigid, Eqn (9.64) reduces to

$$\text{PE}_{g_j} = m_{\text{joint}_j} \mathbf{g}^T\, {}^0\mathbf{O}_j + m_j \mathbf{g}^T\, {}^0\mathbf{p}_{c_j} \tag{9.65}$$

where ${}^0\mathbf{p}_{c_j}$ is the location of the centre of mass of link j.

The gravitational potential energy due to the payload mass m_p is given by

$$\text{PE}_{g_{\text{payload}}} = m_p \mathbf{g}^T\, {}^0\mathbf{p}_{Tool} \tag{9.66}$$

The total potential energy of the system (PE) is given by

$$\text{PE} = \sum_{i=1}^{n} (\text{PE}_{f_j} + \text{PE}_{g_j}) + \text{PE}_{g_{\text{payload}}} \tag{9.67}$$

where n is the number of links.

Once the total kinetic and potential energy is obtained, we can use the Lagrange equations, Eqns (9.53) and (9.54), to derive the equations of motion of a flexible manipulator. The equations of motion can be written in a compact matrix form as

$$\begin{pmatrix} [\mathbf{M}_{rr}] & [\mathbf{M}_{rf}]^T \\ [\mathbf{M}_{rf}] & [\mathbf{M}_{ff}] \end{pmatrix} \begin{pmatrix} \ddot{\mathbf{q}}_r \\ \ddot{\mathbf{q}}_f \end{pmatrix} + \begin{pmatrix} \mathbf{C}_r(\mathbf{q}, \dot{\mathbf{q}}) \\ \mathbf{C}_f(\mathbf{q}, \dot{\mathbf{q}}) \end{pmatrix} + \begin{pmatrix} \mathbf{G}_r(\mathbf{q}) \\ \mathbf{G}_f(\mathbf{q}) \end{pmatrix} + \begin{pmatrix} \mathbf{0} & \mathbf{0} \\ \mathbf{0} & [\mathbf{K}] \end{pmatrix} \begin{pmatrix} \mathbf{q}_r \\ \mathbf{q}_f \end{pmatrix} = \begin{pmatrix} \boldsymbol{\tau} \\ \mathbf{0} \end{pmatrix} \tag{9.68}$$

where the vector of generalized variables $\mathbf{q}$ consists of an $n \times 1$ vector of rigid-joint variables $\mathbf{q}_r$ together with an $N \times 1$ vector of flexible variables $\mathbf{q}_f$. It may be noted that for an n-link manipulator with the number of flexible links $n_f \leq n$, if we consider the assumed modes model with N_j modes for each flexible link, N becomes $2\sum_{j=1}^{n_f} N_j$ in the spatial case and $\sum_{j=1}^{n_f} N_j$ in the planar case. On the other hand, if we consider a finite element model with N_j elements for each flexible link, N becomes $4\sum_{j=1}^{n_f} N_j$ in the spatial case and $2\sum_{j=1}^{n_f} N_j$ in the planar case. In the case of the finite element model the number N is slightly less since in *each* of the flexible links, in the *first* element, δ_{j1} and ϕ_{j1} are set to zero to represent clamped boundary conditions.

The generalized mass matrix for the flexible manipulator system, $[\mathbf{M}(\mathbf{q})]$, is an $(n+N) \times (n+N)$ symmetric and positive definite matrix as it is related to the total kinetic energy of the system. The generalized mass matrix consists of an $n \times n$ (square), symmetric, positive definite sub-matrix associated with the second derivative of the rigid joint variables or $\ddot{\mathbf{q}}_r$ and is denoted by $[\mathbf{M}_{rr}]$. In general, the elements of $[\mathbf{M}_{rr}]$ could be functions of both $\mathbf{q}_r$ and $\mathbf{q}_f$. The $N \times N$ sub-matrix $[\mathbf{M}_{ff}]$ is also a symmetric, positive definite matrix and is associated with the second derivative of the flexible variables or $\ddot{\mathbf{q}}_f$. The sub-matrix $[\mathbf{M}_{rf}]$ is $N \times n$ and represents the coupling between the rigid joint variables and the elastic displacement variables in the link. The $(n+N) \times 1$ vectors, $\mathbf{C}(\mathbf{q}, \dot{\mathbf{q}})$ and $\mathbf{G}(\mathbf{q})$, represent the Coriolis and centripetal terms and the gravity terms and are an extension of the centripetal and Coriolis terms and gravity terms obtained in the case of a rigid manipulator [see Eqn (6.14) in Section 6.3]. Similar to $[\mathbf{M}_{rr}]$ and $[\mathbf{M}_{ff}]$, the centripetal, Coriolis, and gravity terms are also partitioned into those associated with rigid variables and those associated with flexible variables as it is useful for developing model-based control laws (see Section 9.7). Finally, in the equations of motion, Eqn (9.68), the $N \times N$ symmetric, positive definite matrix $[\mathbf{K}]$ is called the flexural stiffness matrix and arises from the strain energy of the flexible links.

The right-hand side of the equations of motion contain an $n \times 1$ vector of input torques (or forces) applied at the joints. It may be mentioned again that zero generalized forces on the right-hand side, corresponding to the flexible variables $\mathbf{q}_f$'s, are a result of the clamped boundary condition used at the actuator end of a flexible link. Finally, the equations of motion as given in Eqn (9.68) do not contain frictional or dissipative terms and, as in rigid manipulators, these have to be added separately.

In a typical FEM-based solution to structural dynamics problems, the $N \times N$ matrices $[\mathbf{M}_{ff}]$ and $[\mathbf{K}]$ are used to obtain natural frequencies and mode shapes. In structural dynamics problems, in contrast to flexible manipulators, there is no gross joint motion $\mathbf{q}_r$ or coupling between the joint motion variables and the flexible variables, $\mathbf{q}_f$, in the form of $[\mathbf{M}_{rf}]$ or in the Coriolis/centripetal term $\mathbf{C}(\mathbf{q}, \dot{\mathbf{q}})$. In flexible manipulators, as mentioned earlier, the motion at the joint can excite vibrations in the flexible links, which in turn can cause motion at the joints. The equation of motion given in Eqn (9.68) captures this inherent coupling between the rigid (joint) and flexible variables.

9.6.3 Symbolic Equations of Motion

The symbolic equations of motion for a flexible manipulator, Eqn (9.68), can be derived using MATHEMATICA or MAPLE once the expressions for the kinetic and potential energies are known. The steps are very similar to those listed in Example 6.1 for a rigid planar 2R manipulator. The main difference in the case of flexible manipulators is in the additional computation of the kinetic and potential energies due to the bending of the flexible links. In the following, we present an algorithm to symbolically obtain the elements of the mass matrix of a flexible-link manipulator. It may be noted that instead of the explicit time derivative of ${}^0\mathbf{p}_j$ [as in Eqn (9.27)] we can use $\partial\,{}^0\mathbf{p}_j/\partial q_i$ to compute the elements of the mass matrix for a flexible-link manipulator.

Algorithm to obtain elements of the mass matrix

1. For $j = 1 \rightarrow n$, symbolically compute the transformation matrices ${}^{j-1}_{\;\;j}[T]$ using Eqns (9.21) and (9.22).
2. For $j = 1 \rightarrow n$, symbolically compute the position vectors ${}^0\mathbf{p}_j$ as per Eqns (9.25) and (9.26). In this step the local position vector $\mathbf{r}_j$ depends on the type of joint j (revolute or prismatic). One of the two discretization methods, discussed in Section 9.5 must be used. At this step the choice of the number of modes or elements in each flexible link is made and the $(n+N)\times 1$ vector of generalized variables $\mathbf{q} = (q_{r_1}, \cdots, q_{r_n}, q_{f_1}, \cdots, q_{f_N})$ is fixed.
3. For $j = 1 \rightarrow n$

   For $i = 1 \rightarrow (n+N)$, compute $\delta_i\,{}^0\mathbf{p}_j = (\partial\,{}^0\mathbf{p}_j/\partial q_i)$

   For $k = i \rightarrow (n+N)$, compute $\delta_k\,{}^0\mathbf{p}_j = (\partial\,{}^0\mathbf{p}_j/\partial q_k)$, and using $\delta_i\,{}^0\mathbf{p}_j$ and $\delta_k\,{}^0\mathbf{p}_j$ compute $M^j_{ik} = \int_0^{l_j} \rho_j A_j (\delta_i\,{}^0\mathbf{p}_j)^T (\delta_k\,{}^0\mathbf{p}_j) dx$

   Compute $M^{\text{link}}_{ik} = M^{\text{link}}_{ik} + M^j_{ik}$

We make the following remarks about this algorithm.

1. This algorithm does not use the time derivative of the position vector as in the case of the rigid manipulators discussed in Chapter 6. In the case of rigid manipulators, the explicit expression for the kinetic energy of the links is used to determine the Lagrangian and the equations

of motion. In the case of flexible manipulators, although a similar procedure can be adopted, this algorithm is more efficient. One can demonstrate, using basic calculus, that using $\partial\, {}^0\mathbf{p}_j/\partial q_i$ and explicit time derivatives are equivalent.

2. The integration to obtain M_{ik}^j is more complicated for assumed modes models as it involves integrating trigonometric functions and is practical only for long slender beams. In the finite element approach, the integration is much easier since it involves at most cubic functions. In addition, the finite element method is widely used in various fields and we can borrow the art of properly assembling element mass (and stiffness) matrices to obtain the mass matrix of a link and, eventually, of the entire flexible manipulator. The finite element approach is more useful when the geometry of the link is more complicated.
3. If a link in the flexible manipulator is rigid, this algorithm needs to be modified. We have to obtain the kinetic energy, compute the quadratic form, extract the mass matrix, and then carefully modify the terms of the mass matrix to include the effect of the rigid link.
4. The contribution of the mass and inertia of the joint and the payload to the mass matrix can be computed by computing the kinetic energy due to the joint and the payload, and, similar to the case of the rigid link, the terms in the mass matrix can be modified. The total generalized mass matrix of a flexible manipulator system is the sum of the contribution of links, joint (mass and inertia), and payload.

Once the elements of the mass matrix are known symbolically, the vector of the Coriolis and centripetal terms $[\mathbf{C}(\mathbf{q}, \dot{\mathbf{q}})]$ can be obtained by taking the partial derivatives of the elements of the mass matrix. The term C_{ij} for the flexible manipulator is similar to that given in Eqn (6.15) for a rigid manipulator except now the summation is from 1 to $n + N$. The gravity term can also be obtained by taking the partial derivative of the gravity portion of the potential energy with respect to q_j. The $N \times N$ stiffness matrix is also computed from the potential energy; only this time we have to take the potential energy due to the bending strain. The strain energy is only a function of the N flexible variables and can be expressed as $(1/2)\mathbf{q}_f{}^T[\mathbf{K}]\mathbf{q}_f$. From the symbolic expression of the strain energy one can collect the coefficients of $(1/2)q_{f_i}q_{f_j}$ and this is the (i, j) element of $[\mathbf{K}]$.

9.7 Control of Flexible-link Manipulators

The dynamic equations of motion of a flexible manipulator, Eqn (9.68), show the coupling between the rigid variables $\mathbf{q}_r$ and the flexible variables $\mathbf{q}_f$. The input torques or forces will not only result in gross motion of the links but the coupling will lead to transverse vibrations. This is shown schematically for a single-link planar, flexible manipulator in the block diagram in Fig. 9.8. In the block diagram, we can observe that $\ddot{\theta}_1$ can excite the flexible dynamics of the flexible manipulator through $[\mathbf{M}_{rf}]$ and the resulting $\ddot{\mathbf{q}}_f$ can in turn influence the rigid dynamics through $[\mathbf{M}_{rf}]^T$. In a multi-link flexible manipulator, there will be additional coupling due to the centripetal/Coriolis terms.

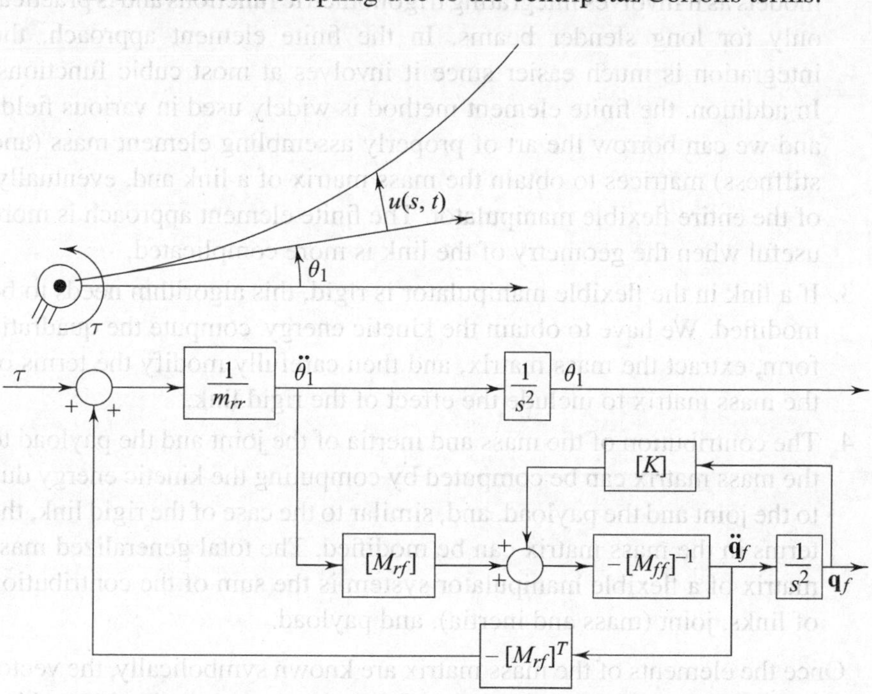

Fig. 9.8 Block diagram of a single-flexible-link manipulator

In contrast to the open-loop block diagram of a rigid, single-link manipulator (see Fig. 8.3 in Chapter 8) or the block diagram of a rigid, single-link manipulator with flexibility at the joint (see Fig. 9.2), the dynamics of a flexible-link manipulator is vastly more complicated. As a result, the control of flexible-link manipulators is vastly complicated by the fact that the control algorithm should not only track the desired motion at the joints but also suppress link vibrations excited during the motion. One approach to control the excited vibrations would be to use external *passive* dampers or to design flexible-links with damping built into the link in the form of visco-elastic or other layers capable of damping out vibrations. Passive dampers have limitations since once designed and built, their characteristics cannot be changed. In this section, we discuss the *active* control of flexible-link manipulators where the input joint torque or force *alone* is used to track a desired joint trajectory and damp out the unwanted vibration of the end-effector. We will use the truncated finite-dimensional models derived[11] in Section 9.6 to develop model-based controllers similar to the model-based controllers discussed in Chapter 8 for rigid manipulators.

9.7.1 Controllability of Flexible-link Manipulators

As seen in the equations of motion of an n-link flexible manipulator, Eqn (9.68), and schematically in Fig. 9.8 for a single-link flexible manipulator, the generalized joint variables $\mathbf{q}_f$ are not directly related to the input joint torques $\boldsymbol{\tau}$. Hence, it is not obvious that the flexible variables can be controlled *only* by joint actuator inputs $\boldsymbol{\tau}$. To analyse this in more detail, we rewrite the equations of motion of a flexible-link manipulator as

$$\begin{aligned}\ddot{\mathbf{q}}_r &= [\mathbf{H}_{rr}]\boldsymbol{\tau} - [\mathbf{H}_{rr}](\mathbf{C}_r + \mathbf{G}_r) - [\mathbf{H}_{rf}]^T(\mathbf{C}_f + \mathbf{G}_f + [\mathbf{K}]\mathbf{q}_f)\\ \ddot{\mathbf{q}}_f &= [\mathbf{H}_{rf}]\boldsymbol{\tau} - [\mathbf{H}_{rf}](\mathbf{C}_r + \mathbf{G}_r) - [\mathbf{H}_{ff}](\mathbf{C}_f + \mathbf{G}_f + [\mathbf{K}]\mathbf{q}_f)\end{aligned} \tag{9.69}$$

where

$$\begin{aligned}[\mathbf{H}_{rr}] &= ([\mathbf{M}_{rr}] - [\mathbf{M}_{rf}]^T[\mathbf{M}_{ff}]^{-1}[\mathbf{M}_{rf}])^{-1}\\ [\mathbf{H}_{rf}]^T &= -[\mathbf{H}_{rr}][\mathbf{M}_{rf}]^T[\mathbf{M}_{ff}]^{-1}\\ [\mathbf{H}_{ff}] &= ([\mathbf{M}_{ff}] - [\mathbf{M}_{rf}][\mathbf{M}_{rr}]^{-1}[\mathbf{M}_{rf}]^T)^{-1}\end{aligned} \tag{9.70}$$

To obtain these equations, we have used the fact that $[\mathbf{M}_{rr}]$ and $[\mathbf{M}_{ff}]$ are square and invertible and, for clarity, the dependence of the matrices on $\mathbf{q}$ and $\dot{\mathbf{q}}$ has been dropped.

If one or more rows of $[\mathbf{H}_{rf}]$ is a zero vector, then the corresponding $\ddot{q}_f$'s are not influenced by $\boldsymbol{\tau}$ and these $\ddot{q}_f$'s *cannot* be *directly* controlled by the joint actuator inputs, $\boldsymbol{\tau}$, alone. It may be noted that $[\mathbf{H}_{rf}]$ is dependent on $\mathbf{q}_r$ and $\mathbf{q}_f$ and hence the controllability of one or more components of $\mathbf{q}_f$ is dependent only on the flexible-manipulator geometry. Physically, if any of the deflection variables induces a moment about the same axis about which a joint input is applied, then that deflection component can be controlled by a proper choice of the joint actuator input. On the other hand, if a joint axis lies in a plane that contains the deflection components, then the induced moment due to the link deflection about the joint axis will be zero and, consequently, these deflection components cannot be controlled by the joint input. It may be mentioned that an ith generalized flexible variable can still

[11] There exists literature on the control of flexible systems where the infinite-dimensional PDE models are directly used for developing control algorithms. These advanced concepts are outside the scope of this textbook. The interested reader is referred to Pazy (1983), Chen et al. (1987), and Morgül (1992) as a starting point for this topic.

be influenced *indirectly* by $[\mathbf{H}_{ff}](\mathbf{C}_f + \mathbf{G}_f + [\mathbf{K}]\mathbf{q}_f)$ as long it is *non-zero*. Due to this, the condition of one or more rows of $[\mathbf{H}_{rf}]$ becoming a zero vector is slightly different from classical uncontrollability; it has been termed as an *inaccessibility* condition. Of course if one or more rows of $[\mathbf{H}_{rf}]$ become zero and $[\mathbf{H}_{ff}](\mathbf{C}_f + \mathbf{G}_f + [\mathbf{K}]\mathbf{q}_f)$ is also zero simultaneously, then the flexible manipulator is uncontrollable at that position.

9.7.2 Model-based Control for Trajectory Following

To develop a model-based control scheme for a flexible-link manipulator system, we start by rewriting the equations of motion, Eqn (9.68), as

$$[\mathbf{M}_{rr}]\ddot{\mathbf{q}}_r + [\mathbf{M}_{rf}]^T\ddot{\mathbf{q}}_f + \mathbf{C}_r(\mathbf{q},\dot{\mathbf{q}}) + \mathbf{G}_r(\mathbf{q}) = \boldsymbol{\tau} \quad (9.71)$$

$$[\mathbf{M}_{rf}]\ddot{\mathbf{q}}_r + [\mathbf{M}_{ff}]\ddot{\mathbf{q}}_f + \mathbf{C}_f(\mathbf{q},\dot{\mathbf{q}}) + \mathbf{G}_f(\mathbf{q}) + [\mathbf{K}]\mathbf{q}_f = \mathbf{0} \quad (9.72)$$

From Eqn (9.72), we can solve for $\ddot{\mathbf{q}}_f$ as

$$\ddot{\mathbf{q}}_f = -[\mathbf{M}_{ff}]^{-1}([\mathbf{M}_{rf}]\ddot{\mathbf{q}}_r + \mathbf{C}_f + \mathbf{G}_f + [\mathbf{K}]\mathbf{q}_f) \quad (9.73)$$

Substituting $\ddot{\mathbf{q}}_f$ in Eqn (9.71), we get after rearrangement

$$([\mathbf{M}_{rr}] - [\mathbf{M}_{rf}]^T[\mathbf{M}_{ff}]^{-1}[\mathbf{M}_{rf}])\ddot{\mathbf{q}}_r + \{\mathbf{C}_r + \mathbf{G}_r - [\mathbf{M}_{rf}]^T[\mathbf{M}_{ff}]^{-1}(\mathbf{C}_f + \mathbf{G}_f + [\mathbf{K}]\mathbf{q}_f)\} = \boldsymbol{\tau} \quad (9.74)$$

Equation (9.74) is similar in form to Eqn (8.28) used to develop model-based control schemes for rigid manipulators and can be thought of as the equation of motion of an equivalent rigid-link manipulator. The equivalent manipulator has a mass matrix given by $([\mathbf{M}_{rr}] - [\mathbf{M}_{rf}]^T[\mathbf{M}_{ff}]^{-1}[\mathbf{M}_{rf}])$ and an equivalent non-linear term given by $(\mathbf{C}_r + \mathbf{G}_r - [\mathbf{M}_{rf}]^T\,[\mathbf{M}_{ff}]^{-1}(\mathbf{C}_f + \mathbf{G}_f + [\mathbf{K}]\mathbf{q}_f))$ and they incorporate the modification that the rigid-body dynamics undergoes due to link vibrations. It may be noted that the control vector $\boldsymbol{\tau}$ and rigid-body degrees of freedom of the flexible manipulator $\mathbf{q}_r$ are both $n \times 1$, and the equivalent mass matrix has full rank (see Exercise 9.7).

Following the concept of control-law partitioning described in section 8.4 for rigid manipulators, we assume that the output of a controller can be written as

$$\boldsymbol{\tau}_{\mathbf{q}_r} = [\alpha]\boldsymbol{\tau}'_{\mathbf{q}_r} + \beta \quad (9.75)$$

where we choose

$$[\alpha] = [\mathbf{M}_{rr}] - [\mathbf{M}_{rf}]^T[\mathbf{M}_{ff}]^{-1}[\mathbf{M}_{rf}]$$
$$\beta = \mathbf{C}_r + \mathbf{G}_r - [\mathbf{M}_{rf}]^T\,[\mathbf{M}_{ff}]^{-1}(\mathbf{C}_f + \mathbf{G}_f + [\mathbf{K}]\mathbf{q}_f) \quad (9.76)$$

Substituting Eqns (9.75) and (9.76) in Eqn (9.74), similar to the case of a rigid manipulator discussed in Section 8.4, we get

$$\tau'_{\mathbf{q}_r} = \ddot{\mathbf{q}}_r \tag{9.77}$$

which represents a unit inertia plant with a new input $\tau'_{\mathbf{q}_r}$. If $\tau'_{\mathbf{q}_r}$ is chosen as

$$\tau'_{\mathbf{q}_r} = \ddot{\mathbf{q}}_{r_d}(t) + [K_p]_{\mathbf{q}_r}\mathbf{e}(t) + [K_v]_{\mathbf{q}_r}\dot{\mathbf{e}}(\mathbf{t}) \tag{9.78}$$

with $\mathbf{e}(t) = \mathbf{q}_{r_d} - \mathbf{q}_r$ and $\mathbf{q}_{r_d}(t)$ denoting a desired joint trajectory then we can write a linear error equation for the unit inertia plant as

$$\ddot{\mathbf{e}}_r(t) + [K_p]_{\mathbf{q}_r}\mathbf{e}_r(t) + [K_v]_{\mathbf{q}_r}\dot{\mathbf{e}}_r(t) = \mathbf{0} \tag{9.79}$$

and by appropriate choice of controller gains $[K_p]_{\mathbf{q}_r}$ and $[K_v]_{\mathbf{q}_r}$, we can ensure that $\mathbf{e}_r(t)$ and $\mathbf{e}_r(t)$ tend to zero asymptotically and the controller can track any desired joint trajectory.

The model-based control law (also called a joint inversion control law) given in Eqns (9.75) and (9.76) yields the closed-loop equations given by

$$\ddot{\mathbf{q}}_r(t) = \tau'_{\mathbf{q}_r} \tag{9.80}$$

$$[\mathbf{M}_{ff}]\ddot{\mathbf{q}}_f + \mathbf{C}_f(\mathbf{q}, \dot{\mathbf{q}}) + \mathbf{G}_f(\mathbf{q}) + [\mathbf{K}]\mathbf{q}_f = -[\mathbf{M}_{rf}]\tau'_{\mathbf{q}_r} \tag{9.81}$$

and the choice of $\tau'_{\mathbf{q}_r}$, given in Eqn (9.78), results in smooth tracking of a desired $\mathbf{q}_{r_d}(t)$. However, this is true as long as the induced flexible vibrations are also stable. The generalized flexible variables $\mathbf{q}_f$ are coupled to control input $\tau'_{\mathbf{q}_r}$ through the non-zero rows of the coupling matrix $[\mathbf{M}_{rf}]$ and are unobservable from the output $\mathbf{q}_r$ (see Fig. 9.8 for a simple one-link planar example). To study the stability of the induced flexible oscillations, we consider the zero dynamics[12] given by

$$\ddot{\mathbf{q}}_f = -[\mathbf{M}_{ff}]^{-1}(\mathbf{C}_f + \mathbf{G}_f + [\mathbf{K}]\mathbf{q}_f) \tag{9.82}$$

where all terms are evaluated for a constant $\mathbf{q}_r^*$ and for $\dot{\mathbf{q}}_r = \mathbf{0}$. The equilibrium points of the system are given by $\dot{\mathbf{q}}_f = \mathbf{0}$ and a $\mathbf{q}_f^*$ which satisfies

$$[\mathbf{K}]\mathbf{q}_f^* + \mathbf{G}_f(\mathbf{q}_r^*, \mathbf{q}_f^*) = \mathbf{0} \tag{9.83}$$

It may be noted that $\mathbf{q}_f^*$ corresponds to the static deflection of the flexible links due to gravity.

---

12 The zero dynamics of a non-linear system describe the dynamic behaviour of the system when inputs are chosen to constrain the outputs of the system to be zero or constant (Isidori 1989).

To study the stability, we choose a Lyapunov function

$$V(\mathbf{q}_f, \dot{\mathbf{q}}_f) = \frac{1}{2}\dot{\mathbf{q}}_f^T[\mathbf{M}_{ff}]\dot{\mathbf{q}}_f + \frac{1}{2}(\mathbf{q}_f^* - \mathbf{q}_f)^T[\mathbf{K}](\mathbf{q}_f^* - \mathbf{q}_f)$$
$$+ \left[V_G(\mathbf{q}_r^*, \mathbf{q}_f) - V_G(\mathbf{q}_r^*, \mathbf{q}_f^*)\right]$$
$$+ (\mathbf{q}_f^* - \mathbf{q}_f)^T\mathbf{G}_f(\mathbf{q}_r^*, \mathbf{q}_f^*) \tag{9.84}$$

where V_G denotes the gravitational potential energy giving rise to the gravity term $\mathbf{G}_f$—the gravity term is the partial derivative of V_G with respect to $\mathbf{q}_f$. We can observe that the Lyapunov function $V(\mathbf{q}_f, \dot{\mathbf{q}}_f)$ is positive and vanishes only at the desired equilibrium state $(\mathbf{q}_f^{*T}, \mathbf{0})^T$. The time derivative of $V(\mathbf{q}_f, \dot{\mathbf{q}}_f)$ along the trajectories of system described by Eqn (9.82) is given by

$$\dot{V} = \dot{\mathbf{q}}_f^T([\mathbf{M}_{ff}]\ddot{\mathbf{q}}_f + \frac{1}{2}[\dot{\mathbf{M}}_{ff}]\dot{\mathbf{q}}_f) - \dot{\mathbf{q}}_f^T[\mathbf{K}](\mathbf{q}_f^* - \mathbf{q}_f)$$
$$+ \dot{\mathbf{q}}_f^T\left[\mathbf{G}_f(\mathbf{q}_r^*, \mathbf{q}_f) - \mathbf{G}_f(\mathbf{q}_r^*, \mathbf{q}_f^*)\right]$$
$$= \frac{1}{2}\dot{\mathbf{q}}_f^T\left([\dot{\mathbf{M}}_{ff}] - 2[\mathbf{C}_{ff}]\right)\dot{\mathbf{q}}_f$$
$$- \dot{\mathbf{q}}_f^T\left[[\mathbf{K}]\mathbf{q}_f^* + \mathbf{G}_f(\mathbf{q}_r^*, \mathbf{q}_f^*)\right] \tag{9.85}$$

where we have used the fact that the velocity-dependent term $\mathbf{C}_f$ can be always written as $\mathbf{C}_f = [\mathbf{C}_{ff}]\dot{\mathbf{q}}_f$ (see Exercise 9.9) such that $[[\dot{\mathbf{M}}_{ff}] - 2[\mathbf{C}_{ff}]]$ is skew symmetric.

From the skew-symmetric property of $[[\dot{\mathbf{M}}_{ff}] - 2[\mathbf{C}_{ff}]]$ and since the equilibrium state satisfies Eqn (9.83), we get $\dot{V} = 0$. This implies that under the model-based control law, given in Eqns (9.75) and (9.76), the zero dynamics of the resulting closed-loop system is *not* asymptotically stable but only *critically* stable. However, if material damping is incorporated in the model, the zero dynamics of the system can become *asymptotically stable*.[13] Fortunately, some amount of damping is always naturally present in the links of the flexible manipulators, although perhaps not to the extent necessary for the precise positioning of the end-effector at the end of the joint trajectory. In the following section, we present a control scheme, based on end-point sensing, which can be used to damp out vibrations once the joint variables reach their desired end positions.

[13] The reasoning is similar to the example of the Lyapunov stability analysis of a damped single-link manipulator discussed in Section 8.10.1.

9.7.3 End Position Vibration Control

The control scheme for end position vibration control is based on sensing the vibration of the end-effector and using the rigid Jacobian matrix of a flexible manipulator. We will first develop the notions of Jacobian matrices for a flexible-link manipulator.

In Section 9.4, we presented the derivation of the 4×4 transformation matrices for a flexible multi-link manipulator. Using these matrices we can obtain the position vector, ${}^0\mathbf{p}_{Tool}$ and the rotation matrix ${}^{0}_{Tool}[R]$, describing the position and orientation of the end-effector or the tool with respect to $\{0\}$. We denote the 6×1 entity representing the position and orientation of the end-effector by $\mathcal{X}$. For a multi-link flexible manipulator, $\mathcal{X}$ will be a function of rigid and flexible variables, namely, $\mathbf{q}_r$ and $\mathbf{q}_f$. We can write, in general,

$$\mathcal{X} = \mathbf{f}(\mathbf{q}_r, \mathbf{q}_f) \tag{9.86}$$

and at a desired end-effector end position and orientation, we can write

$$\mathcal{X}_d = \mathbf{f}(\mathbf{q}_{r_d}, \mathbf{q}_{f_d}) \tag{9.87}$$

where the subscript d denotes the desired end position of the flexible manipulator.

Using Taylor's series expansion about the desired $(\mathbf{q}_{r_d}^T, \mathbf{q}_{f_d}^T)^T$, and assuming small motions, we get up to first order

$$\begin{aligned}\mathcal{X}_d + \delta\mathcal{X} &= \mathbf{f}((\mathbf{q}_{r_d} + \delta\mathbf{q}_r), (\mathbf{q}_{f_d} + \delta\mathbf{q}_f)) \\ &= \mathbf{f}(\mathbf{q}_{r_d}, \mathbf{q}_{f_d}) + \left(\frac{\partial \mathbf{f}}{\partial \mathbf{q}_r}\right)_{\mathbf{q}=\mathbf{q}_d} \delta\mathbf{q}_r + \left(\frac{\partial \mathbf{f}}{\partial \mathbf{q}_f}\right)_{\mathbf{q}=\mathbf{q}_d} \delta\mathbf{q}_f\end{aligned} \tag{9.88}$$

where $\mathbf{q}$ denotes the vector of rigid and flexible variables $({\mathbf{q}_r}^T, {\mathbf{q}_f}^T)^T$. Equation (9.88) can be written in a matrix form as

$$\delta\mathcal{X} = [J_{\mathbf{q}_r}]\delta\mathbf{q}_r + [J_{\mathbf{q}_f}]\delta\mathbf{q}_f \tag{9.89}$$

where $[J_{\mathbf{q}_r}]$ is the $6 \times n$ dimensional Jacobian associated with differential changes in joint variables $\mathbf{q}_r$, and $[J_{\mathbf{q}_f}]$ is the $6 \times N$ dimensional Jacobian associated with differential changes in flexible variables. The Jacobian matrices are similar in concept to the Jacobian matrix derived for rigid serial manipulators (see Section 5.4); however, in the case of flexible manipulators, there are two of them arising from the rigid joint and flexible variables.

The Jacobian $[J_{\mathbf{q}_r}]$ can be split into two parts as

$$[J_{\mathbf{q}_r}] = [J_{\mathbf{q}_r}^r(\mathbf{q}_{r_d})] + [J_{\mathbf{q}_r}^f(\mathbf{q}_{r_d}, \mathbf{q}_{f_d})] \tag{9.90}$$

such that

$$[J^r_{\mathbf{q}_r}(\mathbf{q}_{r_d})] = \left(\frac{\partial \mathbf{f}}{\partial \mathbf{q}_r}\right)_{\mathbf{q}_r=\mathbf{q}_{r_d},\mathbf{q}_f=\mathbf{0}}$$

$$[J^f_{\mathbf{q}_r}(\mathbf{q}_{r_d},\mathbf{q}_{f_d})] = \left(\frac{\partial \mathbf{f}}{\partial \mathbf{q}_r}\right)_{\mathbf{q}_r=\mathbf{q}_{r_d},\mathbf{q}_f=\mathbf{q}_{f_d}}$$

The Jacobian $[J^r_{\mathbf{q}_r}]$ always exists as it is the the conventional Jacobian for rigid-link manipulators (see Section 5.4) and is square for non-redundant robots. The Jacobian $[J^r_{\mathbf{q}_r}]$ will be called the *rigid Jacobian* for the flexible manipulator. In most flexible manipulators, it is difficult to measure or estimate all the flexible variables in all the links, i.e., the full vector $\mathbf{q}_f$, and hence it is not practical to obtain the Jacobian matrix $[J_{\mathbf{q}_f}]$. It is more reasonable to assume that only measurements of the end-effector vibrations, $\mathcal{X}(t)$, are available, and hence, in this section we will use *only* the rigid Jacobian together with measurements of $\mathcal{X}$ and $\dot{\mathcal{X}}$ to develop a controller which can damp out vibrations at the end of the joint motion.

A controller based on measured $\mathcal{X}$ and $\dot{\mathcal{X}}$ with gravity compensation can be written as

$$\boldsymbol{\tau}_{\mathcal{X}} = [J^r_{\mathbf{q}_r}]^T(-[K_p]_{\mathcal{X}}\delta\mathcal{X} - [K_v]_{\mathcal{X}}\dot{\mathcal{X}}) + \mathbf{G}_r(\mathbf{q}_{r_d},\mathbf{q}_{f_d}) \tag{9.91}$$

where $\delta\mathcal{X}$ is the quantity $\mathcal{X} - \mathcal{X}_d$ and is the error between the measured end-effector position and the desired end-effector position.[14] The gain matrices $[K_p]_{\mathcal{X}}$ and $[K_v]_{\mathcal{X}}$ are constant diagonal matrices representing the position and velocity gains, respectively. The vector $\mathbf{q}_{r_d}$ is the final point of the desired joint trajectory and $\mathbf{q}_{f_d}$ is defined by

$$\mathbf{q}_{f_d} = -[\mathbf{K}]^{-1}\mathbf{G}_f(\mathbf{q}_{r_d},\mathbf{q}_{f_d}) \tag{9.92}$$

and corresponds to the static deflection of links under gravity at the desired joint configuration $\mathbf{q}_{r_d}$. For typical fine manipulation or pick-and-place tasks with a flexible manipulator, the desired end-effector velocity $\dot{\mathcal{X}}_d$ at the end of the trajectory following phase will be zero and hence is not considered in the control law. It may be kept in mind that the error $\mathcal{X} - \mathcal{X}_d$ is due to the *flexible* vibrations and is expected to be small. Finally, the control torque $\boldsymbol{\tau}_{\mathcal{X}}$ is applied at the joint although $\mathcal{X} - \mathcal{X}_d$ is a Cartesian error vector. The projection of the Cartesian error to joint torque is done by the transpose of the rigid Jacobian $[J^r_{\mathbf{q}_r}]$ and is similar in concept to the transformation of the Cartesian force to joint torques used for the Cartesian control of rigid manipulators in Section 8.7.

[14] Note that error in $(\cdot)$ has been defined as $(\cdot)_d - (\cdot)$ till now and hence the negative sign in the control law. The error is defined as $\mathcal{X} - \mathcal{X}_d$ to be consistent with Taylor's series expansion in Eqn (9.88).

To investigate the stability of the control scheme given in Eqn (9.91), we note that equilibrium states of the closed-loop system given by Eqns (9.71), (9.72), and under the control scheme given in Eqn (9.91) satisfy the equations

$$\mathbf{G}_r(\mathbf{q}_r, \mathbf{q}_f) = -[J^r_{\mathbf{q}_r}]^T [K_p]_{\mathcal{X}} \delta\mathcal{X} + \mathbf{G}_r(\mathbf{q}_{r_d}, \mathbf{q}_{f_d}) \tag{9.93}$$

$$\mathbf{G}_f(\mathbf{q}_r, \mathbf{q}_f) = -[\mathbf{K}]\mathbf{q}_f \tag{9.94}$$

For small deformations, $\mathbf{G}_f$ depends only on the joint variables $\mathbf{q}_r$ and Eqn (9.94) has a unique solution $\mathbf{q}_f$ for any value of $\mathbf{q}_r \in \Re^n$. Adding $[\mathbf{K}]\mathbf{q}_{f_d} + \mathbf{G}_f(\mathbf{q}_d) = \mathbf{0}$ [from Eqn (9.92)] to the right-hand side of Eqn (9.94) yields

$$\begin{aligned}
[\widehat{\mathbf{K}}] \begin{pmatrix} \mathcal{X}_d - \mathcal{X} \\ \mathbf{q}_{f_d} - \mathbf{q}_f \end{pmatrix} &\triangleq \begin{pmatrix} [J^r_{\mathbf{q}_r}]^T [K_p]_{\mathcal{X}} & \mathbf{0} \\ \mathbf{0} & [\mathbf{K}] \end{pmatrix} \begin{pmatrix} \mathcal{X}_d - \mathcal{X} \\ \mathbf{q}_{f_d} - \mathbf{q}_f \end{pmatrix} \\
&= \begin{pmatrix} \mathbf{G}_r(\mathbf{q}_r, \mathbf{q}_f) - \mathbf{G}_r(\mathbf{q}_{r_d}, \mathbf{q}_{f_d}) \\ \mathbf{G}_f(\mathbf{q}_r, \mathbf{q}_f) - \mathbf{G}_f(\mathbf{q}_{r_d}, \mathbf{q}_{f_d}) \end{pmatrix} \\
&= \mathbf{G}(\mathbf{q}) - \mathbf{G}(\mathbf{q}_d)
\end{aligned} \tag{9.95}$$

It has been shown by De Luca and Siciliano (1993) that the gravity term $\mathbf{G}(\mathbf{q})$ satisfies the inequality

$$\| \mathbf{G}(q_1) - \mathbf{G}(q_2) \| \le c \| q_1 - q_2 \|, \quad \forall q_i \in \Re^n \times \Re^N \quad (i = 1, 2) \tag{9.96}$$

where c is a positive scalar constant dependent on the potential energy and the maximum eigenvalue of $[\mathbf{K}]$. Under the assumption that the minimum eigenvalue[15] of $[\widehat{\mathbf{K}}]$, denoted by $\lambda_{\min}([\widehat{\mathbf{K}}])$, satisfies the inequality

$$\lambda_{\min}([\widehat{\mathbf{K}}]) > c \tag{9.97}$$

we have for $\mathcal{X}_d \neq \mathcal{X}$ and $\mathbf{q}_{f_d} \neq \mathbf{q}_f$,

$$\begin{aligned}
\left\| [\widehat{\mathbf{K}}] \begin{pmatrix} \mathcal{X}_d - \mathcal{X} \\ \mathbf{q}_{f_d} - \mathbf{q}_f \end{pmatrix} \right\| &\ge \lambda_{\min}\left([\widehat{\mathbf{K}}]\right) \left\| \begin{pmatrix} \mathcal{X}_d - \mathcal{X} \\ \mathbf{q}_{f_d} - \mathbf{q}_f \end{pmatrix} \right\| \\
&> c \| \mathbf{q} - \mathbf{q}_d \| \\
&\ge \| \mathbf{c}(\mathbf{q}) - \mathbf{c}(\mathbf{q}_d) \|
\end{aligned} \tag{9.98}$$

where the last inequality follows from Eqn (9.96). This analysis implies that $\mathbf{q} = \mathbf{q}_d$ (hence $\mathcal{X}_d = \mathcal{X}$), $\dot{\mathbf{q}} = \mathbf{0}$ are the *unique* equilibrium states of the closed-loop system represented by Eqns (9.71), (9.72), and under the

[15] The maximum eigenvalue of a square matrix is also called a matrix norm, and physically the maximum and minimum eigenvalue of $[\mathbf{K}]$ 'represents' the maximum and minimum stiffness of the flexible structure.

control scheme given in Eqn (9.91). The condition in Eqn (9.97) will be automatically satisfied if the stiffness matrix $[\mathbf{K}]$ satisfies

$$\lambda_{\min}([\mathbf{K}]) > c \tag{9.99}$$

and that the position gain matrix $[K_p]_{\mathcal{X}}$ is chosen so that the condition

$$\lambda_{\min}\left([J^r_{\mathbf{q}_r}]^T [K_p]_{\mathcal{X}}\right) > c \tag{9.100}$$

is also satisfied. Physically, this analysis implies that we can have an arbitrary equilibrium point $\mathbf{q} = \mathbf{q}_d$ or $\mathcal{X} = \mathcal{X}_d$ and $\dot{\mathbf{q}} = \mathbf{0}$, if the minimum stiffness and the minimum controller position gain are large enough to overcome the static deflection due to gravity.

To analyse the stability of the equilibrium point, we consider the Lyapunov function

$$\begin{aligned} V = {} & \frac{1}{2}\dot{\mathbf{q}}^T[\mathbf{M}(\mathbf{q})]\dot{\mathbf{q}} + \frac{1}{2}(\mathbf{q}_{f_d} - \mathbf{q}_f)^T[\mathbf{K}](\mathbf{q}_{f_d} - \mathbf{q}_f) \\ & + [V_G(\mathbf{q}) - V_G(\mathbf{q}_d)] + (\mathbf{q}_d - \mathbf{q})^T \mathbf{G}(\mathbf{q}_d) \\ & + \frac{1}{2}\delta\mathcal{X}^T[K_p]_{\mathcal{X}}\delta\mathcal{X} \end{aligned} \tag{9.101}$$

which vanishes only at the desired equilibrium states due to Eqns (9.93)–(9.98). It may be noted that V_G denotes the gravitational potential energy giving rise to the gravity term $\mathbf{G}(\mathbf{q})$. The time derivative of V along the trajectories of the closed-loop system, Eqns (9.71)–(9.72) and (9.91) is

$$\begin{aligned} \dot{V} = {} & \dot{\mathbf{q}}^T\left([\mathbf{M}]\ddot{\mathbf{q}} + \frac{1}{2}[\dot{\mathbf{M}}]\dot{\mathbf{q}}\right) - \dot{\mathbf{q}}_f^T[\mathbf{K}](\mathbf{q}_{f_d} - \mathbf{q}_f) \\ & + \dot{\mathbf{q}}^T[\mathbf{G}(\mathbf{q}) - \mathbf{G}(\mathbf{q}_d)] + \dot{\mathcal{X}}^T[K_p]_{\mathcal{X}}\delta\mathcal{X} \end{aligned} \tag{9.102}$$

Using the equations of motion and the skew-symmetry property we get

$$\dot{V} = \dot{\mathbf{q}}_r^T \boldsymbol{\tau}_{\mathcal{X}} - \dot{\mathbf{q}}^T \begin{pmatrix} \mathbf{G}_r(\mathbf{q}_d) \\ [\mathbf{K}]\mathbf{q}_{f_d} + \mathbf{G}_f(\mathbf{q}_d) \end{pmatrix} + \dot{\mathcal{X}}^T[K_p]_{\mathcal{X}}\delta\mathcal{X} \tag{9.103}$$

Using the control law given in Eqns (9.91) and (9.92), we obtain

$$\dot{V} = -\dot{\mathcal{X}}^T[K_v]_{\mathcal{X}}\dot{\mathcal{X}} + (\dot{\mathcal{X}} - [J^r_{\mathbf{q}_r}]\dot{\mathbf{q}}_r)^T([K_p]_{\mathcal{X}}\delta\mathcal{X} + [K_v]_{\mathcal{X}}\dot{\mathcal{X}}) \tag{9.104}$$

and $\dot{V}$ is *strictly* negative if

$$|\,(\dot{\mathcal{X}} - [J^r_{\mathbf{q}_r}]\dot{\mathbf{q}}_r t)^T([K_p]_{\mathcal{X}}\delta\mathcal{X} + [K_v]_{\mathcal{X}}\dot{\mathcal{X}})\,| \;<\; |\,\dot{\mathcal{X}}^T[K_v]_{\mathcal{X}}\dot{\mathcal{X}}\,| \tag{9.105}$$

where $|\cdot|$ denotes the absolute value.

The velocity gains $[K_v]_{\mathcal{X}}$ must satisfy the inequality in Eqn (9.105) for stability, and we can use the inequality to obtain a lower bound on the minimum eigenvalue of $[K_v]_{\mathcal{X}}$. We define $\| (\dot{\mathcal{X}} - [J^r_{\mathbf{q}_r}]\dot{\mathbf{q}}_r) \| = \gamma$, $\| \delta\mathcal{X} \| = \alpha$, $\| \dot{\mathcal{X}} \| = \beta$, $\lambda_{\min}([K_p]_{\mathcal{X}}) = \lambda_p$, and $\lambda_{\min}([K_v]_{\mathcal{X}}) = \lambda_v$, at the end of the trajectory following phase of the links. λ_p is so chosen such that it satisfies the inequality in Eqn (9.100). Then the condition $\dot{V} < 0$ yields

$$\lambda_v > \frac{\gamma\, \lambda_p\, \alpha}{\beta(\beta - \gamma)} \tag{9.106}$$

It should be noted that $\beta \neq 0$ at the end of trajectory following motion of joints as the link vibrations are not driven to zero at this point. The deflected configuration of the arm at the end of the joint trajectory provides the initial conditions to the vibration controller, and therefore the minimum eigenvalue of the velocity feedback gain matrix can be chosen from Eqn (9.106) to ensure stability.

The above analysis shows that $V(t)$ is decreasing as long as $\dot{\mathcal{X}}$ is *not* zero. To show that the end-effector of a flexible-link manipulator will not reach a position where $\dot{\mathcal{X}} = \mathbf{0}$, but $\mathcal{X} \neq \mathcal{X}_d$, we can invoke LaSalle's invariance principle (LaSalle & Lefschetz 1961) and show that $\ddot{\mathbf{q}} = \mathbf{0}$ if and only if $\mathbf{q} = \mathbf{q}_d$, or $\mathbf{q}_r = \mathbf{q}_{r_d}$ and $\mathbf{q}_f = \mathbf{q}_{f_d}$, and hence the equilibrium state of the flexible-link manipulator system is asymptotically stable.

The end position vibration control law given in Eqn (9.91) does not require any feedback from the deflection variables and is similar to a PD controller with gravity compensation [see Section 8.10.2], Eqn (8.92)). The gravity compensation is, however, partial as only the rigid portion of the full gravity term $\mathbf{G}(\mathbf{q})$ is used in the control law.

9.7.4 A Two-stage Control Algorithm

The model-based control law given in Eqns (9.75), (9.76), and (9.78) can achieve asymptotic trajectory following for the rigid-joint variables $\mathbf{q}_r$. It, however, has poor capability to damp out end-effector vibrations at the end of the trajectory following phase. The end position vibration control law given in Eqn (9.91) can be used after the trajectory following phase to damp out end-effector vibrations excited during the trajectory following phase. We can thus have a two-stage controller given by

$$\boldsymbol{\tau} = ([U] - [\mathbf{S}])\boldsymbol{\tau}_{\mathbf{q}_r} + [\mathbf{S}]\boldsymbol{\tau}_{\mathcal{X}} \tag{9.107}$$

where

$$[\mathbf{S}] = \begin{cases} [\mathbf{0}] \text{ null matrix during joint trajectory tracking stage} \\ [U] \text{ identity matrix during end position vibration control} \end{cases}$$

and

$$\boldsymbol{\tau}_{\mathbf{q}_r} = ([\mathbf{M}_{rr}] - [\mathbf{M}_{rf}]^T[\mathbf{M}_{ff}]^{-1}[\mathbf{M}_{rf}])\boldsymbol{\tau}'_{\mathbf{q}_r} + \mathbf{C}_r + \mathbf{G}_r$$
$$-[\mathbf{M}_{rf}]^T\,[\mathbf{M}_{ff}]^{-1}(\mathbf{C}_f + \mathbf{G}_f + [\mathbf{K}]\mathbf{q}_f) \qquad (9.108)$$
$$\boldsymbol{\tau}_{\mathcal{X}} = [J^r_{\mathbf{q}_r}]^T(-[K_p]_{\mathcal{X}}\delta\mathcal{X} - [K_v]_{\mathcal{X}}\dot{\mathcal{X}}) + \mathbf{G}_r(\mathbf{q}_{r_d}, \mathbf{q}_{f_d})$$

A block diagram representation of the two-stage controller is shown in Fig. 9.9.

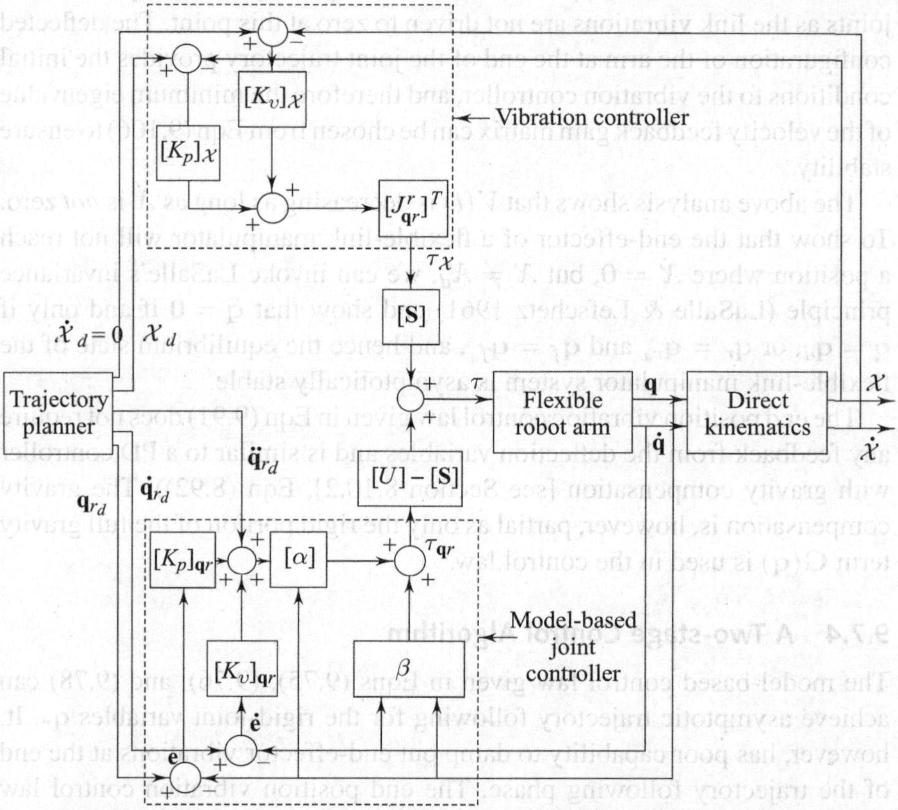

Fig. 9.9 Block diagram of a two-stage controller for a flexible-link manipulator, where $[\alpha] = ([\mathbf{M}_{rr}] - [\mathbf{M}_{rf}]^T[\mathbf{M}_{ff}]^{-1}[\mathbf{M}_{rf}])$ and $\beta = (\mathbf{C}_r + \mathbf{G}_r - [\mathbf{M}_{rf}]^T[\mathbf{M}_{ff}]^{-1}(\mathbf{C}_f + \mathbf{G}_f + [\mathbf{K}]\mathbf{q}_f))$

The development of the two control laws and the two-stage controller assume perfect knowledge of the model parameters and that the models can be computed in real time. Perfect knowledge of model parameters is unrealistic. In the following section we consider the effect of uncertainty in model parameters on the trajectory following control law. The effect of uncertainty in model parameters for the end position vibration controller is not that serious since the only model-based term is the rigid gravity term. If the gravity compensation term $\mathbf{G}_r(\mathbf{q}_{r_d}\mathbf{q}_{f_d})$, is not known exactly, then we will achieve a different asymptotically stable equilibrium state which can be brought arbitrarily close to the desired equilibrium state by a proper choice of $[K_p]_{\mathcal{X}}$ and the error driven portion of the control law.

9.7.5 Effect of Uncertainty in Mass and Stiffness

In the model of a flexible manipulator, the uncertainty in the model parameters can be due to uncertainty in the flexural stiffness of the links, which leads to uncertainty in the stiffness matrix $[\mathbf{K}]$, and uncertainty in mass parameters of the links, which leads to uncertainty in $[\mathbf{M}(\mathbf{q})]$. The uncertainty in stiffness and mass matrices can be considered together as uncertainty in the structural natural frequencies of the model. The natural frequencies of a flexible manipulator at a nominal position determined by $\mathbf{q}_{r_d}$ can be determined from the eigenvalues of the matrix $[\mathbf{M}_{ff}]^{-1}[\mathbf{K}]$, denoted by $[\mathbf{\Omega}]$, as

$$\omega_i^2 = \lambda_i([\mathbf{\Omega}]) = \lambda_i([\mathbf{M}_{ff}]^{-1}[\mathbf{K}]), \qquad i = 1, 2, \ldots, N \tag{9.109}$$

where $\lambda_i(\cdot)$ denotes the ith eigenvalue of a matrix. It should be noted that the matrix $[\mathbf{M}_{ff}]$ depends on the joint variables of the flexible manipulator arm and the natural frequencies vary with the robot configuration.

The manipulator mass and stiffness matrix are derived using one of the discretization methods, AMM or FEM, discussed in Section 9.5. As mentioned in Section 9.5.3, any discretization leads to overestimation of natural frequencies of a system. In addition, due to mechanical joints, invariably present in any manipulator but not always precisely modelled, the actual stiffness in a manipulator is always less than in a model. Hence the natural frequencies in the model are greater than the actual manipulator natural frequencies. To analyse the effect of this *overestimation* of natural frequencies in the trajectory following control law, we rewrite Eqns (9.75) and (9.76) as

$$\begin{aligned}\boldsymbol{\tau}_{\mathbf{q}_r} = {} & ([\mathbf{M}_{rr}] - [\mathbf{M}_{rf}]^T[\mathbf{M}_{ff}]^{-1}[\mathbf{M}_{rf}])\boldsymbol{\tau}'_{\mathbf{q}_r} \\ & + \{\mathbf{C}_r + \mathbf{G}_r - [\mathbf{M}_{rf}]^T([\mathbf{M}_{ff}]^{-1}(\mathbf{C}_f + \mathbf{G}_f + [\widehat{\mathbf{\Omega}}]\mathbf{q}_f)\}\end{aligned} \tag{9.110}$$

where we have used the symbol $[\widehat{\mathbf{\Omega}}]$ to denote the estimated (computed) $[\mathbf{M}_{ff}]^{-1}[\mathbf{K}]$.

For $\tau'_{\mathbf{q}_r}$ chosen as in Eqn (9.78), the closed-loop error equation for the $\mathbf{q}_r$ variable is given by

$$\ddot{\mathbf{e}}(t) + [K_v]_{\mathbf{q}_r}\dot{\mathbf{e}}(t) + [K_p]_{\mathbf{q}_r}\mathbf{e}(t) = -([\mathbf{M}_{rr}] - [\mathbf{M}_{rf}]^T[\mathbf{M}_{ff}]^{-1}[\mathbf{M}_{rf}])^{-1}[\mathbf{M}_{rf}]^T\Delta[\boldsymbol{\Omega}]\mathbf{q}_f \tag{9.111}$$

and the flexible variables $\mathbf{q}_f$ are governed by

$$\ddot{\mathbf{q}}_f + [\mathbf{M}_{ff}]^{-1}(\mathbf{C}_f + \mathbf{G}_f) + ([\boldsymbol{\Omega}] - [\mathcal{M}][\Delta\boldsymbol{\Omega}])\mathbf{q}_f = -[\mathbf{M}_{ff}]^{-1}[\mathbf{M}_{rf}]\tau'_{\mathbf{q}_r} \tag{9.112}$$

where $[\mathcal{M}] = [\mathbf{M}_{ff}]^{-1}[\mathbf{M}_{rf}]\left([\mathbf{M}_{rr}] - [\mathbf{M}_{rf}]^T[\mathbf{M}_{ff}]^{-1}[\mathbf{M}_{rf}]\right)^{-1}[\mathbf{M}_{rf}]^T$ and $[\Delta\boldsymbol{\Omega}] = [\widehat{\boldsymbol{\Omega}}] - [\boldsymbol{\Omega}]$

From Eqn (9.112), we can observe that for a stable response of $\mathbf{q}_f$, the closed-loop *frequency matrix* $([\boldsymbol{\Omega}] - [\mathcal{M}]\Delta[\boldsymbol{\Omega}])$ must be positive definite (Inman 1989). This is intuitively true: consider a simple spring mass system whose equation of motion is given by $\ddot{x} + \omega^2 x = u(t)$. If the natural frequency ω^2 is *negative*, then $x(t)$ will go to infinity. The matrix $([\boldsymbol{\Omega}] - [\mathcal{M}][\Delta\,\boldsymbol{\Omega}])$ is like an equivalent closed-loop natural frequency matrix for the multi-link flexible manipulator and it must be positive definite for $\mathbf{q}_f(t)$ to be bounded. From the definition of $[\boldsymbol{\Omega}]$, we know that it is positive definite. The equivalent mass matrix $([\mathbf{M}_{rr}] - [\mathbf{M}_{rf}]^T[\mathbf{M}_{ff}]^{-1}[\mathbf{M}_{rf}])$ has full rank and is positive definite and, from the property of the rank of product matrices, we can show that $[\mathcal{M}]$ is at least positive-semi-definite. Hence, if $[\Delta\,\boldsymbol{\Omega}] < 0$, the closed-loop frequency matrix is positive definite and the response of $\mathbf{q}_f$ will be stable for a proper choice of gains in $\tau'_{\mathbf{q}_r}$. On the other hand, if $[\Delta\,\boldsymbol{\Omega}] > 0$, the closed-loop frequency matrix may cease to become positive definite and we can get an unstable response of $\mathbf{q}_f$. We can obtain bounds on the uncertainty by considering the maximum eigenvalues of $[\boldsymbol{\Omega}]$, $[\mathcal{M}]$, and $[\Delta\,\boldsymbol{\Omega}]$. Let these be given by $\lambda_{[\boldsymbol{\Omega}]}$, $\lambda_{[\mathcal{M}]}$, and $\alpha\lambda_{[\boldsymbol{\Omega}]}$, respectively. If the natural frequencies are *underestimated* with $\alpha < 0$, then the closed-loop frequency matrix $([\boldsymbol{\Omega}] - [\mathcal{M}]\Delta[\boldsymbol{\Omega}])$ is always positive definite. If natural frequencies are *overestimated* with $\alpha > 0$, then the closed-loop frequency matrix is positive definite only if $\alpha\lambda_{[\mathcal{M}]} < 1.0$.

This analysis implies that the model-based trajectory following controller can become unstable if the model of the flexible arm is *more rigid* than it actually is. As discussed earlier, this situation is entirely possible since the discretization models *always* overestimate the stiffness. This analysis also sets an upper bound on the allowable estimation error in the natural frequencies so as to still obtain stable closed-loop behaviour. More details on this topic and the design of a robust compensator to effectively control flexible manipulators with overestimation of natural frequencies can be found in Theodore (1995) and Theodore and Ghosal (2003). We end this section with representative numerical simulation results of a flexible manipulator arm.

9.7.6 Numerical Simulation of a Flexible Manipulator

We consider a three-DOF manipulator with two flexible links as shown schematically in Fig. 9.10. The manipulator is similar to a PUMA 560 (in terms of configuration), but in these simulations we have assumed that the second and third links are flexible to some extent. We also assume that the flexible links can be modelled as slender beams so that their modelling can be done with the method proposed in Section 9.3. The Denavit–Hartenberg, mass and inertial parameters, and the flexural rigidity *(EI)* values used in the numerical simulation are given in Table 9.2. These numerical values have been chosen arbitrarily. The desired trajectory is chosen to be a smooth (sine profile with zero velocity and acceleration at the start and at the end of trajectory) right-circular helix of radius 25 cm, pitch 2.5 cm, and 3π rotations about the helix axis. The time for the entire trajectory is chosen to be relatively small, 1.0 sec, to excite oscillations and to evaluate the performance of the two-stage controller in trajectory tracking and suppression of tip

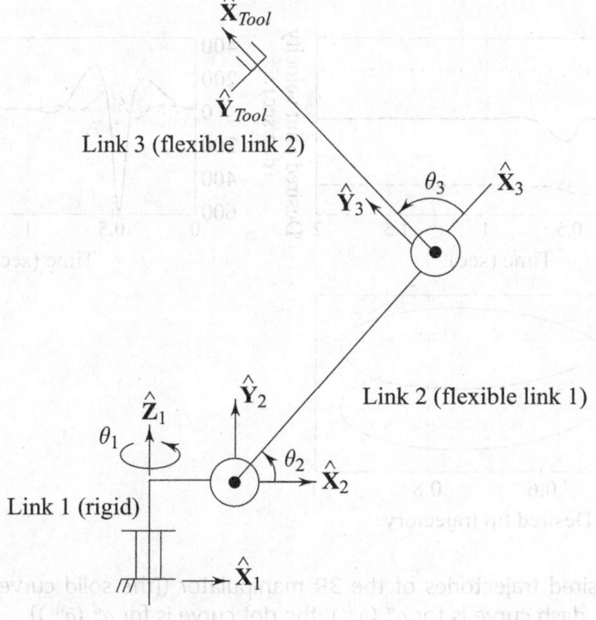

Fig. 9.10 Schematic of a 3R flexible manipulator

Table 9.2 Numerical values of the 3R flexible manipulator

Physical system parameters	Value
Mass of link 1 (m_1)	3.66 kg
Linear mass density of link 2 ($\rho_2 A_2$)	0.331 kg m^{-1}
Linear mass density of link 3 ($\rho_3 A_3$)	0.331 kg m^{-1}
Mass of payload (m_p)	0.1 kg
Length of link 1	0.12 m
Length of link 2	1.0 m
Length of link 3	1.0 m
Rotary inertia of joint 1 (I_{joint_1})	0.4 kg m^2
Rotary inertia of joint 2 (I_{joint_2})	3.275 kg m^2
Rotary inertia of joint 3 (I_{joint_3})	3.275 kg m^2
Flexural rigidity of link 2 [$(EI)_2$]	1165.4916 N m^2
Flexural rigidity of link 3 [$(EI)_3$]	1165.4916 N m^2

oscillations. The desired trajectory of the end-effector and the corresponding joint trajectories (assuming the links are rigid) are shown in Fig. 9.11. After the initial 1.0 sec, the desired tip position is as obtained from the joint trajectory at 1.0 sec and $\dot{\mathcal{X}}_d$ is chosen to be zero. The desired time for damping out tip vibrations is chosen to be 1.0 sec.

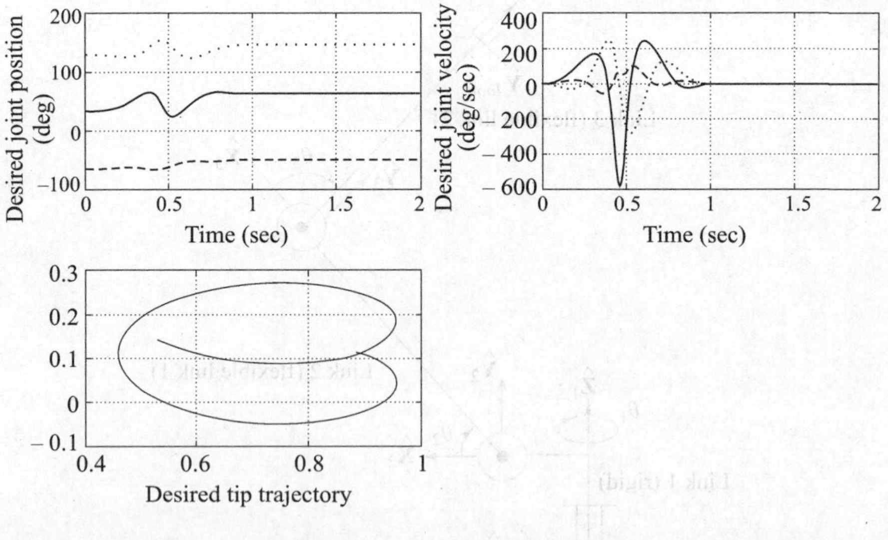

Fig. 9.11 Desired trajectories of the 3R manipulator ([the solid curve is for $q^d_{r_1}$ ($\dot{q}^d_{r_1}$), the dash curve is for $q^d_{r_2}$ ($\dot{q}^d_{r_2}$), the dot curve is for $q^d_{r_3}$ ($\dot{q}^d_{r_3}$)]

The parameters of controller gain matrices are chosen as follows: for the I-stage model-based joint controller, $[K_p]_{\mathbf{q}_r}$ and $[K_v]_{\mathbf{q}_r}$ are diagonal matrices with equal diagonal elements of 64.0 and 32.0, respectively. For the II-stage end position vibration controller, the gain matrices $[K_p]_{\mathcal{X}}$ and $[K_v]_{\mathcal{X}}$ are chosen as diagonal matrices with elements $\{100.0, 100.0, 400.0\}$ and $\{40.0, 40.0, 80.0\}$, respectively. The equations of motion are derived using the finite element formulation with two elements in each of the two flexible links. For the purpose of numerical simulation, we consider a relatively large uncertainty in model parameters—the mass parameters of the manipulator model are under-estimated by 25% and the stiffness parameters of the model are overestimated by 25%. We present the simulation results for two cases.

Case 1 Two-stage control algorithm with no uncertainties in model parameters with the model-based trajectory following control law as given in Eqns (9.75), (9.76), and (9.78).

Case 2 Two-stage control algorithm with the model-based control law as given in Eqns (9.110) and (9.78) and with the above-mentioned uncertainty in model parameters.

Figures 9.12 and 9.14 show the time history plot of the joint position, joint velocity, tip position, and tip velocity errors for the two-stage controller

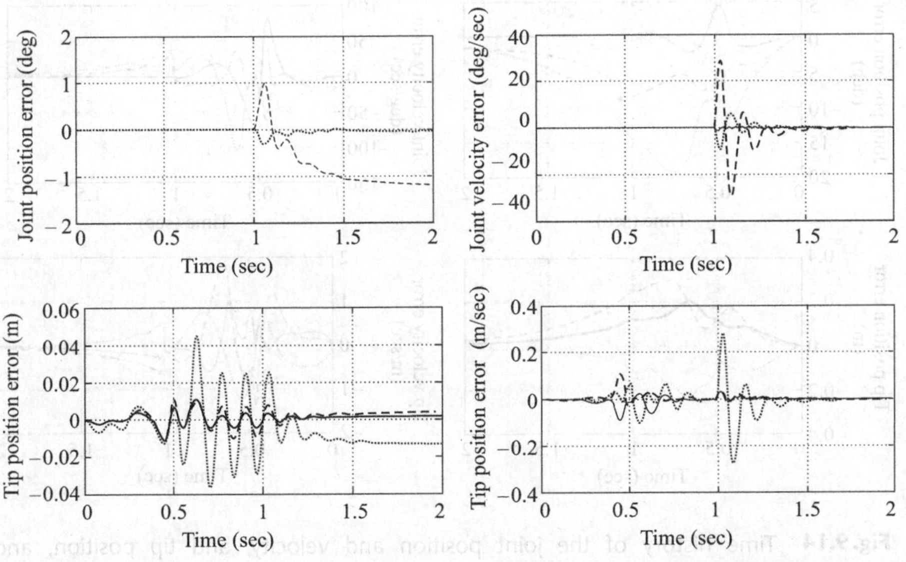

Fig. 9.12 Time history of the joint position, and velocity, and tip position, and velocity errors for two-stage controller [joint error: the solid curve shows $e_1(\dot{e}_1)$, the dash curve shows $e_2(\dot{e}_2)$, the dot curve shows $e_3(\dot{e}_3)$; tip error: the solid curve shows e_x $(\dot{e}_x)$, the dash curve shows $e_y(\dot{e}_y)$, the dot curve shows $e_z(\dot{e}_z)$]

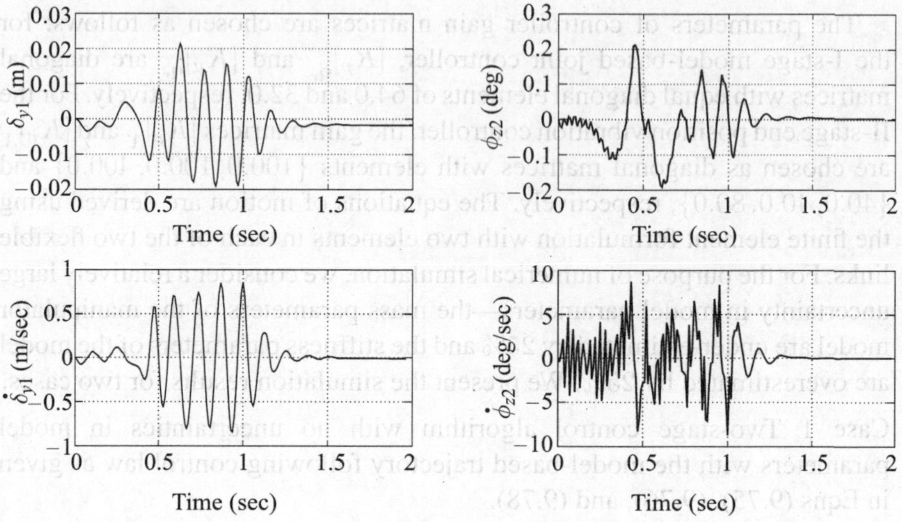

Fig. 9.13 Time history of the elastic deflection variable along the Y direction, at the tip of flexible link 1, and its rate; Time history of the elastic rotation variable about the Z direction, at the tip of flexible link 2, and its rate.

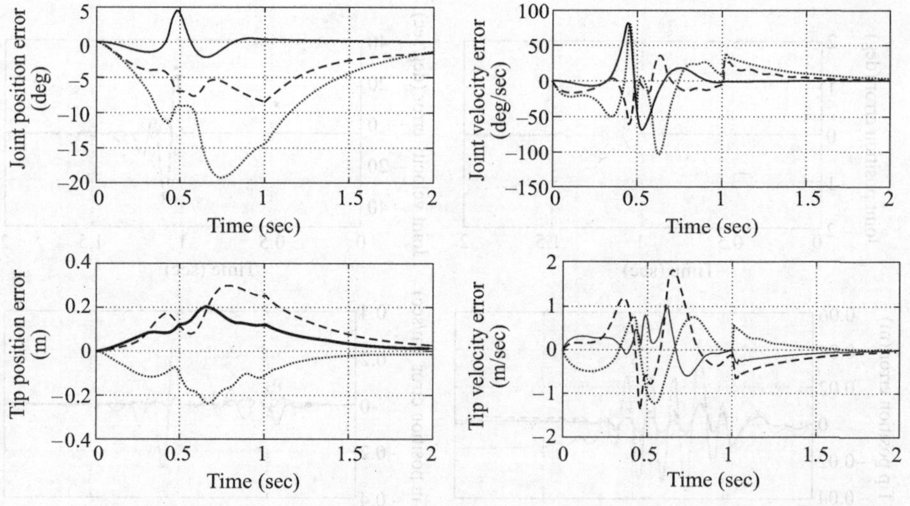

Fig. 9.14 Time history of the joint position and velocity, and tip position, and velocity errors for two-stage controller [joint error: solid curve is for $e_1(\dot{e}_1)$, dash curve is for $e_2(\dot{e}_2)$, dot curve is for $e_3(\dot{e}_3)$; Tip error: solid curve is for $e_x(\dot{e}_x)$, dash curve is for $e_y(\dot{e}_y)$, dot curve is for $e_z(\dot{e}_z)$]

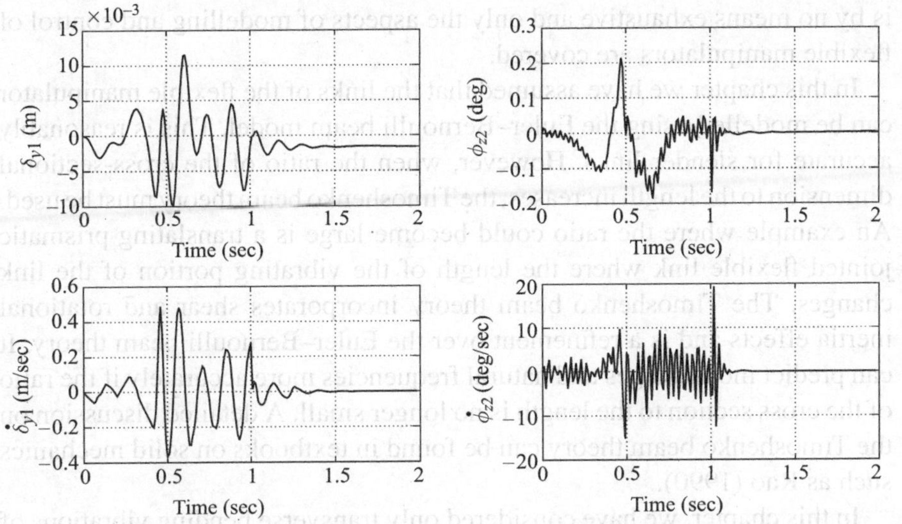

Fig. 9.15 Time history of the elastic deflection variable along the Y direction, at the tip of flexible link 1, and its rate; time history of the elastic rotation variable about the Z direction, at the tip of flexible link 2, and its rate

for cases 1 and 2, respectively. It can be seen from Fig. 9.12, for case 1, that during the trajectory tracking stage (between 0 and 1 sec) though the joint errors are close to zero, the tip errors are in the order of 5 cm, due to the oscillatory response of elastic variables (see Fig. 9.13). These elastic vibrations are then suppressed to zero by the end position vibration control law during the second stage (between 1 and 2 sec), and hence the tip errors are also driven to a lower value of about 1 cm during this stage.

When there are uncertainties in model parameters (case 2), it can be seen from Fig. 9.14 that the joint errors are much larger, of the order of 20°, and the tip errors are of the order of 30 cm. Once the end position vibration controller becomes active between 1 and 2 sec, the joint and tip position errors are again driven to lower levels of about 2° and 3 cm, respectively (see Fig. 9.15). Nevertheless, these errors are much too large to be acceptable and we need a robust compensator. In Theodore and Ghosal (2003) the details of a robust compensator using assumed quantitative bounds on uncertainty are discussed. It is shown that a robust compensator can bring down both joint and tip errors to about the same levels as in case 1.

9.8 Other Topics in Flexible Manipulators

In this section, we list some of the important research topics in the area of flexible manipulators which have not been covered in this chapter. The list is by no means exhaustive and only the aspects of modelling and control of flexible manipulators are covered.

In this chapter we have assumed that the links of the flexible manipulator can be modelled using the Euler–Bernoulli beam model. This is reasonably accurate for slender links. However, when the ratio of the cross-sectional dimension to the length increases, the Timoshenko beam theory must be used. An example where the ratio could become large is a translating prismatic jointed flexible link where the length of the vibrating portion of the link changes. The Timoshenko beam theory incorporates shear and rotational inertia effects and is a refinement over the Euler–Bernoulli beam theory. It can predict mode shapes and natural frequencies more accurately if the ratio of the cross section to the length is no longer small. A detailed discussion on the Timoshenko beam theory can be found in textbooks on solid mechanics such as Rao (1990).

In this chapter, we have considered only transverse bending vibrations of the flexible links. In reality, a flexible link, especially in a multi-link flexible manipulator with a payload, may undergo combined transverse bending and torsional vibrations, and modelling and control of both bending and torsion is a challenging task. For single flexible link, some results on modelling and control of flexible beams undergoing combined bending and torsional vibration can be found in Matsuno et al. (1994).

In this chapter, we have assumed that the deformations in the flexible links of a manipulator are small and that the linear theory of elasticity can be used. Linear elasticity assumes linear relationships between strain and displacement and also between stress and strain. If deformations are large, it is more appropriate to use the non-linear strain displacement relationship, also called *geometrically* non-linear formulations. The use of the non-linear strain displacement relationship leads to coupling between axial and bending modes of vibration (Przemineiecki 1968). Simo and Vu-Quoc (1987) showed that for rotating structures appropriate accounting of the influence of centrifugal force on the bending stiffness requires the use of geometrically non-linear beam theory and the use of the first-order (or linear) beam theory results in a lower bending stiffness, thereby lower natural frequencies. Bakr and Shabana (1986) and Mayo et al. (1995) have used non-linear strain displacement relationships to model beams in multi-body flexible systems. Another classic paper by Kane et al. (1987) discusses the importance of geometrically non-linear formulations for a rotating beam. In a recent work by Chandra Shaker and Ghosal (2006), the authors have used the characteristic speeds, U_g (see Section 9.3.1) and the speed of sound in the material of the link, to obtain conditions under which a geometrically non-linear formulation is desirable.

An alternative to the time domain control method, discussed in this chapter, is to use the frequency-domain-based control. In frequency domain methods the poles and zeros of the flexible manipulators are identified from the local maxima and minima of the manipulators frequency response. These points are easily recognizable since a typical flexible manipulator is very lightly damped. The frequency domain methods are known to be less computationally intensive and less noisy as compared to time domain methods. For more details the reader is referred to Tzes and Yurkovitch (1991). Related to the frequency domain methods are feed-forward control methods based on input preshaping. In the simplest form of input preshaping, an impulse input is shaped into two impulses with the second delayed by one-half period of the vibration frequency to be avoided. The vibration frequencies in a flexible manipulator are configuration dependent and hence non-linear model-based controllers have been used in conjunction with input preshaping for better results. For more details on input shaping methods, the reader is referred to Singer and Steering (1990) and Khorrami et al. (1994, 1995).

In this chapter, we showed that with the model-based controller the flexible dynamics is only critically stable in the absence of material damping and asymptotically stable when damping is present. Most materials used for the flexible links, such as aluminium or steel, have very low material damping. Several researchers have attempted to increase material damping by attaching visco-elastic material to the link material (Alberts et al. 1992). Alternately, one can actively increase damping using piezo-electric elements embedded in the link material and several researchers have demonstrated damping of vibrating cantilever beams using piezo-electric crystals. This is a vast field by itself and the reader is referred to textbooks and research papers on smart materials. A good starting point is the book by Fuller et al. (1996).

Exercises

†9.1 Consider a rigid single-link manipulator with $l_1 = 1.241$ m, $m_1 = 20.15$ kg, the location of the centre of mass $r_1 = 1.2$ m from the rotary joint and the moment of inertia about the centre of mass as 9.6 kg/m^2 moving the horizontal plane. The link is connected to the actuator by a spring of spring constant $K_s = 0.1$ (N m)/rad. Using the theory of Section 9.2 and Chapter 8, design a PD controller such that the link can rotate 150° in 5 sec.

9.2 By referring to Marino and Spong (1986), obtain the feedback linearization scheme for a single rigid link, flexible-joint manipulator moving under gravity.

9.3 In this chapter we have used clamped conditions at the actuator end. What would change in kinematic modelling if free or pinned conditions are used at the actuator end?

†9.4 For a planar 2R manipulator, with both links flexible, derive the symbolic dynamic equations of motion using the assumed mode method. Take two modes for discretizing each flexible link and assume that there is no gravity.

†9.5 For a planar 2R manipulator, with both links flexible, derive the symbolic dynamic equations of motion using the finite element method. Take two elements in each flexible link and assume there is no gravity.

†9.6 Numerically simulate the equations of motion obtained in Exercises 9.3 and 9.4 for the same link, mass and inertial parameters, and actuator torques. Compare and comment on the results.

9.7 Obtain the inverse of the generalized mass

$$\begin{pmatrix} [\mathbf{M}_{rr}] & [\mathbf{M}_{rf}]^T \\ [\mathbf{M}_{rf}] & [\mathbf{M}_{ff}] \end{pmatrix}$$

in terms of the matrices $[\mathbf{M}_{rr}]$, $[\mathbf{M}_{rf}]$, and $[\mathbf{M}_{ff}]$ and verify results in Eqns (9.69) and (9.70).

9.8 Show that the equivalent mass matrix $([\mathbf{M}_{rr}] - [\mathbf{M}_{rf}]^T[\mathbf{M}_{ff}]^{-1}[\mathbf{M}_{rf}])$ used in the trajectory following control law has full rank n.

9.9 Similar to a rigid manipulator (see Sections 6.3 and 8.10.2), show that the flexible Coriolis/centripetal vector $\mathbf{C}_f$ can be written as $[\mathbf{C}_{ff}]\dot{\mathbf{q}}_f$. Also show that $\left[[\dot{\mathbf{M}}_{ff}] - 2[\mathbf{C}_{ff}]\right]$ is skew symmetric.

9.10 From the paper by De Luca and Siciliano (1993), or otherwise, verify the inequality given in Eqn (9.96).

References and Suggested Additional Reading

Alberts, T.E., H. Xia, and Y. Chen 1992, 'Dynamic analysis to evaluate viscoelastic passive damping augmentation for Space Shuttle Remote Manipulator System', *Trans. ASME J. Dyn. Syst.*, vol. 114, pp. 468–75.

Bakr, E.M. and A. A. Shabana 1986, 'Geometrically non-linear analysis of multi-body systems', *Comput. Struct.*, vol. 23, pp. 739–51.

Bayo, E. 1987, 'A finite element approach to control the end point of a single link flexible robot', *J. Robotic. Syst.*, vol. 4, pp. 63–75.

Book, W.J. 1984, 'Recursive Lagrangian dynamics of flexible manipulator arms', *Int. J. Robot. Res.*, vol. 3, pp. 87–101.

Cannon, R.H. and E. Schmitz 1984, 'Initial experiments on end-point control of flexible one-link robot', *Int. J. Robot. Res.*, vol. 3, pp. 62–75.

Cetinkunt, S. and W.-L. Yu 1991, 'Closed-loop behavior of a feedback-controlled flexible arm: A comparative study', *Int. J. Robot. Res.*, vol. 10, pp. 263–75.

Chandra Shaker, M. and A. Ghosal 2006, 'Nonlinear modeling of flexible manipulators using non-dimensional variables', *ASME Trans., J. Nonlinear and Comput. Dyn.*, vol. 1, pp. 123–34.

Chedmail, P., Y. Aoustin, and Ch. Chevallereau 1991, 'Modelling and control of flexible robots', *Int. J. Num. Meth. Eng.*, vol. 32, pp. 1595–619.

Chen, G., M.C. Delfour, A.M. Krall, and G. Payre 1987, 'Modeling, stabilization and control of serially connected beams", *SIAM J. Control Optim.*, vol. 25, pp. 526–46.

De Luca, A. and B. Siciliano 1991, 'Closed-form dynamic model of a planar multi-link lightweight robots', *IEEE Trans. Syst. Man Cyb.*, vol. SMC-21, pp. 826–39.

De Luca, A. and B. Siciliano 1993, 'Regulation of flexible arms under gravity', *IEEE Trans. Robotic Autom.*, vol. RA-9, pp. 463–67.

Fuller, C., S. Elliott, and P. Nelson 1996, *Active Control of Vibration*, Academic Press.

Hu, F.L. and A.G. Ulsoy 1994, 'Dynamic modeling of constrained flexible robot arms for controller design", *Trans. ASME J. Dyn. Syst.*, vol. 116, pp. 56–65.

Inman, D.J. 1989, *Vibration: With Control, Measurement and Stability*, Prentice Hall.

Isidori, A. 1989, *Nonlinear Control Systems: An Introduction*, 2nd edn, Springer-Verlag.

Jonker, J.B. 1990, 'A finite element dynamic analysis of flexible manipulators', *Int. J. Robot. Res.*, vol. 9, pp. 59–74

Kane, T.R., R.R. Ryan, and A.K. Banerjee 1987, 'Dynamics of a cantilever beam attached to a moving base', *AIAA J. Guid. Control*, vol. 10, pp. 139–51.

Khorrami, F., S. Jain, and A. Tzes 1994, 'Experiments on rigid-body-based controllers with input preshaping for a two link flexible manipulator', *IEEE Trans. Robotic. Autom.*, vol. RA-10, pp. 55–65.

Khorrami, F., S. Jain, and A. Tzes 1995, 'Experimental results on adaptive nonlinear control and input preshaping for multi-link flexible manipulators', *Automatica*, vol. 31, pp. 83–97.

LaSalle, J. and S. Lefschetz 1961, *Stability by Liapunov's Direct Method with Applications*, Academic Press.

Li, C.-J. and T.S. Sankar 1993, 'Systematic methods for efficient modeling and dynamics of flexible robot manipulators', *IEEE Trans. Syst. Man Cyb.*, vol. SMC-23, pp. 79–94.

Marino, R.W. and M.W. Spong 1986, 'Nonlinear control techniques for flexible joint manipulators: A single link case study', *Proc. of IEEE Conf. on Robotic Autom.*, San Francisco, pp. 1030–26.

Matsuno, F., T. Murachi, and Y. Sakawa 1994, 'Feedback control of decoupled bending and torsional vibrations of flexible beams', *J. Robotic. Syst.*, vol. 11, pp. 341–53.

Mayo, J., J. Dominguez, and A.A. Shabana 1995, 'Geometrically nonlinear formulation of beams in flexible multi-body systems', *J. Vib. Acoust.*, vol. 117, pp. 501–09.

Meirovitch, L. 1967, *Analytical Methods in Vibrations*, Macmillan Co.

Meirovitch, L. 1986, *Elements of Vibration Analysis*, 2nd edn, McGraw-Hill Book Co.

Morgül, Ö. 1992, 'Dynamic boundary control of a Euler-Bernoulli Beam', *IEEE Trans. Automat. Control*, vol. AC-37, pp. 639–42.

Oakley, C.M. and R.H. Cannon 1989, 'End-point control of a two-link manipulator with a very flexible forearm: Issues and experiments', *Proc. of American Control Conference*, pp. 1381–88.

Pazy, A. 1983, *Semigroups of Linear Operators and Applications to Partial Differential Equations*, Springer-Verlag.

Przemieniecki, J.S. 1968, *Theory of Matrix Structure Analysis*, McGraw-Hill Book Co.

Rao, S.S. 1990, *Mechanical Vibrations*, 2nd edn, Addison-Wesley.

Simo, J.C. and L. Vu-Quoc 1987, 'The role of nonlinear theories in transient dynamic analysis of flexible structures', *J. Sound Vib.*, vol. 119, pp. 487–508.

Singer, N.C. and W.P. Seering 1990, 'Preshaping command inputs to reduce system vibration', *Trans. ASME J. Dyn. Syst.*, vol. 112, pp. 76–82.

Spong, M.W. 1987, 'Modeling and control of elastic joint robots', *Trans. ASME J. Dyn. Syst.*, vol. 109, pp. 310–19.

Stylianou, M. and B. Tabarrok, 'Finite element analysis of a axially moving beam, Part 1: Time integration; Part II: Stability analysis', *J. Sound Vib.*, vol. 178, pp. 433–81.

Sunada, W.H. and S. Dubowsky 1981, 'The application of finite element methods to the dynamic analysis of flexible linkages', *Trans. ASME J. Mech. Des.*, vol. 103, pp. 643–51.

Theodore, R.J. 1995, *Dynamic modeling and control analysis of multilink flexible manipulators*, PhD Thesis, Dept. of Mechanical Engg., Indian Institute of Science, Bangalore.

Theodore, R.J. and A. Ghosal 1995, 'Comparison of the assumed modes and finite elements model for flexible multilink manipulators', *Int. J. Robot. Res.*, vol. 14, pp. 91–111.

Theodore, R.J. and A. Ghosal 1997, 'Modeling of flexible manipulators with prismatic joints', *IEEE Trans. Syst. Man Cyb.—Part B*, vol. 27, pp. 296–305.

Theodore, R.J. and A. Ghosal 2003, 'Robust control of multilink flexible manipulators', *Mech. Mach. Theory*, vol. 38, pp. 367–77.

Tzes, A. and S. Yurkovich 1991, 'Application and comparison of on-line identification methods for flexible manipulator control', *Int. J. Robot. Res.*, vol. 10, pp. 515–27.

Usoro, P.B., R. Nadira, and S.S. Mahil 1986, 'A finite elment/Lagrangian approach to modeling lightweight flexible manipulators', *Trans. ASME J. Dyn. Syst.*, vol. 108, pp. 198–205

10

Modelling and Analysis of Wheeled Mobile Robots*

10.1 Introduction

In the chapters so far, we have looked at modelling and analysis of robots and manipulators which fall under the category of being *stationary* with one member or a link fixed to the ground. Apart from such fixed-base robots, engineers and designers have built a large number of autonomous mobile robots and devices which are capable of locomotion on a surface, and, in the recent past, capable of motion in 3D space. The earliest devices, motivated by the industrial need of transferring parts and finished goods from one location in a factory to another, were automated guided vehicles or AGVs. These vehicles could move around on a flat factory floor, guided by metal or optical strips laid out on the floor. Autonomous devices with legs and treads capable of moving on an uneven terrain have also been built, but a vast majority of mobile robots are wheeled mobile robots or WMRs. WMRs have been used in industrial environments, military and security applications, handling of hazardous material, providing mobility to handicapped persons, and in planetary exploration—the most spectacular example of the last is the Mars Rover of NASA. The science and technology of WMRs spans a large number of topics such as kinematics, dynamics, control, design, sensing, path planning, obstacle avoidance, and intelligence. In this chapter, we look at only the kinematic and dynamic modelling aspects of WMRs. We develop the constraints between the wheel and an uneven terrain and show that a WMR can be modelled and analysed as a parallel manipulator. The chapter focuses on issues such as inverse and direct kinematics, dynamics and simulation of WMRs, and relies heavily on the techniques developed for kinematic and dynamic analyses of parallel manipulators in Chapters 4–6. Kinematic and dynamic analyses and simulation results are presented for a three-wheeled mobile robot. We will follow the treatment in Chakraborty (2003) and Chakraborty and Ghosal (2004; 2005).

According to the definition by Muir and Neuman (1987), a WMR is a 'robot capable of locomotion on a surface solely through the actuation of wheel assemblies mounted on it and in contact with the surface. A wheel assembly is a device which provides or allows relative motion between its mount and the surface on which it is intended to have single-point contact'. As defined, a wheel of a WMR contacts the ground surface at a point.[1] This is unlike a revolute or prismatic joint where the two parts of a joint, having relative motion, have an area contact.[2] The constraints arising out of the point contact between the wheel and the ground surface involve rolling (or sliding) and are called *non-holonomic* constraints. The topic of a wheel modelled as a thin disc rolling or sliding on a flat surface has been extensively studied [see, for example, Muir and Neuman (1987) and Alexander and Maddocks (1989)]. In this chapter, we start with a more general approach of two arbitrary smooth surfaces in contact and develop the contact equations following the treatment in Montana (1988). This is applied to a torus-shaped wheel[3] moving on an uneven terrain, and simulation results are presented in Section 10.2.

A typical WMR is modelled as a planar rigid body that rides on multiple wheels. A rigid plane in contact with a flat surface has three degrees of freedom, namely, the x and y coordinates of a point of interest on the plane and the angular rotation θ about the normal to the plane. Intuitively one might expect that the rigid platform of a WMR would also have three degrees of freedom. This is not so. For a WMR with wheels modelled as thin discs and moving on a flat surface, the two horizontal components of the linear velocity of the centre of mass and the angular speed about the vertical are related by the fact that the linear velocity vector of the centre of mass is tangent to the path traced by the centre of mass at all instants. This constraint gives rise to non-holonomic constraints, which is a distinctive feature of all WMRs.[4] Alexander and Maddocks (1989) have modelled WMRs moving on flat surfaces with an arbitrary number of wheels. They have derived relationships between the linear and angular velocity components of the rigid platform and the steering and driving rates of the wheels. For a three-wheeled mobile robot, the complications arising due to the steering mechanism need

[1] In practical wheels, due to deformation, there is actually a surface contact. This is, however, very difficult to model and analyse and is not considered in this text.

[2] Typical revolute, prismatic, and spherical joints are called *lower pair* joints, whereas the contact between a sphere and a plane or between a wheel and the ground surface is called a *higher pair* joint.

[3] A wheel modelled as a torus is not very unrealistic if the tyre pressure is large. For motorbike wheels, a torus is a reasonable approximation.

[4] In a four-wheeled car, either the front or rear wheels are driven by the engine and the front two wheels are steered. The front two wheels must rotate by different angles during a turn and this is accomplished by steering mechanisms such as an Ackerman steering mechanism. In addition, the angular speeds of the driven wheels, on a common axle, are adjusted by the differential to ensure that the car does not skid or slide while turning. Since there are only two actuators, all the linear and angular velocity components of the body of a car are not independent.

not be considered, and in this text we will consider the simple case of a three-wheeled mobile robot with the two rear wheels driven and the front wheel having steering capability. We show that a three-wheeled mobile robot with torus-shaped wheels and with three degrees of freedom can be modelled as an equivalent parallel manipulator with three closed loops, three actuated joints, and several passive joints. The techniques developed for kinematic and dynamic analyses of a parallel manipulator and closed-loop mechanisms, in earlier chapters, are used to model and analyse the kinematics and dynamics of a three-wheeled mobile robot.

Most researchers have modelled and analysed the motion of WMRs on a flat terrain. In the recent past there has been an increasing interest in off-road vehicles and WMRs for use in rough and uneven terrains and for planetary exploration—a famous example is the Mars Rover of NASA. A three-wheeled mobile robot on an uneven terrain, unlike on a flat surface, has three degrees of freedom and three actuators are required to achieve general motions of the rigid platform. Waldron (1995) has argued that two wheels joined by a common rigid axle cannot roll on an uneven terrain without slip since, in general, there is no instantaneous centre compatible with both the wheels. Slip at the wheel–ground contact point leads to *localization* errors, and in such a situation odometric sensing together with a kinematic model of the WMR is not enough to determine the position and attitude of the WMR with respect to a given coordinate system.[5] Slip at the wheel–ground contact point also wastes power, which is at a premium in tasks such as planetary exploration. To avoid slip on an uneven terrain, Choi et al. (1999) and Choi and Sreenivasan (1999) have suggested the use of a variable-length axle (VLA) where the axle length is changed by an unactuated prismatic joint. At high speeds and at large inclinations, a VLA with prismatic joint can still slip. In this chapter, an alternative to the VLA concept is presented. We model and analyse a three-wheeled mobile robot with torus-shaped wheels, three actuated joints, and passive rotary joints; this allows the distance between the wheel–ground contact points to change *without* changing the axle length. Kinematic and dynamic modelling of a three-wheeled mobile robot capable of traversing smooth uneven terrains without slip are discussed in Sections 10.4 and 10.5, respectively, and numerical simulation results are presented.

A WMR of a given geometry can traverse only certain kinds of uneven terrains with its wheels maintaining single-point contact with the ground—a car with a given wheel diameter, axle length, and other geometrical

[5] One can also use a global positioning satellite (GPS) based sensing scheme (Cooper & Durrant 1994). However, a GPS-based scheme has its own disadvantages such as loss of signal and frequency jamming.

dimensions can get stuck in a ditch or may not cross over a bump! Traversability of an uneven terrain for a WMR is similar to the concept of the workspace of stationary robots. In Section 10.6, we present a brief discussion on the traversability of a *single* torus-shaped wheel on uneven terrains. Finally, in Section 10.7, we present a few related topics in modelling of WMR not covered in this text.

10.2 Motion of a Single Wheel on Uneven Terrain

Most WMR literature model the wheels of a WMR as thin discs[6] which can roll without slipping on a flat surface. The motion of a thin disc rolling without slipping on a flat surface is a very well known problem in mechanics—it is perhaps the best known example of a mechanical system with *non-holonomic* constraints and is discussed in several textbooks in mechanics [see, for example, Goldstein (1980)]. For motion of a WMR on a flat plane, the wheel modelled as a thin disc is quite reasonable since the wheel–ground contact point lies in a vertical plane through the centre of the wheel. In an uneven terrain, a thin disc model of a wheel is not very good since the contact point between a conventional wheel and the ground can be on the lateral surface of the wheel.[7] With this in mind, we will model the wheel of a mobile robot as a torus and allow the contact point to be on the lateral surface of the torus.

10.2.1 Model of a Torus-shaped Wheel

A surface in $\Re^3$ can be represented as a single equation of the form $z = f(x, y)$ or as $f(x, y, z) = 0$. It is more useful to represent a surface in a parametric form as a vector equation

$$(x, y, z)^T = \mathbf{f}(u, v) \tag{10.1}$$

where (x, y, z) are the coordinates of a point ${}^0\mathbf{p}$ on the surface with respect to a coordinate system $\{0\}$. The variables (u, v) are called the parameters and any $(u, v) \in U \subseteq \Re^2$ maps to a point on the surface as shown schematically in Fig. 10.1. The equation of a torus-shaped wheel in parametric form is given by

$$x = r_1 \cos u_w$$

[6] Wheeled mobile robots with omni-directional and ball wheels have also been built. Kinematic analyses of WMRs with omni-directional and ball wheels have been discussed in Muir and Neuman (1987).

[7] In the case of a motorbike negotiating a turn on a curved road, the wheel–ground contact point is on the lateral surface of the wheel.

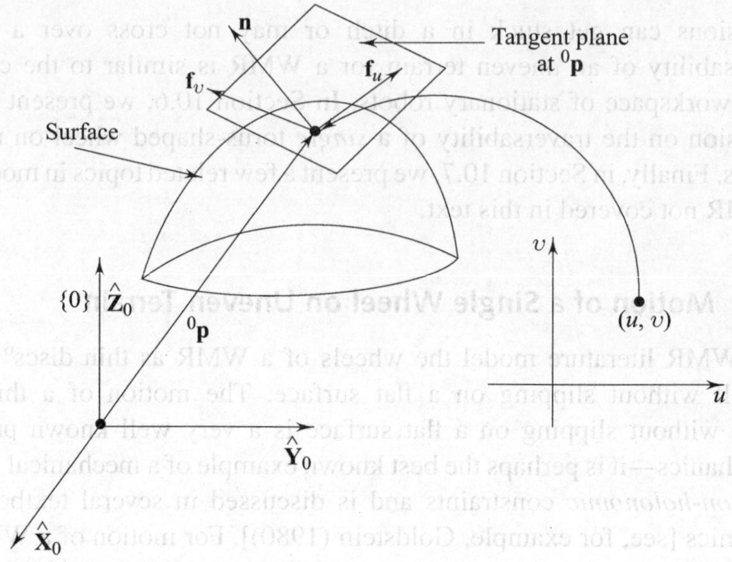

Fig. 10.1 A surface in $\Re^3$

$$y = \cos v_w (r_2 + r_1 \sin u_w) \qquad (10.2)$$
$$z = \sin v_w (r_2 + r_1 \sin u_w)$$

where r_1 and r_2 are the two radii associated with the torus and we have used the subscript w on the parameters u and v to denote a wheel. Figure 10.2 shows a sketch of a torus-shaped wheel on an uneven terrain.

At any point ${}^0\mathbf{p}$ on the surface, we can define tangent vectors $\mathbf{f}_u$ and $\mathbf{f}_v$ as

$$\mathbf{f}_u = \frac{\partial \mathbf{f}}{\partial u}, \quad \mathbf{f}_v = \frac{\partial \mathbf{f}}{\partial v} \qquad (10.3)$$

At any non-singular point, we can define the normal to the surface (and the tangent plane formed by $\mathbf{f}_u$ and $\mathbf{f}_v$) as

$$\mathbf{n} = \frac{\mathbf{f}_u \times \mathbf{f}_v}{|\mathbf{f}_u \times \mathbf{f}_v|} \qquad (10.4)$$

The tangent vectors $\mathbf{f}_u$ and $\mathbf{f}_v$, the tangent plane, and $\mathbf{n}$ are shown schematically in Fig. 10.1.

We assume that $\mathbf{f}_u$ and $\mathbf{f}_v$ are orthogonal.[8] This implies that the vectors $\mathbf{f}_u/|\mathbf{f}_u|$, $\mathbf{f}_v/|\mathbf{f}_v|$, and $\mathbf{n}$ form a right-handed coordinate system at any point on the surface. Based on this, we can define the following:

[8] If $\mathbf{f}_u$ and $\mathbf{f}_v$ are not orthogonal, then we can still obtain an orthogonal set as $\{\mathbf{f}_u/|\mathbf{f}_u|, \mathbf{n} \times \mathbf{f}_u/|\mathbf{f}_u|, \mathbf{n}\}$.

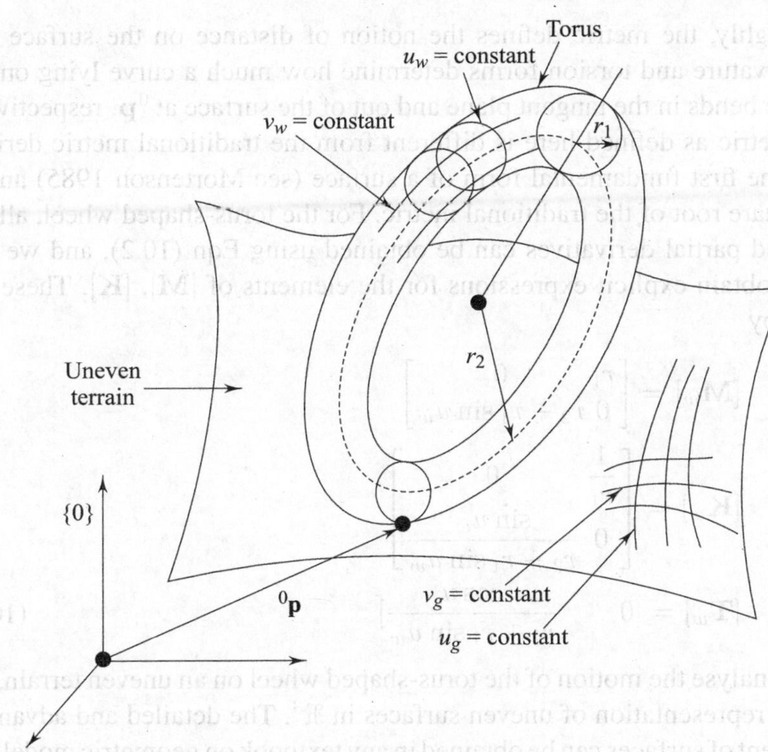

Fig. 10.2 Torus-shaped wheel on uneven terrain

- A metric [M] on the surface as

$$[\mathbf{M}] = \begin{bmatrix} |\mathbf{f}_u| & 0 \\ 0 & |\mathbf{f}_v| \end{bmatrix} \tag{10.5}$$

- A curvature form [K] as

$$[\mathbf{K}] = \begin{bmatrix} \dfrac{\mathbf{f}_u \cdot \mathbf{n}_u}{|\mathbf{f}_u|^2} & \dfrac{\mathbf{f}_u \cdot \mathbf{n}_v}{|\mathbf{f}_u||\mathbf{f}_v|} \\ \dfrac{\mathbf{f}_v \cdot \mathbf{n}_u}{|\mathbf{f}_u||\mathbf{f}_v|} & \dfrac{\mathbf{f}_v \cdot \mathbf{n}_v}{|\mathbf{f}_v|^2} \end{bmatrix} \tag{10.6}$$

- A torsion form [T] as

$$[\mathbf{T}] = \begin{bmatrix} \dfrac{\mathbf{f}_v \cdot \mathbf{f}_{uu}}{|\mathbf{f}_u|^2|\mathbf{f}_v|} & \dfrac{\mathbf{f}_v \cdot \mathbf{f}_{uv}}{|\mathbf{f}_v|^2|\mathbf{f}_u|} \end{bmatrix} \tag{10.7}$$

where $\mathbf{f}_{(\cdot)(\cdot)}$ denotes the second partial derivatives of $\mathbf{f}$, $\partial^2\mathbf{f}/[\partial(\cdot)\partial(\cdot)]$, with respect to u and v, and $\mathbf{n}_{(\cdot)}$ denotes partial derivative of $\mathbf{n}$ with respect to u and v.

Roughly, the metric defines the notion of distance on the surface and the curvature and torsion forms determine how much a curve lying on the surface bends in the tangent plane and out of the surface at ${}^0\mathbf{p}$, respectively. The metric as defined here is different from the traditional metric derived from the first fundamental form of a surface (see Mortenson 1985) and is the square root of the traditional metric. For the torus-shaped wheel, all the required partial derivatives can be obtained using Eqn (10.2), and we can easily obtain explicit expressions for the elements of [**M**], [**K**]. These are given by

$$[\mathbf{M}_w] = \begin{bmatrix} r_1 & 0 \\ 0 & r_2 + r_1 \sin u_w \end{bmatrix}$$

$$[\mathbf{K}_w] = \begin{bmatrix} \dfrac{1}{r_1} & 0 \\ 0 & \dfrac{\sin u_w}{r_2 + r_1 \sin u_w} \end{bmatrix}$$

$$[\mathbf{T}_w] = \left[0 \quad \frac{\cos u}{r_2 + r_1 \sin u_w}\right] \tag{10.8}$$

To analyse the motion of the torus-shaped wheel on an uneven terrain, we need a representation of uneven surfaces in $\Re^3$. The detailed and advanced treatment of surfaces can be obtained in any textbook on geometric modelling or CAD [see, for example, Mortenson (1985)]. In the following section, we discuss the representation of an uneven terrain in $\Re^3$, which is enough for our purpose.

10.2.2 Representation of Uneven Terrain in $\Re^3$

We assume that the uneven terrain is a smooth and differentiable surface in $\Re^3$ with at least C^3 continuity.[9] This is valid when the uneven terrain is considered to be hard and smooth, and it clearly precludes terrains with sand, dirt, and any kind of steps or discontinuities.

For WMRs moving on an uneven terrain, the explicit equations of the uneven terrain are usually not available, although (as we will see in Section 10.2.3) we need to compute explicit expressions for the elements of [**M**], [**K**], and [**T**] of the terrain. Typically n measured data points of the terrain in the form of $(x, y, z)_i$, where $i = 1, 2, ..., n$, are available, and these data come from sensors, such as a laser range finder, located on the WMR or external to the WMR. These data are called a digital elevation model

---

[9] In geometric modelling, C^k continuity implies that derivatives up to and including the kth-order exist. As we will see, for kinematic analysis, C^2 continuity is enough; whereas for dynamics, C^3 continuity is required.

(DEM) of the terrain. The reconstruction of a surface or obtaining $\mathbf{f}(u, v)$ from its DEM is an ill-posed problem, and, generally, a 'correct' surface does not exist. As a result various notions such as the least square error, distance metrics, energy minimization, and smoothness assumptions on the underlying surface, with their own advantages and disadvantages, have been proposed in the literature to reconstruct a surface from the available DEM [for some of the techniques for surface reconstruction, see Edelsbrunner and Mucke (1994), Bajaj et al. (1995), and Amenti et al. (1998)]. Our focus is not on surface modelling techniques. In this text, our only goal is to efficiently compute the partial derivatives up to the required order. As can be seen from the expressions of the metric, curvature and torsion, we need to obtain explicit expressions for the partial derivatives up to the second derivatives and, as we will see later, for dynamics we require up to the third derivatives. Any representation of the terrain which allows us to do this will serve our purpose. Therefore, without loss of generality we will model the uneven terrain as a spline tensor product surface (see Mortenson 1985). We will use bi-cubic and B-spline surfaces depending on the order of the derivatives required, and these are briefly explained next.

A bi-cubic patch is given by the equation

$$\mathbf{f}(u, v) = \sum_{i=0}^{3}\sum_{j=0}^{3} a_{ij}u^i v^j \quad (u, v)\epsilon[0, 1] \tag{10.9}$$

The quantities a_{ij}'s are called algebraic coefficients and there are 16 of them. One can find them from four given corner points and the slopes and twist vectors at these corner points. One can also find them if 16 points on the surface are given. In a bi-cubic patch, a change in any of the algebraic coefficients (or any of the 16 points) changes the whole surface.

An alternative approach is to use non-uniform rational B-splines or NURBS. A NURBS is defined as

$$\mathbf{f}(u, v) = \sum_{i=0}^{m}\sum_{j=0}^{n} \mathbf{p}_{ij}N_{i,k}(u)N_{j,l}(v) \quad (u, v)\epsilon[0, 1] \tag{10.10}$$

where $\mathbf{p}_{ij}$'s are the vertices of a defining polyhedron, $N_{i,k}$ and $N_{j,l}$ are blending functions whose degrees are determined by j and l. A blending function is defined recursively as

$$\begin{aligned} N_{i,1}(u) &= 1 \text{ if } t_i \leq u \leq t_{i+1} \\ &= 0 \text{ otherwise} \\ N_{i,k}(u) &= \frac{(u - t_i)N_{i,k-1}(u)}{t_{i+k-1} - t_i} + \frac{(t_{i+k} - u)N_{i+1,k-1}(u)}{t_{i+k} - t_{i+1}} \end{aligned} \tag{10.11}$$

where t_i's are knot values which relate u to a given control point. These are

$$
\begin{aligned}
t_i &= 0 \text{ if } i \leq k && (10.12)\\
&= i - k + 1 \text{ if } k \leq i \leq n\\
&= n - k + 2 \text{ if } i \geq n
\end{aligned}
$$

with $0 \leq i \leq n + k$. The range of u is $0 \leq u \leq n - k + 2$. In the case of NURBS, the choice of k and l is free and determines how the local shape of the surface changes if one of the control points, $\mathbf{p}_{ij}$, is changed. One can hence have more control over the local shape of the surface in comparison to bi-cubic patches.

We have used the MATLAB spline toolbox to generate the surfaces from assumed synthetic ground data. In our simulations we have used both bi-cubic surfaces and B-spline surfaces based on our continuity requirement. Typical examples of synthetic surfaces representing uneven terrains, generated using MATLAB spline toolbox, are given in Figs 10.3–10.5.

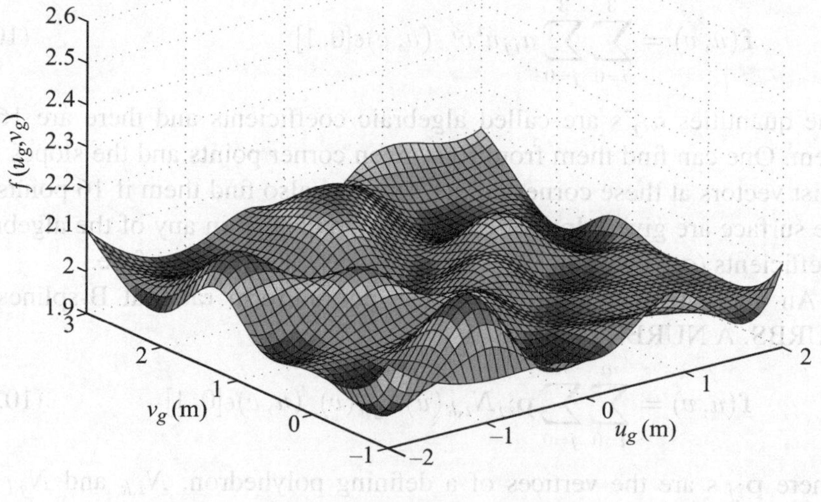

Fig. 10.3 A bi-cubic surface

The bi-cubic or B-spline surfaces can be represented in the form $\mathbf{f}(u_g, v_g)$, where we have used the subscript g to denote the uneven ground or terrain. The function $\mathbf{f}(u_g, v_g)$ has the required continuity. According to the definitions given in Eqns (10.5)–(10.7), we can obtain explicit expressions for the elements of $[\mathbf{M}_g]$, $[\mathbf{K}_g]$, and $[\mathbf{T}_g]$. The explicit expressions can be

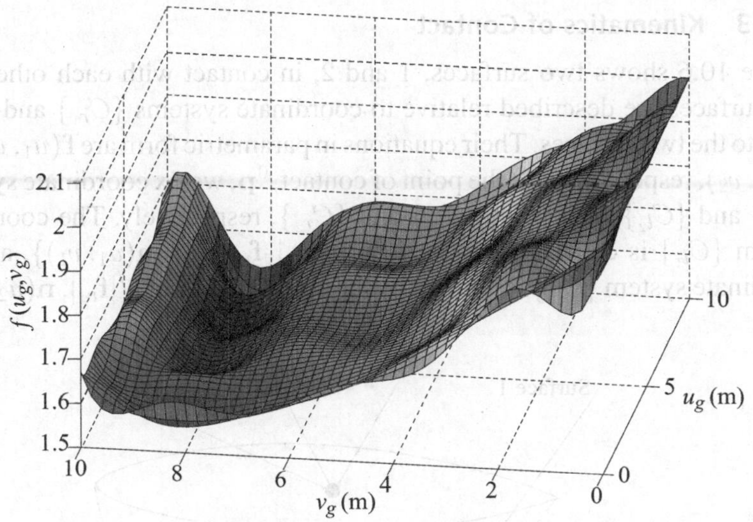

Fig. 10.4 B-spline surface

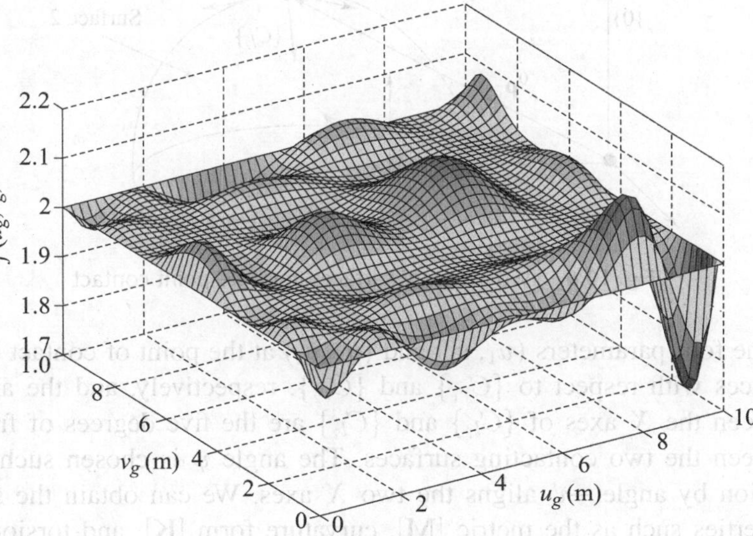

Fig. 10.5 Another B-spline surface

obtained by using a symbolic manipulation software such as MAPLE (Heck 2003) or MATHEMATICA (Wolfram 1999).

To analyse the motion of the torus-shaped wheel on an uneven terrain, we use the differential equations describing the contact between two arbitrary surfaces derived by Montana (1988). This is discussed next.

10.2.3 Kinematics of Contact

Figure 10.6 shows two surfaces, 1 and 2, in contact with each other. The two surfaces are described relative to coordinate systems $\{C_{r_1}\}$ and $\{C_{r_2}\}$ fixed to the two surfaces. Their equations in parametric form are $\mathbf{f}(u_1, v_1)$ and $\mathrm{f}_2(u_2, v_2)$, respectively. At the point of contact, ${}^0\mathbf{p}$, we fix coordinate systems $\{C_{l_1}\}$ and $\{C_{l_2}\}$ relative to $\{C_{r_1}\}$ and $\{C_{r_2}\}$, respectively. The coordinate system $\{C_{l_1}\}$ is defined by the set $\{\mathbf{f}_{u_1}/|\mathbf{f}_{u_1}|, \mathbf{f}_{v_1}/|\mathbf{f}_{v_1}|, \mathbf{n}(u_1, v_1)\}$, and the coordinate system $\{C_{l_2}\}$ is defined by the set $\{\mathbf{f}_{u_2}/|\mathbf{f}_{u_2}|, \mathbf{f}_{v_2}/|\mathbf{f}_{v_2}|, \mathbf{n}(u_2, v_2)\}$.

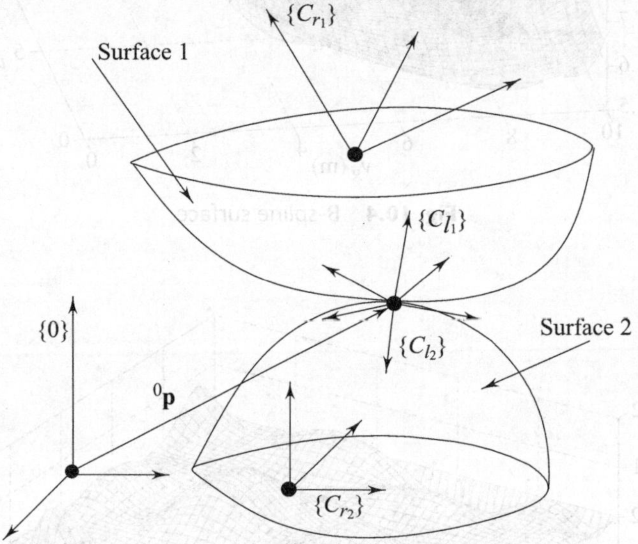

Fig. 10.6 Two arbitrary surfaces in single-point contact

The four parameters (u_1, v_1) and (u_2, v_2) at the point of contact on two surfaces with respect to $\{C_{r_1}\}$ and $\{C_{r_2}\}$, respectively, and the angle ψ between the X axes of $\{C_{l_1}\}$ and $\{C_{l_2}\}$ are the five degrees of freedom between the two contacting surfaces. The angle ψ is chosen such that a rotation by angle $-\psi$ aligns the two X axes. We can obtain the surface properties such as the metric [**M**], curvature form [**K**], and torsion form [**T**] for both the surfaces using Eqns (10.5)–(10.7). The curvature matrix of surface 2 at the point of contact *relative* to surface 1, denoted by $[\mathbf{K}^*]$, is given by

$$[\mathbf{K}^*] = [R_\psi][\mathbf{K}_2][R_\psi]^T \qquad (10.13)$$

where $[R_\psi]$ is the matrix

$$[R_\psi] = \begin{pmatrix} \cos\psi & -\sin\psi \\ -\sin\psi & -\cos\psi \end{pmatrix}$$

The contact equations as derived by Montana (1988) are given by

$$
\begin{aligned}
(\dot{u}_1, \dot{v}_1)^T &= [\mathbf{M}_1]^{-1}([\mathbf{K}_1] + [\mathbf{K}^*])^{-1}[(-\omega_y, \omega_x)^T - [\mathbf{K}^*](v_x, v_y)^T] \\
(\dot{u}_2, \dot{v}_2)^T &= [\mathbf{M}_2]^{-1}[R_\psi]([\mathbf{K}_1] + [\mathbf{K}^*])^{-1}[(-\omega_y, \omega_x)^T + [\mathbf{K}_1](v_x, v_y)^T] \\
\dot{\psi} &= \omega_z + [\mathbf{T}_1][\mathbf{M}_1](\dot{u}_1, \dot{v}_1)^T + [\mathbf{T}_2][\mathbf{M}_2](\dot{u}_2, \dot{v}_2)^T \qquad (10.14) \\
0 &= v_z
\end{aligned}
$$

where ω_x, ω_y, and ω_z are the angular velocity components and v_x, v_y, and v_z are the linear velocity components of $\{C_{l_1}\}$ relative to $\{C_{l_2}\}$. The last equation, $v_z = 0$, is a constraint specifying that contact is always maintained during the relative motion of surface 1 with respect to surface 2.

The first five equations in Eqn (10.14) can be used to find $\dot{u}_1$, $\dot{v}_1$, $\dot{u}_2$, $\dot{v}_2$, $\dot{\psi}$ given ω_x, ω_y, ω_z, v_x, v_y at the point of contact. One can also invert Eqn (10.14) to get

$$
\begin{aligned}
(v_x, v_y)^T &= -[\mathbf{M}_1](\dot{u}_1, \dot{v}_1)^T + [R_\psi][\mathbf{M}_2](\dot{u}_2, \dot{v}_2)^T \\
(\omega_y, -\omega_x)^T &= -[\mathbf{K}_1][\mathbf{M}_1](\dot{u}_1, \dot{v}_1)^T - [R_\psi][\mathbf{K}_2][\mathbf{M}_2](\dot{u}_2, \dot{v}_2)^T \\
\omega_z &= \dot{\psi} - [\mathbf{T}_1][\mathbf{M}_1](\dot{u}_1, \dot{v}_1)^T - [\mathbf{T}_2][\mathbf{M}_2](\dot{u}_2, \dot{v}_2)^T \qquad (10.15) \\
v_z &= 0
\end{aligned}
$$

and use the first five equations in Eqn (10.15) to obtain ω_x, ω_y, ω_z, v_x, v_y for a given $\dot{u}_1$, $\dot{v}_1$, $\dot{u}_2$, $\dot{v}_2$, $\dot{\psi}$ at the point of contact.

The contact equations are analogous to the constraint equations derived for rotary, prismatic, and other joints in Section 2.5. The *key* difference, however, is that the constraints involve linear and angular velocity components and derivatives of u_i, v_i and ψ together with u_i, v_i $(i = 1, 2)$ and ψ. In general, it is not possible to solve for $u_i(t)$, $v_i(t)$ $(i = 1, 2)$, and $\psi(t)$ in the closed form for given linear and angular velocity components. Likewise, it is also not possible to solve in the closed form, position, and orientation of $\{C_{r_1}\}$ and $\{C_{r_2}\}$ given the time derivatives of u_i, v_i $(i = 1, 2)$ and ψ. The first five differential equations in Eqn (10.14) or Eqn (10.15) are called *non-holonomic* due to the fact that they are *non-integrable* (Goldstein 1980) and they, in general, cannot be written purely in terms of position and orientation variables and u_i, $v_i (i = 1, 2)$ and ψ. There are two special cases of contact equations. If $v_x = v_y = 0$, then the surfaces are in *pure rolling* contact. If $\omega_x = \omega_y = 0$, then the surfaces are in *pure sliding* contact.[10] In wheeled mobile robots, the interest is on pure rolling of the wheels as we get the desired situation of no slip.

---

[10] In Montana (1988), $\omega_z = 0$ is also assumed for rolling. This is, however, a special case. The condition $\omega_z \neq 0$ allows rotation about the common normal and point of contact. This is similar to steering and assuming $\omega_z = 0$ removes the possibility of steering. For wheels in mobile robots, undergoing pure rolling, we will assume only $v_x = v_y = 0$.

The contact between the wheel and the ground, with $v_x = v_y = v_z = 0$, has three degrees of freedom, namely, one can have arbitrary ω_x, ω_y, and ω_z. These three degrees of freedom are, however, very different from the three degrees of freedom in a spherical (S) joint discussed in Section 2.5. In the case of an S joint, not only $v_x = v_y = v_z = 0$, but also the x, y, z coordinates of two points on the two parts of the S joint are coincident. As known in literature, non-holonomic constraints restrict only the space of achievable velocities and *not* the positions. In the classic example of a vertical disc rolling without slip on a flat plane, the linear velocity components v_x and v_y are zero, but the disc can achieve any x, y position and can have any rotation θ normal to its plane at the x, y location. Likewise, for the two surfaces in contact and undergoing rolling without slip, there are *five* relative degrees of freedom as far as the position and orientation are concerned but only three degrees of freedom for velocities. It may be noted that the constraint $v_z = 0$ or $z =$ constant is like the holonomic constraints in lower pair joints. It signifies that the wheel cannot lose contact with the plane. Due to the above reasoning, we call the 'equivalent' joint between the wheel and the ground as a *non-holonomic* joint.

For two given surfaces, and given linear and angular velocity components and initial values, Eqns (10.14) can, however, be *integrated numerically* to obtain the evolution of u_1, v_1, u_2, v_2, ψ as functions of time. From the numerical integration we can obtain the motions of the two surfaces which always maintain contact with each other. This is illustrated for the case of the torus-shaped wheel undergoing pure rolling on an uneven terrain next.

10.2.4 Kinematics of a Single Wheel

We denote surface 1 as the wheel and surface 2 as the ground. The coordinate system $\{C_{r_2}\}$ is the same as the fixed coordinate system $\{0\}$. The coordinate system $\{C_{r_1}\}$ is fixed at the wheel centre C and is denoted by $\{w\}$. The coordinate systems $\{C_{l_1}\}$ and $\{C_{l_2}\}$ at the contact point are denoted $\{2\}$ and $\{1\}$, respectively. We also fix coordinate systems $\{3\}$ and $\{4\}$ as shown in Fig. 10.7. The coordinates of the point of contact on the wheel are given in the wheel frame $\{w\}$ by (u_w, v_w) and in the ground frame $\{0\}$ by (u_g, v_g). The 4×4 transformation matrix ${}^0_1[T]$ describes the uneven ground at the wheel–ground contact point and is given by

$$
{}^0_1[T] = \begin{pmatrix} l_1 & m_1 & n_1 & u_g \\ l_2 & m_2 & n_2 & v_g \\ l_3 & m_3 & n_3 & f(u_g, v_g) \\ 0 & 0 & 0 & 1 \end{pmatrix}
$$

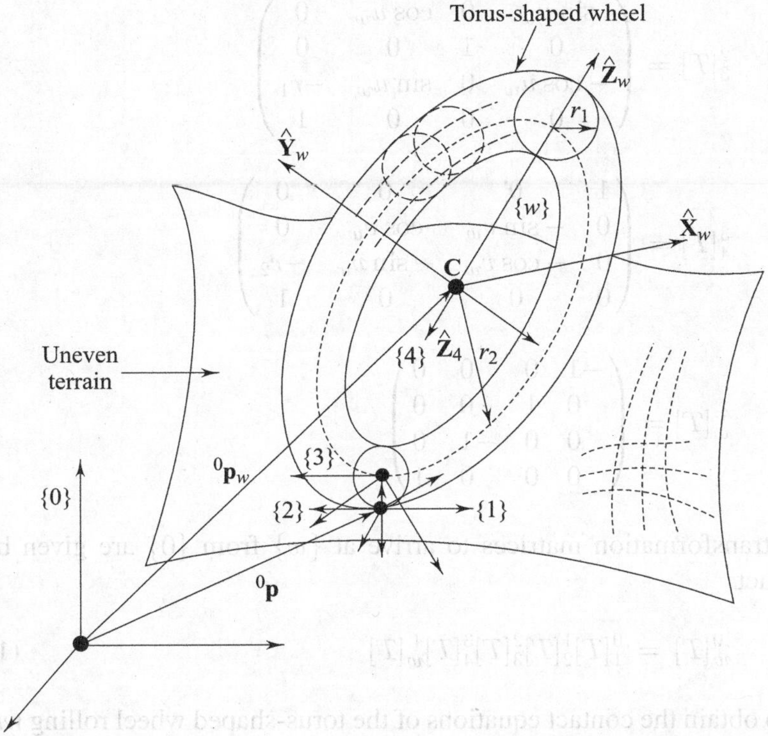

Fig. 10.7 Coordinate systems for the torus-shaped wheel on uneven terrain

where l_i, m_i, n_i $(i = 1, 2, 3)$ are the components of the orthogonal vectors $\{\mathbf{f}_u/|\mathbf{f}_u|, \mathbf{n} \times \mathbf{f}_u/|\mathbf{f}_u|, \mathbf{n}\}$, with $\mathbf{n}$ as defined in Eqn (10.4). The components of ${}^0_1[T]$ can be obtained explicitly in closed-form once we have obtained the closed-form equations of the reconstructed surface as a bi-cubic or as a NURBS (see Section 10.2.2).

The 4×4 transformation matrix ${}^1_2[T]$ describes the ψ rotation of the torus-shaped wheel as discussed in Section 10.2.3. The transformation matrices ${}^2_3[T]$ and ${}^3_4[T]$ describe the tilt and rotation of the torus-shape wheel, and are functions of u_w and v_w, respectively. The final ${}^{\;4}_{w}[T]$ constant matrix ensures that the Z axis of the torus-shaped wheel is initially pointed vertically upwards. The 4×4 matrices ${}^1_2[T]$, ${}^2_3[T]$, ${}^3_4[T]$, and ${}^{\;4}_{w}[T]$ are

$$
{}^1_2[T] = \begin{pmatrix} \cos\psi & -\sin\psi & 0 & 0 \\ -\sin\psi & -\cos\psi & 0 & 0 \\ 0 & 0 & -1 & 0 \\ 0 & 0 & 0 & 1 \end{pmatrix}
$$

$$ {}^2_3[T] = \begin{pmatrix} \sin u_w & 0 & \cos u_w & 0 \\ 0 & 1 & 0 & 0 \\ -\cos u_w & 0 & \sin u_w & -r_1 \\ 0 & 0 & 0 & 1 \end{pmatrix} $$

$$ {}^3_4[T] = \begin{pmatrix} 1 & 0 & 0 & 0 \\ 0 & -\sin v_w & \cos v_w & 0 \\ 0 & -\cos v_w & -\sin v_w & -r_2 \\ 0 & 0 & 0 & 1 \end{pmatrix} $$

$$ {}^4_w[T] = \begin{pmatrix} -1 & 0 & 0 & 0 \\ 0 & 1 & 0 & 0 \\ 0 & 0 & -1 & 0 \\ 0 & 0 & 0 & 1 \end{pmatrix} $$

The transformation matrices to arrive at $\{w\}$ from $\{0\}$ are given by the product

$$ {}^0_w[T] \;=\; {}^0_1[T]^1_2[T]^2_3[T]^3_4[T]^4_w[T] \tag{10.16} $$

To obtain the contact equations of the torus-shaped wheel rolling *without slip* on an arbitrary uneven surface, we set v_x and v_y to zero in Eqn (10.14). The contact equations in this special case are

$$ \begin{aligned} (\dot{u}_w, \dot{v}_w)^T &= [\mathbf{M}_w]^{-1}([\mathbf{K}_w] + [\mathbf{K}^*])^{-1}[(-\omega_y, \omega_x)^T] \\ (\dot{u}_g, \dot{v}_g)^T &= [\mathbf{M}_g]^{-1}[\mathbf{R}_\psi]([\mathbf{K}_g] + [\mathbf{K}^*])^{-1}[(-\omega_y, \omega_x)^T] \\ \dot{\psi} &= \omega_z + [\mathbf{T}_w][\mathbf{M}_w](\dot{u}_w, \dot{v}_w)^T + [\mathbf{T}_g][\mathbf{M}_g](\dot{u}_g, \dot{v}_g)^T \\ 0 &= v_z \end{aligned} \tag{10.17} $$

where the subscript w denotes the wheel, subscript g denotes the uneven ground, and the angular velocity components $\omega_x, \omega_y, \omega_z$ are the three inputs to this three-DOF system. The relative curvature matrix $[\mathbf{K}^*]$ is evaluated as in Eqn (10.13).

With the above preliminaries, we can now numerically simulate the motion of the torus-shaped wheel by following the steps given next.

Step 1: Generate the surface The ground DEM is assumed to be known. For kinematics purposes we need C^2 continuity of the surface, hence we can use a bi-cubic representation of the surface. The surface representation and its derivatives are computed using built-in MATLAB functions (Mathwork 1992).

Step 2: Form contact equations The contact equations, Eqn (10.17), are obtained in symbolic form. We have a system of five first-order, non-linear, ordinary differential equations (ODEs) in five variables u_w, v_w, u_g, v_g, ψ.

Step 3: Obtain initial conditions The initial values of the point of contact on the ground (u_g, v_g) can be chosen freely. The coordinates of the point of contact on the wheel (u_w, v_w) can also be chosen as desired. The heading angle ψ can also be arbitrary. The choice of u_w, v_w, and ψ determines the orientation of the wheel at the starting point.

Step 4: Solve contact equations The contact equations, Eqn (10.17), can be solved with any ODE solver, using initial conditions obtained in step 3 and using specified inputs ω_x, ω_y, ω_z. This gives the evolution of u_w, v_w, u_g, v_g, and ψ in time.

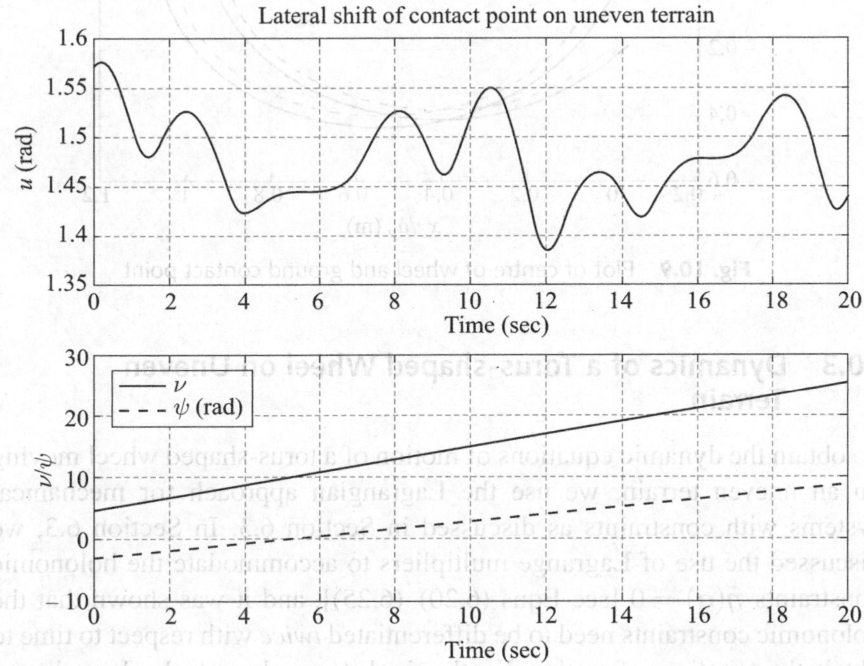

Fig. 10.8 Variation of parameters for a torus-shaped wheel moving on uneven terrain

To illustrate the kinematics of a single torus-shaped wheel, we present numerical results for the synthetically generated surface shown in Fig. 10.3. The geometrical parameters of the wheel used in simulation are $r_1 = 0.05$ m and $r_2 = 0.25$ m. The initial conditions used are $u_g = 0$, $v_g = 0$, $u_w = \pi/2$, $v_w = 3\pi/2$, and $\psi = -\pi$. Figure 10.8 shows the variation of u_w, v_w, and

ψ. It can be seen that the wheel tilts as it rolls on the uneven surface. It can also be seen from Fig. 10.9 that the trace of the wheel centre and the ground contact point are different. For the motion of wheel, modelled as a thin disc, on a flat terrain, the trace of the wheel centre and the ground contact point would be coincident.

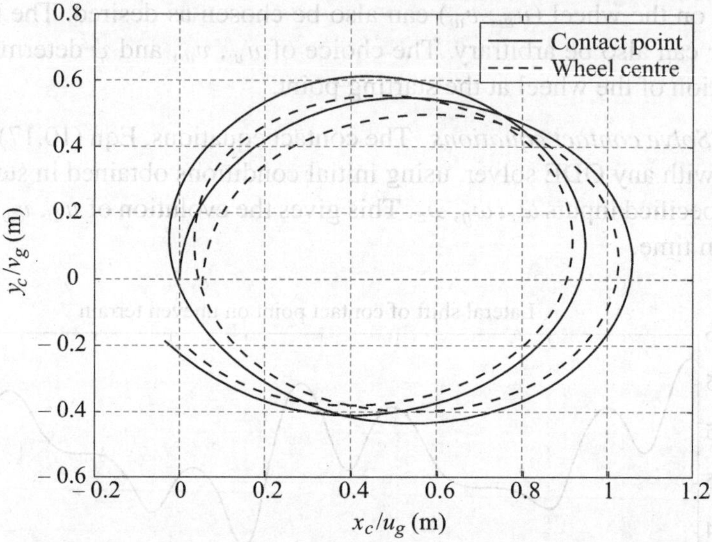

Fig. 10.9 Plot of centre of wheel and ground contact point

10.3 Dynamics of a Torus-shaped Wheel on Uneven Terrain

To obtain the dynamic equations of motion of a torus-shaped wheel moving on an uneven terrain, we use the Lagrangian approach for mechanical systems with constraints as discussed in Section 6.3. In Section 6.3, we discussed the use of Lagrange multipliers to accommodate the holonomic constraints $\eta(\mathbf{q}) = 0$ [see Eqns (6.20)–(6.25)], and it was shown that the holonomic constraints need to be differentiated *twice* with respect to time to obtain the equations of motion. For the single torus-shaped wheel moving on an uneven terrain, the no-slip constraints $v_x = v_y = 0$ are already available in terms of the first derivatives of the generalized coordinates. The two non-holonomic constraints can be written as

$$(v_x, v_y)^T = -[\mathbf{M}_w](\dot{u}_w, \dot{v}_w)^T + [R_\psi][\mathbf{M}_g](\dot{u}_g, \dot{v}_g)^T = (0, 0)^T \tag{10.18}$$

We can rearrange Eqn (10.18) as

$$[\mathbf{\Psi}(\mathbf{q})]\dot{\mathbf{q}} = \mathbf{0} \tag{10.19}$$

where the generalized variables $\mathbf{q}$ are $(u_w, v_w, u_g, v_g, \psi)$.

To obtain the kinetic energy and potential energy of the wheel, required for the Lagrangian formulation, we note that the transformation matrix of $\{w\}$ with respect to $\{0\}$ is given by

$$ {}^{0}_{w}[T] = {}^{0}_{1}[T]\,{}^{1}_{2}[T]\,{}^{2}_{3}[T]\,{}^{3}_{4}[T]\,{}^{4}_{w}[T] = \left(\begin{array}{c|c} {}^{0}_{w}[R] & {}^{0}\mathbf{p_w} \\ \hline 0 & 1 \end{array}\right) \tag{10.20} $$

where the rotation matrix ${}^{0}_{w}[R]$ describes the orientation of the coordinate system $\{w\}$ fixed at the wheel centre, which is located by the vector ${}^{0}\mathbf{p}_w$ from the fixed coordinate system $\{0\}$. From the rotation matrix and its derivative with respect to time, we can obtain the angular velocity matrix of $\{w\}$ with respect to $\{0\}$ [see Eqn (5.3)] as

$$ {}^{0}_{w}[\Omega] \triangleq {}^{0}_{w}[\dot{R}]\,{}^{0}_{w}[R]^T = \begin{pmatrix} 0 & -\Omega_z & \Omega_y \\ \Omega_z & 0 & -\Omega_x \\ -\Omega_y & \Omega_x & 0 \end{pmatrix} \tag{10.21} $$

where $(\Omega_x, \Omega_y, \Omega_z)$ are the components of the angular velocity vector $\boldsymbol{\Omega}$. The linear velocity can be obtained by the differentiation of ${}^{0}\mathbf{p}_w$ and we get

$$ {}^{0}\mathbf{V}_w = {}^{0}\dot{\mathbf{p}}_{\mathbf{w}} \tag{10.22} $$

It may be noted that $\boldsymbol{\Omega}$ and ${}^{0}\mathbf{V}_w$ are *not* the same as $(\omega_x, \omega_y, \omega_z)^T$ and $(v_x, v_y, v_z)^T$ used in wheel–ground contact equations.

The kinetic energy of the wheel is given by

$$ \mathrm{KE} = \frac{1}{2}\boldsymbol{\Omega}^T[I_w]\boldsymbol{\Omega} + \frac{1}{2}m_w\,{}^{0}\mathbf{V}_w^2 \tag{10.23} $$

where m_w is the mass of the wheel and $[I_w]$ is the moment of inertia matrix of the wheel. In the case of the torus-shaped wheel, $[I_w]$ in $\{w\}$ is given by

$$ [I_w] = \begin{pmatrix} \frac{1}{4}m_w(3r_1^2 + 4r_2^2) & 0 & 0 \\ 0 & \frac{1}{8}m_w(5r_1^2 + 4r_2^2) & 0 \\ 0 & 0 & \frac{1}{8}m_w(5r_1^2 + 4r_2^2) \end{pmatrix} $$

The potential energy of the wheel is given by

$$ \mathrm{PE} = m_w g z_{wc} \tag{10.24} $$

where z_{wc} is the height of the centre of the wheel, C, in $\{0\}$ and is obtained from the third component of ${}^{0}\mathbf{p}_w$ (see Fig. 10.7).

Following the Lagrangian formulation in Section 6.3, we can derive the equations of motion of the single torus-shaped wheel in the form

$$ [\mathbf{M}(\mathbf{q})]\ddot{\mathbf{q}} + [\mathbf{C}(\mathbf{q}, \dot{\mathbf{q}})]\dot{\mathbf{q}} + \mathbf{G}(\mathbf{q}) = \boldsymbol{\tau} + [\boldsymbol{\Psi}(\mathbf{q})]^T\boldsymbol{\lambda} \tag{10.25} $$

where $\boldsymbol{\lambda}$ is a 2×1 vector of Lagrange multipliers arising from the non-holonomic constraints of no slip at the wheel–ground contact point.

As shown in Section 6.3, we can solve for $\boldsymbol{\lambda}$ by differentiating the constraint $[\boldsymbol{\Psi}(\mathbf{q})]\dot{\mathbf{q}} = \mathbf{0}$ *once* and obtain equations of motion of the wheel in a form similar to Eqn (6.25). All the terms in the equations of motion, namely, $[\mathbf{M}(\mathbf{q})]$, $[\mathbf{C}(\mathbf{q}, \dot{\mathbf{q}})]$, and $\mathbf{G}(\mathbf{q})$, can be symbolically derived as discussed in Section 6.3.

Once the equations of motion are known, we can simulate the motion of the torus-shaped wheel on an uneven terrain under the action of external forces or moments $\boldsymbol{\tau}$ and for the no-slip constraints of Eqn (10.18). The steps that need to be followed are given next.

Step 1: Generate the surface The uneven terrain is reconstructed from the ground DEM with at least C^3 continuity. For dynamics C^3 continuity is required since $[\dot{\boldsymbol{\Psi}}]$ is required and $[\dot{\boldsymbol{\Psi}}]$ contains third partial derivatives of the surface function. Hence, we use a fourth-degree B-spline to generate the surface. The surface representation and its derivatives are computed using in-built MATLAB functions (Mathwork 1992).

Step 2: Form the equations of motion The equations of motion [Eqn (10.25)] are formed in the manner outlined earlier. We have a set of five second-order ODEs in five variables, u_w, v_w, u_g, v_g, ψ. The equations of motion are derived symbolically using MAPLE (Heck 2003) or MATHEMATICA (Wolfram 1999).

Step 3: Obtain initial condition The initial conditions must be chosen such that these satisfy the non-holonomic constraints of no slip at the wheel–ground contact point. The initial values for u_w, v_w, u_g, v_g, and ψ can be chosen arbitrarily, depending on the initial desired position and orientation of the wheel on the ground. Among the first derivatives, $\dot{u}_w$, $\dot{v}_w$, $\dot{u}_g$, $\dot{v}_g$, and $\dot{\psi}$, we can choose only three. The other two are to be determined from the non-holonomic no-slip constraints given by Eqn (10.18).

Step 4: Solve equations of motion The five second-order equations of motion are solved with an ODE solver using initial conditions obtained in step 3. This gives the evolution of u_w, v_w, u_g, v_g, ψ and their first derivatives $\dot{u}_w$, $\dot{v}_w$, $\dot{u}_g$, $\dot{v}_g$, $\dot{\psi}$ as functions of time.

To illustrate the dynamics of the torus-shaped wheel, we use the surface shown in Fig. 10.4. In this simulation we assume that there is no external applied force or moment acting on the wheel, i.e., $\boldsymbol{\tau} = \mathbf{0}$. The initial conditions used are: $u_g = 4.00103$ m, $v_g = 0.49346$ m, $u_w = 1.5698$ rad, $v_w = (3\pi/2)$ rad, $\psi = -3.141545$ rad, $\dot{u}_w = 0$ rad/sec, $\dot{v}_w = 2$ rad/sec, $\dot{u}_g = 0$ m/sec, $\dot{v}_g = 0.5975$ m/sec, and $\dot{\psi} = 0$ rad/sec. Figure 10.10 shows the variation of u_w, v_w, and ψ as a function of time t. It can be seen that the wheel tilts as it rolls on the uneven surface. Due to this, the trace

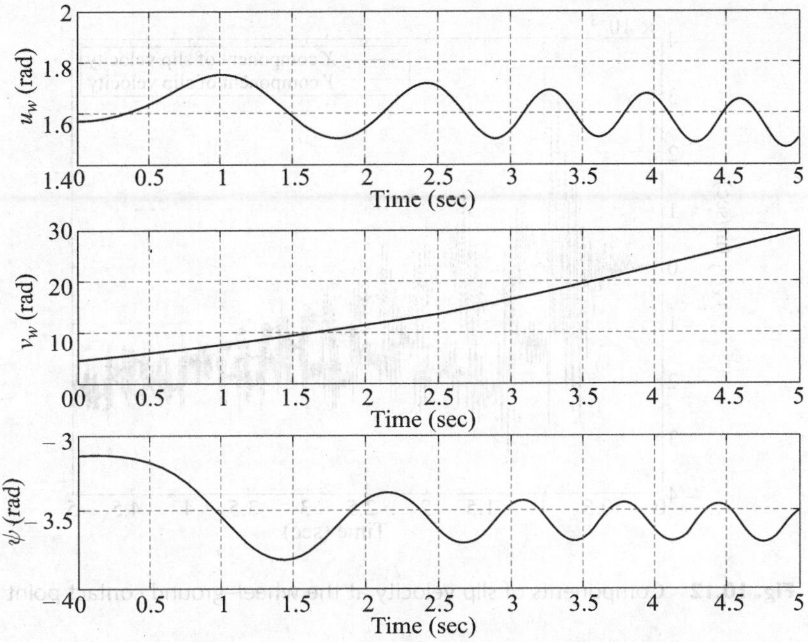

Fig. 10.10 Variation of parameters for a torus-shaped wheel moving on uneven terrain

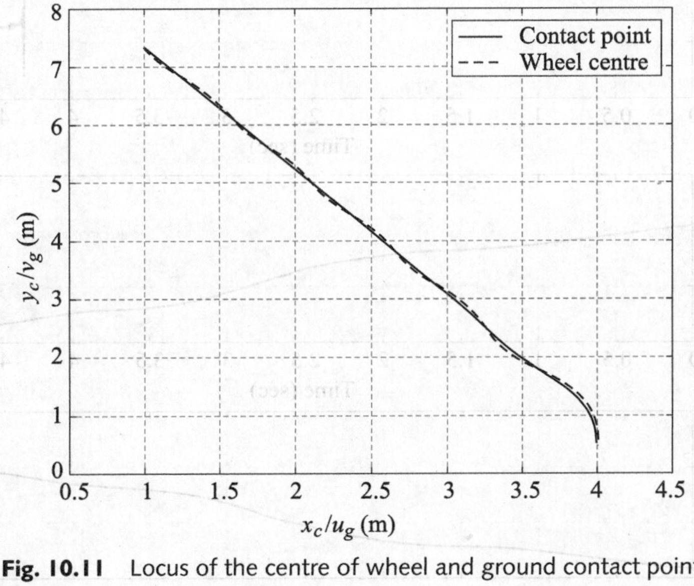

Fig. 10.11 Locus of the centre of wheel and ground contact point

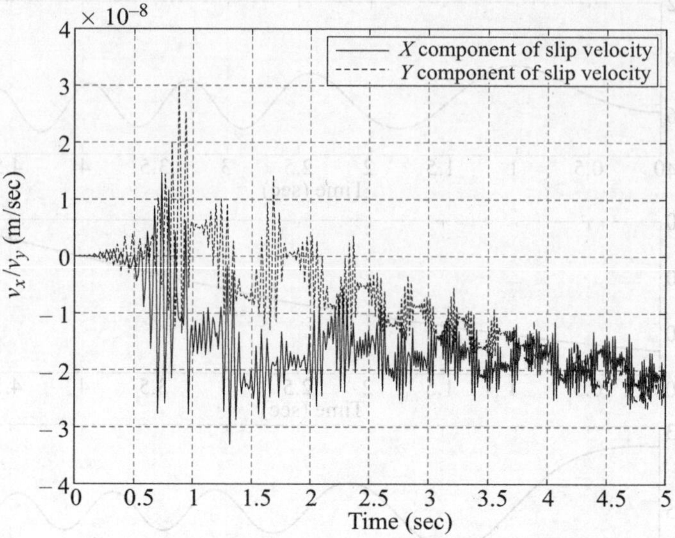

Fig. 10.12 Components of slip velocity at the wheel–ground contact point

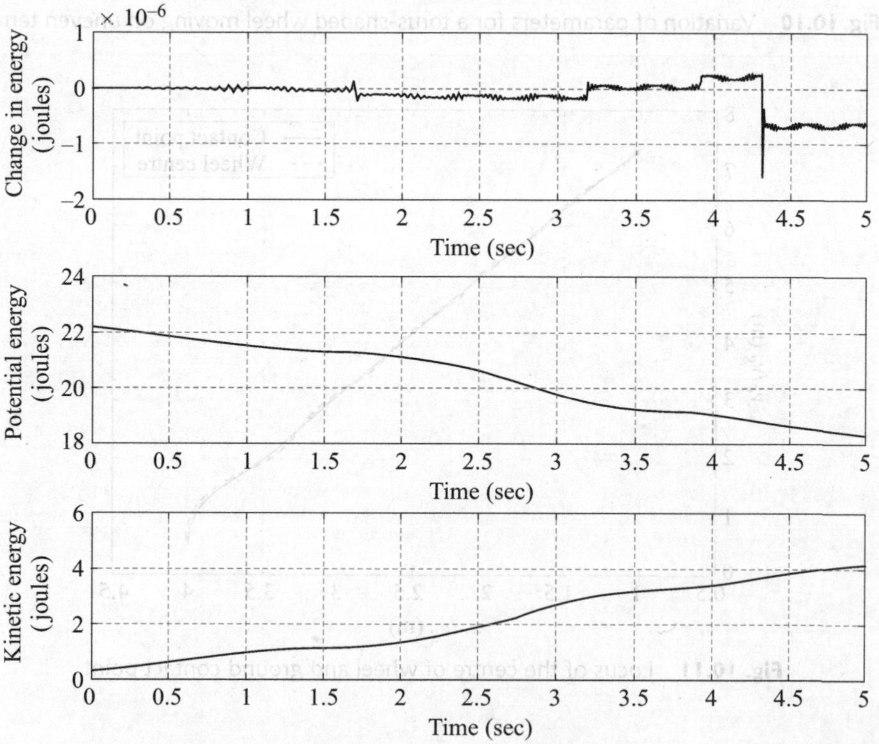

Fig. 10.13 Conservation of energy and the variation of potential and kinetic energies

of the wheel centre and the contact point, as shown in Fig. 10.11, are different. Figure 10.12 shows that the wheel rolls without slip as the velocity component of the wheel contact point at the wheel–ground contact is very small (of the order of 10^{-8} m/sec). The conservation of energy is depicted in Fig. 10.13, which implies that the simulation results are correct. Figure 10.13 also shows the variation in potential and kinetic energies as the wheel moves over the uneven terrain.

In the following section, we will use three torus-shaped wheels connected to a rigid platform with rotary (R) joints to model a three-wheeled mobile robot capable of traversing uneven terrains without slip. We will model the three-wheeled mobile robot as a parallel manipulator and analyse its kinematics.

10.4 Kinematic Modelling of a Three-wheeled Mobile Robot

With three torus-shaped wheels and a rigid platform, we can have two possible architectures: (1) a WMR with the top platform having three degrees of freedom, and (2) a WMR with the top platform having six degrees of freedom. In the first case, the rear wheels are attached to the rigid platform with two rotary (R) joints, with one of the joints being passive and one actuated. The passive joint allows lateral *tilt* of the wheels, where the lateral tilt is the rotation about an axis perpendicular to the axle and lying in the plane of the platform. The actuated joint represents a motor or an actuator for in-plane rotation, or simply *rotation*, of the wheel. The front wheel can be *steered* or rotated about an axis perpendicular to the plane of the top platform and has no lateral tilt capability. The three-DOF WMR can be *instantaneously* modelled as an equivalent spatial parallel manipulator shown schematically in Fig. 10.14.

The degree of freedom for the spatial parallel manipulator can be verified using Grüebler's criterion given in Eqn (3.1) with $\lambda = 6$ (reproduced below for convenience),

$$\text{dof} = 6(N - J - 1) + \sum_{i=1}^{J} F_i$$

For the parallel manipulator shown in Fig. 10.14, we have $N = 8$, $J = 9$, and $\sum_{i=1}^{J} F_i = 15$ from three degrees of freedom for each wheel–ground contact and one degree of freedom for each of the six rotary (R) joints. Hence, the dof of the WMR is three and three of the joint variables should be actuated. We choose rotation at the two rear wheels, θ_1 and θ_2, and the

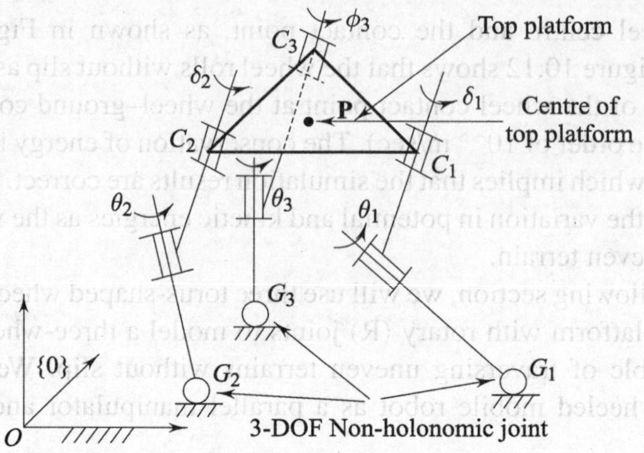

Fig. 10.14 Equivalent instantaneous mechanism for the 3-DOF WMR

steering at the front wheel, ϕ_3, as the actuated variables. The two lateral tilts at the rear wheels, δ_1 and δ_2, and the rotation of the front wheel, θ_3, are taken to be the passive variables which are to be computed.

In addition to the joints provided in the three-DOF WMR, if we also provide a rotary (R) joint in each of the two rear wheels to allow steering of the rear wheels and a joint in the front wheel to allow for lateral tilt in the front wheel, the instantaneous equivalent parallel manipulator is as shown in Fig. 10.15. In this case, $N = 11$, $J = 12$, and $\sum_{i=1}^{J} F_i = 18$, and Grüebler's criterion gives us the degrees of freedom of the top platform as 6. For this

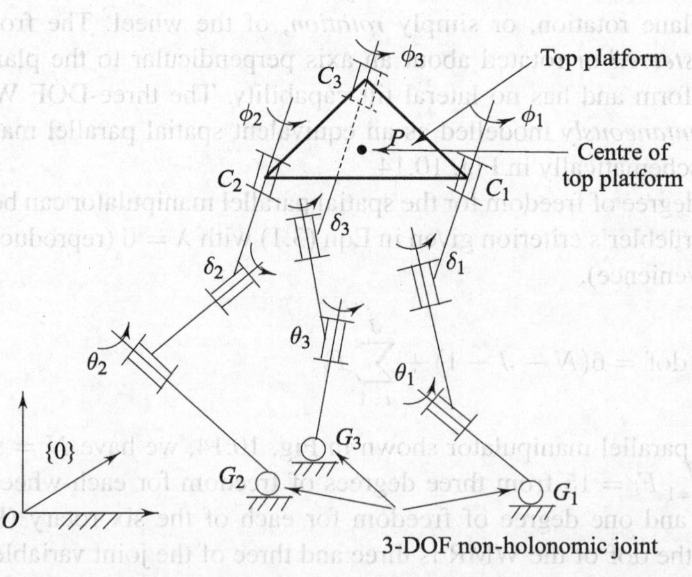

Fig. 10.15 Equivalent instantaneous mechanism for the 6-DOF WMR

case, the WMR would require six actuations. The WMR with the platform having six degrees of freedom is similar to the six-DOF parallel manipulator described in Example 4.3 in Section 4.4 (see also Section 2.8.6) except that it is inverted. The moving top platform in the WMR is actually the fixed base in Example 4.3 and the (fixed) ground for the WMR is analogous to the rigid body connected to the 'fingers' by spherical (S) joints in Example 4.3. The serial chains connecting the wheel to the moving platform are analogous to the 'fingers' in Example 4.3. The key difference is that a wheel–ground contact point, although having three degrees of freedom, is *not* the same as an S joint—the constraint equations at the wheel–ground contact point are non-holonomic, whereas in Example 4.3 the constraint equations between two S joints are holonomic.

In this section, we present the kinematic analysis and simulation results for the 3-DOF WMR configuration. The analysis procedure can be extended to the 6-DOF WMR configuration with actuated variables chosen as the rotation of the three wheels θ_1, θ_2, θ_3 and the three steering angles ϕ_1, ϕ_2, ϕ_3. The lateral tilt of the three wheels, δ_1, δ_2, δ_3, for the 6-DOF configuration would be the passive variables.

10.4.1 Direct and Inverse Kinematics of the 3-DOF WMR

As the WMR is subjected to non-holonomic, no-slip constraints, the kinematics problem can be formulated only in terms of the first derivatives of the kinematic variables. We first state the direct and inverse kinematics problems for the 3-DOF MWR and then present steps to solve them.

Direct kinematics problem Given the actuation rates, $\dot{\theta}_1$, $\dot{\theta}_2$, $\dot{\phi}_3$, and the ground and wheel geometry, find the orientation of the top platform ${}^0_p[R]$ and the position of the centre of the platform or any other point of interest.

Inverse kinematics problem For a given geometry of the ground and wheels, and given any three components out of $V_{p_x}, V_{p_y}, V_{p_z}, \Omega_{p_x}, \Omega_{p_y}, \Omega_{p_z}$, where $V_{p_x}, V_{p_y}, V_{p_z}$ are the components of the linear velocity vector of the centre of the platform (or any other point of interest) and $\Omega_{p_x}, \Omega_{p_y}, \Omega_{p_z}$ are the components of the angular velocity vector of the platform, find the two actuator inputs to the rear wheels $\dot{\theta}_1$ and $\dot{\theta}_2$, and the steering input to the front wheel $\dot{\phi}_3$.

To solve both the problems we proceed as follows.

Step 1: Generate the uneven terrain surface As described in Section 10.2.2, we can use bi-cubic patches to reconstruct the surface from elevation data. From the interpolated surface we can find the metric, curvature, and torsion forms for the ground at the three wheel–ground contact points. We also obtain expressions for the metric, curvature, and torsion forms for the torus-shaped wheels.

Step 2: Form contact equations For *each* wheel we write the five differential equations [see Eqn (10.14)] in terms of the contact variables u_i, v_i, u_{g_i}, v_{g_i}, and ψ_i, where $i = 1, 2, 3$. Since the wheels undergo no-slip motion, we set $v_x = v_y = 0$ for each of the three wheels. It may be noted that ω_x, ω_y, and ω_z in the contact equations for each wheel are the three components of angular velocities of $\{2\}$ with respect to $\{1\}$ (see Section 10.2.3) and are unknown. These are related to the angular velocity of the platform, Ω_{p_x}, Ω_{p_y}, Ω_{p_z}, and the input and passive joint rates. In the fixed coordinate system, $\{0\}$, we can write for each wheel

$$^0(\omega_x, \omega_y, \omega_z)^T = {}^0(\Omega_{p_x}, \Omega_{p_y}, \Omega_{p_z})^T + {}^0\boldsymbol{\omega}_{\text{input}} \tag{10.26}$$

where the vector ${}^0\boldsymbol{\omega}_{\text{input}}$, for each wheel, can be obtained from ${}^0[\dot{R}]_{\text{in}}\,{}^0[R]_{\text{in}}^T$ with

$$^0[R]_{\text{in}} = {}^0_p[R][R(\widehat{\mathbf{Z}}, \phi_i)][R(\widehat{\mathbf{Y}}, \delta_i)][R(\widehat{\mathbf{X}}, \theta_i)] \quad i = 1, 2, \text{ or } 3$$

Note that for the 3-DOF WMR, $\phi_1 = \phi_2 = \delta_3 = 0$. Equation (10.26) couples all five sets of ODEs and we get a set of 15 coupled ODEs in 21 variables—namely, the 15 contact variables u_i, v_i, u_{g_i}, v_{g_i}, ψ_i $(i = 1, 2, 3)$, the three wheel rotations $\theta_1, \theta_2, \theta_3$, the two lateral tilts δ_1, δ_2, and the front wheel steering ϕ_3. Of these, the rates of the three actuated variables θ_1, θ_2, and ϕ_3 are assumed to be known.

Step 3: Obtain the angular and linear velocities of centre of the platform The angular and linear velocities of the centre of the platform can be expressed in terms of the 15 wheel variables u_i, v_i, u_{g_i}, v_{g_i}, ψ_i $(i = 1, 2, 3)$ and their first derivatives with respect to time. If γ, β, α be a Z-Y-X Euler angle parametrization (see Chapter 2, Section 2.2.3) representing the orientation of the platform, we have

$$\begin{aligned}
{}^0\Omega_{p_x} &= \dot{\alpha}\cos\beta\cos\gamma - \dot{\beta}\sin\gamma = f_1(u_i, v_i, u_{g_i}, v_{g_i}, \psi_i, \dot{u}_i, \dot{v}_i, \dot{u}_{g_i}, \dot{v}_{g_i}, \dot{\psi}_i)\\
{}^0\Omega_{p_y} &= \dot{\alpha}\cos\beta\sin\gamma + \dot{\beta}\cos\gamma = f_2(u_i, v_i, u_{g_i}, v_{g_i}, \psi_i, \dot{u}_i, \dot{v}_i, \dot{u}_{g_i}, \dot{v}_{g_i}, \dot{\psi}_i)\\
{}^0\Omega_{p_z} &= \dot{\gamma} - \dot{\alpha}\sin\beta = f_3(u_i, v_i, u_{g_i}, v_{g_i}, \psi_i, \dot{u}_i, \dot{v}_i, \dot{u}_{g_i}, \dot{v}_{g_i}, \dot{\psi}_i), \; i = 1, 2, 3
\end{aligned} \tag{10.27}$$

If x_c, y_c, and z_c denote the coordinates of the centre of the platform in $\{0\}$, the linear velocity of the centre of the platform is given by

$$^0(V_{p_x}, V_{p_y}, V_{p_z})^T \triangleq {}^0(\dot{x}_c, \dot{y}_c, \dot{z}_c)^T = {}^0\mathbf{V}_{w_i} + {}^0(\Omega_{p_x}, \Omega_{p_y}, \Omega_{p_z})^T \times {}^0\mathbf{p}_{c_i} \tag{10.28}$$

where $i = 1, 2, 3$ denote the three wheels, the vector ${}^0\mathbf{p}_{C_i}$ locates the point of attachment of the wheel to the platform from the centre of the platform, and ${}^0\mathbf{V}_{w_i}$ is the velocity of the centre of the wheel given by Eqn (10.22).

Step 4: Form holonomic constraint equations In addition to the contact equations, the distance between the three points C_1, C_2, and C_3 (see Fig. 10.14) must remain constant. These *holonomic* constraint equations can be written as

$$\begin{aligned} \|{}^0\mathbf{p}_{C_1} - {}^0\mathbf{p}_{C_2}\|^2 &= l_{12}^2 \\ \|{}^0\mathbf{p}_{C_2} - {}^0\mathbf{p}_{C_3}\|^2 &= l_{23}^2 \\ \|{}^0\mathbf{p}_{C_3} - {}^0\mathbf{p}_{C_1}\|^2 &= l_{31}^2 \end{aligned} \tag{10.29}$$

where ${}^0\mathbf{p}_{C_i}$ $(i = 1, 2, 3)$ are the position vectors of the centre of the three wheels, C_1, C_2, C_3, respectively, from the origin of the fixed coordinate system $\{0\}$ and l_{ij} is the distance between centres of wheels i and j, respectively. The holonomic constraints in Eqn (10.29) are similar to the spherical–spherical (S-S) pair joint constraint discussed in Section 2.5.6.

Solution of the direct kinematics problem From steps 1, 2, and 4, we have 15 first-order ODEs and 3 algebraic constraint equations for the 18 unknown variables (recall that $\dot{\theta}_1$, $\dot{\theta}_2$, and $\dot{\phi}_3$ are actuated variables and are given). This system of differential algebraic equations (DAEs) can be converted to 18 ODEs in 18 variables by differentiating the three algebraic constraint equations. Using an ODE solver, we solve the set of 18 ODEs numerically, with the initial conditions obtained as outlined below. Once we have obtained $u_i, v_i, u_{g_i}, v_{g_i}, \psi_i\, (i = 1, 2, 3)$ and $\delta_1, \delta_2, \theta_3$, we can obtain the rotation matrix of the platform ${}^0_p[R]$. From Fig. 10.14, the position vector of the centre of the platform ${}^0\mathbf{p}_P$ with respect to $\{0\}$, denoted by (x_c, y_c, z_c), is given by

$$(x_c, y_c, z_c)^T = {}^0\mathbf{p}_{C_i} + {}^0_p[R]\overrightarrow{C_iP} \text{ for } i = 1, 2, 3 \tag{10.30}$$

where $\overrightarrow{C_iP}$ is the vector from point C_i to the centre of the platform P.

Solution of the inverse kinematics problem From steps 1–4 we have 21 first-order ODEs and 3 algebraic constraints for 24 unknowns assuming the two components of the velocity of the centre of the platform $\dot{x}_c$, $\dot{y}_c$, and the angular velocity about the vertical, $\dot{\gamma}$, are given. This set of DAEs is also converted to ODEs and integrated using initial conditions determined as discussed below. The numerical solution gives the 15 contact variables u_i, v_i, u_{g_i}, v_{g_i}, ψ_i $(i = 1, 2, 3)$, the 3 actuated variables $\theta_1, \theta_2, \phi_3$, the 3 passive variables $\delta_1, \delta_2, \theta_3$, and the other 3 platform variables z_c, α, β.

Initial conditions To solve the set of ODEs in the direct or inverse kinematics problem, we have to choose the initial conditions such that these satisfy the holonomic constraint equations. For the direct kinematics, among the 18 variables we can choose $\delta_1 = 0$, $\delta_2 = 0$, $\theta_3 = 0$ at $t = 0$. Moreover, we can also choose v_1, v_2, v_3 to be $3\pi/2$, and the point of contact of wheel 2 in $\{0\}$ obtained from u_{g_2}, v_{g_2}. The other two wheels must also be in contact with the ground. Hence, for these two wheels, we have

$$^0\mathbf{p}_{C_i} + {}^0_w[R]\overrightarrow{C_iG_i} = {}^0\mathbf{p}_{G_i} \quad \text{for } i = 1, 2, 3 \qquad (10.31)$$

where $\overrightarrow{C_iG_i}$ is the vector from point C_i to the ground contact point G_i and $^0\mathbf{p}_{G_i}$ locates the ground contact point in $\{0\}$. The vectors in Eqn (10.31) can be converted to unit vectors. This yields two independent equations for each of the two wheels. In addition, for each of the three wheels, from the definition of the X axis in $\{1\}$ and $\{2\}$ (see Fig. 10.7), we have

$$\cos\psi_i = {}^0\widehat{\mathbf{X}}_{1_i} \cdot {}^0\widehat{\mathbf{X}}_{2_i} \quad i = 1, 2, 3 \qquad (10.32)$$

which gives rise to three equations. In addition, there are three holonomic constraints in Eqn (10.29). This gives us 10 non-linear equations in 10 variables and we can solve them numerically to obtain the required initial conditions.

For the inverse kinematics problem involving 24 ODEs, in addition to those for the direct kinematics problem, we have to obtain the initial values of θ_1, θ_2, ϕ_3, α, β, z_c. We can choose $\theta_1 = 0$, $\theta_2 = 0$, $\phi_3 = 0$, or any other initial desired heading. As we know u_i, v_i, u_{g_i}, v_{g_i}, ψ_i $(i = 1, 2, 3)$, we can obtain the position vector of the centre of the three wheels $^0\mathbf{p}_{C_i}$ $(i = 1, 2, 3)$, hence we get the rotation matrix ${}^0_p[R]$ of the platform. Once the rotation matrix is known, we can obtain the 3-2-1 Euler angle sequence, γ, β, α. The position of the centre of the platform is given by Eqn (10.30) and we have x_c, y_c, and z_c at the initial instant.

10.4.2 Numerical Simulation Results

For the 3-DOF, three-wheeled mobile robot we use the following numerical values for our simulation purposes:
Length of the rear axle $= 2l_a = 1$ m.
Distance of the centre of the front wheel from the middle of the rear axle $= l_s = 1$ m.

The two radii of the torus-shaped wheel are $r_1 = 0.05$ m, $r_2 = 0.25$ m.

The centre of the WMR is assumed to be at $(1/3)l_s$ from the centre of the axle along the line joining the centre of the axle to the centre of the front wheel.

We present simulation results for both direct and inverse kinematics problems.

Results from direct kinematics The synthetic surface used for simulation is shown in Fig. 10.4. The initial conditions used are as follows:

$$u_1 = 1.5816 \text{ rad}, \; v_1 = \frac{3\pi}{2} \text{ rad}, \; u_{g_1} = 4.089 \text{ m},$$
$$v_{g_1} = 0.3917 \text{ m}, \; \psi_1 = -3.1414 \text{ rad}$$
$$u_2 = 1.5560 \text{ rad}, \; v_2 = \frac{3\pi}{2} \text{ rad}, \; u_{g_2} = 3.1 \text{ m},$$
$$v_{g_2} = 0.4 \text{ m}, \; \psi_2 = -3.1413 \text{ rad}$$
$$u_3 = 1.5735 \text{ rad}, \; v_3 = \frac{3\pi}{2} \text{ rad}, \; u_{g_3} = 3.5977 \text{ m},$$
$$v_{g_3} = 1.4097 \text{ m}, \; \psi_3 = -3.1404 \text{ rad}$$
$$\theta_3 = 0 \text{ rad}, \; \delta_1 = 0 \text{ rad}, \; \delta_2 = 0 \text{ rad}$$

The actuator inputs used for simulations are $\dot{\theta}_1 = -1$ rad/sec, $\dot{\theta}_2 = -0.9$ rad/sec, and $\dot{\phi}_3 = 0.005t$ rad/sec. The variation of lateral tilts of the rear wheels is shown in Fig. 10.16. The satisfaction of holonomic constraints is depicted in Fig. 10.17, and the locus of the wheel centres, wheel–ground contact points and the centre of the platform are shown in Fig. 10.18. Figure 10.19 shows that there is virtually no slip at the wheel–ground contact points of the three wheels.

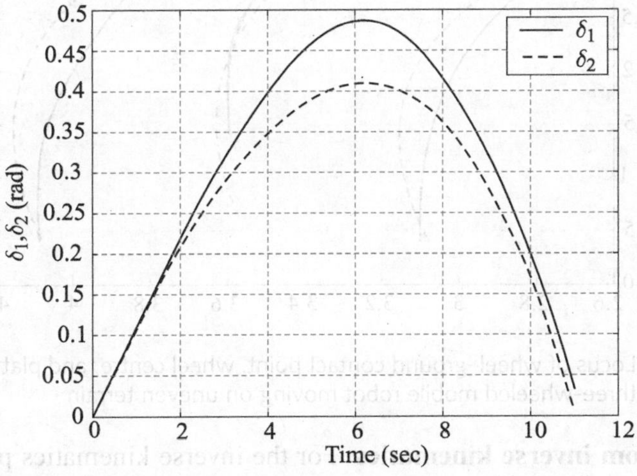

Fig. 10.16 Variation of δ_1 and δ_2 for the three-wheeled mobile robot moving on uneven terrain

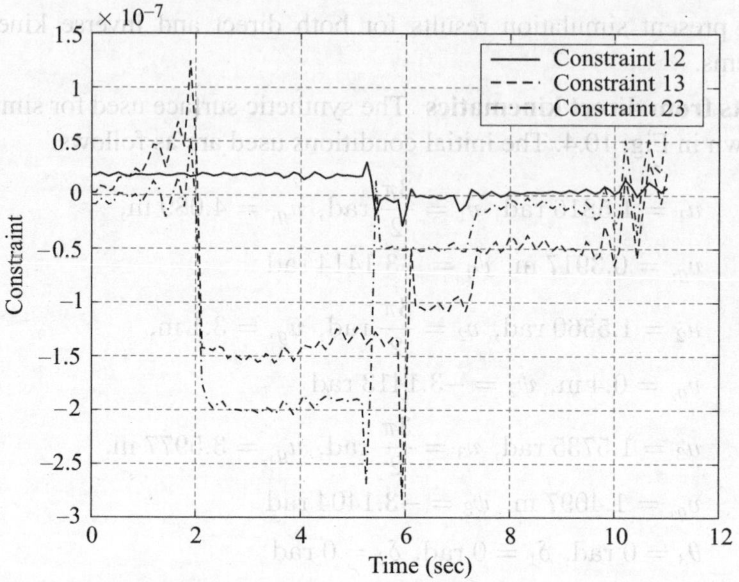

Fig. 10.17 Satisfaction of holonomic constraint

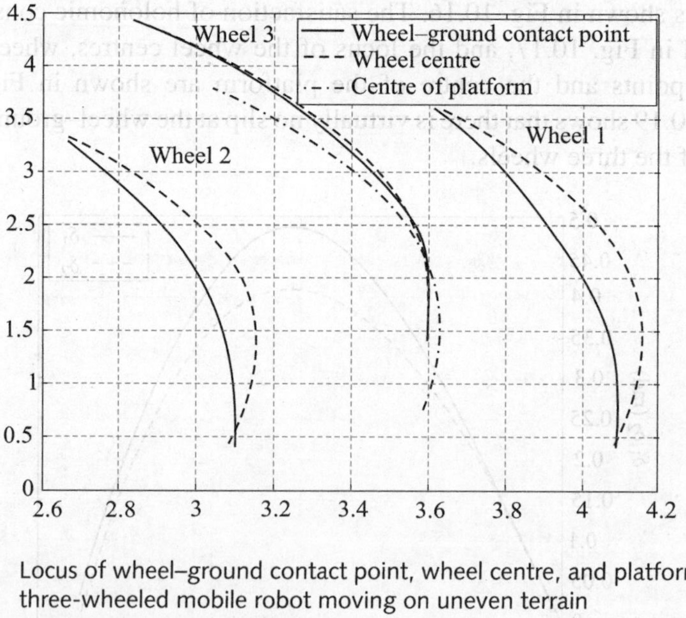

Fig. 10.18 Locus of wheel–ground contact point, wheel centre, and platform centre of three-wheeled mobile robot moving on uneven terrain

Results from inverse kinematics For the inverse kinematics problem we again use the synthetically generated surface shown in Fig. 10.4. The initial conditions used are as follows:

$$u_1 = 1.5801 \text{ rad}, \; v_1 = \frac{3\pi}{2} \text{ rad}, \; u_{g_1} = 3.9904 \text{ m},$$

$$v_{g_1} = 0.4895 \text{ m}, \; \psi_1 = -3.1415 \text{ rad}$$

$$u_2 = 1.5535 \text{ rad}, \; v_2 = \frac{3\pi}{2} \text{ rad}, \; u_{g_2} = 2 \text{ m},$$

$$v_{g_2} = 0.5 \text{ m}, \; \psi_2 = -3.1411 \text{ rad}$$

$$u_3 = 1.5808 \text{ rad}, \; v_3 = \frac{3\pi}{2} \text{ rad}, \; u_{g_3} = 3.4898 \text{ m},$$

$$v_{g_3} = 1.5702 \text{ m}, \; \psi_3 = -3.1414 \text{ rad}$$

$$z_c = 2.22 \text{ m}, \; \alpha = -0.0448 \text{ rad}, \; \beta = -0.0498 \text{ rad}$$

The inputs are $\dot{x}_c = 0.03$ m/sec, $\dot{y}_c = 0.15$ m/sec, and $\dot{\gamma} = -0.005t$ rad/sec.

Figure 10.20 shows the variation in lateral tilt of the two wheels. The satisfaction of the holonomic constraints is shown in Fig. 10.21. The locus of the wheel centres, wheel–ground contact point, and the centre of the platform are shown in Fig. 10.22. Figure 10.23 shows that there is no slip at the wheel–ground contact points of the three wheels.

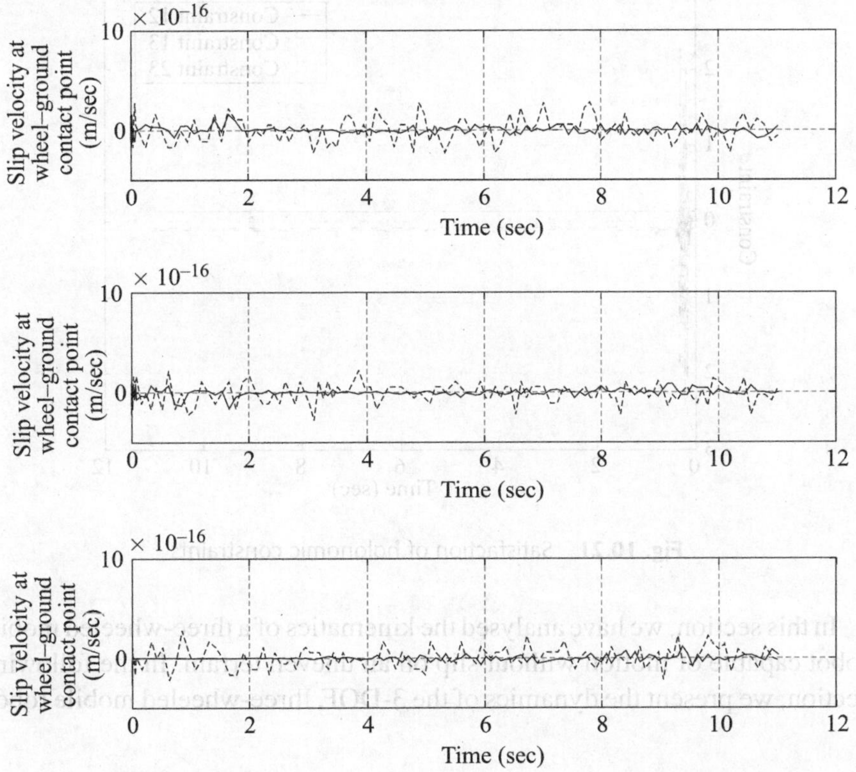

Fig. 10.19 Slip velocities at wheel–ground contact points for three wheels

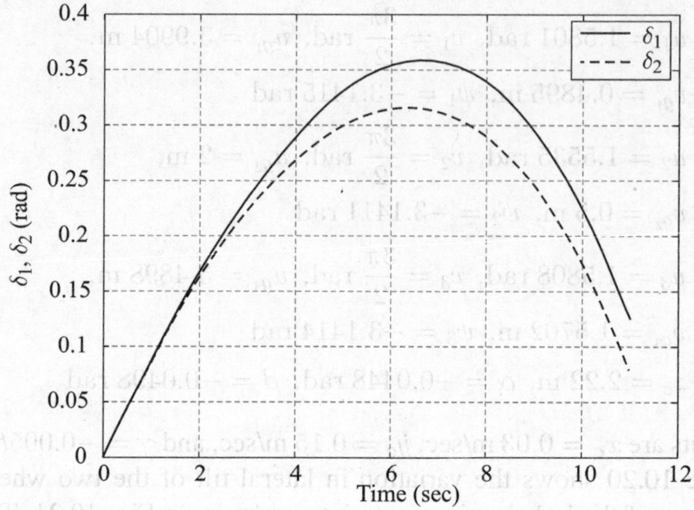

Fig. 10.20 Variation of δ_1 and δ_2 for the three-wheeled mobile robot moving on uneven terrain

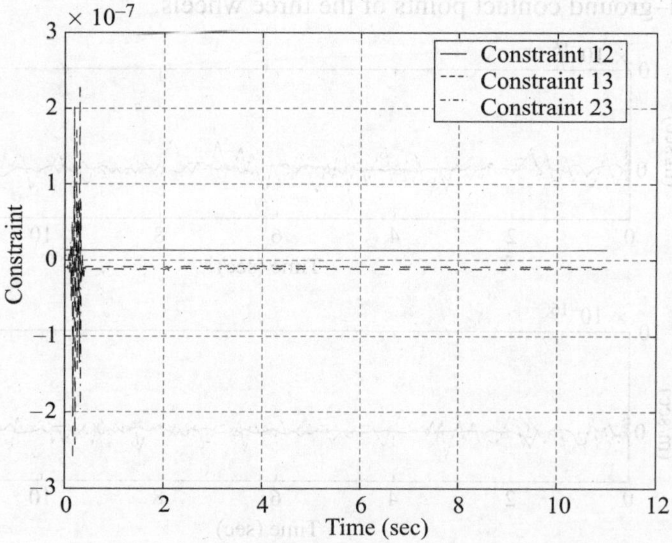

Fig. 10.21 Satisfaction of holonomic constraints

In this section, we have analysed the kinematics of a three-wheeled mobile robot capable of motion without slip on an uneven terrain. In the following section, we present the dynamics of the 3-DOF, three-wheeled mobile robot.

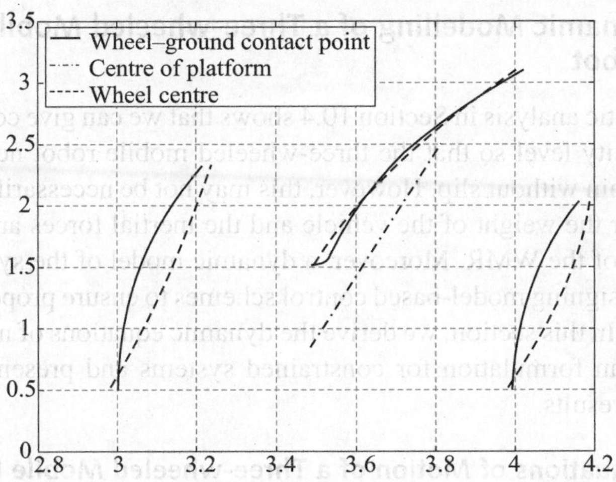

Fig. 10.22 Locus of wheel–ground contact point, wheel centre, and platform centre of three-wheeled mobile robot moving on uneven terrain

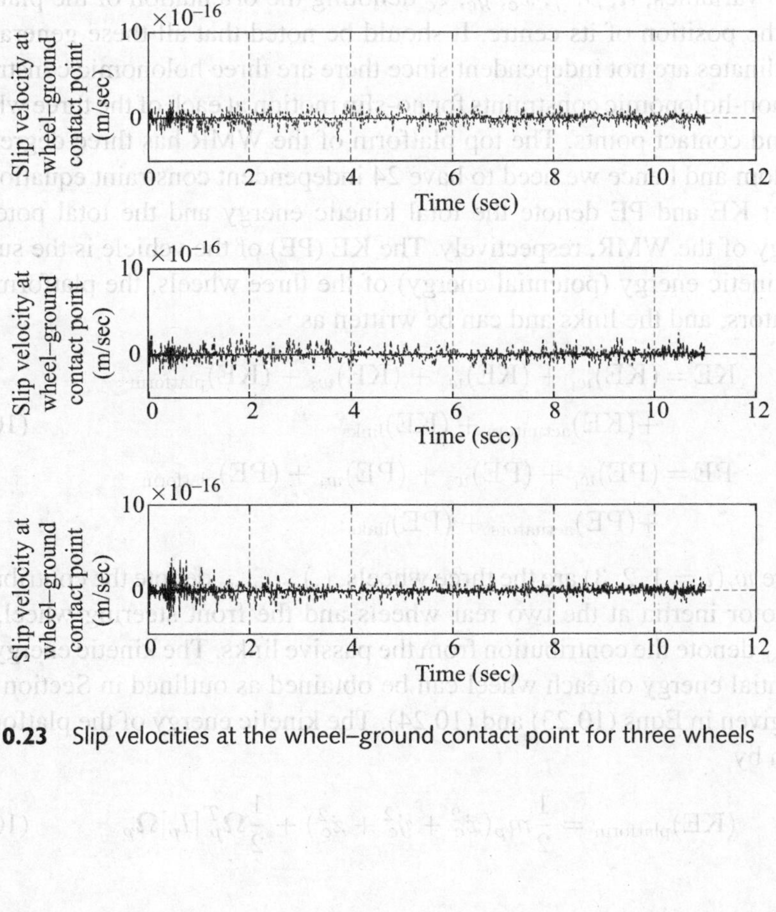

Fig. 10.23 Slip velocities at the wheel–ground contact point for three wheels

10.5 Dynamic Modelling of a Three-wheeled Mobile Robot

The kinematic analysis in Section 10.4 shows that we can give control inputs at the velocity level so that the three-wheeled mobile robot negotiates the uneven terrain without slip. However, this may not be necessarily true when we consider the weight of the vehicle and the inertial forces arising due to the motion of the WMR. Moreover, a dynamic model of the system is also useful in designing model-based control schemes to ensure proper control of the system. In this section, we derive the dynamic equations of motion using a Lagrangian formulation for constrained systems and present numerical simulation results.

10.5.1 Equations of Motion of a Three-wheeled Mobile Robot

As shown in the kinematic model developed in Section 10.4, there are 27 generalized coordinates, namely, 15 contact variables u_i, v_i, u_{g_i}, v_{g_i}, ψ_i $(i = 1, 2, 3)$; 3 passive variables δ_1, δ_2, θ_3; 3 actuated variables θ_1, θ_2, ϕ_3; and 6 variables, α, β, γ, x_c, y_c, z_c denoting the orientation of the platform and the position of its centre. It should be noted that all these generalized coordinates are not independent since there are three holonomic constraints and non-holonomic constraints for no-slip motion at each of the three wheel–ground contact points. The top platform of the WMR has three degrees of freedom and hence we need to have 24 independent constraint equations.

Let KE and PE denote the total kinetic energy and the total potential energy of the WMR, respectively. The KE (PE) of the vehicle is the sum of the kinetic energy (potential energy) of the three wheels, the platform, the actuators, and the links and can be written as

$$\begin{aligned} \text{KE} &= (\text{KE})_{w_1} + (\text{KE})_{w_2} + (\text{KE})_{w_3} + (\text{KE})_{\text{platform}} \\ &\quad + (\text{KE})_{\text{actuators}} + (\text{KE})_{\text{links}} \qquad (10.33) \\ \text{PE} &= (\text{PE})_{w_1} + (\text{PE})_{w_2} + (\text{PE})_{w_3} + (\text{PE})_{\text{platform}} \\ &\quad + (\text{PE})_{\text{actuators}} + (\text{PE})_{\text{links}} \end{aligned}$$

where $w_i (i = 1, 2, 3)$ are the three wheels, $(.)_{\text{actuators}}$ denote the contribution of motor inertia at the two rear wheels and the front steering wheel, and $(.)_{\text{links}}$ denote the contribution from the passive links. The kinetic energy and potential energy of each wheel can be obtained as outlined in Section 10.3 and given in Eqns (10.23) and (10.24). The kinetic energy of the platform is given by

$$(\text{KE})_{\text{platform}} = \frac{1}{2} m_p (\dot{x}_c^2 + \dot{y}_c^2 + \dot{z}_c^2) + \frac{1}{2} \boldsymbol{\Omega}_p^T [I_p] \boldsymbol{\Omega}_p \qquad (10.34)$$

where $\mathbf{\Omega}_p$ is the angular velocity of the platform, m_p is the mass of the platform, $[I_p]$ is the mass moment of inertia matrix of the platform, and $(x_c, y_c, z_c)^T$ is the position vector of the centre of the platform in $\{0\}$. The potential energy of the platform is given by

$$(\text{PE})_{\text{platform}} = m_p g z_c \tag{10.35}$$

The mass of the passive joints allowing lateral tilt is assumed to be lumped with the mass of the platform and that of the actuators is assumed to be lumped with the wheel. Thus their contribution to the potential energy term and the kinetic energy term is included in the kinetic energies of the wheel and platform. As shown in the equivalent parallel manipulator model in Fig. 10.14, the passive joint assembly and the actuators for each wheel form a serial chain from the ground to the platform body. Similar to a serial manipulator, we can obtain the rotational kinetic energy of the passive joint assemblies and the actuators. For the two rear wheels, the angular velocity of the passive joint assembly in a local coordinate system fixed to the joint is

$$\mathbf{\Omega}_{\text{links}_i} = [R(\widehat{\mathbf{Y}}, \delta_i)]^T \mathbf{\Omega}_p + (0, \dot{\delta}_i, 0)^T, \quad i = 1, 2$$

since the passive joint rotation is about the local $\widehat{\mathbf{Y}}$ axis[11] and $\mathbf{\Omega}_p$ is the angular velocity of the platform with respect to the coordinate system attached to the centre of the platform. From the angular velocity, we can obtain the rotational kinetic energy of the passive links for the two rear wheels (wheel 1 and 2) as

$$\text{KE}_{\text{links}_i} = \frac{1}{2}\mathbf{\Omega}^T_{\text{links}_i}[I_{\text{links}_i}]\mathbf{\Omega}_{\text{links}_i}, \quad i = 1, 2$$

where $[I_{\text{links}_i}]$ is the moment of inertia matrix of the passive links. The angular velocity of the two actuators in the two rear wheels is given by

$$\mathbf{\Omega}_{\text{actuators}_i} = [R(\widehat{\mathbf{X}}, \theta_i)]^T \mathbf{\Omega}_{\text{links}_i} + (\dot{\theta}_i, 0, 0)^T, \quad i = 1, 2$$

since the local axis of rotation for the actuators is the $\widehat{\mathbf{X}}$ axis. The rotational kinetic energy of the actuators can be obtained as

$$\text{KE}_{\text{actuators}_i} = \frac{1}{2}\mathbf{\Omega}^T_{\text{actuators}_i}[I_{\text{actuators}_i}]\mathbf{\Omega}_{\text{actuators}_i}, \quad i = 1, 2$$

where $[I_{\text{actuators}_i}]$ is the moment of inertia matrix of the actuators. The angular velocity of the steering actuator in the front wheel (wheel 3) is given by

$$\mathbf{\Omega}_{\text{actuators}_i} = [R(\widehat{\mathbf{Z}}, \phi_i)]^T \mathbf{\Omega}_p + (0, 0, \dot{\phi}_i)^T, \quad i = 3$$

[11] Note that we use the same symbols $\widehat{\mathbf{X}}$, $\widehat{\mathbf{Y}}$, and $\widehat{\mathbf{Z}}$ for representing the unit vectors at all the joints. This is just for notational simplicity, and $\widehat{\mathbf{X}}$, $\widehat{\mathbf{Y}}$, $\widehat{\mathbf{Z}}$ are *not* the same for all the joints.

since the steering is done about the local $\hat{\mathbf{Z}}$ axis. The rotational kinetic energy of the steering actuator is given by

$$\mathrm{KE}_{\text{actuators}_i} = \frac{1}{2}\boldsymbol{\Omega}^T_{\text{actuators}_i}[I_{\text{actuators}_i}]\boldsymbol{\Omega}_{\text{actuators}_i}, \; i = 3$$

The angular velocity of the passive joint and link in the front wheel (wheel 3) allowing in-plane rotation of the wheel is given by

$$\boldsymbol{\Omega}_{\text{links}_i} = [R(\hat{\mathbf{X}}, \theta_i)]^T \boldsymbol{\Omega}_{\text{actuators}_i} + (\dot{\theta}_i, 0, 0)^T, \; i = 3$$

and hence its rotational kinetic energy is given by

$$\mathrm{KE}_{\text{links}_i} = \frac{1}{2}\boldsymbol{\Omega}^T_{\text{links}_i}[I_{\text{links}_i}]\boldsymbol{\Omega}_{\text{links}_i}, \; i = 3$$

Once we have the total kinetic energy and the total potential energy, by following the Lagrangian formulation, we can obtain the components of the mass matrix $[\mathbf{M}(\mathbf{q})]$, the Coriolis and centripetal terms $[\mathbf{C}(\mathbf{q}, \dot{\mathbf{q}})]$, and the gravity term $\mathbf{G}(\mathbf{q})$ symbolically using MAPLE (Heck 2003) or MATHEMATICA (Wolfram 1999).

For no-slip motion on an uneven terrain, the three-wheeled mobile robot is subjected to both non-holonomic and holonomic constraints as discussed in Section 10.4. As mentioned earlier, there are 27 generalized coordinates, but the platform has three degrees of freedom. The required 24 constraint equations are the *same* as the set of inverse kinematics equations obtained in Section 10.4.1. These can be arranged in a matrix form as

$$[\boldsymbol{\Psi}]\dot{\mathbf{q}} = \mathbf{0} \tag{10.36}$$

where $[\boldsymbol{\Psi}]$ is a 24×27 matrix. Following the Lagrangian formulation, the equations of motion of the 3-DOF three-wheeled mobile robot can be expressed in the form

$$[\mathbf{M}(\mathbf{q})]\ddot{\mathbf{q}} + [\mathbf{C}(\mathbf{q}, \dot{\mathbf{q}})]\dot{\mathbf{q}} + \mathbf{G}(\mathbf{q}) = \boldsymbol{\tau} + [\boldsymbol{\Psi}(\mathbf{q})]^T \boldsymbol{\lambda} \tag{10.37}$$

where $[\mathbf{M}(\mathbf{q})]$ is a 27×27 symmetric, positive definite mass matrix, $[\mathbf{C}(\mathbf{q}, \dot{\mathbf{q}})]\dot{\mathbf{q}}$, $\mathbf{G}(\mathbf{q})$ are 27×1 vectors representing the Coriolis/centripetal terms and gravity terms, respectively, and $\boldsymbol{\tau}$ is a 27×1 vector of external forces/torques. It may be noted that only three elements of $\boldsymbol{\tau}$, corresponding to the actuated variables θ_1, θ_2, ϕ_3, are non-zero and all others are zero. The 24×1 vector of the Lagrange multipliers $\boldsymbol{\lambda}$ can be obtained by taking the derivative of the constraint equation, Eqn (10.36), once (see Chapter 6, Section 6.3). It should be noted that the generation of the above equations of motion requires intensive symbolic computation.

On solving the equations of motion on a flat plane with the actuators locked and the rear wheels tilted, it was observed that the wheels fell under the action of gravity. This is contrary to a parallel manipulator which, with its actuators locked, behaves as a structure at a non-singular configuration, and under any external loading there is no motion of the links. This is due to the fact that in our equivalent parallel manipulator model the 3-DOF joints at the wheel–ground contact points are not like spherical joints. The constraints at the wheel–ground contact point are in terms of velocities, and they do not restrict the position variables. The wheels can have motion as long as the no-slip constraints are not violated and, in fact, can tilt laterally till the WMR is lying on the ground.[12] To prevent falling of the wheels under gravity, a flexible coupling between the actuator and the platform body can be used. The flexible coupling allows lateral tilt of the wheels but provides some stiffness similar to a suspension in a car. For our modelling purposes, we have assumed the system to be a simple spring-mass-damper system, and a spring and damping force is added to the right-hand side of the equations of motion. Hence, in the vector $\boldsymbol{\tau}$ corresponding to the generalized variables δ_1 and δ_2, there are two additional terms of the form $k_{s_i}\delta_i + k_{d_i}\dot{\delta}_i (i = 1, 2)$, where k_{s_i} is the stiffness of the flexible coupling and k_{d_i} is the damping coefficient.

10.5.2 Algorithm for Solving the Equations of Motion

The key steps in solving the dynamic equations of motion for the three-wheeled mobile robot are as follows:

Step 1: Generate the surface As before, we assume the terrain DEM is known and we generate a spline representation of the surface. For dynamics, we need the C^3 continuity of the surface, hence we use a fourth-degree B-spline to generate the surface. The surface representation and its derivatives are computed using in-built MATLAB functions (Mathwork 1992).

Step 2: Form equations of motion The equations of motion [Eqn (10.37)] are formed in the manner outlined earlier. We have a set of 27 second-order ODEs in 27 variables which can be solved with any ODE solver using appropriate initial conditions.

Step 3: Obtain initial conditions The initial conditions for the configuration variables are determined in the same manner as for the inverse kinematics problem. For the first derivatives of the generalized coordinates, the initial conditions are chosen such that these satisfy the no-slip non-holonomic

[12] In kinematics literature there is a notion of *form closure* which is applicable to joints or a mechanism. In form closure, when the actuated joints are locked, the joint or the mechanism does not have any motion. In our case, due to the nature of wheel–ground contact equations, we *do not* have form closure. Instead we have what is known as *force closure*. In force closure, forces or moments are required to stop the motion of links even when the actuators are locked.

constraints at the initial instant. We choose three of them arbitrarily and determine the other 24 by solving the 24 constraint equations which are linear in the first derivatives of the configuration variables.

Step 4: Solve equations of motion The 27 second-order equations of motion can be solved with any ODE solver using initial conditions as obtained in step 3. This gives the evolution of u_i, v_i, u_{g_i}, v_{g_i}, ψ_i $(i = 1, 2, 3)$ and its first derivatives $\dot{u}_i$, $\dot{v}_i$, $\dot{u}_{g_i}$, $\dot{v}_{g_i}$, $\dot{\psi}_i$ in time. We also obtain the variation of the passive lateral tilts and their rates, the actuator rates, and the position of the centre of the platform and its orientation as a function of time.

10.5.3 Numerical Simulation Results

To illustrate the dynamics of the three-wheeled mobile robot, we use the synthetically generated uneven terrain shown in Fig. 10.5. The mass and inertia of the various components used in the simulation are as given below:
Mass of the platform (and passive joint assemblies) = m_p = 10 kg
Mass of each wheel (and actuators) = m_w = 1 kg
Mass moment of inertia of each actuator (in SI units)

$$[I_{\text{actuators}_i}] = \begin{pmatrix} 0.1 & 0 & 0 \\ 0 & 0.1 & 0 \\ 0 & 0 & 0.1 \end{pmatrix}, \quad i = 1, 2, 3$$

Mass moment of inertia of passive joint assembly (in SI units)

$$[I_{\text{links}_i}] = \begin{pmatrix} 0.1 & 0 & 0 \\ 0 & 0.1 & 0 \\ 0 & 0 & 0.1 \end{pmatrix}, \quad i = 1, 2, 3$$

The flexible coupling is assumed to be a simple torsional spring and its stiffness k_{s_i} $(i = 1, 2)$ is estimated by assuming that the maximum allowable deflection is $\delta_{\max} = \pi/4$ for a static load of $W/3(r_1 + r_2)$ N m, where $W = (m_p + 3m_w)g$ N is the total weight of the WMR. The damping coefficient k_{d_i} $(i = 1, 2)$ is assumed to be 5% of the stiffness. The numerical values of k_{s_i} and k_{d_i} used in the simulation are $k_{s_i} = 16.24$ (N m)/rad and $k_{d_i} = 0.57$ (N m)/rad.

The initial conditions used for simulation are

$$u_1 = 1.57 \text{ rad}, \; v_1 = \frac{3\pi}{2} \text{ rad}, \; u_{g_1} = 5.008 \text{ m},$$

$$v_{g_1} = 1.5067 \text{ m}, \; \psi_1 = -3.142 \text{ rad}$$

$$u_2 = 1.5648 \text{ rad}, \; v_2 = \frac{3\pi}{2} \text{ rad}, \; u_{g_2} = 4 \text{ m},$$

$$v_{g_2} = 1.5 \text{ m}, \; \psi_2 = -3.1418 \text{ rad},$$

$$u_3 = 1.5818 \text{ rad}, v_3 = \frac{3\pi}{2} \text{ rad},$$
$$u_{g3} = 4.4897\, m, v_{g3} = 2.483 \text{ m}, \psi_3 = -3.1419 \text{ rad},$$
$$\theta_1 = 0 \text{ rad}, \theta_2 = 0 \text{ rad}, \theta_3 = 0 \text{ rad}, \delta_1 = 0 \text{ rad}, \delta_2 = 0 \text{ rad},$$
$$\phi_3 = 0 \text{ rad}, z_c = 2.3088 \text{ m},$$
$$\alpha = -0.0127 \text{ rad}, \beta = 0.0133 \text{ rad}$$

All the initial values of the first derivatives are assumed to be 0. The two driving actuator torques and the steering torque are assumed to be $\tau_1 = -0.35$ N m, $\tau_2 = -0.5$ N m, and $\tau_3 = -0.001t$ N m.

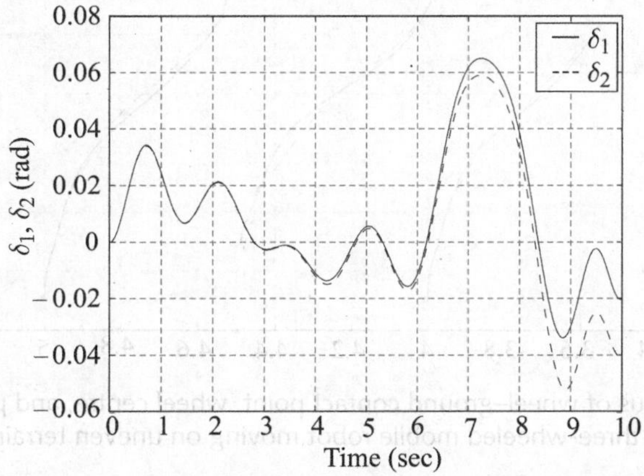

Fig. 10.24 Variation of δ_1 and δ_2 of the three-wheeled mobile robot moving on uneven terrain

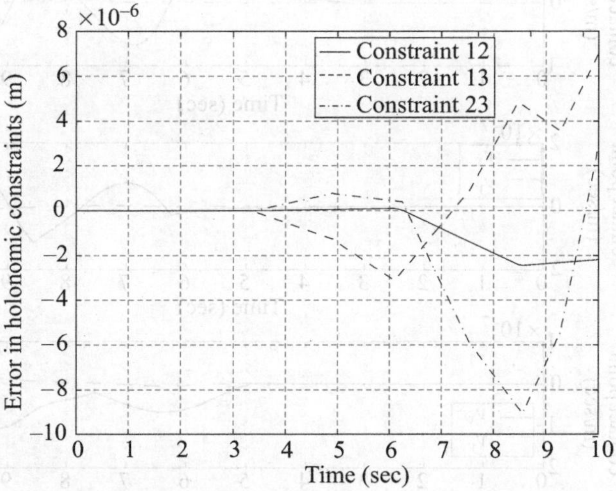

Fig. 10.25 Satisfaction of holonomic constraints in equations of motion

The variation of lateral tilts of the rear wheels is shown in Fig. 10.24. Figure 10.25 shows that the holonomic constraints are satisfied, and the locus of the wheel centres, wheel–ground contact points, and the centre of the platform are shown in Fig. 10.26. Figure 10.27 shows that there is no slip at the wheel–ground contact points of the three wheels.

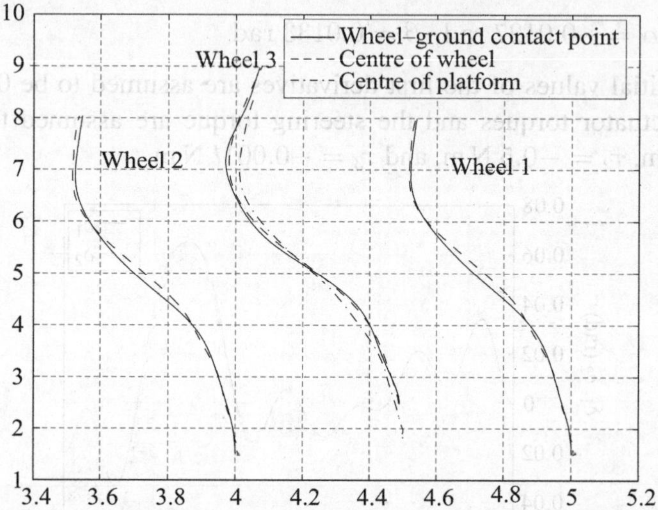

Fig. 10.26 Locus of wheel–ground contact point, wheel centre, and platform centre of the three-wheeled mobile robot moving on uneven terrain

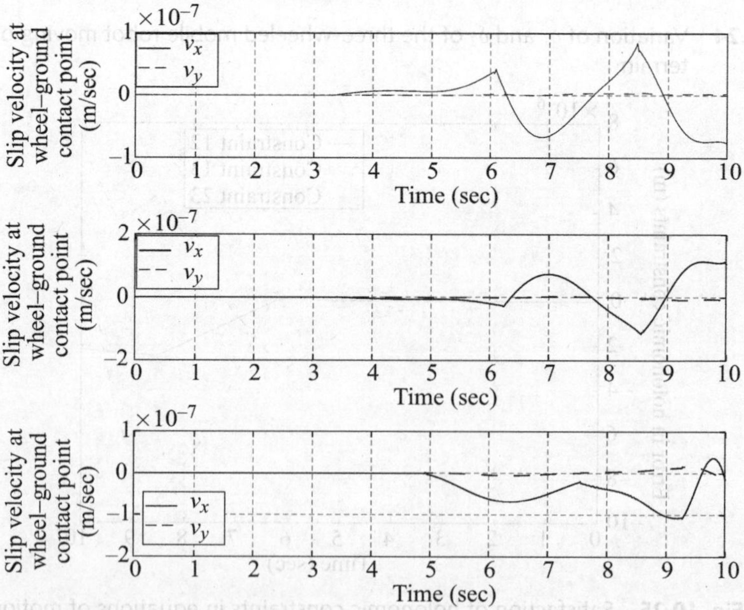

Fig. 10.27 Slip velocities at the wheel–ground contact point

The kinematic and dynamic simulation results show that the three-wheeled mobile robot is capable of traversing uneven terrain, under gravity and subjected to external loading, without slip at the wheel–ground contact points. Intuitively, a mobile robot with a given geometry cannot traverse *all* possible uneven terrains. It is quite likely that the wheels or the rigid platform can get stuck in a depression or bulge in the uneven terrain. Obtaining conditions about uneven terrains which a wheeled mobile robot can traverse is analogous to the notion of obtaining the workspace of a given stationary serial or parallel manipulator. (Recall that the workspace of a serial or parallel manipulator is defined as the set of positions and orientation an end-effector of a manipulator can reach in three-dimensional space.) In the following section, we obtain analytical conditions which can predict whether a *single* torus-shaped wheel can traverse an uneven terrain with *single-point contact* and with no slip.

10.6 Traversability of a Single Wheel

The contact equations, Eqn (10.14), are valid as long as there is single-point contact between the two surfaces, or equivalently, the matrix $([\mathbf{K}_w] + [\mathbf{K}^*])$ is invertible. In this section, we present analytical conditions when multi-point contact between the torus-shaped wheel and a surface may occur and, consequently, a path may not be traversable. We restrict our analysis to the class of surfaces where the DEM data obtained from sensors are unique, i.e., for each point (x_i, y_i) there is a unique z_i giving the height at that point on the surface. The analysis is based on the net curvature at the point of contact and hence is a *local* analysis.

As the first step, we consider planar curves and a circle moving in contact with the planar curve, and then extend it to surfaces and the torus-shaped wheel. The curvature of a planar curve defined by the parametric equation $[u, f(u)]$, at a known point, is given by

$$\kappa = \frac{d^2f/du^2}{[1 + (df/du)^2]^{3/2}}$$

The net curvature or relative curvature of two curves at a point of contact is the difference between the curvatures of the two curves at that point. Figure 10.28 shows different possible cases of contact between a circle or a wheel modelled as a thin disc and an arbitrary curve. From the geometry of plane curves it can be seen that two curves in contact with each other will intersect if the relative curvature at the point of contact is negative [see Fig. 10.28(c), circle 3 and surface curve at point C_3]. However, if the relative curvature at the point of contact is positive, we cannot say whether the

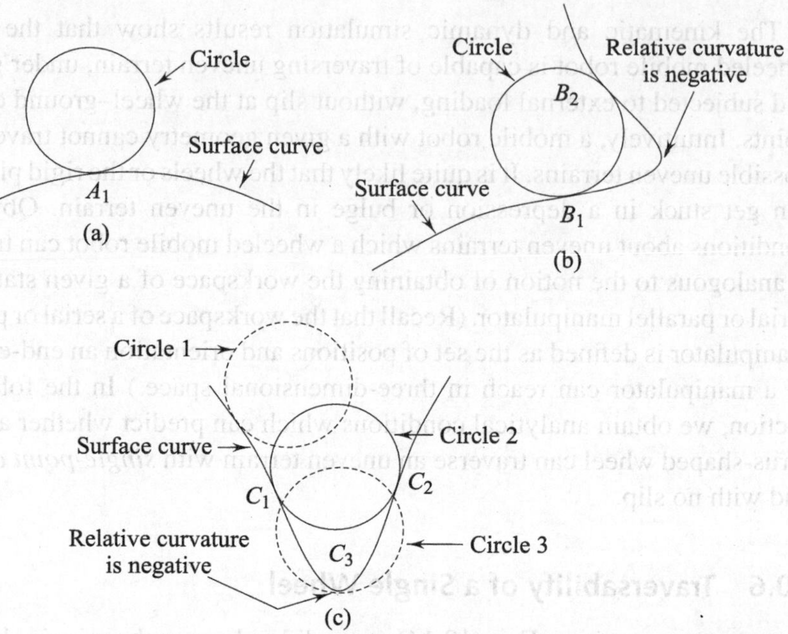

Fig. 10.28 A circle in contact with different curves

two curves have multi-point contact or not. For example, Figs 10.28(a)–(c) all show the circle in contact with different curves with positive relative curvature at the point of contact [A_1 in Fig. 10.28(a), B_1 in Fig. 10.28(b), and C_1 in Fig. 10.28(c)]. But there is no multi-point contact in the case of Fig. 10.28(a), whereas in Figs 10.28(b)[13] and 10.28(c) (see circle C_2) there is multi-point contact. This is because curvature is a local property and we cannot make any conclusions about the global intersection of two curves from it. It should be noted [as shown in Fig. 10.28(c)] that for multi-point contact to occur, there has to be at least one point on the curve where the relative curvature is negative. If a circle is moved along a curve and the relative curvatures at all points of contact are positive, then there will be no multi-point contact [see Fig. 10.28(a)]. On the other hand, if there is a point where the relative curvature is negative [point C_3 in Fig. 10.28(c)], the circle curve cannot pass through it and multi-point contact will occur between the curve and the circle before the circle reaches the point C_3. Thus, from a purely local analysis of net curvature we can have an idea of whether two curves will have multi-point contact or not. We call a point *locally traversable*[14] if

[13] We mention the case represented by Fig. 10.28(b) for the sake of completeness, although in this case the height at certain points along the curve is non-unique.

[14] This is a slightly different use of the terminology, as in the literature the term 'local traversability' has been used when it is possible to pass through the point.

the relative curvature at that point is positive. Whether the circle can actually pass through that point or not depends upon the relative curvature at future points along the path as stated above.

We now extend this idea to the case of the torus-shaped wheel in contact with an uneven terrain. It may be noted that $([\mathbf{K}_w] + [\mathbf{K}^*])$ is the net curvature of the wheel and the ground at the contact point expressed in $\{1\}$ and $\{2\}$ at the contact point (see Fig. 10.7). The relative *principal* curvatures are given by the eigenvalues of $([\mathbf{K}_w] + [\mathbf{K}^*])$. For a given wheel geometry, the eigenvalues are dependent only on ψ and u_w. Based on the discussion on multi-point contact of a circle and a curve, we can argue that the wheel can traverse an uneven terrain at a point (or the point is locally traversable) if the least of the relative principal curvatures is positive. This implies that if we obtain the eigenvalues of the relative curvature matrix at a given point as a function of ψ and u_w, we can find sets of ψ and u_w for which both the eigenvalues would be positive and hence the point will be locally traversable for those sets of ψ and u_w. Alternatively, if one of the eigenvalues is negative or zero for certain values of ψ and u_w, then the surfaces will intersect each other. For the case of a torus-shaped wheel, this implies that the wheel cannot be driven to pass through such a point with those particular values of ψ and u_w and multi-point contact will occur before the wheel reaches that point. A given path is fully traversable, if there exist values of ψ and u_w for which all points along the curve are traversable.

The expression for $([\mathbf{K}_w] + [\mathbf{K}^*])$ for the case of a torus-shaped wheel on an uneven terrain is given by

$$[\mathbf{K}_w] + [\mathbf{K}^*] = \begin{pmatrix} A & B \\ C & D \end{pmatrix} = \begin{pmatrix} (\frac{1}{r_1} + a) - Y & X - c \\ X - b & \left(\dfrac{\sin u_w}{r_2 + r_1 \sin u_w} + d\right) + Y \end{pmatrix} \tag{10.38}$$

where

$$X = (b + c)\sin^2\psi - (a - d)\cos\psi \sin\psi \tag{10.39}$$
$$Y = (a - d)\sin^2\psi + (b + c)\cos\psi \sin\psi$$

and a, b, c, d are the components of the curvature tensor of the ground, $[\mathbf{K}_g]$. The characteristic polynomial of $([\mathbf{K}_w] + [\mathbf{K}^*])$ is given by

$$\lambda^2 - (A + D)\lambda + (AD - BC) = 0 \tag{10.40}$$

where $A + D$ is the trace and $(AD - BC)$ is the determinant of $([\mathbf{K}_w] + [\mathbf{K}^*])$. These are given by

$$A + D = \left(\frac{1}{r_1} + a\right) + \left(\frac{\sin u_w}{r_2 + r_1 \sin u_w} + d\right) \tag{10.41}$$

$$AD - BC = \left[\left(\frac{1}{r_1} + a\right) - Y\right]\left[\left(\frac{\sin u_w}{r_2 + r_1 \sin u_w} + d\right) + Y\right]$$
$$-(X - c)(X - b) \quad (10.42)$$

where X and Y are given by Eqn (10.39). The eigenvalues λ_1 and λ_2 are given by the roots of Eqn (10.40) and we can obtain analytical expressions for them in terms of the wheel radii, ground geometry, ψ, and u_w. At any point, the eigenvalues are *only* functions of u_w and ψ. Hence, we can classify a point into one of the following.

(a) *Always traversable*: For such a point the eigenvalues are positive for all values of u_w and ψ or, more concisely, both the determinant and the trace are positive for any u_w and ψ.

(b) *Always non-traversable:* For such a point the eigenvalues are negative for all values of u_w and ψ or, more concisely, the determinant is positive and the trace is negative for any u_w and ψ.

(c) *May be traversable:* For such a point there are some *specific* values of ψ and u_w for which both the eigenvalues are positive.

We present a few results below illustrating the above analysis.

Flat plane In this case $a = b = c = d = 0$, hence

$$A = \frac{1}{r_1}, \quad B = 0, \quad C = 0, \quad D = \frac{\sin u_w}{r_2 + r_1 \sin u_w}$$

In this case $\lambda_1 = 1/r_1$ and $\lambda_2 = (\sin u_w)/(r_2 + r_1 \sin u_w)$. λ_1 is always positive and $\lambda_2 \geq 0$ ($\lambda_2 = 0$ implies $u_w = 0$, which in turn implies that the torus is lying flat on the ground). Physically, it can be seen that a torus on a flat plane can roll without slip from any place to any other place provided it does not fall flat.

Ruled surface For a ruled surface, $a = b = c = 0$ and $d = (-f_{v_g v_g})/(1 + {f_{v_g}}^2)^{3/2}$. In this special case d is one of the principal curvatures of the ground (the other being 0).[15] The trace $A + D$ and determinant $AD - BC$ are given by

$$A + D = \frac{1}{r_1} + \frac{\sin u_w}{r_2 + r_1 \sin u_w} + d \quad (10.43)$$

$$AD - BC = \frac{1}{2}\left(\frac{d}{r_1} - \frac{d \sin u_w}{r_2 + r_1 \sin u_w}\right)\cos 2\psi$$

[15] Note that we have assumed here that the ruled surface has zero curvature in the direction of u_g. If we had assumed the ruled surface to have zero curvature along v_g, then $a = (-f_{u_g u_g})/(1 + {f_{u_g}}^2)^{3/2}$ and $b = c = d = 0$.

$$+\frac{1}{2}\left(\frac{d}{r_1}+\frac{d\sin u_w}{r_2+r_1\sin u_w}\right)+\frac{\sin u_w}{r_1(r_2+r_1\sin u_w)} \tag{10.44}$$

Using Eqn (10.40), we see that for $d > 0$ both the eigenvalues are positive. For $d < 0$, if both $\lambda_1 > 0$ and $\lambda_2 > 0$ are to be always positive, we obtain $|d| < 1/(r_1 + r_2)$. This implies that if the ruled surface has a positive curvature (convex shape), then it is always navigable. However, if the ruled surface has a negative curvature (concave shape), the wheel can pass through a point if the magnitude of the curvature of ground at that point is less than the smallest curvature of the wheel. If both the eigenvalues are always negative, then $|d| > 1/r_1$. This implies that if the magnitude of the curvature of ground is more than the largest curvature of wheel there will definitely be multi-point contact. For $1/(r_1 + r_2) < |d| < 1/r_1$ there will be sets of ψ and u_w for which the wheel may cross the point. They are those values for which the least of λ_1 and λ_2 is positive. In the sinusoidal ruled surface, whose coordinates are given by $[u_g, v_g, a_0 \sin(\omega_0 v_g)]$, a possible two-point contact resulting from a choice of the largest $|d|$ in between the two curvatures of the torus-shaped wheel is shown in Fig. 10.29. For $a_0 = 1$, $\omega_0 = 3$, initial conditions $u_g = 0$, $v_g = \pi/2\omega_0$, $u_w = \pi/2$, $v_w = 3\pi/2$, $\psi = -\pi$ and inputs $\omega_x = 2$, $\omega_y = 0$, $\omega_z = 0$, the sets of values of ψ and u_w for which $1/(r_1 + r_2) < |d| < 1/r_1$ can be computed (see Fig. 10.30). Although the torus-shaped wheel can get stuck as shown in Fig. 10.29, one can visualise that with a non-zero heading

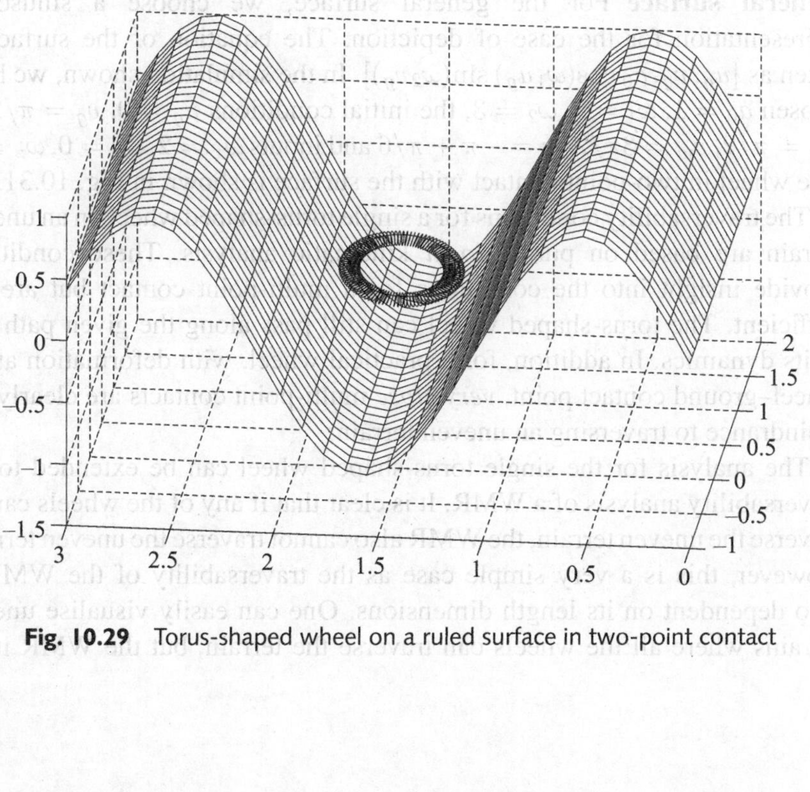

Fig. 10.29 Torus-shaped wheel on a ruled surface in two-point contact

angle ψ the torus-shaped wheel can traverse the ruled surface. The possible values of ψ as a function of u_w which allow the torus-shaped wheel to cross the trench are shown by filled circles (•) in Fig. 10.30, and the $\pm$ values of traversable ψ's can be explained from the symmetry of the problem.

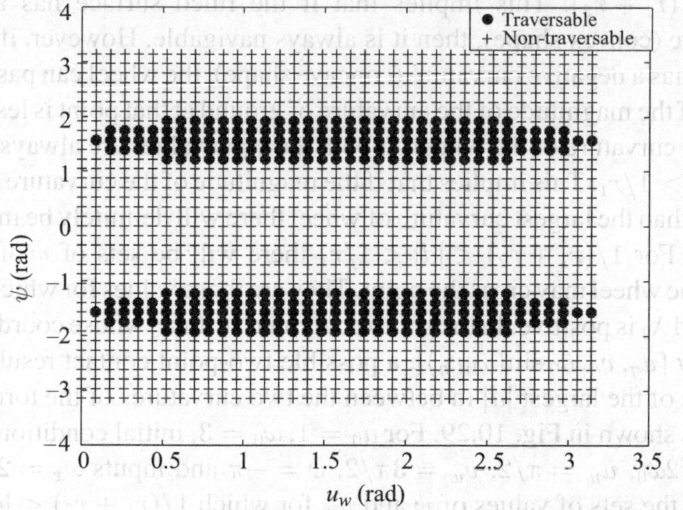

Fig. 10.30 Values of (ψ, u_w) for which the trench can be crossed

General surface For the general surface, we choose a sinusoidal representation for the ease of depiction. The equation of the surface is taken as $[u_g, v_g, a_0 \cos(\omega_1 u_g)\sin(\omega_2 v_g)]$. In the simulation shown, we have chosen $a_0 = 1$, $\omega_1 = 2$, $\omega_2 = 3$, the initial conditions $u_g = 0$, $v_g = \pi/2\omega_0$, $u_w = \pi/2$, $v_w = 3\pi/2$, $\psi = -\pi + \pi/6$ and inputs $\omega_x = 2$, $\omega_y = 0$, $\omega_z = 0$. The wheel in two-point contact with the surface is shown in Fig. 10.31.

The traversability conditions for a single torus-shaped wheel on an uneven terrain are based on purely *local* kinematic analysis. These conditions provide insight into the conditions for a multi-point contact but are not sufficient. The torus-shaped wheel can still pass along the given path due to its dynamics. In addition, for a practical wheel, with deformation at the wheel–ground contact point, *very close* multi-point contacts are clearly not a hindrance to traversing an uneven terrain.

The analysis for the single torus-shaped wheel can be extended to the traversability analysis of a WMR. It is clear that if any of the wheels cannot traverse the uneven terrain, the WMR also cannot traverse the uneven terrain. However, this is a very simple case as the traversability of the WMR is also dependent on its length dimensions. One can easily visualise uneven terrains where all the wheels can traverse the terrain, but the WMR itself

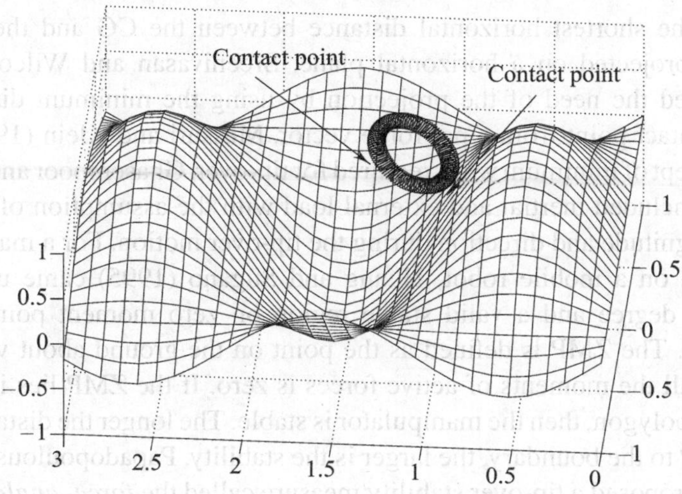

Fig. 10.31 Torus-shaped wheel on an undulated surface in two-point contact

can get stuck at a bulge lying in between the wheels. Obtaining conditions for traversability of a WMR on uneven terrains is a difficult problem due to the (possible) infinite complexity of the uneven terrain and the complicated kinematics of a WMR.

10.7 More on Modelling of Wheeled Mobile Robots

In this section, we present a brief discussion on two topics of relevance to wheeled mobile robots in the area of modelling and analysis. The first topic, related to kinematics and dynamics, is stability—a wheeled mobile robot on an uneven terrain, executing certain manoeuvres, can lose stability and lead to tip-over. The second topic, related to wheel slip, is the loss of traction and sliding on an uneven terrain. Many researchers have worked on both these topics and, in this section, we present the main ideas and concepts. More details and analytical treatment can be found in the literature listed at the end of the chapter.

Tip-over or roll-over instability occurs when a vehicle body undergoes rotation, which results in reduction in the number of wheel–ground contact points. The remaining wheel–ground contact points lie on a line called the *tip-over axis*. If the situation is not controlled, control is lost and the vehicle overturns. For tip-over stability, a low centre of gravity (CG) is desirable. Large weight is stabilizing at low speeds but is destabilizing at high speeds. In order to detect and prevent tip-over, the concept of instantaneous stability margin is useful. For legged vehicles, McGhee and Iswandhi (1979) defined this as the shortest horizontal distance between the CG and the support pattern projected on a horizontal plane. Sreenivasan and Wilcox (1994) eliminated the need of the projection by using the minimum distance of each contact point from a net force vector. Messuri and Klein (1985) used the concept of minimum work required for tip-over. Ghasempoor and Sepehri (1995) included inertial and external load with the assumption of constant load magnitude and direction during the tip-over motion. For a manipulator mounted on a mobile robot, Huang and Sugano (1995) came up with a stability degree and a valid stable region on zero moment point (ZMP) criterion. The ZMP is defined as the point on the ground about which the sum of all the moments of active forces is zero. If the ZMP lies inside the support polygon, then the manipulator is stable. The longer the distance from the ZMP to the boundary, the larger is the stability. Papadopoulous and Rey (1996) proposed a tip-over stability measure called the *force–angle stability measure*. The force–angle stability measure is defined as the minimum of all angles made by the resultant forces through the CG to the tip-over axis normal. The measure is also weighted by the magnitude of the net force vector to take into account the weight of the vehicle. The force–angle measure has a simple graphical interpretation and is relatively easy to compute for mobile robots moving on an uneven terrain and subjected to inertial and external forces.

In the context of sliding or loss of traction, the earliest work was done for automobiles and aircraft landing on wet runways (see, for example, Puleo 1970; Dugoff et al. 1970; Allen & O'Massey 1981; Chang & Lee 1990; Tan & Chin 1991). Wheel slip and traction at the wheel–ground contact point in WMRs have been discussed in Balakrishna and Ghosal (1995). For a wheel modelled as a disc on a flat surface, the authors have defined wheel slip as the ratio

$$\lambda = \frac{\dot{\theta} - \omega^*}{y}$$

where $\omega^* = v/r$ with v as the linear velocity, r is the radius of the wheel, and $\dot{\theta}$ is the angular velocity of the wheel. The value of y is ω^* when $\omega^* > \dot{\theta}$ and is $\dot{\theta}$ when $\omega^* < \dot{\theta}$. For ideal rolling, $v = r\dot{\theta}$ and the wheel slip is zero. When $\lambda = 1$, the wheel rolls at the same place without moving and the wheel skids if $\lambda = -1$. It is shown in Balakrishna and Ghosal (1995) that for a single wheel on a flat surface the traction force at the wheel–ground contact point must be at least C^1 function of $\dot{\theta}$ and v, otherwise the wheel cannot be controlled by a torque acting at its centre. Simulation results for a three-wheeled vehicle with omni-directional wheels and moving on a flat surface are presented in Balakrishna and Ghosal (1995), but very little work has been done for an uneven terrain. A reader interested in skidding and control during braking is referred to literature on automotive braking and control.

Exercises

10.1 Consider two flat planes meeting at an angle and a WMR with two wheels, modelled as thin discs, and a fixed-length (rigid) axle moving on it in a way such that the two wheels are on two different planes. From Waldron (1995) or from first principles, show that the two-wheeled mobile robot with the rigid axle will slip while moving on this piecewise flat terrain.

10.2 From the paper by Montana (1988) or first principles derive the contact equations given in Eqn (10.14).

10.3 Consider a sphere rolling on a flat surface. Derive the contact equations and show that a sphere rolling on a flat surface gives rise to non-holonomic constraints.

10.4 Obtain the kinematic equations of the three-wheeled mobile robot moving on a flat surface.

10.5 On a flat surface, the three-wheeled mobile robot has two degrees of freedom. What is the relationship between the three inputs which will enable the three-wheeled mobile robot to move without slip on a flat surface.

References and Suggested Additional Reading

Alexander, J.C. and J.H. Maddocks 1989, 'On the kinematics of wheeled mobile robots', *Int. J. Robot. Res.*, vol. 8, pp. 15–27.

Allen, R.R. and R.C. O'Massey 1981, 'Longitudinal instability in braked landing gear', *Trans. ASME J. Dyn. Syst.*, vol. 103, pp. 259–65.

Amenti, N., M. Bern, and M. Kamvysselis 1998, 'A new Voronoi-based surface reconstruction algorithm', *SIGGRAPH '98 Proceedings*, pp. 415–21.

Balakrishna, R. and A. Ghosal 1995, 'Modeling of slip for wheeled mobile robots', *IEEE Trans. Robotic Autom.*, vol. 11, pp. 126–32.

Bajaj, C.L., F. Bernardini, and G. Xu 1995, 'Automatic reconstruction of surfaces and scalar fields from 3D scans', *SIGGRAPH '95 Proceedings*, pp. 109–18.

Chakraborty, N. 2003, *Modeling of wheeled mobile robots on uneven terrain*, MSc(Engg) Thesis, Dept. of Mechanical Engg., IISc, Bangalore.

Chakraborty, N. and A. Ghosal 2004, 'Kinematics of wheeled mobile robots on uneven terrain', *Mech. Mach. Theory*, vol. 39, pp. 1273–87.

Chakraborty, N. and A. Ghosal 2005, 'Dynamic modeling and simulation of a wheeled mobile robot for traversing uneven terrain without slip', *Trans. ASME J. Mech. Des.*, vol. 127, pp. 901–09.

Chang, C. and T. Lee 1990, 'Stability analysis of three and four-wheeled vehicles', *JSME Int. J.*, Series III, vol. 33, pp. 567–74.

Choi, B.J. and S.V. Sreenivasan 1999, 'Gross motion characteristics of articulated robots with pure rolling capability on smooth uneven surfaces', *IEEE Trans. Robotic Autom.*, vol. 15, pp. 340–43.

Choi, B.J., S.V. Sreenivasan, and P.W. Davis 1999, 'Two wheels connected by an un-actuated variable length axle on uneven ground: kinematic modeling and experiments', *Trans. ASME J. Mech. Des.*, vol. 121, pp. 235–40.

Cooper, S. and H.F. Durrant Whyte 1994, 'A Kalman filter model for GPS navigation of land vehicles', *Proc. of the Int. Conf. of Intelligent Robots and Systems*, pp. 157–63.

Dugoff, H., P.S. Fancher, and L. Segel 1970, 'An analysis of tire traction properties and their influence on vehicle dynamic performance', *SAE Transactions*, Paper No. 700377.

Edelsbrunner, H. and E.P. Mucke 1994, 'Three-dimensional alpha shapes', *ACM Trans. Graphic*, vol. 19, pp. 43–72.

Ghasempoor, A. and N. Sepehri 1995, 'A measure of machine stability for moving base manipulators', *Proc. of IEEE Conf. on Robotics and Automation*, pp. 2249–54.

Goldstein, H. 1980, *Classical Mechanics*, 2nd edn, Addison-Wesley.

Heck, A. 2003, *Introduction to Maple*, 3rd edn, Springer-Verlag.

Huang, Q. and S. Sugano 1995, 'Manipulator motion planning for stabilizing a mobile manipulator', *Proc. of IEEE Conf. on Intelligent Robotics and Systems*, pp. 467–72.

MATLAB Users Manual, Mathwork Inc., Massachusetts, 1992.

McGhee, R.B. and G.I. Iswandhi 1979, 'Adaptive locomotion of a multi-legged robot over rough terrain', *IEEE Trans. Syst. Man Cyb.*, vol. SMC-9, pp. 76–82.

Messuri, D.A. and C.A. Klein 1985, 'Automatic body regulation for maintaining stability of legged vehicles during rough terrain locomotion', *IEEE Trans. Robotic Autom.*, vol. RA-1, pp. 132–41.

Montana, D.J. 1988, 'The kinematics of contact and grasp', *Int. J. Robot. Res.*, vol. 7, pp. 17–32.

Mortenson, M.E. 1985, *Geometric Modeling*, John Wiley.

Muir, P.F. and C.P. Newman 1987, 'Kinematic modeling of wheeled mobile robots', *J. Robotic Sys.*, vol. 4, pp. 281–329.

Papadopoulous, E.G. and D.A. Rey 1996, 'A new measure of tip-over stability marging for mobile manipulators' *Proc. of IEEE Conf. on Robotics and Automation*, pp. 3111–16.

Puleo, G. 1970, 'Automatic brake proportioning devices', *SAE Transactions*, Paper No. 700375.

Sreenivasan, S.V. and B.H. Wilcox 1994, 'Stability and traction control of an actively actuated micro-rover', *J. Robotic Syst.*, vol. 11, pp. 487–502.

Tan, H. and Y. Chin 1991, 'Vehicle traction control: Variable structure control approach', *Trans. ASME J. Dyn. Syst.*, vol. 113, pp. 223–30.

Waldron, K.J. 1995, 'Terrain adaptive vehicles', *Trans. ASME J. Mech. Des.*, vol. 117B, pp. 107–12.

Wolfram, S. 1999, *The Mathematica Book*, 4th edn, Cambridge University Press.

Index

I

J

K

L

Q

R

T

U

V

W